Sustainable Water Developments
Resources, Management, Treatment,
Efficiency and Reuse

Series Editor

Jochen Bundschuh
University of Southern Queensland (USQ), Toowoomba, Australia
Royal Institute of Technology (KTH), Stockholm, Sweden

ISSN: 2373-7506

Volume 2

Innovative Materials and Methods for Water Treatment: Solutions for Arsenic and Chromium Removal

Editors

Marek Bryjak

*Polymer and Carbon Materials Department, Faculty of Chemistry,
Wrocław University of Technology, Wrocław, Poland*

Nalan Kabay

Chemical Engineering Department, Faculty of Engineering, Ege University, Izmir, Turkey

Bernabé L. Rivas

Polymer Department, University of Concepción, Concepción, Chile

Jochen Bundschuh

*Deputy Vice Chancellor's Office (Research and Innovation) & Faculty of Health,
Engineering and Sciences & International Centre for Applied Climate
Sciences, University of Southern Queensland, Toowoomba, Australia*

CRC Press
Taylor & Francis Group
Boca Raton London New York

CRC Press is an imprint of the
Taylor & Francis Group, an **informa** business

A BALKEMA BOOK

Cover photo
The photograph shows part of an arsenic adsorption system located in southern Arizona. The system became operational in late 2009 and is capable of treating a water flow of up to 2000 gallons per minute (about 454 cubic meters per hour). The well water is split into two separate, equal streams, each of which passes through two arsenic adsorption beds arranged in a lead-lag configuration. Each bed contains 375 cubic feet (about 10.6 cubic meters) of media which equates to a contact time of 2.8 minutes per vessel at full flow. The photo is courtesy of Dr. Paul Sylvester.

Published by:
CRCPress/Balkema
P.O. Box 447, 2300 AK Leiden, The Netherlands
e-mail: Pub.NL@taylorandfrancis.com
www.crcpress.com – www.taylorandfrancis.com

First issued in hardback 2020
First issued in paperback 2021

ISBN-13: 978-1-138-60659-3 (pbk)
ISBN-13: 978-1-138-02749-7 (hbk)

Typeset by MPS Limited, Chennai, India

Library of Congress Cataloging-in-Publication Data

Innovative materials and methods for water treatment: solutions for arsenic and chromium removal / editors, Marek Bryjak, Wrocław University of Technology, Wrocław, Poland, Nalan Kabay, Ege University, Izmir, Turkey, Bernabé L. Rivas, University of Concepción, Concepción, Chile, Jochen Bundschuh, University of Southern Queensland, Deputy Vice Chancellor's Office (Research and Innovation) & Faculty of Health, Engineering and Sciences, Toowoomba, Australia & Royal Institute of Technology (KTH), Stockholm, Sweden.
    pages cm. – (Sustainable water developments; volume 2)
    Includes bibliographical references and index.
    ISBN 978-1-138-02749-7 (hardcover: alk. paper) – ISBN 978-1-315-68260-0
(ebook: alk. paper)   1. Water–Purification–Arsenic removal.
2. Water–Purification–Chromium removal.   I. Bryjak, Marek.
    TD427.A77I553   2015
    628.1′6836–dc23
                                                                    2015021341

Visit the Taylor & Francis Web site at
http://www.taylorandfrancis.com

and the CRC Press Web site at
http://www.crcpress.com

Publisher's Note
The publisher has gone to great lengths to ensure the quality of this reprint but points out that some imperfections in the original copies may be apparent.

About the book series

Augmentation of freshwater supply and better sanitation are two of the world's most pressing challenges. However, such improvements must be done economically in an environmentally and societally sustainable way.

Increasingly, groundwater – the source that is much larger than surface water and which provides a stable supply through all the seasons – is used for freshwater supply, which is exploited from ever-deeper groundwater resources. However, the availability of groundwater in sufficient quantity and good quality is severely impacted by the increased water demand for industrial production, cooling in energy production, public water supply and in particular agricultural use, which at present consumes on a global scale about 70% of the exploited freshwater resources. In addition, climate change may have a positive or negative impact on freshwater availability, but which one is presently unknown. These developments result in a continuously increasing water stress, as has already been observed in several world regions and which has adverse implications for the security of food, water and energy supplies, the severity of which will further increase in future. This demands case-specific mitigation and adaptation pathways, which require a better assessment and understanding of surface water and groundwater systems and how they interact with a view to improve their protection and their effective and sustainable management.

With the current and anticipated increased future freshwater demand, it is increasingly difficult to sustain freshwater supply security without producing freshwater from contaminated, brackish or saline water and reusing agricultural, industrial, and municipal wastewater after adequate treatment, which extends the life cycle of water and is beneficial not only to the environment but also leads to cost reduction. Water treatment, particularly desalination, requires large amounts of energy, making energy-efficient options and use of renewable energies important. The technologies, which can either be sophisticated or simple, use physical, chemical and biological processes for water and wastewater treatment, to produce freshwater of a desired quality. Both industrial-scale approaches and smaller-scale applications are important but need a different technological approach. In particular, low-tech, cost-effective, but at the same time sustainable water and wastewater treatment systems, such as artificial wetlands or wastewater gardens, are options suitable for many small-scale applications. Technological improvements and finding new approaches to conventional technologies (e.g. those of seawater desalination), and development of innovative processes, approaches, and methods to improve water and wastewater treatment and sanitation are needed. Improving economic, environmental and societal sustainability needs research and development to improve process design, operation, performance, automation and management of water and wastewater systems considering aims, and local conditions.

In all freshwater consuming sectors, the increasing water scarcity and correspondingly increasing costs of freshwater, calls for a shift towards more water efficiency and water savings. In the industrial and agricultural sector, it also includes the development of technologies that reduce contamination of freshwater resources, e.g. through development of a chemical-free agriculture. In the domestic sector, there are plenty of options for freshwater saving and improving efficiency such as water-efficient toilets, water-free toilets, or on-site recycling for uses such as toilet flushing, which alone could provide an estimated 30% reduction in water use for the average household. As already mentioned, in all water-consuming sectors, the recycling and reuse of the respective wastewater can provide an important freshwater source. However, the rate at which these water efficient technologies and water-saving applications are developed and adopted depends on the behavior of individual consumers and requires favorable political, policy and financial conditions.

Due to the interdependency of water and energy (water-energy nexus); i.e. water production needs energy (e.g. for groundwater pumping) and energy generation needs water (e.g. for cooling), the management of both commodities should be more coordinated. This requires integrated energy and water planning, i.e. management of both commodities in a well-coordinated form rather than managing water and energy separately as is routine at present. Only such integrated management allows reducing trade-offs between water and energy use.

However, water is not just linked to energy, but must be considered within the whole of the water-energy-food-ecosystem-climate nexus. This requires consideration of what a planned water development requires from the other sectors or how it affects – positively or negatively – the other sectors. Such integrated management of water and the other interlinked resources can implement synergies, reduce trade-offs, optimize resources use and management efficiency, all in all improving security of water, energy, and food security and contributing to protection of ecosystems and climate. Corresponding actions, policies and regulations that support such integral approaches, as well as corresponding research, training and teaching are necessary for their implementation.

The fact that in many developing and transition countries women are disproportionately disadvantaged by water and sanitation limitation requires special attention to this aspect in these countries. Women (including schoolgirls) often spend several hours a day fetching water. This time could be much better used for attending school or working to improve knowledge and skills as well as to generate income and so to reduce gender inequality and poverty. Absence of in-door sanitary facilities exposes women to potential harassment. Moreover, missing single-sex sanitation facilities in schools and absence of clean water contributes to diseases. This is why women and girls are a critical factor in solving water and sanitation problems in these countries and necessitates that men and women work side by side to address the water and wastewater related operations for improvement of economic, social and sustainable freshwater provision and sanitation.

Individual volumes published in the series span the wide spectrum between research, development and practice in the topic of freshwater and related areas such as gender and social aspects as well as policy, regulatory, legal and economic aspects of water. The series covers all fields and facets in optimal approaches to the:

- Assessment, protection, development and sustainable management of groundwater and surface water resources thereby optimizing their use.
- Improvement of human access to water resources in adequate quantity and good quality.
- Meeting of the increasing demand for drinking water, and irrigation water needed for food and energy security, protection of ecosystems and climate and contribution to a socially and economically sound human development.
- Treatment of water and wastewater also including its reuse.
- Implementation of water efficient technologies and water saving measures.

A key goal of the series is to include all countries of the globe in jointly addressing the challenges of water security and sanitation. Therefore, we aim for a balanced selection of authors and editors originating from developing and developed countries as well as for gender equality. This will help society to provide access to freshwater resources in adequate quantity and of good quality, meeting the increasing demand for drinking water, domestic water and irrigation water needed for food security while contributing to socially and economically sound development.

This book series aims to become a state-of-the-art resource for a broad group of readers including professionals, academics and students dealing with ground and surface water resources, their assessment, exploitation and management as well as the water and wastewater industry. This comprises especially hydrogeologists, hydrologists, water resources engineers, wastewater engineers, chemical engineers and environmental engineers and scientists.

The book series provides a source of valuable information on surface water but especially on aquifers and groundwater resources in all their facets. As such, it covers not only the scientific and technical aspects but also environmental, legal, policy, economic, social, and gender aspects of groundwater resources management. Without departing from the larger framework of integrated groundwater resources management, the topics are centered on water, solute and heat transport in aquifers, hydrogeochemical processes in aquifers, contamination, protection, resources assessment and use.

The book series constitutes an information source and facilitator for the transfer of knowledge, both for small communities with decentralized water supply and sanitation as well as large industries that employ hundreds or thousands of professionals in countries worldwide, working in the different fields of freshwater production, wastewater treatment and water reuse as well as those concerned with water efficient technologies and water saving measures. In contrast to many other industries, suffering from the global economic downturn, water and wastewater industries are rapidly growing sectors providing significant opportunities for investments. This applies especially to those using sustainable water and wastewater technologies, which are increasingly favored. The series is also aimed at communities, manufacturers and consultants as well as a range of stakeholders and professionals from governmental and non-governmental organizations, international funding agencies, public health, policy, regulating and other relevant institutions, and the broader public. It is designed to increase awareness of water resources protection and understanding of sustainable water and wastewater solutions including the promotion of water and wastewater reuse and water savings.

By consolidating international research and technical results, the objective of this book series is to focus on practical solutions in better understanding groundwater and surface water systems, the implementation of sustainable water and wastewater treatment and water reuse and the implementation of water efficient technologies and water saving measures. Failing to improve and move forward would have serious social, environmental and economic impacts on a global scale.

The book series includes books authored and edited by world-renowned scientists and engineers and by leading authorities in economics and politics. Women are particularly encouraged to contribute, either as author or editor.

Jochen Bundschuh
(Series Editor)

Editorial board

Table of contents

List of contributors

Anna Bastrzyk — Department of Chemical Engineering, Faculty of Chemistry, Wrocław University of Technology, Wrocław, Poland

Ulker Beker — Chemical Engineering Department, Yildiz Technical University, Istanbul, Turkey

Nicole Blute — Hazen and Sawyer, Los Angeles, California, USA

Michał Bodzek — Silesian University of Technology, Faculty of Energy and Environmental Engineering, Gliwice, Poland & Institute of Environmental Engineering of the Polish Academy of Sciences, Zabrze, Poland

Nupur Bose — Department of Environment and Water Management, A N College, Patna, Bihar, India

Eugenio Bringas — Department Chemical and Biomolecular Engineering, University of Cantabria, Cantabria, Spain

Marek Bryjak — Polymer and Carbon Materials Department, Faculty of Chemistry, Wrocław University of Technology, Wrocław, Poland

Jochen Bundschuh — Deputy Vice Chancellor's Office (Research and Innovation) & Faculty of Health, Engineering and Sciences & International Centre for Applied Climate Sciences – University of Southern Queensland, Toowoomba, Queensland, Australia

Cristian Campos — Polymer Department, University of Concepción, Concepción, Chile

Tülin Deniz Çiftçi — Chemistry Department, Faculty of Science, Ege University, İzmir, Turkey

Virginia S.T. Ciminelli — Department of Metallurgical and Materials Engineering, Universidade Federal de Minas Gerais, Belo Horizonte, Minas Gerais, Brazil

David R. Contreras — Faculty of Chemical Sciences, University of Concepción, Chile

Lorena Cornejo — Laboratory of Environmental Research on Arid Zones, LIMZA, EUIIIS, University of Tarapacá, Chile & Environmental Resources Area, Centro de Investigaciones del Hombre en el Desierto, CIHDE, Chile

José Luis Cortina — Departament d'Enginyeria Química, Universitat Politècnica de Catalunya, Barcelona, Spain & Water Technology Center, CETaqua, Barcelona, Spain

Alessandra Criscuoli — Institute on Membrane Technology (ITM-CNR), Rende (CS), Italy

Alberto Figoli — Institute on Membrane Technology (ITM-CNR), Rende (CS), Italy

María José Gallardo — Center for Optics and Photonics, University of Concepción, Concepción, Chile

Sofía Garrido — Instituto Mexicano de Tecnología del Agua, Col. Progreso, Jiutepec, Morelos, Mexico

Ashok Ghosh — Department of Environment and Water Management, A N College, Patna, Bihar, India

Oriol Gibert	Departament d'Enginyeria Química, Universitat Politècnica de Catalunya, Barcelona, Spain & Water Technology Center, CETaqua, Barcelona, Spain
Emür Henden	Chemistry Department, Faculty of Science, Ege University, İzmir, Turkey
Jan Hoinkis	Karlsruhe University of Applied Sciences, Karlsruhe, Germany
Evgenia Iakovleva	LUT Chemtech, Lappeenranta University of Technology, Lappeenranta, Finland
Luděk Jelínek	Department of Power Engineering, Institute of Chemical Technology, Prague, Czech Republic
Malgorzata Jemiola-Rzeminska	Faculty of Biochemistry and Biotechnology, Jagiellonian University, Krakow, Poland
Nalan Kabay	Chemical Engineering Department, Faculty of Engineering, Ege University, Izmir, Turkey
Krystyna Konieczny	Silesian University of Technology, Faculty of Energy and Environmental Engineering, Gliwice, Poland
Stanisław Koter	Faculty of Chemistry, Nicolaus Copernicus University, Toruń, Poland
Tomasz Koźlecki	Department of Chemical Engineering, Faculty of Chemistry, Wrocław University of Technology, Wrocław, Poland
Marta I. Litter	Remediation Technologies Division, Environmental Chemistry Department, Chemistry Management, National Atomic Energy Commission, Buenos Aires; National Scientific and Technique Research Council (CONICET); Institute of Research and Environmental Engineering, National University of General San Martín, Argentina
Marjatta Louhi-Kultanen	LUT Chemtech, Lappeenranta University of Technology, Lappeenranta, Finland
Héctor D. Mansilla	Faculty of Chemical Sciences, University of Concepción, Concepción, Chile
Priyanka Mondal	Department of Chemistry, University of Kalyani, Kalyani, West Bengal, India
Michal Němeček	Department of Power Engineering, Institute of Chemical Technology, Prague, Czech Republic
Syouhei Nishihama	Department of Chemical Engineering, University of Kitakyushu, Japan
Inmaculada Ortiz	Department Chemical and Biomolecular Engineering, University of Cantabria, Cantabria, Spain
Helena Parschová	Department of Power Engineering, Institute of Chemical Technology, Prague, Czech Republic
Izabela Polowczyk	Department of Chemical Engineering, Faculty of Chemistry, Wrocław University of Technology, Wrocław, Poland
Bernabé L. Rivas	Polymer Department, University of Concepción, Concepción, Chile
Peter Roberts	Consultant, Melbourne, VIC, Australia
Juan Saiz	Department Chemical and Biomolecular Engineering, University of Cantabria, Cantabria, Spain
Ana María Sancha	División de Recursos Hídricos y Medio Ambiente, Facultad de Ciencias Físicas y Matemáticas, Universidad de Chile, Santiago de Chile, Chile
Julio Sánchez	Polymer Department, Faculty of Chemistry, University of Concepción, Concepción, Chile

Susana Sánchez	Faculty of Chemical Sciences, University of Concepción, Concepción, Chile
Sergio Santoro	REQUIMTE/CQFB, Department of Chemistry, Universidade Nova de Lisboa, Caparica, Portugal
Sudipta Sarkar	Indian Institute of Technology-Roorkee, Uttarakhanda, India
Arup K. SenGupta	Lehigh University, Bethlehem, Pennsylvania, USA
Mika Sillanpää	LUT Chemtech, Lappeenranta University of Technology, Lappeenranta, Finland
Esra Bilgin Simsek	Chemical&Process Engineering Department, Yalova University, Yalova, Turkey
Ryan C. Smith	Lehigh University, Bethlehem, Pennsylvania, USA
Kazimierz Strzalka	Faculty of Biochemistry and Biotechnology, Jagiellonian University, Krakow, Poland
Mario Suwalsky	Faculty of Chemical Sciences, University of Concepción, Concepción, Chile
Paul Sylvester	Consultant, Waltham, Massachusetts, USA
Leandro Toledo	Polymer Department, Faculty of Chemistry, University of Concepción, Concepción, Chile
Justyna Ulatowska	Department of Chemical Engineering, Faculty of Chemistry, Wrocław University of Technology, Wrocław, Poland
Bruno F. Urbano	Polymer Department, University of Concepción, Concepción, Chile
Cesar Valderrama	Departament d'Enginyeria Química, Universitat Politècnica de Catalunya, Barcelona, Spain
Fernando Villena	Faculty of Biological Sciences, University of Concepción, Concepción, Chile
Jorge Yáñez	Faculty of Chemical Sciences, University of Concepción, Concepción, Chile
Kazuharu Yoshizuka	Department of Chemical Engineering, University of Kitakyushu, Kitakyushu, Japan

Editors' foreword

Climate change, water and energy are global challenges certain to attract attention from policy makers, since the world population will reach 9 billion by 2050. Growing demand for water, driven by the manufacturing and electricity sectors, will strengthen the impact of water on climate change and energy over time (The Economist, 2013).

According to international water authorities, one third of the world population faces some form of water scarcity. Population growth, global warming, widespread mismanagement and increasing demand for energy could lead to a global water crisis, according to the United Nations World Water Development Report (2012). In the last century, the use of water has grown more than two-fold compared to the rate of population increase, and, although there is no global water scarcity so far, an increasing number of regions claim chronic shortage of water. The Food and Agriculture Organization of the United Nations has reported that 1,800 million people will be living in countries or regions with absolute water scarcity, and two-thirds of the world population could be under stress conditions by 2025.

On the other hand, the presence of toxic substances such as heavy metals, organic compounds, radioactive and biological contaminants in water impact the quality of life. Therefore, more stringent regulations are needed that restrict dangerous processes releasing such substances into the environment. In addition, utilization of clean technologies with the slogan "zero discharge" should be accepted by policy makers, stakeholders and the industrial sectors.

The idea for this book was suggested during the international project "CHILTURPOL2", supported by European Commission-Marie Sklodowska-Curie Actions (MC-IRSES) entitled **"Innovative Materials and Methods for Water Treatment"**, as one of its dissemination activities. The book is designed to provide an overview of treatment processes for removal from water of the most toxic elements, arsenic and chromium. It brings together the experiences of different experts in the preparation of selective materials and in running innovative processes for removal of arsenic and chromium from groundwater, wastewater and other contaminated water sources. The reader will get a comprehensive review of materials and methods useful for making water safe. The book discusses various production techniques for sorbents and membranes that are now commercially employed or are in the developmental stage and will be commercialized during the next decades.

The book is divided into four sections, three of which (i) **"Introduction to the problem of arsenic and chromium contaminations"**, (ii) **"Innovative materials for removal of arsenic and chromium"** and (iii) **"Innovative methods for removal of arsenic and chromium"** comprise the technical content of the book. Most of them originate from research in university or corporate laboratories with many of the authors having long, distinguished careers in such work. The last section, **"Case studies: Latin America and the United States of America examples"**, is different. It deals with presentation of the systems that were commissioned in Latin America and the United States of America.

The book will serve as a reference for graduate students, studying in the field of water and wastewater treatment technologies, and for water authorities, scientists, and engineers working in the water industry who wish to learn, select and adapt the best ways to remove the toxic elements from water through the use of innovative materials and technologies.

Marek Bryjak
Nalan Kabay
Bernabé Rivas
Jochen Bundschuh
(editors)
October 2015

About the editors

Professor Marek Bryjak is working at Wrocław University of Technology where he graduated in 1977, received his PhD in 1982 and was awarded a DSc degree in 2001 for studies on application of polymers to separation processes. He carried out a post-doc fellowship at the Centre for Surface Science, Lehigh University USA in 1989–1991 and has visited the following universities: University of Calabria, Minho University, Ege University, Hacepette University, Imperial College, Stellenbosch University, Kitakyushu University, University of Concepcion, University of Santiago, Torino University, Loughborough University, Bratislava University, Institute of Chemical Technology Prague. He was the Head of Department of Specialty Polymers and the Department of Polymer and Carbon Materials. He was the President of the Membrane Section, Polish Chemical Society. He is a member of the European Center for Innovation and Technology and the Center of Advanced Materials and Nanotechnology. Professor Bryjak has authored and co-authored about 90 scientific papers, 6 chapters and edited 4 books. He has supervised about 80 PhD and MSc theses, has given about 60 invited lectures and participated in 15 international and domestic projects. His scientific interest is focused on development of methods for formation of polymer membranes and their surface modification by plasma treatment, preparation and evaluation of new separation materials, and implementation of these materials to water technology.

Professor Nalan Kabay has been working at the Chemical Engineering Department, Faculty of Engineering of Ege University, Turkey since 1994. She graduated from Ege University in 1983. She received her MS degree at Ege University in 1985 and her PhD from Kumamoto University, Japan in 1992. She worked as a post-doc between 1998–1999 at the National Institute for Research in Inorganic Materials-NIRIM, Japan. She several times visited the Chemical Engineering Departments of Loughborough University and Imperial College, London UK as a visiting scientist between 1996–2008. She is former Vice-Dean of the Engineering Faculty of Ege University (2003–2009), former member of the steering committee of the Engineering Research Group at the Turkish Scientific and Technical Research Council (TUBITAK) (2004–2007). Prof. Kabay has authored/co-authored 115 SCI papers, 8 book chapters and acted as guest editor of special issues of 7 SCI journals, and edited 3 books. She is a member of the editorial board for 3 SCI journals and 1 national journal. She has been involved in 15 international and 30 national projects. She has given many lectures at different institutions and conferences as an invited scientist in Japan, UK, Germany, Poland, the Netherlands, India, Australia, Israel, Russia, Ukraine, Spain, S. Africa, S. Korea, UAE and Chile. She was awarded the TUBITAK-Science Promotion Award, the Canon Foundation in Europe Award in 2001 and the SCI-IEX Award in 2012. She is an Honorary Member of the Japan Ion Exchange Society. She has supervised and co-supervised 8 PhD and 40 MS students. Her main interests are water and wastewater treatment by membrane processes (NF, RO, ED, EDI, MBR), desalination of seawater and geothermal water, ion exchange, boron separation, preparation of solvent impregnated resins, elimination of heavy metals and toxic organic compounds from water, and wastewater reuse.

Prof. Bernabé L. Rivas graduated as a biochemist (1975) and received his PhD degree (1980) from the University of Concepción in Chile. He was a post-doc at the University of Mainz (1983–1984) and the University of Tübingen (1989–1991) in Germany. He was Dean of the Faculty of Chemistry, Vice-rector for Research and Development at the University of Concepción where he holds the position of Vice-President now. He was awarded the Municipal Award of Sciences, Municipality of Concepción and the Regional Award of Sciences "Pascual Binimelis", Regional Government, Macromolecules Division of Chilean Chemical Society Award. Prof. Rivas has authored/co-authored 360 publications in ISI journals, 20 book chapters and 560 contributing papers in meetings and symposia proceedings. He has given more than 100 invited lectures at different institutions and conferences in Brazil, Costa Rica, Egypt, France, Germany, Italy, Japan, Mexico, Peru, Spain, Turkey and South Korea. He was Editor of the Journal of the Chilean Chemical Society, a member of the Editorial Boards of Polymer International and the Open Macromolecules Journal. He was advisor for more than 20 PhD and MS theses. His main research interests are the preparation of ion exchange resins, metal separations by ion exchange resins, polymer enhanced ultrafiltration, separation of arsenic and polymer-clay composite materials for metal separations.

Professor Jochen Bundschuh (1960, Germany), finished his PhD on numerical modeling of heat transport in aquifers in Tübingen in 1990. He is working in geothermics, subsurface and surface hydrology and integrated water resources management, and connected disciplines. From 1993 to 1999, he served as an expert for the German Agency of Technical Cooperation (GTZ – now GIZ) and as a long-term professor for the DAAD (German Academic Exchange Service) in Argentina. From 2001 to 2008 he worked within the framework of the German governmental cooperation (Integrated Expert Program of CIM; GTZ/BA) as adviser in mission to Costa Rica at the Instituto Costarricense de Electricidad (ICE). Here, he assisted the country in evaluation and development of its huge low-enthalpy geothermal resources for power generation. Since 2005, he has been an affiliate professor of the Royal Institute of Technology, Stockholm, Sweden. In 2006, he was elected Vice-President of the International Society of Groundwater for Sustainable Development ISGSD. From 2009–2011 he was visiting professor at the Department of Earth Sciences at the National Cheng Kung University, Tainan, Taiwan.

Since 2012, Dr. Bundschuh has been a professor in hydrogeology at the University of Southern Queensland, Toowoomba, Australia working on the wide field of water resources and low/middle enthalpy geothermal resources, water and wastewater treatment and sustainable and renewable energy solutions. In November 2012, Prof. Bundschuh was appointed president of the newly established Australian Chapter of the International Medical Geology Association (IMGA).

Dr. Bundschuh is author of the books "Low-Enthalpy Geothermal Resources for Power Generation" (2008) (CRC Press/Balkema, Taylor & Francis Group) and "Introduction to the Numerical Modeling of Groundwater and Geothermal Systems: Fundamentals of Mass, Energy and Solute Transport in Poroelastic Rocks". He is editor of 16 books and editor of the book series "Multiphysics Modeling", "Arsenic in the Environment", "Sustainable Energy Developments" and the recently established series "Sustainable Water Developments" (all CRC Press/Balkema, Taylor & Francis Group). Since 2015, he has been editor in chief of the Elsevier journal "Groundwater for Sustainable Development".

Acknowledgements

The editors would like to thank the authors of the various chapters for their sincere co-operation and their great contributions in the completion of the book. Our appreciation is directed to Ron Inglis for the time and efforts he devoted to correct the English of the manuscripts. Special thanks are given to Peter Roberts for sharing his knowledge on implementation of new technologies in the water industry. The editors and authors thank also the technical people of CRC Press/Balkema, Taylor & Francis Group, for their cooperation and the excellent typesetting of the manuscript.

Part I

*Introduction to the problem of arsenic
and chromium contaminations*

CHAPTER 1

Occurrence and toxicity of arsenic and chromium

Bernabé L. Rivas, Bruno F. Urbano, Marek Bryjak, Sudipta Sarkar, Stanislaw Koter,
Krystyna Konieczny, Eugenio Bringas, Evgenia Iakovleva & Izabela Polowczyk

1.1 INTRODUCTION

Groundwater which constitutes 97% of global freshwater (Howard *et al.*, 2006) provides water
to rivers during periods of no rainfall. It is therefore an essential resource requiring protection
because two billion people depend directly upon aquifers for drinking water, and 40% of the
world's food is produced by irrigated agriculture that depends basically on groundwater. Globally,
the dependence on groundwater sources is enormous; in 2000, twelve of the twenty-three cities of
the world which had a population higher than 10 million inhabitants made significant use of local
groundwater. For example, China has more than 500 cities, and two-thirds of the water supply for
these is drawn from aquifers (Morris *et al.*, 2003). The use of groundwater for domestic supply
is even more extensive in small towns and rural communities. This is well illustrated in eastern
China, where the Huang-Huai-Hai aquifer system supplies nearly 160 million people, and it is
estimated that almost one-third of drinking water supply in rural communities in Asia comes from
groundwater. In the USA, more than 95% of the rural population depends on aquifers to deliver
their consumption of water.

The presence of a large number of chemical and biological pollutants appears in drinking
water. The pollution of water resources caused by an increasing population occurs via the use and
disposal of chemical, agricultural, and animal products and industrial waste (Gupta and Suhas,
2009; Kanmani *et al.*, 2012; Malaviya and Singh, 2011; Meng *et al.*, 2001; Qi and Donahoe,
2008; Zagorodni, 2007). Water contamination sources can be classified into three categories:

- Natural pollution;
- Waste disposal practices;
- Nonpoint sources due to human activities.

The major percentage of inorganic substances found in groundwater is derived from natural origins. However, significant amounts are also a result of human activities. Thus, the
removal/remediation of these substances is necessary (Gupta and Suhas, 2009; Kanmani *et al.*,
2012). Over the last few years, discarding solid and/or liquid waste products containing heavy
metals emanating from industrial processes has received much attention, and consequently, legislation for the protection of the environment has gradually become more strict (García-Delgado,
2007; WHO, 2004). A variety of methods have been applied for the removal of arsenic (As) and
chromium (Cr) from water and wastewater purification, including chemical, physicochemical,
and biological/biochemical treatment. The removal process includes different types of technologies, which are dependent on the speciation form of the As and Cr. Common chemical treatment
methods include pH adjustment for the control of different species.

1.1.1 *General aspects of arsenic and chromium*

Arsenic is the 33nd element of the periodic system and belongs to group 15, which includes
nitrogen, phosphorus, antimony, and bismuth. Arsenic is a ubiquitous element that occurs naturally
in the earth's crust. More than 245 minerals contain As. Although the source of As is geological,

human activities such as mining, burning fossil fuels, and pesticide application have increased the presence of As in different sources, such as groundwater and landfills (Mandal and Suzuki, 2002). Arsenic pollution has a diverse effect on the living organism; for example, in humans, long exposure to As species results in damage to the skin, lungs, liver, kidney, heart, gastrointestinal tract, and other organs.

Arsenic can also be found in four valence states, -3, 0, $+3$, and $+5$, and a variety of arsenical compounds exist. In fact, As occurs naturally in over 200 different mineral forms; for example, in its elemental form, and in organoarsenicals, arsenides, arsenosulfides, arsenites, silicates, and arsenates. The concentration of As varies depending the source and region. For example, in seawater, the As concentration is usually less than $2.0\,\mu g\,L^{-1}$ (Ng, 2005), while on the surface and in groundwater, the concentration is between 1 and $10\,\mu g\,L^{-1}$ (without any contamination) (Bissen and Frimmel 2003a; 2003b). However, depending on the region, high As concentration can be found in countries such as Argentina, Bangladesh, Cambodia, Canada, Chile, China, India, Mexico, Nepal, Pakistan, Taiwan, Vietnam, and the United States, where more than 100 million people are at risk due to As-contaminated groundwater. This situation is particularly complicated in West Bengal, India and Bangladesh, where the As concentration is within the range of 10–196 and $9.0–28\,mg\,L^{-1}$, respectively (Mandal and Suzuki, 2002).

In addition to the natural existence of As in water, human activities have contaminated water sources with increasing As concentrations, such as the application of arsenical pesticides (Qi and Donahoe, 2008), dusts from the burning of fossil fuels (Han et al., 2003), and the disposal of industrial wastes (Sullivan et al., 2010). Moreover, the natural presence of As in groundwater often originates from the mobilization of natural deposits in rocks, sediments, and soils, as well as from geothermal water. However, the mining activities of humans via the utilization of natural resources have released As into the air, water and soil. The level of As in soil is mainly dependent on the concentration in the parent rock and on human activities. This concentration ranges from 0.1 to $40\,mg\,kg^{-1}$, but varies considerably among geographic regions.

Arsenic in the air exists as a mixture of arsenite and arsenate, which is predominantly absorbed in particulate matters, and the most polluted areas are exposed to arsenical pesticides. However, As in water is the most concerning issue because it is the main route of exposure to humans. Since 1963, the World Health Organization (WHO) has recommended a maximal level of $50\,\mu g\,L^{-1}$ of As in drinking water. However, due the increasing evidence demonstrating the link between exposure to low As concentrations and cancer risk, the WHO lowered the recommendation level to $10\,\mu g\,L^{-1}$ (WHO, 2004). Despite the recommendation of the WHO, some countries such as Bangladesh, Chile, China and India still have maximum level of As concentrations of $50\,\mu g\,L^{-1}$.

The As(V) ions are thermodynamically more stable than As(III), and theoretical calculation predicts a ratio of 10^{26}:1 of As(V) and As(III), respectively. However, due to biological species in water, this ratio can decreases to 0.1:1. Under oxidizing conditions, in aerobic environments, the most stable species are arsenates as arsenates oxyanions (H_3AsO_4, $H_2AsO_4^-$, $HAsO_4^{2-}$, AsO_4^{3-}), which are mainly absorbed into clays. In contrast, under reducing conditions, arsenites are the major species. Several forms of As may be found in nature, and the more abundant species include As(III), As(V), methylarsenic acid (MMA), and dimethylarsenic acid (DMA), which undergo acid-base equilibria. Thus, different major and minor species will be present depending on the pH – see Section 1.2.

Chromium is the seventh most abundant element on earth and is ranked fourth among the most biologically important elements. While Cr exists in oxidation states between -2 and $+6$, only Cr(III) and Cr(VI) (in the form of chromate, CrO_4^{2-}) are considered to be environmentally and biologically important species. Cr(III) occurs naturally in the environment, with a concentration of approximately $100\,mg\,g^{-1}$ in the earth's crust. This species of Cr is fairly immobile in soils because it tends to form insoluble oxides and hydroxides at pH > 4.

Chromium is widely used in industry. The Cr(VI) compounds are used in the metallurgical industry for chrome alloy and metal production and in chrome plating. In the chemical industry, it is used as an oxidizing agent and in the production of other Cr compounds (Dhal et al., 2013). About 80–90% of leather is tanned using Cr chemicals, out of which approximately 40% of the

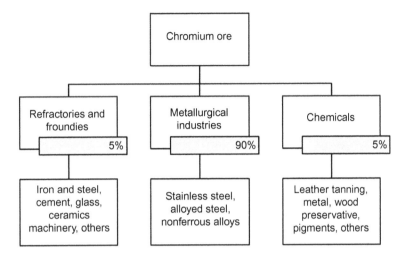

Figure 1.1.   Use of chromium in different industries.

Cr used is discharged in the waste effluent containing Cr(VI) and Cr(III) ions (Saha and Orvig, 2010). The Cr(III) salts are used less widely and are employed in textile dying, in ceramics, the glass industry, and photography. In Figure 1.1, a schematic abstract of the use of Cr in different industries is shown.

Chromium is widely distributed in rocks, freshwater, and seawater in the form of Cr(VI). The fate of Cr in soil and/or water is partly dependent on the redox potential and pH (Choppala *et al.*, 2013). In most soils, Cr is present predominantly as Cr(III) (Barrera-Díaz *et al.*, 2012; Saeedi *et al.*, 2013). Under reducing conditions, Cr(VI) is reduced to Cr(III) by redox reactions with aqueous inorganic species, electron transfer at mineral surfaces, reactions with non-humic organic substances such as carbohydrates and proteins, or reduction by soil humic substances (Barrera-Díaz *et al.*, 2012; Madhavi *et al.*, 2013). The speciation of Cr(VI) in aqueous solutions is shown in Section 1.2. As previously mentioned, Cr(VI) compounds are soluble and thus mobile in the environment. However, Cr(VI) oxyanions are readily reduced to trivalent forms via electron donors such as organic matter or reduced inorganic species, which are ubiquitous in soil, water and atmospheric systems (Barrera-Díaz *et al.*, 2012; Choppala *et al.*, 2013; Dhal *et al.*, 2013; Rengaraj *et al.*, 2001).

Hexavalent chromium ion is classified as a known human carcinogen by inhalation routes of exposure (IARC 1990; USEPA 1998). Increased risks of lung cancer for workers in various industries (electroplating, chrome pigment, mining, leather tanning, and chrome alloy production) due to exposure to airborne Cr(VI) has been well established by long-term epidemiological studies. In 2008, the National Toxicology Program published studies that concluded that oral ingestion of Cr(VI) caused an increase in the incidence of cancer in rats and mice. As a result of the Integrated Risk Information System (IRIS) the United States Environmental Protection Agency (USEPA) released a draft toxicological review of Cr(VI) human health effects (USEPA, 2010d). The USEPA has proposed to consider Cr(VI) as a possible carcinogen and to lower the oral reference dose for non-cancer effects by a factor of three. The Environmental Working Group reported occurrence of trace levels of Cr(VI) in 31 of 35 tap waters tested in the USA so far (Environmental Working Group, 2010). The current USEPA prescribed maximum contaminant level, (MCL) for total Cr is $100\,\mu g\,L^{-1}$ where total Cr is the sum of Cr(III) and Cr(VI). This MCL for total Cr is based on non-cancer health effects and does not account for the Cr(VI)-specific carcinogenicity which was recently established through animal studies and is still being studied widely. The proposed changes to IRIS suggest that the USEPA needs to review the existing MCL for total Cr. USEPA is

Table 1.1.   Regulatory limits of hexavalent and total chromium set by different
enforcing agencies.

Agency	Contaminant	Drinking water limit [$\mu g\,L^{-1}$]	Type of limit*
USEPA	Total Cr	100	MCL
EU Drinking Standards	Total Cr	50	MCL
Canada	Total Cr	50	MCL
India	Cr(VI)	50	MCL
State of California	Total Cr	50	MCL
	Cr(VI)	10	MCL
	Cr(VI)	0.02	PHG

*MCL = Maximum Contaminant Level; PHG = Public Health Goal

presently considering whether or not they will introduce an MCL specifically for Cr(VI) (USEPA, 2011a) for which evaluation of the existing total MCL for Cr could play a pivotal role. Based on the proposed changes to IRIS, a Cr(VI) specific MCL in the low part per billion to sub-part per billion range may likely get enforced.

The State of California has its own total MCL for Cr 50 $\mu g\,L^{-1}$ established in 1977 (California Department of Public Health, 2012). In 2011, the California Office of Environmental Health Hazard Assessment (OEHHA) established a non-enforceable Public Health Goal (PHG) for Cr(VI) as 0.02 $\mu g\,L^{-1}$ (California OEHHA, 2011). The PHG is a concentration standard set corresponding to a lifetime cancer risk level in excess of one in one million caused due to exposure to Cr(VI) in drinking water. The availability of a final PHG enables California Department of Public Health (CDPH) to proceed with setting a primary drinking water standard. In June 2014, the CDPH announced that the regulations establishing the drinking water standard for Cr(VI) have been approved by the Office of Administrative Law (OAL) and according to the law, the drinking water standard for Cr(VI) for the State of California became 10 $\mu g\,L^{-1}$ effective July 1, 2014. California Department of Public Health (CDPH, 2014). In the same line, the State of New Jersey has also considered whether to propose a state MCL for Cr(VI), with a health-based MCL estimated at 0.07 $\mu g\,L^{-1}$ (New Jersey Drinking Water Quality Institute, 2010). Table 1.1 compares regulatory levels for Cr that have been set by various agencies throughout the world. In India, the water quality is enforced on the concentration of Cr(VI) where the MCL is set at 50 $\mu g\,L^{-1}$. At present, Cr(VI) is not regulated in Europe or Canada, but the World Health Organization (WHO) published a Cr(VI)-specific regulatory recommendation setting the regulatory limit on total Cr at 50 $\mu g\,L^{-1}$.

1.1.2   *Arsenic toxicity*

Several studies have reported the toxic and carcinogenic effects in humans and animals after exposure to specific metal ions, such as copper, cadmium, chromium, mercury, and arsenic, among others. The coordinative chemistry, redox properties, molecular structure and characteristic reactivity of these metal ions enable its participation in the biological process, breaking down the mechanism of essential processes (Valko *et al.*, 2005).

Arsenic in the environment can be found in organic and inorganic compounds, and its absorption in humans can occur via inhalation, ingestion and skin absorption (Fig. 1.2). The highest levels of As in humans are normally found in hair, nails, and skin, while it exhibits lower levels in other human organs, such as the upper gastrointestinal tract, epididymis, thyroid, and skeleton (Fowler *et al.*, 2007). The main route of As excretion is through the kidneys. In fact, the presence of As in urine is the most reliable indicator of exposure. Inorganic As in the urine represents the unchanged

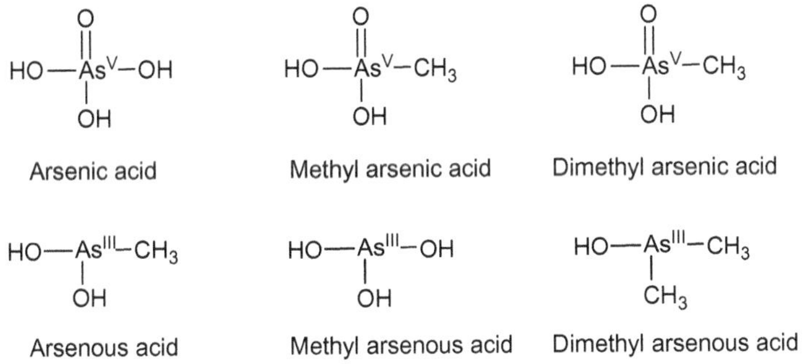

$$HO-\overset{\overset{\displaystyle O}{\|}}{\underset{\underset{\displaystyle OH}{|}}{As^V}}-OH \qquad HO-\overset{\overset{\displaystyle O}{\|}}{\underset{\underset{\displaystyle OH}{|}}{As^V}}-CH_3 \qquad HO-\overset{\overset{\displaystyle O}{\|}}{\underset{\underset{\displaystyle OH}{|}}{As^V}}-CH_3$$

Arsenic acid   Methyl arsenic acid   Dimethyl arsenic acid

$$HO-\underset{\underset{\displaystyle OH}{|}}{As^{III}}-CH_3 \qquad HO-\underset{\underset{\displaystyle OH}{|}}{As^{III}}-OH \qquad HO-\underset{\underset{\displaystyle CH_3}{|}}{As^{III}}-CH_3$$

Arsenous acid   Methyl arsenous acid   Dimethyl arsenous acid

Figure 1.2.   Structures of some toxicologically relevant inorganic and organic arsenic species.

part of As, while the MMA and DMA represent the methylation process (Kurttio *et al.*, 1998). The length of time for As elimination is 2 to 4 months once intake has ceased.

The species of As that are more rapidly absorbed are more toxic, while those that are most rapidly eliminated are less toxic. Arsenite and arsenate forms are highly water-soluble. Exposure to As can result in several health problems in humans, such as cardiovascular and peripheral vascular disease, diabetes, and even various forms of cancer, such as skin, lung, and bladder, among others. Arsenic is considered a non-typical carcinogen because its mechanism is not classified either as an initiating or promoting carcinogenic agent and because it most likely acts by enhancing the effect of other carcinogenic agents (Kenyon and Hughes, 2001). In fact, there is no animal model of carcinogenesis using inorganic As compound alone as a carcinogen; however, there are animal models that use other defined carcinogens combined with As. These studies have revealed the effect of As as a cancer promoter (Wang *et al.*, 2002).

The mechanisms of As toxicity are not well understood; however, some insights regarding its mechanisms of action have been identified. Arsenate oxyanions exhibit a tetrahedral structure that can replace phosphate ions in many biochemical reactions because of their similarity in molecular structure and properties. Arsenate can react with glucose and gluconate to form glucose-6-arsenate and 6-arsenogluconate, respectively, which have activities similar to those of compounds such as glucose-6-phosphate and 6-phosphogluconate (Lagunas, 1980). In addition, phosphate can be replaced in the sodium pump and the anion-exchange transport system of the human red blood cell (Kenney and Kaplan, 1988). However, trivalent arsenicals exhibit a high affinity with thiol-containing molecules such as glutathione (GSH) and cysteine, which may explain the inhibition of enzymatic reactions as well as its accumulation in keratin-rich tissues, such as hair, skin, and the epithelium of the upper gastrointestinal tract (Aposhian and Aposhian, 2006). The binding of As(III) to critical thiol groups, such as enzymes, receptors or coenzymes, may inhibit important biochemical events, which result in toxicity.

Organic arsenicals, methyl arsenic acid, and dimethyl arsenic acid have each been shown to generate oxidative stress via the inhibition of mitochondrial respiration, which results in the formation of reactive oxygen species. This subsequently causes DNA mutations and leads to the development of cancer (Liu *et al.*, 2005). The trivalent species methylarsenous acid exhibits the highest cytotoxicity, while dimethylarsenous acid has shown effects similar to those of inorganic trivalent As and pentavalent arsenicals, which are significantly less cytotoxic than their trivalent analogs (Styblo *et al.*, 2000). An important process performed in the organism is the methylation of inorganic arsenic and the reduction of arsenate to arsenite. This former process is of great practical importance because the methylation of inorganic As is considered the main detoxification pathway of inorganic As (Tseng, 2009). This process occurs in the liver where the methyl groups are transferred from S-adenosylmethionine to As in its trivalent form. However, both methylated and dimethylated trivalent arsenicals have been identified as intermediates in

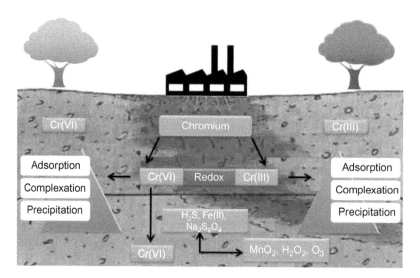

Figure 1.3.   Schematic illustration of chromium dynamics in nature.

the metabolic process. These studies have revealed that MMA and DMA are highly cytotoxic, genotoxic, and potent inhibitors of enzymatic activities compared to As(V) (Valko *et al.*, 2005).

1.1.3   *Chromium toxicity*

In the +3 oxidation state, Cr is also considered essential for glucose metabolism in humans and animals at concentrations of 50–200 mg day^{-1} (Dayan and Paine, 2001). However, sources of hexavalent Cr in the environment are nearly entirely anthropogenic in origin and include fossil fuel combustion, metal, alloy and wood industries, and petrochemical cooling towers (Barrera-Díaz *et al.*, 2012; Madhavi *et al.*, 2013). The Cr waste produced by these industries is commonly disposed of via burial or discharge into lagoons, and leakage from these disposal sites into the groundwater is prevalent. Cr(VI) is generally considered to be toxic and carcinogenic, and its high mobility in soils causes great environmental concern (Barrera-Díaz *et al.*, 2012; Lee and Tiwari, 2012) (Fig. 1.3). Specific oxides and organic compounds are known to convert Cr(III) back to Cr(VI), and recognition of this type of redox chemistry has resulted in the assessment that all Cr compounds are potentially carcinogenic. The carcinogenicity, toxicity and widespread contamination of Cr in populated areas have earned both species of this transition metal positions on the short list of USEPA priority pollutants (Ellis *et al.*, 2002; WHO, 2004).

Exposure to Cr(VI) can cause a range of adverse health effects in humans and animals. Excessive skin contact with chromate has been shown to cause skin ulcers and dermatitis, while inhalation of chromate compounds can result in the ulceration and perforation of the nasal septum and sinonasal cancer (Dayan and Paine, 2001; Langard and Costa, 2007). The concentrations of chromates are on the order of 50–70 mg kg^{-1} body weight, which is considered to be fatal in humans (Langard and Costa, 2007). The biological pathway of Cr toxicity is intricately governed by its speciation or oxidative state. Although Cr(VI) does not directly interact with biological ligands at the physiological level (Dayan and Paine, 2001); once Cr(VI) enters the blood stream, it is taken up by erythrocytes via carboxylate, sulfate and phosphate carrier systems and reduced to Cr(III) (Koedrith *et al.*, 2013). When Cr is present in cells, it can stimulate damage to the three-dimensional structure of DNA via oxygen radical formation, which causes DNA strand breaks, as well as DNA-DNA and DNA-protein cross-links. This subsequently leads to gene mutations and chromosomal abnormalities (Dayan and Paine, 2001). Cellular reducing agents, namely, thiol-containing compounds such as glutathione, cysteine and lipoic acid, appear to be a requirement for such reactions, suggesting that Cr(V) and Cr(IV) intermediates play an important role in the genotoxicity of Cr (Dayan and Paine, 2001; Langard and Costa, 2007). The mechanism involved in the carcinogenic effect of Cr is currently undetermined but likely involves DNA damage due to the presence of reactive Cr intermediates, such as Cr(V) and Cr(IV) in cells (Dayan and Paine, 2001; Koedrith *et al.*, 2013).

1.2 ARSENIC AND CHROMIUM IN AQUEOUS SOLUTIONS

For particular methods of As or Cr removal it is important to know in what conditions the components containing As or Cr are ionized. When discussing the content of species in different oxidation states (As(III) and As(V), Cr(III) and Cr(VI)) the diagrams Eh-pH, where Eh is the potential of reduction for half-reaction, are useful (Schweitzer and Pesterfield, 2010).

1.2.1 *Arsenic speciation*

According to Sharma and Sohn (2009) As can occur in inorganic and organic compounds. In natural waters As occurs mostly as oxyanions of trivalent arsenite (As(III)) or pentavalent arsenate (As(V)) (Smedley and Kinniburgh, 2002). Depending on pH and redox potential (Eh), the following forms are found also: arsenous acid (H_3AsO_3, $H_2AsO_3^-$, $HAsO_3^{2-}$), arsenic acid (H_3AsO_4, $H_2AsO_4^-$, $HAsO_4^{2-}$), as well as the methyl- and dimethyl derivatives. Figure 1.4 presents Eh as the function of pH for different As forms.

At pH values below 6, which correspond to most wastewaters, As is present in forms (V) and (III) (Fig. 1.4). As(V) can be transformed to As(III) when the value of redox potential is 0.25 V or below.

In natural waters As occurs mostly as oxyanions of trivalent arsenite (As(III)) or pentavalent arsenate (As(V)) (Smedley and Kinniburgh, 2002). Although the equilibria involving the As species can be numerous (Couture *et al.*, 2010), below only the basic set of equilibria for As(III) and As(V) is shown (the pK_a values are from Sharma and Sohn, 2009):

a) As(III):

$$H_3AsO_3 \rightleftarrows H_2AsO_3^- + H^+ \qquad pK_{a,1} = 9.2 \qquad (1.1)$$

$$H_2AsO_3^- \rightleftarrows HAsO_3^{2-} + H^+ \qquad pK_{a,2} = 12.1 \qquad (1.2)$$

$$HAsO_3^{-2} \rightleftarrows AsO_3^{3-} + H^+ \qquad pK_{a,3} = 12.7 \qquad (1.3)$$

b) As(V):

$$H_3AsO_4 \rightleftarrows H_2AsO_4^- + H^+ \qquad pK_{a,1} = 2.3 \qquad (1.4)$$

$$H_2AsO_4^- \rightleftarrows HAsO_4^{2-} + H^+ \qquad pK_{a,2} = 6.8 \qquad (1.5)$$

$$HAsO_4^{-2} \rightleftarrows AsO_4^{3-} + H^+ \qquad pK_{a,3} = 11.6 \qquad (1.6)$$

According to Ibáñez and Cifuentes (2004) in the acidic solutions also the AsO^+ cation is present. They calculated the As speciation using the thermodynamic data from the work of Wagman *et al.* (1982). From these data it results that the thermodynamic constant of the reaction forming AsO^+:

$$H_3AsO_3 + H^+ \rightleftarrows AsO^+ + 2H_2O \qquad (1.7)$$

equals 0.496. The pH dependence of ionic forms of As(III) and As(V) calculated according to Equations (1.1)–(1.7) is shown in Figure 1.5a,b.

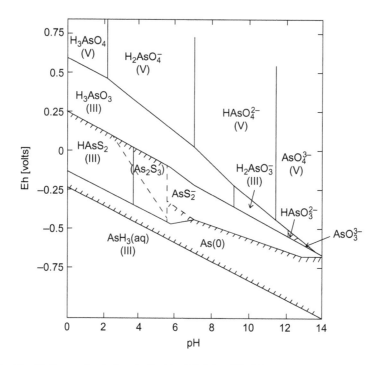

Figure 1.4.   Eh-pH diagram for As ions in liquid at 25°C (modified from Ferguson and Gavis, 1972).

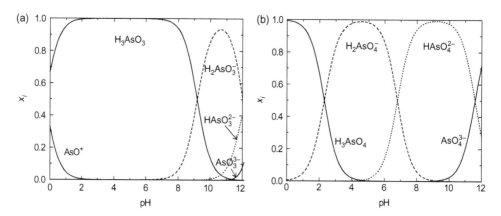

Figure 1.5.   Dependence of the molar fraction, $x_i = n_i/n_{As,tot}$, of various forms of As(III) (a) and As(V) (b) on pH, calculated using the pK_a values given in the text (approximate calculations – the activity coefficients were not taken into account).

Regarding methylarsenic and dimethylarsenic acids the acid-base equilibria are as follows (Sharma and Sohn, 2009):

$$CH_3AsO_3H_2 \rightleftarrows CH_3AsO_3H^- + H^+ \qquad pK_{a1} = 4.1 \qquad (1.8)$$

$$CH_3AsO_3H^- \rightleftarrows CH_3AsO_3H^- + H^+ \qquad pK_{a2} = 8.7 \qquad (1.9)$$

$$(CH_3)_2AsO_2H \rightleftarrows (CH_3)_2AsO_2^- + H^+ \qquad pK_{a1} = 6.2 \qquad (1.10)$$

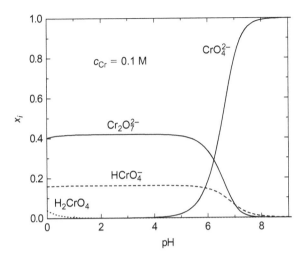

Figure 1.6.   Molar fraction of different forms of Cr(VI), $x_i = n_i/n_{tot,Cr}$ vs. pH.

1.2.2   *Chromium speciation*

The chemistry of Cr in water is not so simple. It can occur in various oxidation states. Because of their stability only Cr(III) and (VI) are discussed here. According to Cotton (Cotton and Wilkinson 1972), Cr(III) forms complexes $[CrX_6]^{3-}$ with $X = H_2O$, F^-, CN^-, SCN^-, the amino-aquo complexes $[CrAm_{6-n}(H_2O)_n]^{3+}$ ($n = 0$–4, 6), and many other complexes. Cr(III) forms hexacoordinated complexes of octahedral geometry. Two hydrolysis reactions of Cr^{3+} are possible (Baes and Messmer, 1976):

$$Cr^{3+} + H_2O \rightleftarrows CrOH^{2+} + H^+ \qquad pK_{h,1} = 3.4\text{–}3.8 \qquad (1.11)$$

$$Cr^{3+} + 2H_2O \rightleftarrows Cr(OH)_2^+ + 2H^+ \qquad pK_{h,2,c} = 9.7 \qquad (1.12)$$

From the values of hydrolysis constants it results that for pH < 3 the dominant ion is Cr^{3+}.

Depending on the solution composition, other Cr(III) complexes should be taken into account. This is clearly demonstrated by Molik *et al.* (2004). Apart of that Cr(III) forms polymeric species also (Rao *et al.*, 2002):

$$pCr^{3+} + qH_2O \rightleftarrows Cr_p(OH)_q^{+(3p-q)} + qH^+ \qquad (1.13)$$

At pH 3.5–4 hydrated Cr_2O_3 precipitates (Schweitzer and Pesterfield, 2010).

Regarding Cr(VI) it is present in oxyanions which take part in the following equilibria (Cotton and Wilkinson, 1972):

$$H_2CrO_4 \rightleftarrows HCrO_4^- + H^+ \qquad pK_{a,1} = -0.61 \qquad (1.14)$$

$$HCrO_4^- \rightleftarrows CrO_4^{2-} + H^+ \qquad pK_{a,2} = 5.9 \qquad (1.15)$$

$$Cr_2O_7^{2-} + H_2O \rightleftarrows 2HCrO_4^- \qquad pK = 2.2 \qquad (1.16)$$

The distribution of the Cr(VI) forms expressed as the molar ratio, $x_i = n_i/n_{tot,Cr}$, is a function of pH (Fig. 1.6). It is a simplified picture because the presence of polychromates ($Cr_3O_{10}^{2-}$ and $Cr_4O_{13}^{2-}$) at low pH values should be also taken into account (Sarmaitis *et al.*, 1996; Schweitzer and Pesterfield, 1996).

According to the calculations made by Vallejo *et al.* (2000), based on the data of (Sarmaitis *et al.*, 1996), the content of $Cr_3O_{10}^{2-}$, in a solution of pH $= 1$ and $c_{Cr,tot} = 0.1$ M, is 13%, and it increases to 28% for $c_{Cr,tot} = 1$ M at the same pH. At pH > 2 the content of polychromates $(Cr_3O_{10}^{2-}, Cr_4O_{13}^{2-})$ decreases practically to zero.

REFERENCES

Aposhian, H.V. & Aposhian, M.M. (2006) Arsenic toxicology: five questions. *Chemical Research in Toxicology*, 19, 1–15.

Baes, Ch.F. & Messmer, R.E. (1976) *The hydrolysis of cations*. J. Wiley & Sons, New York, NY.

Barrera-Díaz, C.E., Lugo-Lugo, V. & Bilyeuc, B. (2012) A review of chemical, electrochemical and biological methods for aqueous Cr(VI) reduction. *Journal of Hazardous Materials*, 223–224, 1–12.

Bissen, M. & Frimmel, F.H. (2003a) Arsenic – a review. Part I: Occurrence, toxicity, speciation, mobility. *Acta Hydrochimica et Hydrobiologica*, 31, 9–18.

Bissen, M. & Frimmel, F.H. (2003b) Arsenic – a review. Part II: Oxidation of arsenic and its removal in water treatment. *Acta Hydrochimica et Hydrobiologica*, 31, 97–107.

California Department of Public Health. (2014) *First drinking water standard for hexavalent chromium now final*. Available from: http://www.cdph.ca.gov/Pages/NR14-053.aspx [accessed 15th July 2014].

California Office of Environmental Health Hazard Assessment (OEHHA). (2011) Final technical support document on public health goal for hexavalent chromium in drinking water. Available from: http://www.oehha.ca.gov/water/phg/072911Cr6PHG.html [accessed 29th July 2011].

Choppala, G., Bolan, N. & Park, J.H. (2013) Chapter two: Chromium contamination and its risk management in complex environmental settings. *Advances in Agronomy*, 120, 129–172.

Cotton, F.A. & Wilkinson, G. (1972) *Advanced inorganic chemistry. A comprehensive text*. Interscience Publishers, J. Wiley & Sons, New York, NY.

Couture, R.-A., Gobeil, C. & Tessier, A. (2010) Arsenic, iron and sulfur co-diagenesis in lake sediments. *Geochimica et Cosmochimica Acta*, 74, 1238–1255.

Dayan, A.D. & Paine, A.J. (2001) Mechanisms of chromium toxicity, carcinogenicity and allergenicity: review of the literature from 1985 to 2000. *Human & Experimental Toxicology*, 20, 439–451.

Dhal, B., Thatoi, H.N., Das, N.N. & Pandey, B.D. (2013) Chemical and microbial remediation of hexavalent chromium from contaminated soil and mining/metallurgical solid waste: a review. *Journal of Hazardous Materials*, 250–251, 272–291.

Ellis, A.S., Johnson, T.M. & Bullen, T.D. (2002) Chromium isotopes and the fate of hexavalent chromium in the environment. *Science*, 295, 2060.

Environmental Working Group (2010) Chromium-6 is widespread in US tap water. Available from: http://www.ewg.org/chromium6-in-tap-water [accessed 7th July 2011].

Ferguson, J.F. & Gavis, J. (1972) A review of the arsenic cycle in natural waters. *Water Research*, 6, 1259–1274.

Fowler, B.A., Selene, C.H., Chou, J., Jones, R.L. & Chen, C.J. (2007) Arsenic. In: Nordberg, G.F., Fowler, B.A., Nordberg, M. & Friberg, L. (eds.) *Handbook on the toxicology of metals*. Academic Press, Inc. p. 367.

García-Delgado, M. (2007) Seasonal and time variability of heavy metal content and of its chemical forms in sewage sludges from different wastewater treatment plants. *Science of the Total Environment*, 382, 82–92.

Gupta, V.K. & Suhas. (2009) Application of low-cost adsorbents for dye removal – a review. *Journal of Environmental Management*, 90, 2313–2342.

Han, F.X., Su, Y., Monts, D.L., Plodinec, M.J., Banin, A. & Triplett, G.E. (2003) Assessment of global industrial-age anthropogenic arsenic contamination. *Naturwissenschaften*, 90, 395–401.

Howard, G., Bartram, J., Pedley, S., Schmoll, O., Chorus, I. & Berger, P. (2006) Groundwater and public health. In: *Protecting groundwater for health*. World Health Organization, Cornwall, UK.

IARC (1990) *IARC Monographs on the Evaluation of Carcinogenic Risks to Humans*: Volume 49: *Chromium, nickel and welding*. International Agency for Research on Cancer, Lyon, France.

Ibáñez, J.P. & Cifuentes, L. (2004) On the kinetics of Cu, As and Sb transport through cation and anion-exchange membranes in acidic electrolytes. *Canadian Metallurgical Quarterly*, 43, 439–448.

Kanmani, P., Aravind, J. & Preston, D. (2012) Remediation of chromium contaminants using bacteria. *International Journal of Environmental Science & Technology*, 9, 183–193.

Kenney, L.J. & Kaplan, J.H. (1988) Arsenate substitutes for phosphate in the human red cell sodium pump and anion exchanger. *The Journal of Biological Chemistry*, 263, 7954–7960.

Kenyon, E.M. & Hughes, M.F. (2001) A concise review of the toxicity and carcinogenicity of dimethylarsinic acid. *Toxicology*, 160, 227–236.

Koedrith, P., Kim, H., Weon, J.-I. & Seo, Y.R. (2013) Toxicogenomic approaches for understanding molecular mechanisms of heavy metal mutagenicity and carcinogenicity. *International Journal of Hygiene and Environmental Health*, 216 (5), 587–598.

Kurttio, P., Komulainen, H., Hakala, E., Kahelin, H. & Pekkanen, J. (1998) Urinary excretion of arsenic species after exposure to arsenic present in drinking water. *Archives of Environmental Contamination and Toxicology*, 34, 297–305.

Lagunas, R. (1980) Sugar-arsenate esters: thermodynamics and biochemical behavior. *Archives of Biochemistry and Biophysics*, 205, 67–75.

Langard, S. & Costa, M. (2007) Chromium. In: *Handbook on the toxicology of metals*. Academic Press, Burlington, MA. pp. 487–510.

Liu, S.-X., Davidson, M.M., Tang, X., Walker, W.F., Athar, M., Ivanov, V. & Hei, T.K. (2005) Mitochondrial damage mediates genotoxicity of arsenic in mammalian cells. *Cancer Research*, 65, 3236–3242.

Lee, S.M. & Tiwari, D. (2012) Organo and inorgano-organo-modified clays in the remediation of aqueous solutions: an overview. *Applied Clay Science*, 59–60, 84–102.

Madhavi, V., Bhaskar Reddy, A.V., Reddy, K.G., Madhavi, G. & Prasad, T.N.K.V. (2013) An overview on research trends in remediation of chromium. *Research Journal of Recent Science*, 2, 71–83.

Malaviya, P. & Singh, A. (2011) Physicochemical technologies for remediation of chromium-containing waters and wastewaters. *Critical Reviews in Environmental Science & Technology*, 41, 1111–1172.

Mandal, B.K. & Suzuki, K.T. (2002) Arsenic around the world: a review. *Talanta*, 58, 201–235.

Meng, X., Korfiatis, G.P., Christodoulatos, C. & Bang, S. (2001) Treatment of arsenic in Bangladesh well water using a household co-precipitation and filtration system. *Water Research*, 35, 2805–2810.

Molik, A., Siepak, J., Świetlik, R. & Dojlido, J.R. (2004) Identification of chromium species in tanning solutions. *Polish Journal of Environmental Studies*, 13, 311–314.

Morris, B.L., Lawrence, A.R.L., Chilton, P.J.C., Adams, B., Calow, R.C. & Klinck, B.A. (2003) Groundwater and its susceptibility to degradation: a global assessment of the problem and options for management. Early Warning and Assessment Report Series. United Nations Environment Programme, Nairobi, Kenya.

New Jersey Drinking Water Quality Institute, Testing Subcommittee Meeting Minutes. (2010) Available from: http://www.state.nj.us/dep/watersupply/pdf/minutes100224.pdf [accessed 7th July 2011].

Ng, J.C. (2005) Environmental contamination of arsenic and its toxicological impact on humans. *Environmental Chemistry*, 2, 146–160.

Qi, Y. & Donahoe, R.J. (2008) The environmental fate of arsenic in surface soil contaminated by historical herbicide application. *Science of the Total Environment*, 405, 246–254.

Rao, L., Zhang, Z., Friese, J.I., Ritherdon, B., Clark, S.B., Hess, N.J. & Rai, D. (2002) Oligomerization of chromium(III) and its impact on the oxidation of chromium(III) by hydrogen peroxide in alkaline solutions. *Journal of the Chemical Society, Dalton Transactions*, 2, 267–274.

Rengaraj, S., Yeon, K.-H. & Moon, S.-H. (2001) Removal of chromium from water and wastewater by ion exchange resins. *Journal of Hazardous Materials*, 87, 273–287.

Saeedi, M., Li, L.V. & Gharehtapeh, A. (2013) Effect of alternative electrolytes on enhanced electrokinetic remediation of hexavalent chromium in clayey soil. *International Journal of Environmental Research*, 7, 39–50.

Saha, B. & Orvig, C. (2010) Biosorbents for hexavalent chromium elimination from industrial and municipal effluents. *Coordination Chemistry Reviews*, 254, 2959–2972.

Sarmaitis, R., Dikinis, V. & Rezaite, V. (1996) Equilibrium in solutions of chromic acid. *Plating and Surface Finishing*, 83, 53–57.

Schweitzer, G.K. & Pesterfield, L.L. (2010) *The aqueous chemistry of the elements*. Oxford University Press, Inc., New York, NY.

Sharma, V.K. & Sohn, M. (2009) Aquatic arsenic: toxicity, speciation, transformations, and remediation. *Environment International*, 35, 743–759.

Smedley, P.L. & Kinniburgh, D.G. (2002) A review of the source, behavior and distribution of arsenic in natural waters. *Applied Geochemistry*, 17, 517–568.

Styblo, M., Del Razo, L.M., Vega, L., Germolec, D.R., LeCluyse, E.L., Hamilton, G.A., Reed, W., Wang, C., Cullen, W.R. & Thomas, D.J. (2000) Comparative toxicity of trivalent and pentavalent inorganic and methylated arsenicals in rat and human cells. *Archives of Toxicology*, 74, 289–299.

Sullivan, C., Tyrer, M., Cheeseman, C.R. & Graham, N.J.D. (2010) Disposal of water treatment wastes containing arsenic – a review. *Science of the Total Environment*, 408, 1770–1778.

Tseng, C.H. (2009) A review on environmental factors regulating arsenic methylation in humans. *Toxicology and Applied Pharmacology*, 235, 338–350.

USEPA, IRIS (1998) Chromium. Available from: http://www.epa.gov/ncea/iris/subst/0144.htm [accessed 26th July 2011].

USEPA, IRIS (2010) Toxicological review of hexavalent chromium (external review draft). EPA/635/R-10/004A.

USEPA (2011) Chromium in drinking water. Available from: http://water.epa.gov/drink/info/chromium/index.cfm [accessed 24th July 2011].

Valko, M., Morris, H. & Cronin, M.T.D. (2005) Metals, toxicity and oxidative stress. *Current Medicinal Chemistry*, 12, 1161–1208.

Vallejo, M.E., Persin, F., Innocent, C., Sistat, P. & Pourcelly, G. (2000) Electrotransport of Cr(VI) through an anion-exchange membrane. *Separation and Purification Technology*, 21, 61–69.

Wagman, D.D., Evans, W.H., Parker, V.B., Schumm, R.H., Halow, I., Bailey, S.M., Churney, K.L. & Nuttall, R.L. (1982) The NBS tables of chemical thermodynamics, selected values for inorganic and C1 and C2 organic substances in SI units. *Journal of Physical and Chemical Reference Data*, 11 (Suppl. 2), 392.

Wang, J.P., Qi, L., Moore, M.R. & Ng, J.C. (2002) A review of animal models for the study of arsenic carcinogenesis. *Toxicology Letters*, 133, 17–31.

WHO (2004) Guidelines for drinking-water quality. World Health Organization, Geneva, Switzerland.

Zagorodni, A.A. (2007) *Ion exchange materials properties and applications*. Elsevier BV, Amsterdam, The Netherlands.

CHAPTER 2

The clean water problem – example of arsenic contamination in the Gangetic Plain of India

Ashok Ghosh & Nupur Bose

2.1 INTRODUCTION

The problem of safe drinking water is one of the major challenges faced by the human population. With the growth of human population to around seven billion, an estimated 780 million persons (WHO/UNICEF, 2012) do not have access to clean drinking water, while 2.6 billion persons – half the developing world, have no access to water for sanitation purposes. As an increasing number of surface fresh water sources either dry up or are heavily polluted, communities are exploiting the groundwater reserves to meet their water demands. Groundwaters have relatively limited or no harmful microbial content, and are rich in minerals, and, therefore, they are regarded as the safest option for natural fresh water.

However, especially in developing regions, increasing dependence on groundwater sources has led to its overexploitation and even to accessing contaminated shallow aquifers. Since the last four decades, groundwater contaminations have emerged as a serious health threat to the consumers of such contaminated water. These contaminants include arsenic, lead, iron, fluoride, nitrate, mercury and others, with arsenic (As) leading as the major contaminant found in almost all the major countries of the world.

Arsenic is a metalloid that is widely present in the earth-crust as oxides or sulfides or as a salt of iron, sodium, calcium, copper and others (Singh *et al.*, 2007). The most abundant As ore mineral is arsenopyrite (FeAsS). The biogeochemical cycle is complex involving the transformation of As to different oxidation states and interactions with solid and organic materials. Arsenic may cycle between the atmosphere, soil, water and sediments (Smith *et al.*, 2003). However, the vast majority of As reserves are associated with the regolith materials. Inorganic As is found in compounds with oxygen, sodium, potassium, copper, chlorine, iron and sulfur. Organic As is formed when As in plants and animals combine with carbon and hydrogen. Trivalent As (As(III)) is highly toxic and has acute, sub-acute and chronic effects on the human body. It is a known carcinogen (Toor and Tahir, 2009). Inorganic As has been assigned to Group I of human carcinogens by the International Agency for Research on Cancer because there is sufficient evidence, from epidemiological studies, to support a causal association between inorganic-As and cancer.

Sources of As in groundwater are either geogenic (due to changes in the hydrogeological environment that promote the leaching of As from the rock strata to the aquifer water) or anthropogenic (due to mining, industrial or other human activities). The As content of the crustal materials vary considerably but generally sedimentary materials contain greater Arsenic concentrations than igneous materials. Arsenic concentrations reported in igneous materials typically range from 1.5 to 3.0 mg As kg^{-1} whereas as concentrations in sedimentary materials range from 1.7 to 400 mg As kg^{-1} (Bhumbla and Keefer, 1994). Of the two, the impact of geogenic As poisoning on the affected population is greater in intensity and spread – a fact that has challenged, with its multi-dimensional facets, the hitherto accepted mitigation norms. 80% of the As contaminated water samples from the shallow aquifers in Bihar have tested positive for trivalent arsenic (As(III)) (Mukherjee *et al.*, 2006), this type being the most life threatening (Smith *et al.*, 2000).

2.2 ORIGIN OF ARSENIC IN GROUNDWATER

Geological stratigraphy, hydrogeology and sedimentary geochemistry determine the occurrence and mobility of As in groundwater. Regarding the sources of As, there are over 200 minerals by which As may be released (Smedley and Kinniburgh, 2002). Approximately 60% of natural As minerals are arsenates, 20% sulfides and sulfosalts, and the remaining 20% are arsenides, arsenites, oxides, alloys, and polymorphs of elemental As (Plant, 2004).

The presence of Arsenic in groundwater has been commonly related to the dissolution of sulfidic minerals, (Smedley and Kinniburgh, 2001) dissolution or desorption from iron oxyhydroxides (Bhattacharya *et al.*, 1997; Nickson *et al.*, 1998) and/or upflow of geothermal water (Welch, 2000). In some cases, natural weathering and leaching of As from geological formations containing As, like arsenopyrite give rise to high As values (Naidu and Nadebaum, 2003). Arsenic occurs widely in aquifers of deltaic sediments (Acharya *et al.*, 1999), near zones of orogeny, and in deep sandy aquifer layers as fluvial deposits. It is introduced into the aquifer sediments in soluble state and gets adsorbed on Fe-rich clastic grains and authigenic siderite concentrations. Oxidation occurs either by infiltration of oxygenated groundwaters (Mueller *et al.*, 1998), or by lowering of groundwater table into a stratigraphic zone of As-rich sulfides (Schreiber, 2000). The adsorption process and its consequent desorption are stated to be controlled by microbial activity within the concerned aquifers. Sediments containing 1 to 20 $\mu g\,kg^{-1}$ of As can give rise to high dissolved As of $>50\,\mu g\,L^{-1}$ by one or both of two possible causes – an increase of pH of over 8.5 or the onset of reductive Fe dissolution (Smedley and Kinniburgh, 2002). The presence of solutes can also decrease or prevent the adsorption of arsenate and arsenite ions onto fine-grained clays, like Fe oxides. Adsorption to hydrous Al and Mn oxides may also be important if these oxides are present in quantity. Finally, As may also be sorbed to the edges of clays and on the surface of calcite. Arsenic sorption by minerals is the main reason for low and non-toxic concentrations of As found in most natural waters (Smedley and Kinniburgh, 2002). Besides these minerals, As can also be encountered in the crystal structure of rock-forming minerals (pyrite, chalcopyrite, and galena) by the substitution of As for Fe or Al and in sulfide minerals, as a substitute for sulfur (Plant *et al.*, 2004; Smedley and Kinniburgh, 2002).

As far as the different sedimentary rocks and unconsolidated depositions are concerned, sands and sandstones tend to have the lowest arsenic concentrations, reflecting the low As concentrations of their dominant minerals, which are quartz and feldspar. Clays, fine sediments and shales, on the other hand, and especially those that are rich in sulfide minerals, organic matter, secondary Fe oxides, and phosphates are expected to have high As concentrations (Plant *et al.*, 2004). In black shales, these high concentrations are related to their high pyrite content. In addition, all Fe and Mn-rich sediments show high As content. The interaction between minerals and water comprises the key steps towards As release in groundwater. Thorough research on this topic has shown that sediments in the groundwater release As after flooding and anaerobic conditions have occurred. Redox potential (Eh) and pH are the most important factors controlling As speciation in natural waters (Plant *et al.*, 2004; Smedley and Kinniburgh, 2002). From a purely chemical point of view, the two main triggering mechanisms for As release in groundwater are:

- The development of high pH (>8.5) conditions in semi-arid or arid environments usually as a result of the combined effects of mineral weathering and high evaporation rates. This pH change leads either to the desorption of adsorbed As (especially pentavalent As) species) and a range of other anion-forming elements (V, B, F, Mo, Se and U) from mineral oxides, especially Fe oxides, or it prevents them from being adsorbed.
- The development of strongly reducing conditions at near-neutral pH values, leading to the desorption of As from mineral oxides and to the reductive dissolution of Fe and Mn oxides, also leading to As release. Iron (II) and As(III) are relatively abundant in these types of groundwater and SO_4 concentrations are small (typically $1\,mg\,L^{-1}$ or less). Large concentrations of phosphate, bicarbonate, silicate and possibly organic matter can enhance desorption of As because of competition for adsorption sites.

Regarding As transport in groundwater, there are few observations of As transport in aquifers, and its rate of movement is poorly understood. The transport of As, as that of many other chemicals, is closely related to adsorption-desorption reactions. Arsenate and arsenite have different adsorption isotherms. They, therefore, travel through aquifers at different velocities, and tend to be separated (Smedley and Kinniburgh, 2002). Various studies indicate a rather low partition coefficient (Kd) or retardation factor of As species. Factors controlling the partition coefficients are also poorly understood, and they involve the chemistry of groundwater, and the surface chemistry and stability of the solid phases (Plant *et al.*, 2004).

2.3   SPATIAL CHARACTERISTICS OF ARSENIC CONTAMINATED GROUNDWATER

Geographical locations of As contaminated groundwater sources can be categorized into: geogenic As derived from organic-rich or black shales, alluvial sediments of Holocene period, closed basins of arid and semi-arid climates in volcanic regions, strongly reducing aquifers usually composed of alluvial sediments having low sulfate concentrations (Nordstrom, 2002); and, As contaminations resulting from anthropogenic activities, as in mining and certain industrial processes. Till 2009, there were about 70 countries which reported cases of As contaminated drinking water sources that threatened the lives of an estimated 150 million people. Of this around 110 million persons resided in south and southeast Asia (Ravenscroft *et al.*, 2009). By 2012, 68 countries had been identified with geogenic As contaminated groundwater till 2012 (Murcott, 2012). Countries with reducing environments accountable of presence of As in groundwater include India, Bangladesh, Taiwan, north China, Taiwan, Vietnam, Hungary and Romania. The arid oxidizing environments of Mexico, Chile and Argentina have led to release of As in the human environment, while mixed oxidizing and reducing environments have resulted in mobilization of As in south western USA. Geothermal spring waters in parts of USA, Japan, New Zealand, Iceland, France, Turkey and other areas also have high As content. Lastly, large quantities of As leaches onto the fresh water sources in areas where sulfide mineralization and mining activities abound, as in Thailand, Ghana, USA, Greece and South Africa, and countries of South America (Smedley and Kinniburgh, 2002).

2.4   AREAS OF GEOGENIC ARSENIC

Naturally occurring As in groundwater is most commonly released from Fe oxide. Reducing conditions in which As sorbs to metallic oxides like Fe(III) causes elevated As concentrations in aquifers. However in anaerobic conditions, Fe(III) reducing microorganisms can couple the reduction of solid-phase Fe(III) with the oxidation of organic matter, releasing As to the groundwater. Bacteria thriving in As geochemical environments convert As into species with different solubility, mobility, bioavailability and toxicity, by reduction, oxidation, and methylation (Silver and Phung, 2005; Stolz and Oremland, 1999).

In Africa, the As contaminated groundwater sources originate in the areas of the Achaean gold reserves and the salt pans of Holocene lakes in Botswana; the phosphogenic province of late Cretaceous-early Paleogene era in Egypt; areas of gold-arsenic mineralization in Ghana; limestone and black-shale aquifers in Nigeria; the crystalline basement rocks, sedimentary formations as well as the fine-grained clays of Kalahari Formations in Zambia; and the gold bearing strata of central Zimbabwe. Countries like Ethiopia, Sudan, and Morocco have known arsenic contaminated drinking water sources (Ravenscroft, 2007).

Very few studies have been undertaken on the As contaminated ground water of Latin American and South American countries, but an emerging crisis of As contaminated water resource in the developing countries of this continent have been widely recognised. The first reports of As contaminated groundwater came from Argentina between 1913–1917 (Goyenechea, 1917). At least 14 Latin American countries have reported arsenic in their groundwater deposits in the last 10–15 years. Mexico, Guatemala, El Salvador, Honduras, Nicaragua, and Costa Rica

(Bundschuh *et al.*, 2011). In Argentina, Peru, Chile, Bolivia and Peru arsenic-rich sediment deposits have rendered more than 14 million people vulnerable to arsenic poisoning through drinking and irrigation waters. The As sources are predominantly volcanic in origin located in the Andean Cordilleras (Bundschuh *et al.*, 2011), although the Tertiary loess deposits in the Pampas, the Tertiary-Quaternary fluvial and Aeolian deposits in the Gran Chaco Plains, and the lacustrine deposits of Bolivia also carry sufficient As bearing minerals. Speciation studies have indicated that As III is the predominant form of geogenic As that plagues the lives of the vulnerable population. Bundschuh and others have identified three major processes by which As mobilization occurs in Latin America as – by sulfide oxidation in mineralized areas; by formation of secondary As minerals; and, by As remobilization from metal oxides and oxyhydroxides (Bundschuh *et al.*, 2011).

The largest identified area having high As concentration in groundwater is the 1 million km^2 Chaco-Pampean Plain (Bhattacharya *et al.*, 2004; Bundschuh *et al.*, 2004; 2005; Nicolli *et al.*, 2010). About eight million inhabitants are at risk from this contamination (Bunschuh *et al.*, 2011). However, apart from geogenic sources, anthropogenic causes also contribute to the contamination in Mexico, especially from mine wastes and smelter fumes (Romero *et al.*, 2006; 2008; Talavera Mendoza *et al.*, 2006). In Chile, arsenic is found in the Atacama Desert, in the river sediments originating in the Andes, and also in the Holocene lacustrine deposits. Similar conditions in Peru have resulted in usage of As contaminated water for irrigation and animal husbandry (Bundschuh *et al.*, 2011), thereby increasing the infiltration of this As in the human environment. In Peru, geogenic As is due to natural mobilization in geological sources, while in Uruguay, As is derived from the volcanic ash deposits of the Raigon Formation.

In United States of America, As contaminated groundwater occurs in both reducing and oxidizing environments in 36 states. Major contamination sites include the Basin and Range area in Arizona, San Joaquin Valley in California and south Carson Desert in Nevada (Welsh *et al.*, 2000). In Europe, As abounds in the highly mineralized areas of Spain, France, Germany, Switzerland, Sweden, Finland, Hungary, Italy, Poland, Portugal and Romania. in countries such as Hungary, Romania, Slovakia, Serbia, Croatia, Greece, Italy and Spain, elevated arsenic concentrations have been detected requiring special treatment to reduce the As to acceptable levels (van Halam *et al.*, 2009).

The rich mineral deposits of Australia are a natural source of arsenic contamination, in groundwater of coastal sand dune environment, and in western Australia. However, mining activities, arsenic solutions used for cattle dips, and use of pesticides have led to serious health implications in areas like Victoria and northern New South Wales (Smith *et al.*, 2005) In New Zealand, As is found in the Waikato river basin and Ohakuri Lake (Murcott, 2012).

Asia has a large number of geogenic As contaminated sources of groundwater, mostly in its eastern and southern segments. In east and south Asia, parts of China and Taiwan, Bangladesh, India, Indonesia, Vietnam and Cambodia, have high As contaminations and are also densely populated.

In 2004, 21 out of 30 provinces of China were identified with As contaminated groundwater of >50 µg L^{-1}. Most of the contaminations cluster in northwestern and northeastern parts of the country. The Holocene sediments of the deep artesian aquifers in the arid provinces of Inner Mongolia and Xinjiang Uighur Autonomous region, in China promote mobilization of geogenic As under strongly reducing conditions. Anaerobic Holocene aquifers also abound in Shanxi and Taiwan, the remaining areas of China coming under anthropogenic As contamination. However, the spatial extent of the geogenic As hotspots is larger, covering 137 towns and 942 villages, in Inner Mongolia, Xinjiang and Shanxi, while the intensity of provinces reeling under anthropogenic As is greater, reaching >4000 µg L^{-1} in northeast China. In Taiwan, reductive aquifer systems having Holocene and Pleistocene clay deposits leach out As in the southwestern ChiaNan Plains (Lin *et al.*, 2006). Groundwater As is termed as "the persistent contaminant" in the Red River Delta in Vietnam and the Mekong Basin in Cambodia (Berg *et al.*, 2007). Both Holocene and Pleistocene deposits account for high As levels here, as over extraction of groundwater extends the downward shift of the reducing environment and mobilizes As even in deeper aquifers. Geogenic

As has also been detected in the peat swamps of Sumatra. These lowlands are drained by the Calik, the Banuyasin and the Musi rivers, having a thin veneer of Holocene alluvium, which promotes leaching of As into the aquifers (Murcott, 2012).

Countries of south Asia that possess similar geological environs have reported widespread areas of geogenic As contaminations in their fresh water sources. These include Bangladesh, India, Myanmar, Nepal and Pakistan (Bhattacharya *et al.*, 2011).

In Bangladesh, the principal sources of As exposure are through drinking water drawn from contaminated aquifers in Holocene deposits under reducing conditions. There was a rapid increase in the utilization of groundwater in the decade of the 1970s. By the nineties, symptoms of arsenicosis resulting from consumption of contaminated groundwater were established (Smith *et al.*, 2002). The levels of As vary from 0.5 to over 3000 $\mu g\,L^{-1}$ (Smedley and Kinniburgh, 2002). Presently, 59 out of 64 districts, located in southern and eastern parts are affected. Maximum concentration of As hotspots lie along the Meghna and Brahmaputra rivers, and in the east the depth of the tube wells reach up to 80 m. The worst affected districts are Chandpur and Munshiganj, followed by large parts of Comilla, Noakhali, Faridpur, Madaripur, Gopalganj and others. Here a number of sources of high As concentrations often exceed 2000 $\mu g\,L^{-1}$ (Murcott, 2012). Over 12,000 persons have already become victims of arsenicosis. According to World Bank estimates, over 28 million to 35 million persons are exposed to As poisoning at present (Rahman, 2002; World Bank, 2005).

In India, confirmed As contaminated areas are located extensively in the northern fluvial plains and in the southern highlands. The northern occurrences in the Indo-Ganga-Brahmaputra plains abound with confirmed groundwater contaminations by geogenic As. However, because of its extent and severity, most of the studies have been done within a relatively small area of the Middle and Lower Gangetic Plains, and the Assam Plain.

The initial discovery of this health hazard occurred in late 1970s, which then became the focus of national and international interest and concern by the decade of the 1980s, when an increasing number of arsenicosis patients were identified in areas having high As levels in the groundwaters in West Bengal. The proximity of Bangladesh and the Indian state of West Bengal within the single hydrogeological unit of the Bengal Delta Plains paved the way for several collaborative studies on this phenomenon. Primary investigations revealed that the Holocene sediments originating in the Himalayas and deposited in the Lower Gangetic valley by its drainage network, along with marine transgressions in the geological past have formed a complex fluvial-estuarine-marine environment that promote leaching and mobilization of As in extensive layers of often unconfined aquifers (Chakraborti *et al.*, 1996). However, several isolated cratons of the Indian peninsular region have also been confirmed as "primary or secondary" contributory sources of As in groundwater These are the Gondwana coal seams in Rajmahal basin in eastern India; Bihar mica-belt in eastern India; pyrite-bearing shale from the Proterozoic-aged Vindhyan range containing in central India; and the Sone river valley gold belt (Bhattacharya *et al.*, 2015). In West Bengal, seven districts – Murshidabad, 24 parganas (North), 24 parganas (South), Malda, Nadia, Burdwan, and Hoogly, are the most severely affected districts. The As content in groundwater in Bengal Basin varies from 50 to 3700 $\mu g\,L^{-1}$, with an average of 200 $\mu g\,L^{-1}$, against the WHO guidelines of 10 $\mu g\,L^{-1}$ of As permissible in drinking water (Stuben *et al.*, 2003). In Bihar, aquifers utilized for pumping out drinking water have high As content ($<50\,\mu g\,L^{-1}$) in 18 districts (www.bih.gov.in) located in the fluvial plains (Ghosh *et al.*, 2008). Almost all the contaminated drinking water sources are located in the rural habitation sites. A recent study revealed various contamination levels in shallow irrigation bore well water in 36 districts (Bose *et al.*, 2015). The Bihar contaminations extend westward into the state of Uttar Pradesh into Ballia, Ghazipur, Lakhimpur Kheri and Varanasi districts, thereby covering a substantial part of the Middle Ganga Plains as well. Other parts of north India with geogenic As contaminations include Punjab, Haryana, Himachal Pradesh, and Uttarakhand (Bhattacharya *et al.*, 2014). In the seven states of northeast India, over 4000 groundwater samples were tested. Geogenic As was detected in 17 out of 23 districts of Assam. Maximum concentrations were found in Jorhat, Dhemaji, Golaghat, and Lakhimpur districts. In the Tripura Plains, three districts of West Tripura, North Tripura and Dhalai registered numerous As contaminated groundwater sources. In Manipur, very high readings of 7980–9860 μg L^{-1} of As have been confirmed the single Kakching Block of Thoubal district. In Arunachal Pradesh, six districts – Dibang Valley, West Kameng, East Kameng, Papum Pare, Tirap and Lower Subansiri, lying along the border with Assam, had a large number of As contaminated aquifers, with the highest reading of 618 μg L^{-1} in Dibang Valley. Similarly in Nagaland, two districts (Mon and Mokokchung districts) situated near Assam's Jorhat district had As in a number of their groundwater sources (Bhattacharya *et al.*, 2014). In Chhattisgarh state, As is present in about 30 villages and towns of Rajnandgaon district. The elevated concentration of As in the water sources of the villages and towns vary between 45 and 6000 μg L^{-1} putting an approximate population of 30,000 at risk of chronic As exposure. Geogenic As has also been detected in river Shivnath, a tributary of river Mahanadi in the vicinity of Rajnandgaon varying between 100–300 μg L^{-1}. It is dominated by inorganic As. (Bhattacharya *et al.*, 2011). Levels in then contaminated sediments lie between 200 mg g^{-1} and 10.75 mg g^{-1}, with a mean level of 68 mg g^{-1} and standard deviation of 41.81. A direct result of this contamination has been noted in the form of lower fish productivity in Shivnath river. The nearby Bagdai river also shares similar contamination characteristics with As concentration of 100 μg L^{-1} (Pandey, 2004).

In Myanmar, As was detected in the groundwater hand pumps in Ayeyerwady River delta in the last decade. Although the magnitude of the contaminations is not yet confirmed, around 45% of the samples had a contamination range of 200–500 μg L^{-1} (Tun, 2003).

Pakistan has elevated As content in the groundwater of Bahawalpur, Multan, Muzaffargarh, Piokara and Sahiwal districts. High readings of 250–500 μg L^{-1} have been confirmed. In Punjab district, all the eleven districts have varying elevated As levels in the groundwater. This district is drained by the Indus and its tributaries, and is geologically the western extension of the fluvial plains on the Himalayan fore deep. Hence, the Indus Basin, has similar Holocene sediments where reducing conditions operate to mobilize the As into the aquifers (Toor and Tahir, 2009).

In west Asia, Afghanistan has recorded high As levels in well waters in its eastern and western parts. Although mining activities account for the leaching of As into most of the wells, geogenic As of up to 500 μg L^{-1} has been detected from the sources of urban water supply in Ghazni.

2.5   AREAS OF ANTHROPOGENIC ARSENIC

In the USA, anthropogenic As contamination is found in 30 states, where coal-based industrial productions, chemical industry, wood production, pesticides, metal smelting and wastewater treatment predominate. In Canada 40 anthropogenic activities, particularly in New Brunswick, Alberta, Ontario, Manitoba and British Columbia are causing As contamination in groundwater sources. Latin American countries like Mexico, Cuba and Guatemala, and South American countries of Argentina, Chile, Brazil and Uruguay account of As contaminated aquifers arising from mining and smelting activities, although insufficient data exists regarding the impacts on populations.

In 30 European countries coal-based plants, thermo-electric power plants, wood and pulp industries, cement and smelting units contribute to substantial release of As in the groundwater (Murcott, 2012).

Asia is also subject to substantial anthropogenic As contaminations. In western Asia and China mineral-rich regions have evolved due to geological complexities along the major lines of colliding lithospheric plates. Mining thus is an important economic activity. Mixing of the mine tailings and wastes with the well water often results in leaching of the As into these wells Hence mining activities have resulted in much of elevated As levels in the groundwater sources in Afghanistan, Armenia and Azerbaijan mainly of copper, gold, iron and other arsenopyrite bearing minerals.. The Kurdistan Province of Iran also has proven contaminations of As due to mining activities. In China, coal mining, mining of gold and other arsenopyrite bearing mineral, and petroleum production have resulted in high As levels in groundwater (Murcott, 2012). In India, copper mining activities in Rajasthan account for high As levels in the districts of Udaipur and Jhunjhunu. In Jharkhand, there are many open pit mines from where sulfides bearing copper and lead ores are mined. These ores contain As in trace amounts, which, if mobilized, may significantly contaminate the groundwater resources. In addition, India is the third largest hard coal producer in the world; the coal mining area covers some 855 km^2 and the total coal mines is 572 in 2004. Hence, coal mines are also a potential source of As emission and the average concentration of As in Indian coal ranges from 0.15–40 mg kg^{-1}. India produces over 100 million tons of coal fly ashes and the major part is dumped in the close vicinity of the plant sites. Concern has been raised due to leaching of As during coal washing, combustion and ash. Fertilizers and various pesticides, insecticides, herbicides, and fungicides often contain high concentration of As and their widespread use are known to cause considerable groundwater contamination especially in the agricultural states of India such as documented in Punjab, Andhra Pradesh, Haryana, Karnataka, Tamilnadu, West Bengal, and Uttar Pradesh. Arsenic has been recently been reported from areas adjoining the gold mining sites of India – the Kolar Gold Fields located in the Mangalur greenstone belt in Karnataka. Surveys have also indicated the occurrence of As in the drinking water wells in Gulbarga and Raichur districts, respectively, where a large number of communities within Shorapur Taluka in Gulbarga and Raichur districts revealed elevated As concentrations above 50 µg L^{-1} together with an additional 14 villages in Gulbarga and 39 in Raichur with As concentrations above 10 µg L^{-1} being used for drinking purposes (Bhattacharya *et al.*, 2014).

As more sources of As in the environment are detected, the types of As contamination and number of occurrences among drinking water sources (Murcott, 2012), have now been categorized as follows:

- Anthropogenic-related As – 54 countries/territories.
- Coal-related As – 28 countries/territories.
- Geogenic-related As – 68 countries/territories.
- Mining-related As – 74 countries/territories.
- Petroleum-related As – 17 countries/territories.
- Volcanogenic-related As – 35 countries/territories.

2.6 ARSENIC CONTAMINATED GROUNDWATER IN GANGETIC PLAINS OF BIHAR, INDIA

The eastern Indian state of Bihar is a densely populated, and an important agricultural area, that occupies a large part of the Mid Ganga fluvial plains. Bihar is located in the northeastern part of India (Latitude: 24°20′10″ to 27°31′15″ S and longitude 82°19′50″ to 88°17′40″ E). The state of Bihar has a geographical area of 94163 km^2. It is an entirely land-locked state that lies mid-way between the humid West Bengal in the east and the sub humid Uttar Pradesh in the west which provides it with a transitional position in respect of climate, economy and culture. It is bounded by Nepal in the north and by the state of Jharkhand in the south (Government of Bihar). It is the third most populated state of India, with a population of over 83,000,000 of which about 83% lives in rural areas and 58% is below 25 years old. The capital city of Bihar is Patna. Geographically, the Bihar plain is divided into two unequal halves by the river Ganga which flows through the middle from west to east: these geographical units are the northern and central Ganga Plains and the southern Chotanagpur Plateau region. These geographic units occupy 30, 24 and 46% of the area, respectively. Normal average rainfall is of about 1203.5 mm, 70% of which comes from the SW monsoon. The total annual replenishable groundwater resource in the state is 27.42 billion m^3 with an average state of development of about 39% (CGWB report, 2010).

Bihar is an agro-based economy where agriculture contributes to 42% of state GDP, and engages 81% of the workforce. Hence, the stakeholders aim for agricultural intensification with expanded irrigation facilities. There is a predominance of bore well irrigation in Bihar where 100 ft ($\approx$30.5 m) is the average depth of bore wells in this fluvial plain. With 40% population under "BPL" the option is for agricultural intensification with expanded irrigation facilities. In 2009, 2,855,202 hectares out of the total irrigated area of 4,566,834 hectares (62.52%) in Bihar

(Dept. of Water Resources, Bihar) came from bore wells (both private and state-owned). These provided more than 50% of the total irrigation in 30 districts. Bihar Ground Water Irrigation Scheme (BIGWIS) was launched in 2012 to provide irrigation to 9,280,000 hectares of agricultural land of the State, under which more than 464,000 tube well units of shallow tube wells with pump sets were installed by 2012 (NABARD Report, 2014). An important component in the Special Plan for Bihar under the 10th and 11th Five Year Plans was the Million Shallow Tube Well Program, in order to cope with agricultural drought in the dry season and in event of late outbreak of monsoon rainfall. More than 4,07,758 units of private shallow tube wells had already been sunk with pump sets till 2007 under this scheme.

From 2004 till 2012, more than 36,000 drinking water tube wells have been tested, of which 30% have As levels of more than $50\,\mu g\,L^{-1}$ (Ghosh *et al.*, 2006; 2012). Initially up to 2005, samples of over 27,000 drinking water tube well waters were tested in the flood plains of Patna, Bhojpur, Vaishali and Bhagalpur districts (Ghosh *et al.*, 2006). Of this, 7500 samples had elevated As levels of $>50\,\mu g\,L^{-1}$, the highest being $1861\,\mu g\,L^{-1}$ in Bhojpur, $724\,\mu g\,L^{-1}$ in Patna, $150\,\mu g\,L^{-1}$ in Vaishali and $608\,\mu g\,L^{-1}$ in Bhagalpur districts. An estimated 700,000 persons were identified as vulnerable to As poisoning (Bose *et al.*, 2007). Till 2014, the entire belt of central Bihar was recognized as a zone of maximum arsenic contaminated drinking water sources (Fig. 2.1).

Trivalent As has been identified in <80% of contaminated irrigation bore wells water samples collected from a belt from the banks of the Ganga in Bhagalpur district up to the foothill districts

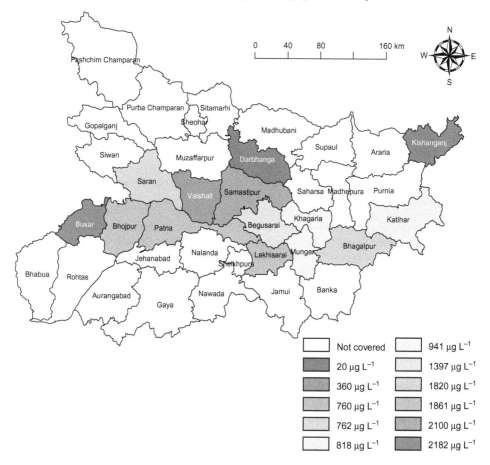

Figure 2.1.   District-level maximum arsenic content in groundwater of Bihar. Source: Singh *et al.* (2014).

in north Bihar (Mukherji *et al.*, 2012). The population vulnerable to As poisoning directly and indirectly through contaminated groundwater usage approximates to 25% of 103,805,000 persons as per current information on shallow aquifer contamination (Ghosh *et al.*, 2007).

2.7 ARSENIC IN FOOD CHAIN

Ingestion of geogenic As by the human body occurs directly through drinking water, and indirectly through the food chain. Long-term use of As-contaminated water for irrigation may result in elevated As concentration in soils and plants (Huq *et al.*, 2003; Ullah, 1998; Panaullah *et al.*, 2003). Significant quantities of water from shallow aquifers are being used for irrigating rice fields, so that the dimensions of As contamination now extend across the water-soil-crop route within the geosphere. Paddy crop cultivation takes place in arsenic-contaminated soils under anaerobic conditions, at which Arsenic is highly available for plant uptake (Meharg and Rahman, 2003; Mondal and Polya, 2008). Roychoudhury *et al.* (2005) analyzed total-As in food coming from an As-affected area of Murshidabad district in West Bengal (Jalangi and Domkal blocks) and the highest concentration was found in cooked rice as $330\,\mu g\,kg^{-1}$. Arsenic uptake and accumulation in rice plant from irrigation water may differ depending on cultivars used. Hence, As in irrigation water poses the greatest threat to soils, crops, the food chain and consequently to human health of mass communities.

Since rice is the staple food, any adverse effects on nutrient content of rice due to As contaminated irrigation water would only enhance the malnutrition problem (Alam and Rahman, 2003). A field study confirmed the presence of As in shallow aquifers extending from the Indo-Nepal border in the north to the Ganga banks, along the Raxaul-Patna line in western Bihar, and the Forbesganj-Bhagalpur line in eastern Bihar. In all, 36 of the 38 districts of Bihar had As contaminated irrigation bore well water. Irrigation bore well water samples with elevated As were located on shallow aquifers of maximum depth of 80 feet ($\approx$24.4 m) in North Bihar Plains, and in aquifers up to 300 feet ($\approx$91.4 m) in the south Bihar Plains (Fig. 2.2). Arsenic contamination levels in

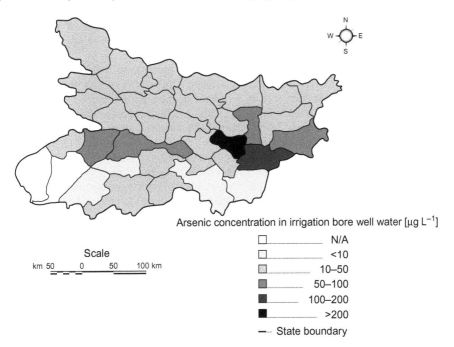

Figure 2.2. Range of arsenic content in water of irrigation bore wells of Bihar. Source: Bose *et al.* (2015).

most of these water sources were frequently above $50 \, \mu g \, L^{-1}$, maximum being $857 \, \mu g \, L^{-1}$ in Khagaria district (Bose *et al.*, 2015).

For the first time, an As contaminated surface water source was detected in a north bank inlet of the river Ganga in Bhagalpur, known as the *Jamunia Chharan* or *Jamunia Dhar*. This water is used for irrigating the adjacent fields in the dry season. Repetitive tests of the Jamunia water samples from different spots revealed the contamination levels up to $66 \, \mu g \, L^{-1}$. Agricultural products, mainly rice, maize, sugarcane, leafy vegetables, and root vegetables tested positive for As levels of over $10 \, \mu g \, g^{-1}$ (this being the WHO recommended permissible limit of As limit in potable water). Onsite sugarcane juice extract tested 19 to $21 \, \mu g \, L^{-1}$ of As content in Bhagalpur. Rice plants and grain samples had 13 to $15 \, \mu g \, g^{-1}$ As. A maize plant sample had over $110 \, \mu g \, g^{-1}$ As content in Chota Sasaram village of Udwantnagar Block in Bhojpur District. The exact cause of the Jamunia Chharan contamination is yet to be studied.

2.8   HEALTH IMPACTS OF ARSENIC

Arsenic contamination has been recognized as one of the major threats to human health by the WHO, the EU and EPA. After microbial contamination, arsenic contamination is ranked as "second among their top priorities in the control of drinking water quality (http://www.epa.gov/safewater/arsenic.html). The World Health Organization (WHO) has classified As among the most powerful carcinogens. In the case of water intended for human consumption, the value recommended in the WHO guidelines has been adopted in both the European legislation (Directive 98/83/EC, Council meeting of 3 November 1998) and the French legislation (Decree 2001-1220 of 20 December 2001, enforced in December 2003) in the form of a 'maximum admissible concentration' and a 'quality threshold' set at $10 \, mg \, L^{-1}$ instead of $50 \, mg \, L^{-1}$, as in 1989" (Lievremont *et al.*, 2009).

Arsenic is highly toxic in its inorganic form and damages multicellular life forms, with the exception of certain microbes with use As compounds as respiratory metabolites (Grund *et al.*, 2005). There are major negative health impacts due to ingestion of As. However, estimates for the population at risk from As poisoning are still confined to a few geographical areas in the developing world, due to paucity of primary data, monitoring and evaluation of contaminated tube wells.

Ingestion of low dose via food or water is the main pathway of this metalloid into the organism, where absorption takes place in the stomach and intestines, followed by release into the bloodstream. In chronic poisoning, arsenicosis then converted by the liver to a less toxic form, from where it is eventually largely excreted in the urine. Only very high exposure can, in fact, lead to appreciable accumulation in the body. Minor alternative pathways of entry are known through inhalation and dermal exposure (Caroli *et al.*, 1996).

Arsenic is consumed mainly in two forms, arsenite (AsIII) and arsenate (AsV). The mechanism of toxicity depends on the As species and their valence state (Sengupta *et al.*, 2006). Mechanism of As(III) toxicity is essentially due to its ability to bind with the sulfhydryl groups present in various essential compounds, e.g., glutathione (GSH), cysteine, etc. (Scott *et al.*, 1993). The binding with function groups (e.g., thiol group) on any receptor or enzyme leads to their inactivation (Aposhian, 1989).

The absorption takes place mainly through ingestion of water, food, beverage, medicine, and sometimes, swallowing of the inhaled particulate matter with As that is cleared by mucociliary escalator (Gomez-Caminero *et al.*, 2001). Following absorption, As undergoes metabolism through repeated reduction and oxidative methylation. It is widely accepted that methylated metabolites of inorganic As are less reactive and less genotoxic; metabolism is regarded as a bio-inactivation mechanism. Following metabolism, As is rapidly cleared from blood, and only 0.1% of the As remains in the plasma 24 hours after dosing. Urine is the most common route of elimination. As much as 45% to 75% of the dose is excreted in the urine within a few days to a week (Vahter and Norin, 1980). The trivalent state of arsenic, As(III), is widely distributed by virtue of its binding with sulfhydryl groups in keratin filament and has a tendency to accumulate in the skin, hair, nails, and mucosae of the oral cavity, esophagus, stomach, and the small intestine (Lindgren *et al.*, 1982). On the other hand, arsenate, As(V), is the predominant form deposited in the skeleton because of its ability to replace phosphate in the apatite crystal in bones; as a result of this it is retained there for a longer time.

Arsenicosis is a disease that needs chronic exposure (almost 10 years; range, 5–20 years) before its symptoms visible or are detected (Das and Sengupta, 2008). Chronic exposure to low levels of As causes different skin lesions in the form of melanosis, leuco-melanosis and keratosis. A few years of continued exposure to low levels of inorganic arsenicals causes different skin lesions, and after a latency period of 20–30 years, internal cancers, particularly of the bladder and lung, can appear (Byrd *et al.*, 1996). However, it is not clear from the literature how much ingestion of As causes what types of skin lesions (Foster *et al.*, 2002; Tondel *et al.*, 2001). Cancer risks from inorganic arsenicals in drinking water have been proved and reported (Hsueh *et al.*, 1995).

Moreover, non-malignant health effects such as diabetes, (Lai *et al.*, 1999; Rahman *et al.*, 1999) peripheral neuropathy, cardiovascular diseases ischemic heart disease, hypertensive heart disease and bronchitis (Abernathy *et al.*, 1999; Rahman *et al.*, 1999) can result from As exposure. If As builds up to higher toxic levels, organ cancers, neural disorders, and organ damage – often fatal – can result.

During 1998–2000, investigations were done on relations between lung function, respiratory symptoms, and As in drinking water among 287 study participants, including 132 with As-caused skin lesions, in West Bengal, India. The source population involved 7683 participants who had been surveyed for As-related skin lesions in 1995–1996. Respiratory symptoms were increased among men with As-caused skin lesions (versus those without lesions), particularly "shortness of breath at night", "morning cough" in smokers and "shortness of breath ever" in nonsmokers. In this study, consumption of As-contaminated water was associated with respiratory symptoms and reduced lung function in men, especially among those with As-related skin lesions (von Ehrenstein *et al.*, 2005).

In West Bengal further between 2001 and 2003, pregnancy outcomes and infant mortality among 202 married women in West Bengal, India were studied. Exposure to high concentrations of As ($\geq 200\,\mu g\,L^{-1}$) during pregnancy was associated with a sixfold increased risk of stillbirth. Arsenic-related skin lesions were found in 12 women who had a substantially increased risk of stillbirth (von Ehrenstein *et al.*, 2006).

Dose-response analysis suggested Type 2 Diabetes Mellitus (T2DM) risk increased by 13% for every $100\,\mu g\,L^{-1}$ increment of As in drinking water. Significant association of T2DM risk with As in urine was also found. This meta-analysis indicates that long-term As exposure might be positively associated with T2DM risk (Wang *et al.*, 2014).

Being a systemic toxin, As has been associated with the risk of skin and several internal cancers (of lung, bladder, liver, and kidney), as well as cardiovascular, respiratory and neurological diseases (Chen *et al.*, 2004; 2009; 2011; Hafeman *et al.*, 2005; Parvez *et al.*, 2010).

Typically, these diseases appear after decades of chronic exposure (Marshall *et al.*, 2007). Evidence suggests that As-related skin lesions typically occur at high As exposure levels (Smith and Steinmaus, 2009). Researchers have long been considering skin lesions as a clinical marker of As susceptibility. Dermatological manifestations, typically associated with chronic As exposure, are predictive of internal cancers among Taiwanese decades after the cessation of exposure. It was found that found that individuals with hyperkeratosis and/or skin cancer had elevated risks of lung and urothelial cancers after adjustments for As exposure and relevant covariates. These elevated risks were evident among individuals exposed to low, medium, or high levels of As (Ahasan and Steinmaus, 2013).

Susceptibility to As related disease varies greatly from person to person. Although some of the factors responsible for these differences in susceptibility have been identified, including diet (Zablotska *et al.*, 2008), genetics (Ghosh *et al.*, 2008; Pierce *et al.*, 2012), or coexposures like smoking (Ferreccio *et al.*, 2000; Melkonian *et al.*, 2011), many of the differences in susceptibility remain unexplained.

Table 2.1.   Summary of health problems due to ingestion of arsenic contaminated water. Source: Singh *et al.* (2014).

Sample Number	Diseases	As in drinking water [μg L^{-1}]	Country/region	References
1	Spontaneous pregnancy loss	10 to 1474		
2	Respiratory complications	216	Bangladesh	Bloom *et al.* (2010)
3	Diabetes	20 to 400		Islam *et al.* (2007)
4	Immunological system	216	Bangladesh	Gonzalez *et al.* (2007)
5	Skin cancer	>100	Cordoba, Argentina	Islam *et al.* (2007)
6	Hepatic damage	200 to 2000	West Bengal, India	Guha Mazumdar *et al.* (1988)
7	Recognizable signs of As toxicity	>100	Cordoba, Argentina	Astolfi *et al.* (1981)
8	Skin lesions	<50	West Bengal, India	Guha Mazumdar *et al.* (1988)
9	Arsenic dermatosis	200	West Bengal, India	Chakraborty and Saha (1987)
10	Neurological disorders	100 to 2000	Bangladesh	Rahman *et al.* (2009)
11	IQ of children	>50		Waserman *et al.* (2004)
12	Melanosis/keratosis	>50	Bihar, India	Chakraborty *et al.* (2003)

Although not directly mutagenic, As is genotoxic, inducing effects including deletion mutations, oxidative DNA damage, DNA strand breaks, sister chromatid exchanges, chromosomal aberrations, aneuploidy, and micronuclei (Basu *et al.*, 2001; Hei *et al.*, 1998), amplification, transforming activity, and genomic instability (Rossman, 2003). These genotoxic effects of As are observed in vitro in mammalian cells and in vivo in laboratory animals and humans (Basu *et al.*, 2001; Rossman, 2003).

Chromosomal aberrations in lymphocytes were observed in workers exposed to As (Beckman, 1977). Trivalent arsenicals, both inorganic and organic, are more potent genotoxins than the pentavalent arsenicals (Kligerman *et al.*, 2003). The mechanism of genotoxic action of As may result from generation of ROS, inhibition of DNA repair, and altered DNA methylation that may lead to genomic instability (Rossman, 2003). The multiple health impact due to ingestion of As contaminated drinking water has been summarized in Table 2.1 (Singh *et al.*, 2014).

2.9   ARSENIC MITIGATION

The existing challenges in mitigating As contamination of water used for human consumption are manifold. Such contaminations plague economically important regions in various countries, these being either agricultural regions, or mining belts and industrial nodes. The current areas of geogenic As contaminations are mostly located in developing countries of south, southeast and east Asia. Abundance of water wealth in the fluvial plains of these areas spawned ancient civilizations with agro-based economies. Presently, with the surface water sources becoming polluted or scarce, there has been an increasing reliance on the abundant groundwater resources for obtaining drinking water amongst such communities. These traditionally densely populated zones grow monsoon-dependent, water-intensive cereals and crops like rice, maize, and sugarcane. Because monsoon rainfall tends to be erratic and uncertain, food production targets are now being met by tapping the same aquifers. The socio-economic profiles of these developing economies create further complexities in implementing solutions to the problems associated with elevated As levels in groundwater.

Hence, in view of the toxic effects of As on the human body, the occurrence and mobilization of As in these aquifers raises two queries in context of As mitigation:

- How to remove the harmful As from the drinking water obtained from underground sources?
- How to minimize, and even stop, the groundwater As from infiltrating into the food chain, while maintaining the vital bore well irrigation programs under food security policies?

It would not be incorrect to state that the state of West Bengal in India and Bangladesh have witnessed the largest number of As mitigation strategies being experimented in laboratory and then undergoing field trials to test their efficacy. However, no single mitigation technique has been identified as "the most preferred option" amongst all the mitigation strategies, on account of diverse socio-economic responses of the stakeholders, and/or the repercussions of the negative effects of a particular adopted strategy.

2.10 MITIGATION STRATEGIES

Many efforts have been undertaken in an attempt to provide relief to individuals consuming As contaminated water. There has been an inundation of technology options for As mitigation both at national and international levels. Some technology options have been adopted without rigorous testing and have failed when applied in the field, leading to severe social consequences. Others have met with partially positive results.

The various As mitigation options tested and implemented in India and Bangladesh are grouped as follows:

- Alternate sources of drinking water, as through rain water harvesting, restoration of open wells, supply of piped water from "safe" sources including treated river water and dug wells.
- Arsenic filtration technologies.
- Bioremediation through phytoextraction, microbial action and vermifiltration.

Rain water harvesting (RWH) by individual households can provide safe drinking water. However, its major drawbacks are microbial growth and decomposition. High costs of construction and maintenance are also major deterrents of RWH in the impoverished, As-contaminated Ganga-Meghna-Brahmaputra plains. Community members are hesitant, and often refuse to consume water that has been collected from rooftops and stored for future use. If constructed in the flood-plains, which hold most of the spatio-temporal concentration of groundwater As as in the Indian state of Bihar, RWH units frequently get flooded during the monsoon. The viability of this method hence becomes questionable along these river plains.

In a number of instances, old existing wells have been renovated, treated and covered for re-use. Again due to lack of water quality monitoring and maintenance, these wells became polluted and defunct. Bihar's deep aquifers are relatively As-free, being as little as $3.5\,\mu g\,L^{-1}$ at depths of 120–130 m (Saha *et al.*, 2010). In Bangladesh and parts of West Bengal in India, elevated As levels have been noted in even deep tube wells, the contaminated water having gained access to the deep aquifers due to improper installation of the pumps.

Piped water supply is frequently touted as the solution to the As problem. In Mohjampur that lies in the western district of Bhojpur in Bihar, a river water treatment plant s been constructed that supplies treated Ganga water to over 20 As affected villages in the area. This region already suffers from energy deficit, and the treatment plant is presently run on diesel fuel. A second such unit is being constructed in Vaishali district of Bihar. The high capital and maintenance costs of such systems are likely to emerge as sustainability issues in near future, as is already evident in Bangladesh.

Dug wells are used by thousands of villagers in Bangladesh. Although the shallowest aquifers tapped by these wells are typically low in As, a full-scale return to this traditional technology is hampered by concerns regarding the microbial quality of the water and the need for regular maintenance (Ahmed *et al.*, 2005). Moreover, in Bihar, in the middle Ganga Plains, As contaminations abound in shallow aquifers (Ghosh *et al.*, 2006). Of considerable success has been the modified dug well design by using of tube well hand pumps to withdraw water, evolved under Project Well, and implemented in 19 administrative blocks of 24 Parganas district in Bengal, and the dug wells tap water of the near-surface unconfined aquifer within the oxidized zone, where the As concentrations is well below $0.05\,mg\,L^{-1}$ (Smith *et al.*, 2003). About 250 dug wells have been installed under Project Well by Aqua Welfare Society (AWS). This project is sponsored by an international NGO, Blue Planet Network. The source of water for these dugwells is rain water. The cost for maintenance depends upon the nature of the breakdown of the unit. The well water is treated with theoline (contains 10% chlorine) three times a month, and the treated water is tested in the state government laboratories. About 0.22 L of theoline is used for a single treatment of a dug well, the cost of which works out to US$ 0.05 only (at conversion rate of US$ 1 = INR 60, year 2012). Each dug well provides drinking water for 20–25 families. This means that around 100 people make use of each well daily. Out of 250 wells constructed, approximately 1–5 wells have not been functioning properly due to improper installation. Around 20 families make use of 1 well. The distance between their houses and the well being short, almost everyone has access to the dug well of their area.

Participation, capacity building and community management serve to create a basis of solidarity, and mutual recognition and cooperation between different families, learning to take democratic decisions in an open process of deliberation. Lack of community participation is also a major hindrance to operation, maintenance and monitoring of these schemes. At present, the As mitigation programs in Bihar are mostly run by government authorities. A few are being implemented by NGOs, but none are being run by community members. In the neighboring state of West Bengal and in Bangladesh, some of the mitigation units are community-driven, implemented and maintained. Dug wells under Project Well were based on creation of awareness about the As problem, thereby creating a demand for clean water. This was followed by community involvement, and training in monitoring of dug well water for As and microbial contaminations. The objective of making the system sustainable at the village level using indigenous labor and materials was thereby achieved on a local scale (Smith *et al.*, 2003).

The spatially erratic nature of arseniferrous aquifers, the varying time-scale of exposure of As poisoning, increasing emergence of cohorts of affected population, along with differences in perception of the As problem among the government institutions have led to severe logistical issues in water supply from alternate As free water sources. Hence, need-based, As filtration units have also been tested and implemented. These include domestic and large community-based filters (Fig. 2.3). However, proper disposal of saturated media, filter sludge and backwash remain critical issues (Brouns *et al.*, 2013).

Presently, in south Asian As affected areas, two types of remediation are being implemented:

- *In situ* remediation: remediation of Arsenic from the aquifer system itself;
- *Ex situ* remediation: treatment of the water by pumping it off the aquifer first and then using filtration techniques (physical/ chemical/bioremediation methods) – *implemented in the Ganga-Brahmaputra Plains of south Asia.*

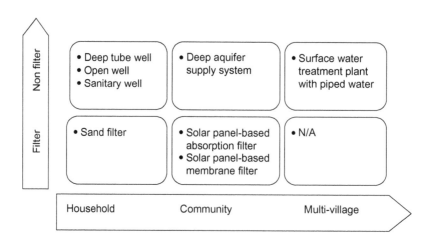

Figure 2.3.   Common mitigation strategies.

Table 2.2. A comparative assessment of two arsenic filter units.

Properties	Reverse Osmosis Schenker®Smart-60	Adsorption-based well head treatment unit
Investment	High initial investment	Relatively less investment
Iron in feed	Problematic	Helpful
Removal	Greatest range	Iron and arsenic
Cleaning	Not frequent	Frequent
Arsenic removal	98–99%	90–95%
Power	Use of electricity	Not required
Supervision	Skilled supervision required	Not required
Speciation in arsenic	Equally efficient in removing both forms	More efficient in removing As(V)
removal mechanism	Pressure driven	Gravity driven
Waste	Toxic waste water produced	Toxic sludge produced
Chemical interference	No interference	Silicates, phosphates can interfere
Maintenance	Can be challenging depending upon field conditions	Relatively less challenging
Flexibility	Flexible in operation	Not flexible
Cost of water (US$ L^{-1})*	0.01	0.003
Self cleaning	Cross-flow	Manual cleaning
Use of chemicals	No use of chemicals	Media chemicals
Recovery	Low water recovery	High recovery
Minerals in treated water	Permeate water readjustment required	Minerals retained
pH adjustment	Not required	Requirement possible
Monitoring of operation	Automatic visualization and logging	Manual

*Currency conversion rate: US$ 1 = Rs. 60 (2012–2013).

In the physical and social milieu of Bihar, the challenges to any As mitigation technology include annual inundations, annual migrations of the As affected, flood affected, impoverished population, energy shortage in rural areas, difficulties in maintenance and monitoring of various As treatment units, social conflicts and exclusions, sludge and backwash disposal, media replacement in filter units.

Based on these shortcomings, three filtration units, one domestic and two community-based were tested in an As affected village in Patna district and a model drawn up based upon their efficacy. The domestic gravity-based As water filter (Aquapal, Sweden) yielded 30,000 L clean water, after which media-saturation process set in, with increasing amounts of As in filtered water. A sludge disposal problem remained. Along with high cost and maintenance issues, this filter was found to be unsuitable for domestic use among the rural inhabitants.

A comparative study of Schenker®smart 60 small-scale double pass reverse osmosis facility and a well head HAIX As removal unit was done. Although the cost per liter of clean water was Rs. 0.59 L^{-1} or U$ 0.010 and Rs. 0.16 L^{-1} or US$ 0.003, respectively, the HAIX-based unit was more cost effective as it is operated manually (Table 2.2). However, community participation in a collectivist, status oriented and hierarchical country however is not easy. Due to the limitations put on and resistance towards community participation by government hierarchies, religious authorities and local rural elites and even in part by the local (male) villagers themselves, there is a risk of the whole range of participatory models to become ineffective.

Researches on bioremediation of As in groundwater have made significant progress in identifying hyper-accumulating plants and microbes. Researchers are also exploring phytoextraction methods that are cost-effective and environment-friendly, to remove As (Cunningham et al., 1996; Terry and Banuelos, 2000). Successful application of phytoextraction to As-contaminated soils depends on many factors, among which are plant biomass and its As concentration. Plants must be able to produce sufficient biomass while accumulating a high concentration of As. In addition, it is important to understand the availability and phytotoxicity of As to the plant itself. Water hyacinth had a removal rate of $600\,mg\,As\,ha^{-1}\,day^{-1}$ and a removal recovery of 18%, under the conditions of the assay. The removal efficiency of water hyacinth was higher due to the biomass production and the more favorable climatic conditions (Alvarado *et al.*, 2008). *Pteris cretica* (moonlight fern) and *Pteris vittata* (Chinese brake fern) are also effective hyperaccumulators of As in water (Baldwin and Butcher, 2007). Similarly, microbes have been isolated that could either use arsenite as an electron donor or arsenate as an electro acceptor under aerobic and anaerobic growth conditions (Anderson and Cook, 2004).

However, a recent report on "the top 10 biotechnologies for improving human health" gave high priority to bioremediation by plants and microorganisms but, while emphasizing the potential of these techniques, it concluded by commenting on the potential environmental risks (Daar *et al.*, 2002). Bioremediation by bacteria, and later by recombinant bacteria, seemed a promising, environmentally correct way of depolluting contaminated sites, thus stimulating a great deal of basic research into the genetics and biochemistry of biodegradation. Despite this, little has been achieved in real environmental cleanup situations and, at present, there are few real applications for recombinant bacteria in the field (Watanabe, 2001). However, As contamination continues, and contaminated sites are still costly to clean up by standard means.

ACKNOWLEDGEMENTS

The authors thank Professor (Dr.) Prosun Bhattacharya, Royal Institute of Technology, Stockholm, Sweden for his valuable inputs for this chapter.

REFERENCES

Abernathy, C.O., Liu, Y.P., Longfellow, D., Aposhian, H.V., Beck, B., Fowler, B., Goyer, R., Menzer, R., Rossman, T., Thompson, C. & Waalkes, M. (1999) Arsenic: health effects, mechanisms of actions, and research issues. *Environmental Health Perspectives*, 107 (7), 593–597.

Acharyya, S.K., Chakraborty, P., Lahiri, S., Raymahashay, B.C., Guha, S. & Bhowmick, A. (1999) Arsenic poisoning in the Ganga Delta. *Nature*, 401, 545.

Ahasan, H. & Steinmaus, C. (2013) Invited commentary: use of arsenical skin lesions to predict risk of internal cancer – implications for prevention and future research. *American Journal of Epidemiology*, 177 (3), 213–216.

Ahmed, M.F., Shamsuddin, S.A.J., Mahmud, S.G., Rashid, H.U., Deere, D. & Howard, G. (2005) Risk assessment of arsenic mitigation options (RAAMO). Arsenic Policy Support Unit (APSU), Dhaka, Bangladesh. Available from: www.apsu-bd.org/ [accessed July 2015].

Alam, M.Z. & Rahman, M.M. (2003) Accumulation of arsenic in rice plant from arsenic contaminated irrigation water and effect on nutrient content. *Water Science and Technology*, 42, 132–135.

Alvarado, S., Guedez, M., Lue-Meru, M., Nelson, G., Alvaro, A., Jesus, A. & Gyula, Z. (2008) Arsenic removal from waters by bioremediation with the aquatic plants water hyacinth *(Eichhornia crassipes)* and lesser duckweed *(Lemna minor)*. *Bioresource Technology*, 99 (17), 8436–8440.

Anderson, C.R. & Cook, G.M. (2004) Isolation and characterization of arsenate-reducing bacteria from arsenic-contaminated sites in New Zealand. *Current Microbiology*, 48, 341–347.

Aposhian, H.V. (1989) Biochemical toxicology of arsenic. In: Hodgson, E., Bend, J.R. & Philpot, R.M. (eds.) *Reviews in biochemical toxicology*. Volume 10. Elsevier Science, New York, NY.

Baldwin, P.R. & Butcher, D.J. (2007) Phytoremediation of arsenic by two hyperaccumulators in a hydroponic environment. *Microchemical Journal*, 85, 297–300.

Basu, A., Mahata, J., Gupta, S. & Giri, A.K. (2001) Genetic toxicology of a paradoxical) human carcinogen, arsenic: a review. *Mutation Research*, 488, 171–194.

Beckman, G., Beckman, L. & Nordenson, I. (1977) Chromosome aberrations in workers exposed to arsenic. *Environmental Health Perspectives*, 19, 145–146.

Berg, M., Stengel, C., Trang, P.T.K., Viet, P.H., Sampson, M.L., Leng, M., Samreth, S. & Fredericks, D. (2007) Magnitude of arsenic pollution in the Mekong and Red River deltas – Cambodia and Vietnam. *Science of the Total Environment*, 372 (2–3), 413–425.

Bhattacharya, P., Chatterjee, D. & Jacks, G. (1997) Occurrence of arsenic-contaminated groundwater in alluvial aquifers from delta plains, eastern India: options for safe drinking supply. *International Journal of Water Research Management*, 13, 79–82.

Bhattacharya, P., Welch, A.H., Ahmad, K.M. & Gunnar, J. (2004) Arsenic in groundwater of sedimentary aquifers. *Applied Geochemistry*, 19 (2), 163–167.

Bhattacharya, P., Mukherjee, A. & Mukherjee, A.B. (2011) Arsenic in groundwater of India. In: Nriagu, J.O. (ed.) *Encyclopedia of environmental health*. Volume 1. Elsevier, Burlington. pp. 150–164.

Bhattacharya, P., Claesson, M., Bundschuh, J., Sracek, O., Fagerberg, J., Jacks, G., Martin, R.A., Storniolo, A. & Thir, J.M. (2006) Distribution and mobility of arsenic in Rio Dulce alluvial aquifers in Santiago del Estero province, Argentina. *Science of the Total Environment*, 358 (1–3), 97–120.

Bhattacharya, P., Mukherjee, A. & Mukherjee, A.B. (2014) Groundwater arsenic in india: source, distribution, effects and mitigation. *Reference Module in Earth Systems and Environmental Sciences*. Elsevier.

Bhumbla, D.K. & Keefer R.F. (1994) Arsenic mobilization and bioavailability in soils. In: Nriagu, J.O. (ed.) *Arsenic in the environment*. Part 1: *Cycling and characterization*. Wiley, New York. pp. 51–82.

Bose, N., Ghosh, A.K., Singh, S.K. & Singh, A. (2015) Impact of arsenic contaminated irrigation water on some edible crops in the fluvial plains of Bihar. In: Ramanathan, Al., Johnston, S., Mukherjee, A. & Nath, B. (eds.) *Safe and sustainable use of arsenic-contaminated aquifers in the Gangetic plain: a multidisciplinary approach*. Cham, Switzerland, Springer International Publishing. pp. 255–264.

Brouns, M., Janssen, M. & Wong, A. (2013) Dealing with arsenic in rural Bihar, India. TU Delft, Delft, The Netherlands. Available from: http://www.indiawaterportal.org/sites/indiawaterportal.org/files/dealing_with_arsenic_in_bihar_india_-_third_version_merged.pdf [accessed July 2015].

Brown, K.G. & Chen, C.J. (1995) Significance of exposure assessment to analysis of cancer risk from inorganic arsenic in drinking water in Taiwan. *Risk Analysis*, 15 (4), 475–484.

Bundschuh, J., Farias, B., Martin, R, Storniolo, A. Bhattacharya, P., Cortes, J., Bonorino, G. & Albouy, R. (2004) Groundwater arsenic in the Chaco-Pampean plain, Argentina: case study from Robles County, Santiago del Estero Province. *Applied Geochemistry*, 19 (2), 231–243.

Bundschuh, J. and Garcia, M.E. (2008) Rural Latin America – a forgotten part of the global ground-water arsenic problem? In: Bhattacharya, P., Ramanathan, Al., Bundschuh, J., Chandrasekharan, D. & Mukherjee, A.B. (eds.) *Groundwater for sustainable development: problems, perspectives and challenges*, CRC Press, Boca Raton, FL. pp. 311–321.

Bundschuh, J., Litter, M.I., Parvez, F., Román-Ross, G., Nicolli, H.B., Jean, J.-S., Liu, C.-W., López, D., Armienta, M.A., Guilherme, L.R.G., Cuevas, A.H., Cornejo, L., Cumbal, L. & Toujaguez, R. (2012) One century of arsenic exposure in Latin America: a review of history and occurrence from 14 countries. *Science of the Total Environment*, 429, 2–35.

Byrd, D.M., Roegner, M.L., Griffiths, J.C., Lamm, S.H., Grumski, K.S., Wilson, R. & Lai, S. (1996) Carcinogenic risks of inorganic arsenic in perspective. *International Archives of Occupational and Environmental Health*, 68 (6), 484–494.

Caroli, F., Torre, L.A., Petrucci, F. & Violante, N. (1996) Element speciation in bioinorganic chemistry. In: Caroli, S. (ed.) *Chemical Analysis Series*. Volume 135. pp. 445–463.

CGWB Report (2010) State profile, ground water scenario of Bihar. Central Ground Water Board, Ministry of Water Resources, Government of India. Available from: http://cgwb.gov.in/gw_profiles/st_Bihar.htm [accessed July 2015].

Chakraborti, D., Das, D., Samanta, B.K., Mandal, B.K., Roy Chowdhury, T., Chanda, C.R., Chowdhury, P.P. & Basu, G.K. (1996) Arsenic in groundwater in six districts of West Bengal, India. *Environmental Geochemistry and Health*, 18, 5–15.

Chen, C.L., Hsu, L.I. & Chiou, H.Y. (2004) Ingested arsenic, cigarette smoking, and lung cancer risk: a follow-up study in arseniasis-endemic areas in Taiwan. *JAMA*, 292 (24), 2984–2990.

Chen, Y., Parvez, F. & Gamble, M. (2009) Arsenic exposure at low to moderate levels and skin lesions, arsenic metabolism, neurological functions, and biomarkers for respiratory and cardiovascular diseases: review of recent findings from the Health Effects of Arsenic Longitudinal Study (HEALS) in Bangladesh. *Toxicology and Applied Pharmacology*, 239 (2), 184–192.

Chen, Y., Graziano, J.H. & Parvez, F. (2011) Arsenic exposure from drinking water and mortality from cardiovascular disease in Bangladesh: prospective cohort study. *BMJ*, 342, d2431.

Cunningham, S.D., Anderson, T.A., Schwab, A.P. & Hsu, F.C. (1996) Phytoremediation of soils contaminated with organic pollutants. *Advances in Agronomy*, 56, 55–114.

Daar, A.S., Thorsteinsdottir, H., Martin, D.K., Smith, A.C., Nast, S. & Singer, P.A. (2002) Top ten biotechnologies for improving health in developing countries. *Nature Genetics*, 32, 229–232.

Das, N.K. & Sengupta, S.R. (2008) Arsenicosis: diagnosis and treatment. *Indian Journal of Dermatology, Venereology, and Leprology*, 74, 571–581.

Ferreccio, C., Gonzalez, C. & Milosavjlevic, V. (2000) Lung cancer and arsenic concentrations in drinking water in Chile. *Epidemiology*, 11 (6), 673–679.

Foster, F., Craun, G. & Brown, K.G. (2002) Detection of excess arsenic-related cancer risks. *Environmental Health Perspectives*, 110, 12–13.

Ghosh, A., Singh, S.K., Bose, N., Upadhyay, A. & Roy, N.P. (2006) Arsenic hotspots detected in the State of Bihar (India): a serious health hazard for estimated human population of 5.5 lakhs. In: Anji Reddy, M. (ed.) *Proceedings of ICEM '05*. I BS Publ., India.

Ghosh, A.K., Bose, N., Kumar, R., Bruining, H., Lourma, S., Donselaar, M.E. & Bhatt, A.G. (2012) Geological origin of arsenic groundwater contamination in Bihar, India. In: Ng, J.C., Noller, B.N., Naidu, R., Bundschuh, J. & Bhattacharya, P. (eds.) *Understanding the geological and medical interface of arsenic. Proceedings of the 4th International Congress on Arsenic in the Environment, As 2012, 22–27 July 2012, Cairns, Australia*. CRC Press, Boca Raton, FL. pp. 85–87.

Ghosh, P., Banerjee, M., Giri, A.K. & Ray, K. (2008) Toxicogenomics of arsenic: classical ideas and recent advances. *Mutation Research*, 659, 293–301.

Gomez-Caminero, A., Howe, P., Hughes, M., Kenyon, E., Lewis, D.R., Moore, M., Ng, J., Aitio, A. & Becking, G. (2001) Environmental health criteria 224: arsenic and arsenic compounds. *Pollutants and Ecotoxicology*. World Health Organization, Geneva, Switzerland.

Grund, S.C., Hanusch, K. & Wolf, H.U. (2005) Arsenic and arsenic compounds. *Ullmann's encyclopedia of industrial chemistry*. Wiley, Weinheim, Germany.

Hafeman, D.M., Ahsan, H., Louis, E.D., Siddique, A.B., Slavkovich, V., Cheng, Z. & van Geen, A. (2005) Association between arsenic exposure and a measure of subclinical sensory neuropathy in Bangladesh. *Journal of Occupational and Environmental Medicine*, 47, 778–784.

Hei, T.K., Liu, S.X. & Waldren, C. (1998) Mutagenicity of arsenic in mammalian cells: role of reactive oxygen species. *Proceedings of the National Academy of Sciences of the United States of America*, 95, 8103–8107.

Hsueh, Y.M., Cheng, G.S., Wu, M.M., Yu, H.S., Kuo, T.L. & Chen, C.J. (1995) Multiple risk factors associated with arsenic-induced skin cancer: effects of chronic liver disease and malnutritional status. *British Journal of Cancer*, 71, 109–114.

Huq, I. & Naidu, R. (2003) Arsenic in groundwater of Bangladesh: contamination in the food chain. In: Ahmad, M.F. (ed.) *Arsenic contamination: Bangladesh perspective*. BUET, Dhaka, Bangladesh.

Kligerman, A.D., Doerr, C.L., Tennant, A.H., Harrington-Brock, K., Allen, J.W., Winkfield, E., Poorman-Allen, P., Kundu, B., Funasaka, K. & Roop, B.C. (2003) Methylated trivalent arsenicals as candidate ultimate genotoxic forms of arsenic: induction of chromosomal mutations but not gene mutations. *Environmental and Molecular Mutagenesis*, 42, 192–205.

Lai, M.S., Hsueh, Y.M., Chen, C.J., Shyu, M.P., Chen, S.Y., Kuo, T.L., Wu, M.M. & Tai, T.Y. (1999) Ingested inorganic arsenic and prevalence of diabetes mellitus. *American Journal of Epidemiology*, 139, 484–492.

Lievremont, D., Bertin, P.N. & Lett, M. (2009) Arsenic in contaminated waters: biogeochemical cycle, microbial metabolism and biotreatment processes. *Biochimie*, 91, 1229–1237.

Lin, Y.B., Lin, Y.P., Liu, C.W. & Tan, Y.C. (2006) Mapping of spatial multi-scale sources of arsenic variation in groundwater on ChiaNan floodplain of Taiwan. *Science of the Total Environment*, 370, 168–181.

Lindgren, A., Vahter, M. & Dencker, L. (1982) Autoradiographic studies on the distribution of arsenic in mice and hamster administered As-arsenite or -arsenate. *Acta Pharmacologica et Toxicologica*, 51, 253–365.

Marshall, G., Ferreccio, C., Yuan, Y., Bates, M.N., Steinmaus, C., Selvin, S., Liaw, J. & Smith, A.H. (2007) Fifty-year study of lung and bladder cancer mortality in Chile related to arsenic in drinking water. *Journal of the National Cancer Institute*, 99, 920–928.

Meharg, A.A. & Rahman, M.M. (2003) Arsenic contamination of Bangladesh paddy field soils: implications for rice contribution to arsenic consumption. *Environmental Science & Technology*, 37, 229–234.

Melkonian, S., Argos, M. & Pierce, B.L. (2011) A prospective study of the synergistic effects of arsenic exposure and smoking, sun exposure, fertilizer use, and pesticide use on risk of premalignant skin lesions in Bangladeshi men. *American Journal of Epidemiology*, 173, 183–191.

Mondal, D. & Polya, D.A. (2008) Rice is a major exposure route for arsenic in Chakdaha block, Nadia district, West Bengal, India: a probabilistic risk assessment. *Applied Geochemistry*, 23, 2986–2997.

Mueller, S.H., Goldfarb, R.J. & Farner, G.L. (1998) A seasonal study of arsenic and groundwater geochemistry in Fairbanks, Alaska. In: Piestrzynski, A. (ed.) *Mineral deposits at the beginning of the 21st century*. Swets and Zeitlinger Publishers, Lisse, The Netherlands. pp. 1043–1046.

Mukherjee, A., Sengupta, M.K., Hossaun, M.A., Ahamed, S., Das, B., Nayak, B., Lodh, D., Rahman, M.M. & Chakraborty, D. (2006) Arsenic contamination in groundwater: a global perspective with emphasis on the Asian scenario. *Journal of Health, Population and Nutrition*, 24, 142–163.

Mukherjee, A., Scanlon, B.R., Fryar, A.E., Saha, D., Ghosh, A., Chowdhuri, S. & Mishra, R. (2012) Solute chemistry and arsenic fate in aquifers between the Himalayan foothills and Indian craton (including central Gangetic plain): influence of geology and geomorphology. *Geochimica et Cosmochimica Acta*, 90, 283–302.

Murcott, S. (2012) *Arsenic contamination in the world: an international sourcebook*. IWA Publishing, London, UK.

Naidu, R. & Nadebaum, P. (2003) Geogenic arsenic and associated toxicity problems in the groundwater-soil-plant-animal-human continuum. In: Skinner, H.C.W. & Berger, A.R. (eds.) *Geology and health – closing the gap*. Oxford University Press, New York, NY.

National Bank for Agriculture and Rural Development, India (NABARD) (2014) Functions. Dept. of Refinance. Available from: www.nabard.org/english/Refinance2.aspx [accessed 10th January 2015].

Nickson, R.T., McArthur, J.M., Burgess, W.G., Ahmed, K.M., Ravenscroft, P. & Rahman, M. (1998) Arsenic poisoning of Bangladesh groundwater. *Nature*, 395, 338.

Nicolli, H.B., Bundschuh, J., Garcia, J.W., Falcon, C.M. & Jean, J. (2010) Sources and controls for the mobility of arsenic in oxidizing groundwaters from loess-type sediments in arid/semi-arid dry climates – evidence from the Chaco–Pampean plain (Argentina). *Water Research*, 44 (19), 5511–5846.

Nordstrom, D.K. (2002) Worldwide occurrences of arsenic in groundwater. *Science*, 296, 2143–2145.

Panaullah, G.M., Ahmad, Z.U., Rahman, G.K., Jahirudding, M. & Miah, M.A.M. (2003) The arsenic hazard in the irrigation water-soil-plant system in Bangladesh: a preliminary assessment. *Proceedings of the 7th Int. Conference Biogeochemistry of Trace Elements, Uppsala*. pp. 104–105.

Pandey, P.K. (2004) Sediment contamination by arsenic in parts of central-east India: an analytical study on its mobilization. *Current Science*, 86, 10.

Parvez, F., Chen, Y. & Brandt-Rauf, P.W. (2010) A prospective study of respiratory symptoms associated with chronic arsenic exposure in Bangladesh: findings from the Health Effects of Arsenic Longitudinal Study (HEALS). *Thorax*, 65, 528–533.

Pierce, B.L., Kibriya, M.G. & Tong, L. (2012) Genome-wide association study identifies chromosome 10q24.32 variants associated with arsenic metabolism and toxicity phenotypes in Bangladesh. *PLoS Genetics*, 8, 1002522.

Plant, J.A., Kinniburgh, D.G., Smedley, P.L., Fordyce, F.M. & Klinck, B.A. (2004) Arsenic and selenium. In: Sherwood, B. (ed.) *Environmental geochemistry*. Elsevier, London, UK.

Rahman, M., Tondel, M., Ahmad, S.A., Chowdhury, I.A., Faruquee, M.H. & Axelson, O. (1999) Hypertension and arsenic exposure in Bangladesh. *Hypertension*, 33, 74–78.

Rahman, M.H. (2002) Water resources management and water-related diseases: Bangladesh experience. *Proceedings of International Conference on Environmental Threats to the Health of Children: Hazards and Vulnerability*, 3–7 March 2002, Bangkok, Thailand, p. 15.

Ravenscroft, P. (2007) Predicting the global extent of arsenic pollution of groundwater and its potential impact on human health – a report. UNICEF, New York, NY.

Ravenscroft, P., Brammer, H. & Richards, K. (2009) Arsenic pollution: a global synthesis. *RGS-IBG Book Series*. Wiley-Blackwell, London, UK.

Rossman, T.G. (2003) Mechanisms of arsenic carcinogenesis. An integrated approach. *Mutation Research*, 533, 37–65.

Roychoudhury, T., Tokunaqa, H., Uchino, T. & Ando, M. (2005) Effect of arsenic-contaminated irrigation water on agricultural land soil and plants in West Bengal, India. *Chemosphere*, 58, 799–810.

Saha, D., Sahu, S. & Chandra, P. (2011) Arsenic – safe alternate aquifers and their hydraulic characteristics in contaminated areas of Middle Ganga Plain, Eastern India. *Environmental Monitoring and Assessment*, 175 (1–4), 331–348.

Schreiber, M.E., Simo, J.A. & Freiberg, P.G. (2000) Stratigraphic and geochemical controls on naturally occurring arsenic in groundwater, eastern Wisconsin, USA. *Hydrogeology Journal*, 8, 161–176.

Scott, N., Hatalid, K.M., MacKenzie, N.E. & Carter, D.E. (1993) Reaction of arsenic (III) and arsenic(V) species with glutiathione. *Chemical Research in Toxicology*, 6, 102–106.

Sengupta, M.K., Hossain, M.A., Mukherjee, A., Ahamed, S., Das, B., Nayak, B., Pal, A. & Chakraborti, D. (2006) Arsenic burden of cooked rice: traditional and modern methods. *Food and Chemical Toxicology*, 44, 1823–1829.

Silver, S. & Phung, L.T. (2005) A bacterial view of the periodic table: genes and proteins for toxic inorganic ions. *Journal of Industrial Microbiology and Biotechnology*, 32, 587–605.

Singh, N., Kumar, D. & Sahu, A.P. (2007) Arsenic in the environment: effects on human health and possible prevention. *Journal of Environmental Biology*, 28, 359–365.

Singh, S.K., Ghosh, A.K., Kumar, A., Kumar, C., Tiwari, R.R., Oarwez, R., Kumar, N. & Imam, M.D. (2014) Groundwater arsenic contamination and associated health risks in Bihar, India. *International Journal of Environmental Research*, 8, 49–60.

Smedley, P.L. & Kinniburgh, D.G. (2002) A review of the source, behaviour and distribution of arsenic in natural waters. *Applied Geochemistry*, 17, 517–568.

Smith, A.H. & Steinmaus, C.M. (2009) Health effects of arsenic and chromium in drinking water: recent human findings. *Annual Review of Public Health*, 30, 107–122.

Smith, A.H., Lingas, E.O. & Rahman, M. (2000) Contamination of drinking-water by arsenic in Bangladesh: a public health emergency. *Bulletin of World Health Organization*, 78 (9).

Smith, E., Smith, J. & Naidu, R. (2005). Distribution and nature of arsenic along former railway corridors of South Australia. *Science of the Total Environment*, 163 (1–3), 175–182.

Smith, M.M.H., Hore, T., Chakraborty, P., Chakraborty, D.K., Savarimuthu, X. & Smith, A.H. (2003) A dugwell program to provide arsenic-safe water in West Bengal, India: preliminary results. *Journal of Environmental Science and Health*, Part-A, A38, 289–299.

Stolz, J.F. & Oremland, R.S. (1999) Bacterial respiration of arsenic and selenium. *FEMS Microbiology Reviews*, 23, 615–627.

Stuben, D., Bernera, Z., Chandrasekharam, D. & Karmakar, J. (2003) Arsenic enrichment in groundwater of West Bengal, India: geochemical evidence for mobilization of As under reducing conditions. *Applied Geochemistry*, 18, 1417–1434.

Talavera Mendoza, O., Hernández, M.A.A., Abundis, J.G. & Mundo, N.F. (2006) Geochemistry of leachates from the El Fraile sulfide tailings piles in Taxco, Guerrero, southern Mexico. *Environmental Geochemistry and Health*, 28 (3), 243–255.

Terry, N. & Banuelos, G. (2000) *Phytoremediation of contaminated soil and water*. Lewis Publishers, Boca Raton, FL.

Tondel, M., Rahman, M., Magnuson, A., Chowdhury, I.A., Faruquee, M.H. & Ahmad, S.A. (2001) The relationship of arsenic levels in drinking water and the prevalence rate of skin lesions in Bangladesh. *Environmental Health Perspectives*, 107, 727–729.

Toor, I.A. & Tahir, S.N. (2009) Study of arsenic concentration levels in Pakistani drinking water. *Polish Journal of Environmental Studies*, 18, 907–912.

Tun, T.N. (2003) Arsenic contamination of water sources in rural Myanmar. *29th WEDC International Conference, Abuja, Nigeria*. Available from: http://wedc.lboro.ac.uk/resources/conference/29/Tun.pdf [accessed July 2015].

Ullah, S.M. (1998) Arsenic contamination of groundwater and irrigated soil of Bangladesh. *International Conference on As Pollution of Groundwater in Bangladesh: Causes, Effects and Remedies*. Dhaka Community Hospital, Dhaka, Bangladesh. pp. 8–12.

Vahter, M. & Norin, H. (1980) Metabolism of As-labelled trivalent and pentavalent inorganic arsenic in mice. *Environmental Research*, 21, 446–457.

Van Halam, D., Bakker, S.A., Amy, G.L. & van Dijk, J.C. (2009) Arsenic in drinking water: a worldwide water quality concern for water supply countries. *Drinking Water Engineering and Science*, 2, 29–34.

von Ehrenstein, O.S., Guha Majumdar, D.N., Yuan, Y., Samanta, S., Balmes, J., Sil, A., Ghosh, N., Hira-Smith, M., Haque, R., Purushothamam, R., Lahiri, S., Das, S. & Smith, A.H. (2005) Decrements in lung function related to arsenic in drinking water in West Bengal, India. *American Journal of Epidemiology*, 162, 533–541.

von Ehrenstein, O.S., Guha Majumdar, D.N., Hira-Smith, M., Ghosh, N., Yuan, Y., Windham, G., Ghosh, A. Haque, R., Lahiri, S., Kalman, D., Das, S. & Smith, A.H. (2006) Pregnancy outcomes, infant mortality, and arsenic in drinking water in West Bengal, India. *American Journal of Epidemiology*, 163, 662–669.

Wang, W., Xie, Z., Lin, Y. & Zhang, D. (2014) Association of inorganic arsenic exposure with type 2 diabetes mellitus: a meta-analysis. *Journal of Epidemiology & Community Health*, 68, 176–184.

Watanabe, M.E. (2001) Can bioremediation bounce back? *Nature Biotechnology*, 19, 1111–1115.

Welch, A.H., Westjohn, D.B., Helsel, D.R. & Wanty, R.B. (2000) Arsenic in ground water of the United States: occurrence and geochemistry. *Groundwater*, 38, 589–604.

WHO (2012) Estimated with data from WHO/UNICEF Joint Monitoring Programme (JMP) for Water Supply and Sanitation. Progress on Sanitation and Drinking-Water. World Health Organisation, Geneva, Switzerland.

World Bank Report (2005) Study: arsenic contamination of groundwater in South and East Asia: towards a more operational response. World Bank, Washington, DC.

Zablotska, L.B., Chen, Y., Graziano, J.H., Parvez, F., van Geen, A., Howe, G.R. & Ahsan, H. (2008) Protective effects of B vitamins and antioxidants on the risk of arsenic related skin lesions in Bangladesh. *Environmental Health Perspectives*, 116, 1056–1062.

CHAPTER 3

Human health through traces of contaminants in water: structural effects induced by Cr(III) and Cr(VI) on red cells

Mario Suwalsky, Susana Sánchez, María José Gallardo, Fernando Villena, Malgorzata Jemiola-Rzeminska & Kazimierz Strzalka

3.1 INTRODUCTION

In the course of an *in vitro* system search for the toxicity screening of chemicals with biological relevance, different cellular models have been applied to examine their adverse effects. The cell membrane is a diffusion barrier which protects the cell interior. Therefore, its structure and functions are susceptible to alterations as a consequence of interactions with chemical species. With the aim to better understand the molecular mechanisms of the interaction of chemical species with cell membranes we utilized human erythrocytes and molecular models of the erythrocyte membrane. Erythrocytes were chosen because although less specialized than many other cell membranes they carry on enough functions in common with them such as active and passive transport, and the production of ionic and electric gradients, to be considered representative of the plasma membrane in general. The molecular models consist of bilayers of dimyristoylphosphatidylcholine (DMPC) and dimyristoylphosphatidylethanolamine (DMPE), representative of phospholipid classes located in the outer and inner monolayers of the human erythrocyte membrane, respectively (Boon and Smith, 2002; Devaux and Zachowsky, 1994).

Chromium exists in many oxidation states, of which only the hexavalent Cr(VI) and the trivalent Cr(III) ions are stable under environmental conditions (Pesti *et al.*, 2000). Chromium toxicity has been almost exclusively linked to the hexavalent form (Barceloux, 1999; Grevatt, 1998; Lamson and Plaza, 2002; Repetto *et al.*, 1996). Water containing more than $0.05 \, \text{mg} \, \text{L}^{-1}$ of Cr(VI) is considered to be toxic (Pesti *et al.*, 2000), whereas the exposure to $250 \, \mu\text{M}$ induced a significant decrease of goldfish isolated hepatocytes (Krumschnabel and Nawaz, 2004). Cr(VI) salts are well known human carcinogens, but the results from *in vitro* studies are often conflicting (Blasiak and Kowalik, 2000). On the other hand, its mechanistic toxicity has not been completely understood; although, a large number of Cr(VI) studies demonstrated that it induces oxidative stress, DNA damage, apoptotic cell death and altered gene expression (Bagchi *et al.*, 2002). Once absorbed into the bloodstream, the water-soluble chromate or dichromate ions readily enter red blood cells through the phosphate and sulfate anion-exchange carrier pathway, though a portion may remain in plasma for an extended period (Wiegand *et al.*, 1985). Since Cr(III) is unable to cross the red cell membrane by this pathway, it may enter red blood cells but only with very low efficiency (O'Flaherty, 1996). Cr(VI) is quickly reduced by glutathione, ascorbic acid, or cysteine to the kinetically much more stable Cr(III) (Gagelli *et al.*, 2002), which may interact with cellular macromolecules, including DNA. About 90% of cellular Cr is present as Cr(III) (Pesti *et al.*, 2000). Despite such toxic features, Cr(III) is essential in mammals that require it for normal carbohydrate and lipid metabolism (Gagelli *et al.*, 2002; Lamson and Plaza, 2002). It is amazing that very few studies are available in the literature that directly address the toxicity of Cr(III) (Grevatt, 1998).

In an attempt to further elucidate the effect of Cr(III) and Cr(VI) on cell membranes, the present work examined their influence on the morphology of intact human erythrocytes by means of fluorescence spectroscopy, defocusing microscopy (DM) and scanning electron microscopy (SEM) while isolated unsealed human erythrocyte membranes (IUM) were also studied by fluorescence spectroscopy. The capacity of Cr(III) and Cr(VI) to perturb the multibilayer structures of DMPC and DMPE was evaluated by X-ray diffraction and differential scanning calorimetry (DSC). These systems and techniques have been used in our laboratories to determine the interaction with and the membrane-perturbing effects of other metallic cations (Suwalsky *et al.*, 2010a; 2010b; 2012; 2013).

3.2   MATERIALS AND METHODS

3.2.1   *Scanning electron microscope studies of human erythrocytes*

Blood (0.1 mL) was obtained from healthy volunteers by aspiration into tuberculin syringes containing $10\,\mu L$ of heparin ($5000\,UI\,mL^{-1}$) in 0.9 mL phosphate buffered saline ($10\,mM\,Na_2PO_4$, $147\,mM\,NaCl$, $3\,mM\,KCl$, pH 7.5, PBS), from which the following samples were prepared: (i) control, by mixing 0.1 mL blood stock with 0.9 mL PBS, and (ii) Cr(III) and Cr(VI) in a range of concentrations (0.1–$0.5\,mM$ and 0.1–$1.0\,mM$, respectively) by mixing the blood with PBS containing $CrCl_3$ and K_2CrO_4 in adequate concentrations, respectively. Samples were incubated at 37°C for 1 h. They were then fixed overnight at 5°C by adding $20\,\mu L$ of each sample to plastic tubes containing 1 mL of 2.5% glutaraldehyde in saline, reaching a final fixation concentration of about 2.4%. The fixed samples were directly placed on Al stubs, air dried at 37°C for 30 min to 1 h and gold-coated for 3 min at 10^{-1} Torr ($\approx 13.3\,Pa$) in a sputter device (Edwards S150, Sussex, England). Resulting specimens were examined in a Jeol SEM (JSM 6380 LB, Japan).

3.2.2   *Defocusing microscopy measurements of human erythrocytes*

RBCs were obtained from healthy donors under no pharmacological treatment. The cells were centrifuged (1000 rpm for 10 min) and washed three times with PBS 1X pH 7.4 with 1 mg/mL of bovine serum albumine (BSA). RBC solution was prepared diluting the washed blood 20 times in a solution of PBS and BSA. $CrCl_3$ and K_2CrO_4 solutions were prepared in the same preparation of PBS and BSA. In order to carry out the analysis, 1.7 mL of the RBC solution were placed in an acrylic cuvette, and visualized at the microscope. The experiment was carried out in an inverted optical microscope Nikon Eclipse Ti-U, where the light source is provided by a halogen lamp with a transmission filter centered at 650 ± 5 nm to avoid physical damage on RBC. The objective is mounted on a C-focus system (Mad City Labs, Madison, USA) for an auto correction of the focus positioning and for a nanometric control of the focal plane position. The visualization was performed through a Retiga EXI Fast 12-BIT CCD camera (Surrey, Canada) which is calibrated to determine the equivalence between the gray level at the observed image and the intensity at the image plane (Agero *et al.*, 2003; 2004; Etcheverry *et al.*, 2012; Mesquita *et al.*, 2006), The computational software permits a simultaneous control of the C-focus and the nanopositioning devices, as well as the CCD camera (Etcheverry *et al.*, 2012). To make three-dimensional reconstructions two images of a morphologically normal erythrocyte were captured in the defocus positions $+1$ and $-1\,\mu m$ with the software and then the concentration of chromates was increased and the procedure was repeated.

3.2.3   *Fluorescence measurements of human erythrocyte membranes*

Human blood was obtained from healthy volunteers. Samples were collected in EDTA vacutainers and stored at 4°C. The erythrocyte preparation protocol followed is a modified version from that already published (Sánchez, 2012). Briefly, erythrocytes were centrifuged and the plasma layer removed. PBS buffer was added to restore the sample to the original hematocrit and the sample was kept at 37°C. From this stock solution the blood was diluted to hematocrit 1.2% into 1 mL of PBS buffer containing either 1 mM $CrCl_3$ or 2 mM K_2CrO_4 (control sample contained only PBS buffer). Samples were incubated at 37°C for 1 h with occasional gentle mixing. After incubation, the erythrocytes were rinsed with PBS by centrifugation (25°C, 1000 g, 15 min) 3–4 times. Samples were diluted to hematocrit 0.2% in PBS containing 1 µM laurdan, incubated 15 min at 37°C and observed under the microscope in a thermostatized stage at 37°C. Laurdan is a membrane probe having a large excited-state dipole moment, which results in its ability to report the extent of water penetration into the bilayer because of the dipolar relaxation effect (Weber and Farris, 1979). Water penetration has been correlated with lipid packing and membrane fluidity (Parasassi *et al.*, 1997). Laurdan is located within a lipid bilayer; its spectrum will sense the environment and shift according to the water content. Laurdan fluorescence shifts were quantitatively evaluated using the generalized polarization (*GP*) concept, which is related to the lipid polar headgroup organization in lipid bilayers, which is defined by the expression:

$$GP = I_{440} - I_{490}/I_{440} + I_{490} \tag{3.1}$$

where I_{440} and I_{490} are the emission intensities at 440 nm and 490 nm, respectively. A full discussion of the use and mathematical significance of *GP* can be found in the literature (Bagatolli *et al.*, 2003; Parasassi *et al.*, 1995). Images were collected on a scanning two photon fluorescence microscope at the Laboratory for Fluorescence Dynamics (University of California at Irvine, CA) as previously described (Sánchez *et al.*, 2007). A mode-locked titanium sapphire laser (Mira 900; Coherent, Palo Alto, CA) pumped by a frequency doubled Nd:vanadate laser (Verdi Coherent, Palo Alto, CA) set to 780 nm was used as the two-photon excitation light source. A two-channel detection system was attached for *GP* image collection (Sánchez *et al.*, 2007). The fluorescence was split in two channels using a Chroma Technology 470DCXR-BS dichroic beam splitter and two interference filters Ealing 490 and Ealing 440 (Chroma Technology, Bellows Falls, VT) which were placed in the emission paths to further isolate the two desired regions of the emission spectrum (440 ± 10 nm and 490 ± 10 nm). Two simultaneous 256 × 256 pixel images (emission centered at 490 and 440 nm) were obtained from the erythrocytes. The two intensity images were used to create a *GP* image by applying the *GP* formula (3.1)) pixel by pixel. Corrections for the wavelength dependence of the emission detection system were accomplished through the comparison of the *GP* value of a known solution (laurdan in DMSO) (Sánchez *et al.*, 2002) taken on a steady-state fluorometer PC1 (ISS Inc., Illinois, USA) and then in the microscope. From the resulting *GP* image two sets of data were obtained: the average *GP* value of each erythrocyte (center of the histogram of number of pixels versus *GP*) and the shape of each cells separated in discocytes or echinocytes. Three samples were used: control (incubated in PBS buffer), a sample incubated with 1 mM Cr(III) and a third one incubated with Cr(VI). From each plate, 10 images were obtained and the analysis was performed using Sim FCS (Laboratory for Fluorescence Dynamics) and Image J software (National Institute of Health, Bethesta, USA).

3.2.4   *X-ray diffraction studies of phospholipid multilayers*

The capacity of Cr(III) and Cr(VI) to perturb the structures of DMPC and DMPE multilayers was determined by X-ray diffraction. Synthetic DMPC (lot 140PC-224, MW 677.9), DMPE (lot 140PE-54, MW 635.9) from Avanti Polar Lipids, ALA, USA), $CrCl_3$ and K_2CrO_4 from Merck (Germany) were used without further purification. About 2 mg of each phospholipid were introduced into 2 mm diameter special glass capillaries (Glas-Technik & Konstruktion, Berlin, Germany) which were then filled with 150 µL of (i) distilled water and (ii) aqueous solutions of the Cr salts in a range of concentrations. Specimens were incubated at 30°C and 60°C with DMPC and DMPE, respectively, centrifuged for 10 min at 2000 rpm and X-ray diffracted with Ni-filtered $CuK\alpha$ from a Bruker Kristalloflex 760 (Karlsruhe, Germany). Specimen-to-film distances were 8 and 14 cm, standardized by sprinkling calcite powder on the capillary surface. The relative reflection intensities were obtained in a MBraun PSD-50M linear position-sensitive detector system (Garching, Germany); no correction factors were applied. The experiments were performed at 19 ± 1°C, which is below the main phase transition temperature of both DMPC and DMPE. Each experiment was repeated three times.

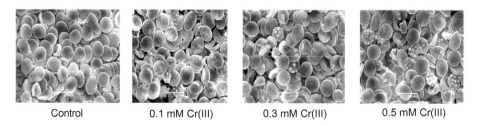

| Control | 0.1 mM Cr(III) | 0.3 mM Cr(III) | 0.5 mM Cr(III) |

Figure 3.1.   Scanning electron microscopy (SEM) images of the effects of Cr(III) on the morphology of human erythrocytes.

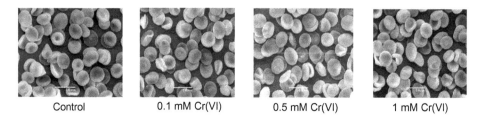

| Control | 0.1 mM Cr(VI) | 0.5 mM Cr(VI) | 1 mM Cr(VI) |

Figure 3.2.   Scanning electron microscopy (SEM) images of the effects of Cr(VI) on the morphology of human erythrocytes.

3.2.5   *Differential scanning calorimetry*

Appropriate amounts of lipids (DMPC or DMPE) dissolved in chloroform were gently evaporated to dryness under a stream of gaseous nitrogen until a thin film on the wall of the glass test tube was formed. To remove the remnants of moisture, the samples were subsequently exposed to vacuum for 1 h and then dry lipid films were suspended in distilled water. $CrCl_3$ and K_2CrO_4 were added in the concentration range of 0.05 mM to 10.0 mM. The multilamellar liposomes (MLV) were prepared by vortexing the samples at the temperature above gel-to-liquid crystalline phase transition of the pure lipid (about 30°C for DMPC and 60°C for DMPE). DSC experiments were performed using a NANO DSC Series III System with Platinum Capillary Cell (TA Instruments, New Castle, USA). The calorimeter was equipped with the original data acquisition and analysis software. In order to avoid bubble formation during heating mode the samples were degassed prior to being loaded by pulling a vacuum of 0.3–0.5 atm. (≈304 hPa) on the solution for a period of 10–15 min. Then the sample cell was filled with about 400 μL of MLV suspension and an equal volume of buffer was used as a reference. The cells were sealed and thermally equilibrated for about 10 min below starting temperature of the run. All measurements were carried out on samples under 0.3 MPa pressure. Data were collected in the range of 5–40°C (DMPC) and 30–70°C (DMPE) at the scan rate 1°C min^{-1} both for heating and cooling. Scans of buffer as a sample and a reference were also performed to collect the apparatus baseline. For the check of the reproducibility each sample was prepared and recorded at least three times. Each data set was analyzed for thermodynamic parameters with the software package supplied by TA Instruments (New Castle, USA).

3.3   RESULTS

3.3.1   *Scanning electron microscope studies of human erythrocytes*

SEM examinations of human erythrocytes incubated with $CrCl_3$ in the range 0.1–0.5 mM indicated that Cr(III) induced echinocytosis (Fig. 3.1). In that altered condition, red blood cells lost their normal profile and presented a spiny configuration with blebs in their surfaces. The extent of these shape changes was dependent on the Cr(III) concentrations. As can be observed in Figure 3.2,

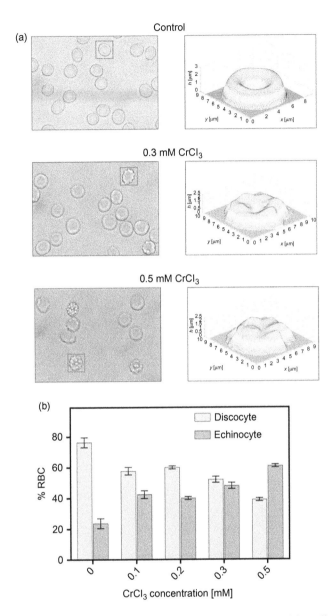

Figure 3.3.   Effects of Cr(III) in human erythrocytes. (a) Defocused images and three dimensional reconstructions of untreated erythrocytes (control) and incubated with 0.3 and 0.5 mM of CrCl$_3$.
(b) Population distribution of RBCs at different CrCl$_3$ concentrations (0–0.5 mM).

Cr(VI) did not induce any significant change to the erythrocyte shape even at a concentration as
high as 1 mM.

3.3.2   *Defocusing microscopy measurements of human erythrocytes*

The erythrocyte morphology was also analyzed through three dimensional reconstructions using
defocusing microscopy (DM). Figure 3.3 shows DM experiments of human red blood cells
exposed to different concentration of CrCl$_3$. Cr(III) induced the transformation of the normal

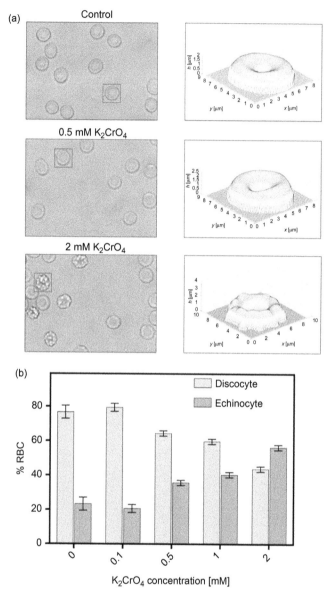

Figure 3.4.   Effects of Cr(VI) in human erythrocytes. (a) Defocused images and three dimensional reconstructions of untreated erythrocytes (control) and incubated with 0.5 and 2 mM K_2CrO_4.
(b) Population distribution of RBC with different K_2CrO_4 concentrations (0–2 mM).

discoid morphology of the erythrocytes to echinocytes (Fig. 3.3a) at low Cr(III) concentration
(0.1 mM) reaching a maximum effect at 0.5 mM. At this concentration 60% of the erythrocytes
population was affected with echinocytosis (Fig. 3.3b). On the other hand, Cr(VI) did not induce
significant changes in the erythrocyte population at low concentration (0.1–0.5 mM). It was
observed that there was a significant increase in the echinocyte transformation only at high
concentration (2 mM) of Cr(VI) where more than 50% of the cell had the echinocyte shape
(Fig. 3.4).

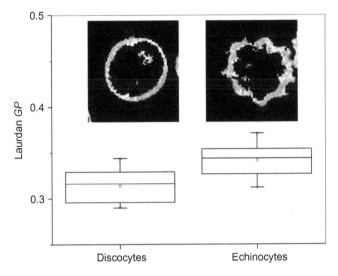

Figure 3.5. Laurdan *GP* values for discocytes and echinocytes. Blood was diluted to hematocrit 0.2% in PBS containing 1 µM laurdan, incubated 15 min at 37°C and observed under the microscope in a thermostatized stage at 37°C. From the laurdan *GP* image (methods), shape and *GP* value for each erythrocyte was obtained. The plot is the results of 20 cells of each type. Box-charts show minimum, median, 25th and 75th percentile values.

3.3.3 *Fluorescence measurements of human erythrocyte membranes*

The effect of Cr(III) and Cr(VI) on the membrane of red blood cells was examined by measuring laurdan generalized polarization (*GP*). In blood samples, and depending on several factors such as pH and temperature, erythrocytes with different shapes were found (Stasiuk *et al.*, 2009). The most common shapes found were discocytes (biconcave discs) and echinocytes (spiculated spheres). To study the effect of any agent on the membrane fluidity using laurdan *GP*, the shape factor has to be considered since the laurdan *GP* value of the discocytes is smaller than that of the echinocytes (discocytes membrane are more fluid) (Best *et al.*, 2002). Figure 3.5 shows laurdan *GP* values for each shape in a control sample incubated in PBS buffer. The image was obtained in a two-photon microscope at the equatorial plane; therefore; discocytes appear as a circle and echinocytes like stars. The median *GP* values extracted from the box plot were 0.32 and 0.34 for discocytes and echinocytes, respectively. In order to study the effect of the Cr ions, erythrocytes were individually analyzed and the data were organized into two groups: echinocytes and discocytes. Samples with the same hematocrit (1.2%) were incubated for 1 h at 37°C with either 1 mM Cr(III) or 2 mM Cr(VI). The control sample consisted in a sample of the same characteristics but containing only PBS. Figure 3.6a shows a representative laurdan *GP* image for each condition (control, Cr(III) 1 mM and Cr(VI) 2 mM).

In the control sample both discocytes and echinocytes can be easily distinguished; however in the samples incubated with Cr(III) and Cr(VI) mainly echinocytes were observed. Figure 3.6b shows the analysis from 10 images analyzed per group. The percentage of discocytes (white) and echinocytes (gray) in the control was 59 and 41%, respectively. In the samples incubated with Cr(III) and Cr(VI) all the erythrocytes were echinocytes. In the case of Cr(VI) twice the concentration was needed in order to get 100% of the cells in the spiculated shape.

The laurdan *GP* parameter from each cell was determined and the plot is shown in Figure 3.6c. For the control sample, the *GP* values for discocytes and echinocytes are separately plotted. The average *GP* value of the cells incubated with Cr(III) and Cr(VI), all of them echinocytes, was similar to the value of echinocytes present in the control.

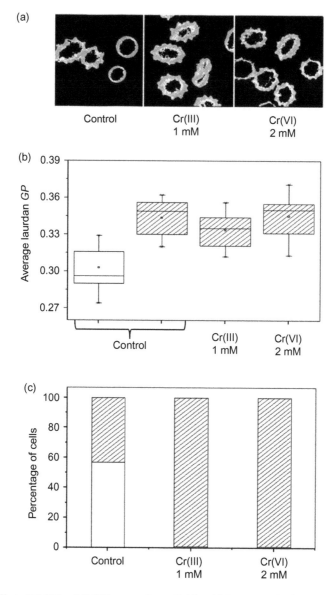

Figure 3.6.   Effect of Cr(III) and Cr(VI) on membrane fluidity. (a) Representative laurdan *GP* image for the three conditions studied, (b) Percentage of discocytes (white) and echinocytes (diagonal lines) in each sample and (c) average *GP* values for the eythrocytes. Echinocytes and discocytes were analyzed separately. In control, both discocytes (white) and echinocytes (diagonal lines) were present; however, in the samples incubated with the chromium ions only echinocytes (diagonal lines) were observed.

3.3.4   *X-ray diffraction studies of phospholipid multilayers*

Figure 3.7a exhibits the results obtained by incubating DMPC with water and CrCl$_3$. As expected, water altered the structure of DMPC, as its bilayer width (phospholipid bilayer width plus the layer of water) increased from about 55 Å in its dry crystalline form (Suwalsky, 1996) to 64.5 Å

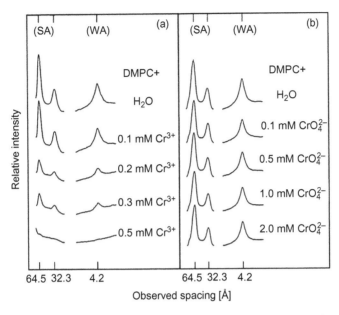

Figure 3.7.   Microdensitograms from X-ray diffraction diagram of DMPC in water and aqueous solutions of (a) Cr(III) and (b) Cr(VI); (SA) small-angle and (WA) wide-angle reflections.

when immersed in water, and its small-angle reflections, which correspond to DMPC polar terminal groups, were reduced to only the first two orders of the bilayer width. On the other hand, only one strong reflection of 4.2 Å showed up in the wide-angle region which corresponds to the average distance between fully extended acyl chains organized with rotational disorder in hexagonal packing. These results were indicative of the fluid state reached by DMPC bilayers. Figure 3.7a discloses that after exposure to 0.2 mM Cr(III) there was a considerable weakening of the small and wide-angle lipid reflection intensities (indicated as (SA) and (WA) in the figure, respectively) which at 0.5 mM practically disappeared. From these results it can be concluded that Cr(III) produced a significant structural perturbation of DMPC bilayers. On the other hand, Figure 3.7b shows that Cr(VI) did not induce any significant change to DMPC, even at higher concentrations that those used with Cr(III). Figure 3.8a shows the results of the X-ray diffraction analysis of DMPE bilayers incubated with water and $CrCl_3$. As reported elsewhere, water did not significantly affect the bilayer structure of DMPE (Suwalsky, 1996). Figure 3.8a also shows that increasing concentrations of Cr(III) caused a gradual but slight weakening of DMPE reflection intensities, all of which still remained at the highest assayed Cr(III) concentration. On the other hand, Figure 3.8b shows that Cr(VI) did not induce any significant change to DMPE bilayers, even at higher concentrations that those used with Cr(III). From these results it can be concluded that Cr(III) induced stronger structural perturbations than Cr(VI) to both phospholipid bilayers.

3.3.5   *Differential scanning calorimetry*

All heating profiles of the binary systems of phospholipids/$CrCl_3$ and phospholipids/K_2CrO_4 multilamellar vesicles were referred to those formed entirely by lipids. Figure 3.9 reports the typical heating curves obtained of DMPC liposomes containing different concentrations of $CrCl_3$ and K_2CrO_4. The first heating profile corresponds to the values of thermal parameters obtained for fully hydrated DMPC bilayers in the absence of any additives, where the pretransition temperature ($T_p = 14.98°C$) and the main phase transition temperature ($T_m = 24.00°C$) are in agreement with the previously published data (Huang and Lee, 1999; Koynova and Caffrey, 1998; Marsh,

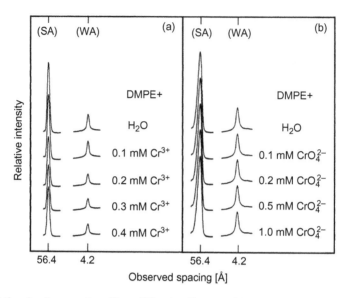

Figure 3.8.   Microdensitograms from X-ray diffraction diagram of DMPE in water and aqueous solutions of (a) Cr(III) and (b) Cr(VI); (SA) small-angle and (WA) wide-angle reflections.

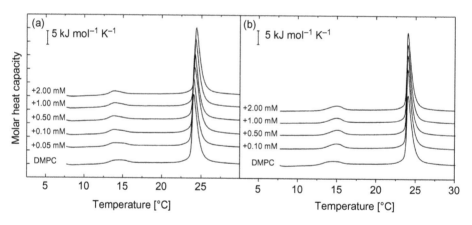

Figure 3.9.   Representative DSC curves obtained for multilamellar DMPC liposomes containing different CrCl$_3$ (a) and K$_2$CrO$_4$ (b) concentrations. Scans were obtained at a heating rate of 1°C min^{-1}.

1991). The highly cooperative main transition ($P_\beta \rightarrow L_\alpha$ phase transition) with an enthalpy change (ΔH) of 24.28 kJ mol^{-1} corresponds to the gel-to-liquid-crystal transition, while the smaller one ($L_\beta \rightarrow P_\beta$ phase transition) with a ΔH of 3.61 kJ mol^{-1} is called pretransition. On the basis of the measured heating curves, the thermodynamic parameters referred to as essential for the description of interactions between phospholipid and guest molecules were calculated: the phase transition temperatures and the calorimetric enthalpy as well as transition entropy (Table 3.1). The first parameter, T_m, is an expression of changes in membrane fluidity, i.e. the passage of a certain number of aliphatic chains from an ordered to a disordered state. The analysis of the variation of T_m values with increasing amounts of CrCl$_3$ and K$_2$CrO$_4$ in DMPC multibilayers is shown in Figure 3.10. It is clear that neither CrCl$_3$ nor K$_2$CrO$_4$ SA has any significant perturbing

Table 3.1. Thermodynamic parameters of the pretransition and main phase transition of pure, fully hydrated DMPC multilamellar liposomes, DMPC/CrCl₃ and DMPC/K₂CrO₄ mixtures determined from heating and cooling scans collected at a heating (cooling) rate of 1°C min⁻¹. The accuracy for the main phase transition temperature and enthalpy was ±0.01°C and ±0.8 kJ mol⁻¹, respectively.

	Pretransition heating			Main transition heating			Pretransition cooling			Main transition cooling		
Conc [mM]	ΔH [kJ mol⁻¹]	ΔS [J mol⁻¹ K⁻¹]	T_m [°C]	ΔH [kJ mol⁻¹]	ΔS [J mol⁻¹ K⁻¹]	T_m [°C]	ΔH [kJ mol⁻¹]	ΔS [J mol⁻¹ K⁻¹]	T_m [°C]	ΔH [kJ mol⁻¹]	ΔS [J mol⁻¹ K⁻¹]	T_m [°C]
DMPC												
1.00	3.61	1.25	14.98	24.28	8.17	24.00	1.77	0.63	9.34	24.19	8.16	23.26
+CrCl₃												
0.005	3.18	1.11	14.32	23.60	7.94	24.03	1.86	0.66	9.28	23.04	7.77	23.27
0.01	3.16	1.10	13.95	23.05	7.75	24.03	1.17	0.41	9.25	22.29	7.52	23.27
0.03	3.22	1.12	13.85	25.51	8.58	24.17	0.82	0.29	9.28	23.47	7.92	23.34
0.05	3.00	1.05	13.81	22.87	7.69	24.16	0.70	0.25	9.45	21.60	7.28	23.37
1.00	3.05	1.06	13.95	23.13	7.78	24.27	0.30	0.11	9.59	21.24	7.16	23.47
2.00	2.82	0.98	13.91	22.15	7.45	24.37	0.10	0.09	9.86	19.51	6.58	23.58
+Kr₂CrO₄												
0.01	3.41	1.18	14.96	24.18	8.14	24.01	1.66	0.59	9.40	24.31	11.90	23.27
0.05	3.35	1.16	15.05	23.31	7.85	24.01	0.9	0.35	9.39	24.52	11.84	23.29
1.00	3.37	1.17	14.96	23.31	7.85	24.01	0.48	0.13	9.36	24.81	11.52	23.27
2.00	3.27	1.13	14.96	23.22	7.81	24.01	0.09	0.03	9.31	24.94	11.37	23.29

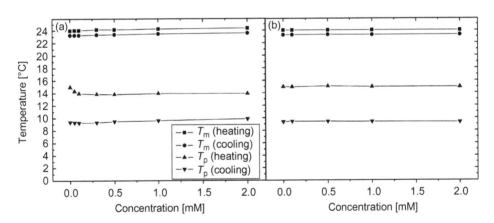

Figure 3.10. A plot of phase transition temperature of DMPC multilamellar liposomes determined for cooling and heating scans as a function of CrCl₃ (a) and K₂CrO₄ (b) content.

effect on the membrane fluidity. The second thermodynamic parameter presented in Figure 3.11 is the enthalpy change associated with the phase transition, which was calculated by integrating the peak area after baseline adjustment and normalization to the amount of sample analyzed. The total enthalpy change associated with the lipid chain melting is related to molecular packing of the acyl chain. Thus, compounds entering the bilayer should lower the stability of the lipid matrix with a decrease in both T_m and ΔH of the gel to liquid crystal transition phase. However, the incorporation of CrCl₃ and K₂CrO₄ does not significantly influence the ΔH in the concentration range from 0.05 to 2 mM.

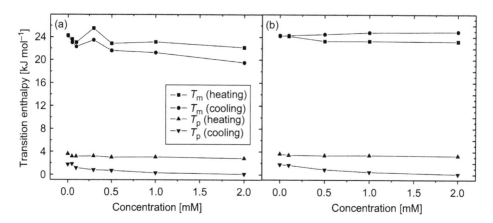

Figure 3.11.   A plot of transition enthalpy of DMPC multilamellar liposomes determined for heating and cooling scans as a function of CrCl$_3$ (a) and K$_2$CrO$_4$ (b) content.

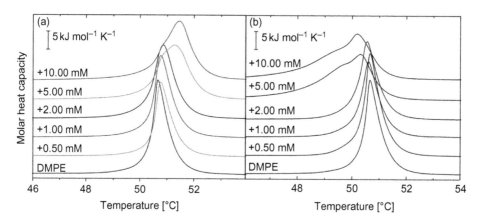

Figure 3.12.   Representative DSC curves obtained for multilamellar DMPE liposomes containing different CrCl$_3$ (a) and K$_2$CrO$_4$ (b) concentrations. Scans were obtained at a heating rate of 1°C min^{-1}.

Figure 3.12 shows the set of representative heating profiles obtained for DMPE liposomes. In the thermal range of 30–70°C, the pure DMPE bilayers exhibit a strong and sharp main-transition at 50.64°C, with an enthalpy change of 22.25 kJ mol^{-1}, arising from the conversion of gel to liquid-crystal phase. The transition is reversible and the shape of the peak is roughly symmetrical. Thermodynamical parameters found for DMPE are concurrent with the literature data (Lewis and McElhaney, 1993). The effectiveness in perturbations of DMPE thermotropic phase transition exerted by examined compounds was further analyzed in terms of thermodynamic parameters. Table 3.2 presents values of temperature, enthalpy and entropy for DMPE/CrCl$_3$ and DMPE/K$_2$CrO$_4$ systems, determined on the basis of heating and cooling scans. In contrast to phospholipids containing choline group, DMPE bilayers are affected by the presence of CrCl$_3$ and K$_2$CrO$_4$ as the phase transition temperature underwent a shift in the concentration-dependent manner to the higher and lower values in presence of CrCl$_3$ and K$_2$CrO$_4$, respectively (Fig. 3.13). Nevertheless, the significant changes are observed for liposomes with high CrCl$_3$ and K$_2$CrO$_4$ contents (>2 mM). As can be seen from Figure 3.14, transition enthalpy (ΔH) is severely altered only under cooling for high K$_2$CrO$_4$ content in the liposomes.

Table 3.2. Thermodynamic parameters of the phase transition of pure, fully hydrated DMPE multilamellar liposomes, DMPE/CrCl$_3$ and DMPE/Kr$_2$CrO$_4$ mixtures determined from heating and cooling scans collected at a heating (cooling) rate of 1°C min^{-1}. The accuracy for the main phase transition temperature and enthalpy was ±0.01°C and ±0.8 kJ mol^{-1}, respectively.

Compound		Heating				Cooling		
	conc [mM]	ΔH [kJ mol^{-1}]	ΔS [J mol^{-1}K^{-1}]	T_m [°C]		ΔH [kJ mol^{-1}]	ΔS [J mol^{-1}K^{-1}]	T_m [°C]
DMPE								
	1.00	22.25	6.9	50.64		28.55	8.9	49.36
+CrCl$_3$								
	0.05	21.88	6.8	50.74		23.19	7.2	49.46
	1.00	18.09	5.6	50.74		26.74	8.3	49.53
	2.00	22.82	7.1	50.84		26.69	8.3	49.70
	5.00	22.99	7.1	51.29		27.24	8.4	50.52
	10.00	24.82	7.7	51.39		24.76	7.6	50.94
+Kr$_2$CrO$_4$								
	0.05	26.12	8.1	50.64		31.09	9.6	49.29
	1.00	21.12	6.5	50.61		25.60	8.0	49.09
	2.00	22.13	6.8	50.54		24.70	7.7	49.05
	5.00	28.07	8.9	50.30		46.24	14.4	48.20
	10.00	27.98	8.7	50.19		42.11	13.1	48.16

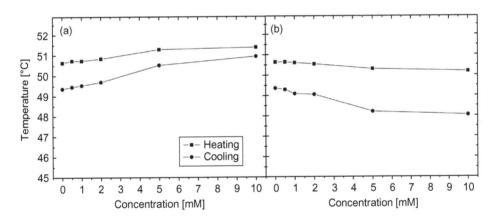

Figure 3.13. A plot of phase transition temperature of DMPE multilamellar liposomes determined for heating and cooling scans as a function of CrCl$_3$ (a) and K$_2$CrO$_4$ (b) content.

3.4 DISCUSSION

It is amazing that after so many studies on Cr toxicity, and particularly on the knowledge that Cr(VI) anions enter the erythrocytes readily (Fernandez *et al.*, 2000; Kitagawa *et al.*, 1982; Ottenwaelder *et al.*, 1988; Wells *et al.*, 2003; Wiegand *et al.*, 1985) there are practically no reports on the structural effects induced by Cr compounds to the erythrocyte membrane. In order to understand these effects, Cr(III) under the form of CrCl$_3$, and Cr(VI) as K$_2$CrO$_4$ were incubated with intact erythrocytes and bilayers built-up of DMPC and DMPE, phospholipid classes present in the outer and inner monolayers of the erythrocyte membrane, respectively. Analysis by X-ray diffraction indicated that 0.2 mM and higher concentrations of Cr(III) induced

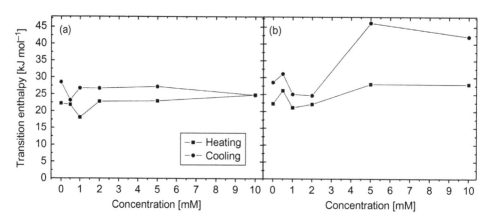

Figure 3.14.   A plot of transition enthalpy of DMPE multilamellar liposomes determined for heating and cooling scans as a function of CrCl$_3$ (a) and K$_2$CrO$_4$ (b) content.

structural perturbations of the polar head group and to the hydrophobic acyl regions of DMPC, whose bilayer structure completely collapsed with 0.5 mM Cr(III). On the other hand, Cr(VI) in a range of concentration up to 1 mM, did not induce any significant effect to DMPC. Interestingly, DSC studies did not reveal significant changes in thermotropic behavior of DMPC upon CrCl$_3$ or K$_2$CrO$_4$ addition. When X-ray diffraction analyses were performed with DMPE, Cr(III) also induced higher structural perturbations than Cr(VI). However, these effects were much milder than those observed with DMPC. These results can be explained on the basis of the different nature of DMPC and DMPE head groups, and of Cr^{3+} and CrO$_4^{2-}$ ions. Chemically the two lipids only differ in their terminal amino groups, these being $^+$N(CH$_3$)$_3$ in DMPC and $^+$NH$_3$ in DMPE. Moreover, both molecular conformations are very similar in their dry crystalline phases: their acyl chains are mostly parallel and extended with the polar groups lying perpendicularly to them; however, DMPE molecules pack tighter than those of DMPC. This effect, due to the DMPE smaller polar group and higher effective charge, makes for a very stable multilayer arrangement that is not significantly perturbed by the presence of water (Suwalsky, 1996). On the other hand, the gradual hydration of DMPC bilayers leads to water filling the highly polar interbilayer spaces. Consequently, there is an increase in its bilayer width from 54.5 Å when dry up to 64.5 Å when fully hydrated at a temperature below that of its main transition. This condition promoted the incorporation of Cr ions into DMPC highly polar interbilayer space and the ensuing molecular perturbation of the phospholipid bilayer structure by Cr^{3+} ions. This interaction can be largely understood on the basis of its exceptionally small Cr^{3+} radius (0.62 Å) having thus a large effective charge. In this context, we postulate electrostatic interactions of Cr(III) ions with the negatively charged choline phosphate groups. On the other hand, the much bulkier tetrahedrically configurated CrO$_4^{2-}$ anions (Belagyi et al., 1999) can hardly interact with the also bulky $^+$N(CH$_3$)$_3$ group of DMPC. Cr(III) ions can also interact with DMPE phosphate groups disrupting its packing arrangement, which could not be achieved by CrO$_4^{2-}$ ions as discussed above. Worth mentioning is the opposite effect CrCl$_3$ and K$_2$CrO$_4$ have on the temperature of DMPE phase transition. The meaningful changes were observed at concentration higher than 2 mM and T_m was shifted of (+0.75) and (−0.45) for 10 mM of CrCl$_3$ and K$_2$CrO$_4$, respectively. Moreover, the shape of the peak corresponding to the phase transition was modified by the appearance of additional shoulder.

According to the bilayer couple hypothesis (Sheetz and Singer, 1974), shape changes are induced in red cells due to the insertion of foreign species in either the outer or the inner monolayer of the erythrocyte membrane. Thus, spiculated shapes (echinocytes) are observed in the first case whereas cup shapes (stomatocytes) are produced in the second due to the differential expansion of the corresponding monolayer. Given the extent of the interaction of Cr(III) with DMPC, class of lipid preferentially located in the outer monolayer of the erythrocyte membrane, echinocytes were expected of erythrocytes incubated with $CrCl_3$. SEM examination of specimens showed that in fact Cr(III) induced a change of shape of the normal biconcave erythrocytes. Although some echinocytes were observed with 0.1 mM Cr(III), the number of equinocytes considerably increased with higher concentrations, reaching an about 50% of the cells with 0.5 mM $CrCl_3$. On the other hand, no equinocytes were observed in erythrocytes incubated with 1 mM Cr(VI).

Defocusing microscopy is a simple, accurate and real-time method to detect alterations in human erythrocytes. Observed results indicate different membrane interactions of Cr ions depending of its oxidation state. Cr(III) induced the formation of spicules even at low concentrations (0.1 mM $CrCl_3$) and Cr(VI) only induced echinocytosis at considerable higher concentrations (2 mM K_2CrO_4). These results are in agreement with SEM experiments where same results were obtained.

In the blood stream of healthy individuals, three types of erythrocytes are normally found, which are not pathological: echinocytes, stomatocytes and discocytes. The cell volume in these three configurations remains unaltered (Reinhart and Chien, 2009) and they are in an equilibrium designed to protect the erythrocytes from lysis caused by factors such as basic pH and ionic amphiphiles (Stasiuk *et al.*, 2009). Laurdan *GP* value gives a parameter related with the fluidity of the plasma membrane and it has been reported that there are differences between the *GP* value in membranes from erythrocytes in either discocyte or echinocytes shapes, being a larger *GP* value for the echinocytes (Best *et al.*, 2002) interpreted as an increase in membrane order (Parasassi *et al.*, 1991). The *GP* values in erythrocytes were measured with different shapes (Fig. 3.5) and data confirmed the previous observations. To study the effect of Cr ions on the erythrocyte membrane data were analyzed considering the two main populations present in the samples (discocytes and echinocytes), separately measuring laurdan *GP* changes within each population. Data indicate that Cr(III) or Cr(VI) do not induce significant changes on the laurdan *GP* parameter of the plasma membrane but it does induce changes on the shape of the erythrocytes. These results indicate that there are not significant changes on the molecular dynamics and/or water content at the glycerol backbone level of the phospholipid polar head groups (where laurdan locates) in the presence of either Cr(III) or Cr(VI) until 2 mM. An important issue with fluorescence measurements with erythrocytes is discerning whether the observed results represent the behavior of the outer membrane leaflet, the inner leaflet, or an average of the two. Laurdan does cross the cell membrane and presumably reports behavior from both leaflets (Stott *et al.*, 2008). Changes in the shape indicate an interaction of the ions with the outer monolayer of the erythrocytes. In a previous report by Suwalsky *et al.* (2008) changes in laurdan *GP* were reported when unsealed erythrocytes membranes incubated with 0.5 mM Cr(III) produced a change of 0.13 *GP* units and not changes were observed with Cr(VI) until 0.1 mM. The differences between the two systems (unsealed membranes and whole erythrocytes) make the comparison difficult: in the unsealed membranes the ionic species can access the membrane from both sides of the bilayers contrary to the whole erythrocytes where just the outer membrane can be reached. On the other hand, the spatial resolution given by the microscopy techniques, such as *GP* imaging, allows correlating the changes in *GP* value with changes in the shape of the red blood cells; in cuvette experiments, however, the only information obtained is the *GP* value from the membrane as an average of all the vesicles in solution. Changes in shape due to the presence of the Cr species were also reported by Suwalsky *et al.* (2008). They reported shape changes from discocytes to echinocytes when the red blood cells where incubated with 0.5 mM Cr(III); however, the same concentration of Cr(VI) did not induced shape changes. Higher concentrations of the Cr ions were used in order to obtain similar response (100% changes in shape and same *GP* value). The concentration of Cr(VI) (2 mM) was double the concentration of Cr(III) (1 mM); therefore, data confirm the higher effectiveness of Cr(III) to produce shape changes in human erythrocytes.

Although it is generally reported that Cr(VI) is highly toxic while Cr(III) is relatively innocuous, there are however others reporting just the opposite. For instance, it has been found that Cr(III) exhibited stronger toxic effects in some plant species than Cr(VI), in humans can decrease immune system activity, and in animal cells have been shown to interact with various targets, including microfilaments, mitochondria, lysosome and nucleus, being often the final outcome necrosis and in some cases necrosis (Speranza *et al.*, 2007). On the other hand, Cr(III) under the form of Cr acetate hydroxide inflicted plasma membrane damage and suppressed mitochondrial function of human fibroblasts. Authors concluded that plasma and mitochondria membranes are important targets for this compound toxicity (Rudolf and Cervinka, 2005). It has also been reported that the lethal dose of $CrCl_3$ subcutaneously injected is $0.8\,g\,L^{-1}$ in dogs and $0.5\,g\,L^{-1}$ for rabbits (Lamson and Plaza, 2002).

3.5   CONCLUSIONS

In conclusion, these results indicate that interaction of Cr ions with phospholipid bilayers perturb the bilayer structure. The lipid bilayer being the major permeability barrier of the membrane, the structural perturbation induced by Cr ions will affect its permeability. It might also affect the functions of ion channels, receptors and enzymes immersed in the membrane lipid moiety. These findings may provide a new insight into the possible mechanism for the toxicity of Cr ions at the level of the erythrocyte membranes.

ACKNOWLEDGEMENTS

This work was supported by grants from FONDECYT (1130043 and 3140167) and PIA-CONICYT (PFB0824). DSC measurements were carried out with equipment purchased thanks to the financial support of the European Regional Development Fund within the framework of the Polish Innovation Economy Operational Program (contract POIG.02.01.00-12-167/08, project Malopolska Center of Biotechnology).

REFERENCES

Agero, U., Monken, C.H., Ropert, C., Gazzinelli, C.R.T. & Mesquita, O.N. (2003) Cell surface fluctuations studied with defocusing microscopy. *Physical Review E*, 67, 051904.

Agero, U., Mesquita, L.G., Neves, B.R.A., Gazzinelli, R.T. & Mesquita, O.N. (2004) Defocusing microscopy. *Microscopy Research and Technique*, 65, 159–174.

Bagatolli, L.A., Sanchez, S.A., Hazlett, T. & Gratton, E. (2003) Giant vesicles, laurdan, and two-photon fluorescence microscopy: evidence of lipid lateral separation in bilayers. *Methods in Enzymology*, 360, 481–500.

Bagchi, D., Stohs, S.J., Downs, B.W., Bagchi, M. & Preuss, H.G. (2002) Cytotoxicity and oxidative mechanisms of different forms of chromium. *Toxicology*, 180, 5–22.

Barceloux, D.G. (1999) Chromium. *Clinical Toxicology*, 37, 173–194.

Belayi, J., Pas, M., Raspor, R., Pesti, M. & Pali, Y. (1999) Effect of hexavalent chromium on eukaryotic plasma membrane studied by EPR spectroscopy. *Biochimica et Biophysica Acta*, 1421, 175–182.

Best, K.B., Ohran, A.J., Hawes, A.C., Hazlett, T.L., Gratton, E., Judd, A.M. & Bell, J.D. (2002) Relationship between erythrocyte membrane phase properties and susceptibility to secretory phospholipase A2. *Biochemistry*, 41, 13,982–13,988.

Blasiak, J. & Kowalik, J. (2000) A comparison of the in vitro genotoxicity of tri- and hexavalent chromium. *Mutation Research*, 469, 135–145.

Boon, J.M. & Smith, B.D. (2002) Chemical control of phospholipid distribution across bilayer membranes. *Medicinal Research Reviews*, 22, 251–281.

Devaux, P.F. & Zachowsky, A. (1994) Maintenance and consequences of membrane phospholipids asymmetry. *Chemistry and Physics of Lipids*, 73, 107–120.

Etcheverry, S., Gallardo, M., Solano, P., Suwalsky, M., Mesquita, O.N. & Saavedra, C. (2012) Real-time study of shape and thermal fluctuations in the echinocyte transformation of human erythrocytes using defocusing microscopy. *Journal of Biomedical Optics*, 17, 106–113.

Fernandes, M.A.S., Geraldes, C.F.G.C., Oliveira, C.R. & Alpoim, M.C. (2000) Chromate-induced human erythrocytes haemoglobin oxidation and peroxidation: influence of vitamin E, vitamin C, salicylate, deferoxamine, and *N*-ethylmaleimide. *Toxicology Letters*, 114, 237–243.

Gagelli, E., Berti, F., D'Amelio, N., Ggelli, N., Valensin, G., Bovalini, L., Paffetti, A. & Trabalzini, L. (2002) Metabolic pathways of carcinogenic chromium. *Environmental Health Perspectives*, 110 (Suppl. 5), 733–738.

Glionna, G., Oliveira, C.K., Siman, L.G., Moyses, H.W., Prado, D.M.U., Monken, C.H. & Mesquita, O.N. (2009) Tomography of fluctuating biological interfaces using defocusing microscopy. *Applied Physics Letters*, 94, 193701.

Grevatt, P.C. (1998) Toxicological review of trivalent chromium. EPA, CAS 16065-83-1, 1–44.

Huang, C. & Li, S. (1999) Calorimetric and molecular mechanics studies of the thermotropic phase behavior of membrane phospholipids. *Biochimica et Biophysica Acta*, 1422, 273–307.

Kitagawa, S., Seki, H., Kametani, F. & Sakurai, H. (1982) Uptake of hexavalent chromium by bovine erythrocytes and its interaction with cytoplasmic components: the role of glutathione. *Chemico-Biological Interactions*, 40, 265–274.

Koynova, R. & Caffrey, M. (1998) Phases and phase transitions of the phosphatidylcholines. *Biochimica et Biophysica Acta*, 1376, 91–145.

Krumschnabel, G. & Nawaz, M. (2004) Acute toxicity of hexavalent chromium in isolated teleost hepatocytes. *Aquatic Toxicology*, 70, 159–167.

Lamson, D.W. & Plaza, S.M. (2002) The safety and efficacy of high dose chromium. *Alternative Medicine Review*, 7, 218–235.

Lewis, N.A.H. & McElhaney, R.N. (1993) Calorimetric and spectroscopic studies of the polymorphic phase behaviour of a homologous series of *n*-saturated 1,2-diacyl phosphatidylethanolamines. *Biophysical Journal*, 64, 1081–1096.

Marsh, D. (1991) Analysis of the chain length dependence of lipid phase transition temperatures: main and pretransitions of phosphatidylcholines; main and non-lamellar transitions of phosphatidylethanolamines. *Biochimica et Biophysica Acta*, 1062, 1–6.

Mesquita, L.G., Agero, U. & Mesquita, O.N. (2006) Defocusing microscopy: an approach for red blood cell optics. *Applied Physics Letters*, 88, 133901.

O'Flaherty, E.J. (1996) A physiologically based model of chromium kinetics in the rat. *Toxicology and Applied Pharmacology*, 138, 54–64.

Ottenwaelder, H., Wiegand, H.J. & Bolt, H.M. (1998) Uptake of ^{51}Cr(VI) by human erythrocytes: evidence for a carrier-mediated transport mechanism. *Science of the Total Environment*, 71, 561–566.

Parasassi, T. & Gratton, E. (1995) Membrane lipid domains and dynamics as detected by laurdan fluorescence. *Journal of Fluorescence*, 8, 365–373.

Parasassi, T., De Stasio, G., Ravagnan, G., Rusch, R.M. & Gratton, E. (1991) Quantitation of lipid phases in phospholipid vesicles by the generalized polarization of laurdan fluorescence. *Biophysical Journal*, 60, 179–189.

Parasassi, T., Gratton, E., Yu, W.M., Wilson, P. & Levi, M. (1997) Two-photon fluorescence microscopy of laurdan generalized polarization domains in model and natural membranes. *Biophysical Journal*, 72, 2413–2429.

Pesti, M., Gazdag, Z. & Belágyi, J. (2000) In vivo interaction of trivalent chromium with yeast plasma membrane, as revealed by EPR spectroscopy. *FEMS Microbiology Letters*, 182, 375–380.

Reinhart, W.H. & Chien, S. (1986) Red cell morphology in stomatocyte-echinocyte transformation: roles of cell geometry and cell shape. *Blood*, 66, 1110–1118.

Repetto, G., del Peso, A., Salguero, M., Garfia, A., Sanz, P. & Repetto, M. (1996) Comparative effects of three chromium compounds no mouse neuroblastoma cells cultured in vitro. *Toxicology Letters*, 88, 41–42.

Rudolf, E. & Cervinka, M. (2005) The role of biomembranes in chromium (III)-induced toxicity in vitro. *Alternatives to Laboratory Animals*, 33, 249–259.

Sánchez, S.A., Bagatolli, L.A., Gratton, E. & Hazlett, T. (2002) A two-photon view of an enzyme at work: *Crotalus atrox* venom PLA$_2$ interaction with single-lipid and mixed-lipid giant unilamellar vesicles. *Biophysical Journal*, 82, 2232–2243.

Sánchez, S.A., Tricerri, M.A. & Gratton, E. (2007) Interaction of high density lipoprotein particles with membranes containing cholesterol. *Journal of Lipid Research*, 48, 1689–1700.

Sánchez, S.A., Tricerri, M.A. & Gratton E. (2012) Laurdan generalized polarization fluctuations measures membrane packing micro-heterogeneity in vivo. *Proceedings of the National Academy of Sciences of the United States of America*, 109, 7314–7319.

Sheetz, M.P. & Singer, S.J. (1974) Biological membranes as bilayer couples. A molecular mechanism of drug-erythrocyte induced interactions. *Proceedings of the National Academy of Sciences of the United States of America*, 71, 4457–4461.

Speranza, A., Ferri, P., Battistelli, M., Falcieri, E., Crinelli, R. & Scoccianti, V. (2007) Both trivalent and hexavalent chromium strongly alter in vitro germination and ultrastructure of kiwifruit pollen. *Chemosphere*, 66, 1165–1174.

Stasiuk, M., Kijanka, G. & Kozubek, A. (2009) Transformations of erythrocytes shape and its regulation. *Postepy Biochemii*, 55, 425–433.

Stott, B.M., Vu, M.P., McLemore, C.O., Lund, M.S., Gibbons, E., Brueseke. T.J., Wilson-Ashworth, H.A. & Bell, J.D. (2008) Use of fluorescence to determine the effects of cholesterol on lipid behavior in sphingomyelin liposomes and erythrocyte membrane. *Journal of Lipid Research*, 49, 1202–1215.

Suwalsky, M. (1996) Phospholipid bilayers. In: Salamone, J.C. (ed.) *Polymeric materials encyclopedia.* Volume 7. CRC Press, Boca Raton, FL. pp. 5073–5078.

Suwalsky, M., Castro, R., Villena, F. & Sotomayor, C.P. (2008) Cr(III) exerts stronger structural effects than Cr(VI) on the human erythrocyte membrane and molecular models. *Journal of Inorganic Biochemistry*, 102, 842–849.

Suwalsky, M., Villena, F. & Sotomayor, C.P. (2010a) Mn^{2+} exerts stronger structural effects than the Mn-citrate complex on the human erythrocyte membrane and molecular models. *Journal of Inorganic Biochemistry*, 104, 55–61.

Suwalsky, M., González, R., Villena, F., Aguilar, L.F., Sotomayor, C.P., Bolognin, S. & Zatta, P. (2010b) Human erythrocytes and neuroblastoma cells are affected in vitro by Au(III) ions. *Biochemical and Biophysical Research Communications*, 397, 226–231.

Suwalsky, M., Fierro, P., Villena, F., Aguilar, L.F., Sotomayor, C.P., Jemiola-Rzeminska, M., Gul-Hinc, S., Ronowska, A. & Szutowicz, A. (2012) Human erythrocytes and neuroblastoma cells are in vitro affected by sodium orthovanadate. *Biochimica et Biophysica Acta*, 1818, 2260–2270.

Suwalsky, M., Fierro, P., Villena, F., Gallardo, M.J., Jemiola-Rzeminska, M., Strzalka, K., Gul-Hinc, S., Ronowska, A., Zysk, M. & Szutowicz, A. (2013) Effects of sodium metavanadate on in vitro neuroblastoma and red blood cells. *Archives of Biochemistry and Biophysics*, 535, 248–256.

Weber, G. & Farris, F.J. (1979) Synthesis and spectral properties of a hydrophobic fluorescent probe: 6-propionyl-2-(dimethylamino)naphthalene. *Biochemistry*, 18, 3075–3078.

Wells, I.C., Claasen, J.P. & Anderson, R.J. (2003) A test for adequacy of chromium nutrition in humans-relation to type 2 diabetes mellitus. *Biochemical and Biophysical Research Communications*, 303, 825–827.

Wiegand, H.J., Ottenwalder, H. & Bolt, H.M. (1985) Fast uptake kinetics in vitro of $^{51}Cr(VI)$ by red blood cells of man and rat. *Archives of Toxicology*, 57, 31–34.

CHAPTER 4

Implementation of innovations as challenges for water treatment

Peter Roberts

4.1 INTRODUCTION

In the "real world" of industry and commerce, to have discovered an innovative material or method is only part of the way to success, implementation is often the larger and more difficult part.

This book is about removal of unwanted solutes – Cr and As – from water, by "chemical" (e.g. adsorption or ion exchange (IX)) or "physical" (e.g. membranes) means or combinations thereof. However, this chapter deals with administration rather than technology and is designed to help readers understand the sometimes complex pathway between a technical discovery and a commercial operating plant. The content of the chapter is based on the personal experiences of the author, an Australian chemical engineer with over 45 years experience in water treatment, none of it in research and much of it directly in the provision of capital plant for power stations, heavy industry and drinking water applications. In particular, he has substantial experience, both as an employee and independent contractor or consultant, in working on specifications, bids, contracts and start-ups for both end-users of plant and plant suppliers. He also has considerable experience in the preparation of organizational standards covering the commercial, contractual and administrative sections of all specifications.

The chapter is divided into the following sections:

- Development and marketing of innovations.
- Initiating an implementation project.
- Internal approval.
- Method of procurement.
- Writing a specification for water treatment plant.
- Writing the "plant specific" specification.
- Items to be specifically addressed.
- Guarantees.
- Evaluation of bids.
- Contract administration.
- Commissioning and performance evaluation.
- Evaluation of a specification.

4.1.1 *Definitions and terms*

4.1.1.1 *General*
In this chapter:

- The terms "water" and "feed water" are independent of source and end use and hence include what may be commonly known as "wastewater".
- The term "wastewater" means water lost in the treatment process.
- The terms "the author" and "Specification author" mean the author of a specification.
- The term "this author" means the author of this chapter.

4.1.1.2   *Materials*

In the water treatment industry, new materials of construction have been adopted where technically or economically advantageous, an example being the widespread adoption of fiberglass reinforced plastic (FRP) for the construction of pressure vessels in IX applications during the last 30 years. However, this and similar advances are not specific to water treatment.

It is exchanger materials, whether synthetic polymers or processed naturally occurring minerals, that are essentially specific to the "chemical" removal of solutes. Membrane materials are the counterparts for "physical" removal. For simplicity, in this chapter "material" is defined as an IX resin, other substance with IX or adsorption properties or a membrane material.

4.1.1.3   *Methods*

Similarly, the industry has been quick to utilize the massive advances in instrumentation and control systems technology that have occurred over the last 30 years. However this is also not industry specific.

What is specific is the manner in which the materials are arranged and the manner by which the desired process chemistry outcomes are achieved by the interaction between the feed water, materials and chemicals used in the process. This can be described as "plant chemical design" and in this chapter "methods" refers primarily to such designs.

Innovative methods often patented and plants that utilize the innovation are given a name by which they become commonly known, even after the patents have expired.

4.2   DEVELOPMENT AND MARKETING OF INNOVATIONS

Water treatment innovations can originate in a variety of sources, including universities and research institutes. However, in the water treatment field, in most cases such originators can be expected to come to a commercial agreement that results in:

- Development and marketing of material innovations being predominantly by the material manufacturer. In the case of synthetic polymer-based IX resins, there are only a limited number of manufacturers and resin manufacturing plants in a few major countries of the world. The number of membrane manufacturers is similarly limited and additionally they tend to specialize either in membranes of relatively large pore size such as microfiltration (MF) and ultrafiltration (UF) or fine pore size such as nanofiltration (NF) and reverse osmosis (RO).
- Development and marketing of methods innovations being by large, usually multi-national, water/wastewater treatment contractors. There are more of these contractors than there are material manufacturers but they are still reasonably limited in number.

Prospective end-users of innovations may extend development work by laboratory or pilot plant work.

4.2.1   *Initiating an implementation project*

If there is an indication that the adoption of an innovation would be of benefit to the end-user then if the innovation is confined to a change in an existing plant, such as a resin or membrane change, there should be little problem in the end-users organization in initiating an implementation project. The manager responsible for plant operation would normally have sufficient authority.

However, if the plant as a whole is to be new, irrespective of whether it is a "stand alone" retrofit or is to form part of a new industrial or public infrastructure project then it is probable that more people and organizations will be involved in the implementation. Firstly, the end-user will need to decide between the two alternatives:

- making incorporation of the innovation into the new plant mandatory
  or
- simply allowing the innovation to be incorporated into the new plant

If the new plant is to be procured by competitive bidding between suppliers then the second of the above alternatives would be the normal selection. If the innovation is patented, then making it mandatory may cause at least some possible suppliers to decline to bid.

4.2.2 *Internal approval*

In most cases there will be a project manager within the end-users organization, formally appointed or de facto. That person should be familiar with the history and major technical aspects of the innovation and would normally be the primary contact with its developers and/or originators.

Usually at some point in time that person will need to request funding approval from the appropriate level of his organization's management.

Questions that can reasonably be expected from management are:

• Why is in-plant implementation of this innovation necessary?

In the technical area covered by this book at least a partial answer to that question could well be "because the law or government regulations require a change from our current practice."

• What are the alternatives?
• Why implement this particular innovation now?
• How much will the implementation cost?
• How accurate is the cost estimate? – this question could be one of the most difficult questions to answer, especially if the request is to incorporate an innovation in a new plant where design is still at an early stage.
• Will the implementation of the innovation cause any side-effects?

In the industrial field, the effect on operating costs is a possible side-effect that usually needs to be specifically addressed. In a public infrastructure project, operating costs could be at least as important a consideration as capital cost.

• What are the risks, technical or other, that could lead to the innovation failing to achieve its objective?
• If the innovation does fail, what is the organization's "fall back" strategy or position?

In general, the more radical the innovation, the greater the importance of these last two questions.

4.2.3 *Method of procurement*

Once internal approval is obtained to implement the innovation, a decision has to be made as to the general method of procurement. One or more of the options below could be utilized:

• Option 1: Issue a purchase order on the manufacturer or local supply agent of the innovative material, plus, if required,
• Option 2: Issue an internal works order on the end-users engineering section or on a maintenance services type contractor partner.

Option 1 alone would be appropriate where the change is strictly material only, e.g. a new resin or membrane is being tried in an existing plant.

However, as materials manufacturers are in general not organized to perform mechanical/electrical work on industrial sites, if the material change requires some consequential changes to methods and equipment,(but those changes are themselves not innovative), a second order for the mechanical/electrical work would normally be appropriate. This contractor would not have to guarantee the outcome of the innovation as a whole.

Typically, these orders would include some of the end-users commercial/contractual conditions, but in a simplified form, as appropriate for relatively straightforward projects.

• Option 3: Issue a specification to specialist water/wastewater contractors, against which they will bid for detail design, manufacture and setting into operation of the plant. Installation of the plant on the site may or may not be included. After evaluation of the bids a purchase order is issued or a contract is formed.

This option is appropriate for a new plant, irrespective of size and technical complexity since as well as the major multinational contractors there are local companies in most countries. However, where the innovation is a method which involves extensive alterations to existing plant, it could also be advantageous to have a specialist contractor involved in the project.

- Where a plant including innovations is to be part of a major new industrial undertaking, e.g. a large power station, petrochemical plant or public infrastructure project, there may be a main contract for the entire undertaking between its owner/end-user and a main contractor, who may be one of the world's large multi-national engineer/constructors.

The engineer/constructor is likely to have his own standard specifications for mechanical plant in general or for various types of water treatment plant, which may be inappropriate in at least some degree for a plant including a particular innovation. This adds another layer of administrative complexity in having the innovation accepted.

In this kind of situation it is important that the engineer/constructor be contacted as early as possible in the project. The importance of the innovation to the overall project will largely determine the level of the contact but it should be formal (i.e. a signed letter or, if permissible, an electronically signed email).

- Large public infrastructure projects, typified by projects to increase or upgrade drinking water supplies, are increasingly being provided under BOO (build, own and operate) or BOOT (build, own, operate and transfer) arrangements. In such arrangements the contractor is usually an organization set up specifically for the project, whose ownership is likely to include at least one large financial institution such as a bank. In both BOO and BOOT arrangements, the contractor will recover the money outlaid during the build (including design) stage by receiving payments over many years as the plant produces water for public use. In a BOOT scheme, the contract has a definite duration (typically 20–30 years) and includes the condition that at the end of the contract period ownership is transferred, normally to the end-user of the water.

This author believes that because of the long term nature and large scale of BOO and BOOT schemes, there would usually be unacceptable financial risks in basing the entire scheme on an innovative technology until that technology has been suitably proven in smaller projects (and thus can no longer be classed as "innovative"). However, existing BOO or BOOT plants could well provide an opportunity to trial new technologies on a small side stream basis, as usually the operating contractor will have direct financial incentive to adopt any technology that can reduce his cost of producing water.

4.2.4   Writing a specification for water treatment plant – general

Typically, a specification for water treatment plant will have four sections or groups of sections, viz.

4.2.4.1   Contractual, commercial and administrative

There are international standard conditions of contract and national standards in many countries.

These standards are intended to be sufficient. However, because they must be general enough to have widespread applicability, most larger organizations, particularly those that are owned by governments, have either their own standards or extra conditions of contract supplementing an international or national standard. Major projects may have project specific standard commercial and/or administrative sections as well.

A Specification author will probably be directed to include such sections in the overall specification but have little or no authority to directly change clauses. However, the author should read them thoroughly and critically. In particular project standards, because they are typically written primarily to cover the main plant of the project (e.g. boilers, turbines, nuclear reactors in power stations, processing plants in the oil/gas/petro-chemicals industry) may contain clauses which are incompatible with, or inappropriate for, the requirements of auxiliary plant such as water treatment plant.

It is best to resolve any such incompatibilities and clarify any ambiguities or other difficulties at the specification writing stage. Deferring their resolution to the bid evaluation or contract stage is all too likely to result in additional costs to the end-user and/or project delays.

4.2.4.2 *General technical*

Structural, mechanical and electrical standards and codes, issued by standards bodies or professional associations are legally recognized in most countries and can be regarded as unchangeable for individual plants or projects.

However, in addition large industrial users of plant are likely to have additional standard requirements particular to their own industry or organization. Thus, a water treatment plant Specification author is also likely to be directed to include such sections in the overall specification and again they need to be read and any incompatibilities or ambiguities resolved. Where the organization standards contain large parts that are clearly irrelevant to the particular water treatment plant, this author recommends obtaining approval to delete those clauses from the particular specification.

4.2.4.3 *Plant specific*

This will be wholly or at least largely written by the Specification author.

4.2.4.4 *Schedules and bid conditions*

The specification documents will typically also contain some standard bid conditions, and financial and technical schedules that bidders have to provide to aid evaluation of their bid. The schedules provided by the successful bidder will form part of the contract.

The technical schedules will normally be prepared by the Specification author. These are commonly issued in a tabular format, with the Specification author listing the information he needs for properly comparing bids, e.g. quantities of resins, diameter and height of tanks and pressure vessels, with an adjacent space for the bidder to provide the information. This author recommends that there be a technical schedule "departures from the specification" where the bidder must give details, with reference to the particular specification clauses, of all items in which his bid differs from the requirements of the specification.

The author should also at least participate in the finalization of the bid conditions and financial schedules. Bid conditions normally contain many clauses which are standard throughout the organization and which may not be altered for individual specifications. However, if, as is common, they contain clauses covering "information to be provided with the bid" the Specification author should be able to add in requests for specific items pertaining to the plant. Particularly where the end-user/specification issuer is a government or government owned organization, this author also prefers that there be a bid condition clause, cross-referenced to the above "departures from the specification" clause, that formally states that any departure in the body of the bid, but not listed in the schedule, will not apply and that the bidder will not be entitled to any financial compensation for compliance with the specification in a subsequent contract.

Financial schedules are plant specific; however, the Specification author may have to design them from an organization or project standard template.

4.2.5 *Writing the "plant specific" specification*

From an overall viewpoint, the two "design philosophy" extremes are

(i) *"Design, manufacture and deliver" or "design, manufacture and install"*
   This philosophy is inherent in BOO and BOOT schemes but is not limited to them.

At this extreme the entire design is left to the bidder/contractor. For a plant to be provided under this philosophy, the "plant specific" specification can be very short, at least in theory little more than the required net output of the plant and the quality of the treated water, plus details of the feed water quality and any specific constraints which will affect the cost of the plant, e.g. plant wastewater cannot be fed into gravity drains but must be collected and pumped to disposal. Another typical situation, relatively common, is where bulk storage facilities for chemicals are to be provided as part of the plant but their size may be governed by other usages or by delivery constraints (minimum quantity and/or maximum frequency) rather than by usage in the plant itself. If such a circumstance applies, either full details will need to be included in the specification or tank numbers and/or sizes will need to be directly specified.

This philosophy may have superficial attractions, but it can be risky, as the end-user's ability to control the details of the plant he has to operate long-term is limited. In this author's experience, many plant operating problems can be traced to inadequacies in the original specification.

It may be thought that writing a minimal specification saves time but this is not necessarily the case. What really matters is the total time prior to a bid being accepted and if such a specification results in a wide diversity of bids, the time taken in bid evaluation may be prolonged.

(ii) *Separated design*

At the other extreme the complete design, including drawings, is carried out by the end-user or other body, all items of equipment are specified in detail and the package is issued for bids. It is common in civil engineering, building and large municipal sewage treatment but rare in industrial water treatment.

In deciding the plant design philosophy, this author recommends keeping firmly in mind:

- "It is not the duty of the bidder/contractor to provide the best plant for the job required, it is his job to design and provide the minimum cost plant which meets the specification."

  and a major corollary,

- "The end-user cannot expect anything that is primarily about the long term reliability of the plant or its "user friendliness" to be provided by the contractor unless it is directly specified."

These two statements are based upon the "commercial reality" of a contract gained by competitive bidding. The degree of instrumentation and flexibility of the control system are two particular design areas where the above are especially relevant.

However, as regards instrumentation, over-specifying can be as bad or even worse than under-specifying, as unless the end-user has the skills to properly maintain an instrument, it is better not to include it in the plant.

The flexibility of the control system and the user friendliness of the plant are closely related, particularly in the area of operation of the plant in manual mode. In this context "manual mode" refers to the forced operation of a single item or group of items of plant such as valves, pumps etc. by specific action at the operator interface – normally a screen. Specification authors should consider this aspect in detail, if possible in conjunction with the long term operators of the plant.

Both IX (e.g. start, regeneration) and membrane processes (e.g. start, stop, backwash, clean, etc.) normally involve sequential operations of varying degrees of complexity. In normal "auto" mode initiation of these sequences may be automatic or at the operator interface and they then proceed automatically to conclusion. This author's basic method of obtaining a flexible and user friendly control system was to have a clause in the specification requiring facilities to be provided such that, in manual mode, the plant could be placed into any configuration that formed part of any automatic sequence without operating individual pump/air blower etc. "start/stop" controls nor individual valve "open/shut" controls. Furthermore, once the plant was placed into a specific configuration the plant would remain in that configuration until it was either manually stopped or reached the same sequence end-point as in "auto" mode (the time-out of a timer in most steps of an IX regeneration sequence) whereupon it would shut down. Other clauses required the ability to change from auto to manual mode and vice-versa at any time without immediately affecting the physical configuration of the plant and the necessity of maintaining all essential plant safety interlocks operational in manual mode.

In practice most specifications for plant lie somewhere between the two extremes detailed above.

Plants that utilize innovative materials and/or methods as discussed in this book may be significantly different from previous practice. Furthermore, even a large, experienced contractor may not be fully familiar with the innovation and may protect himself by declining to guarantee process performance. Thus it is perhaps likely that the appropriate design philosophy would be closer to the "separated design" than most plants, e.g. all or most of the "plant chemical design" would be done at the specification preparation stage and included in it.

4.2.6   *Items to be specifically addressed*

Specifications for water treatment and many other types of mechanical/electrical plants normally include different clauses as outlined in the following.

4.2.6.1   *Work by the contractor*
These formally state which functional areas such as design, manufacture, delivery, site installation, commissioning and maintenance under warranty are included in the contract and list the major mechanical and electrical items in the plant.

Four items that often need to be addressed on an individual case basis are:

- *Associated civil and building works*: Suppliers of specialist plant would in general have to sub-contract these works and this author believes that they should be provided separately from the plant under most circumstances.
- *Control system software*: Nowadays owners and/or engineer/constructors of major projects often prefer to centralize control systems over the entire project and have a single software provider. If the plant is a retrofit, the owner could similarly have an existing on-going contract for software covering the entire manufacturing operation.

In such cases, clauses covering commissioning of the plant will need to take into account that two contractors will be involved.

Note that it should not be assumed that this software provider has any water treatment knowledge, especially if the plant uses the kind of innovative technologies described in this book. Establishing the control system design philosophy and providing full details of how the plant is to work is the responsibility of the plant contractor, ideally in conjunction with the plant Specification author.

To minimize delays in plant commissioning, this author recommends that a thorough off-plant control system simulation be carried out. If the plant contractor has designed the software, the simulation would be to demonstrate to the end user that the plant will work in accordance with the contractor's detailed design and the specification. This author recommends that the performance of the simulation be listed as a specific item under "work by the contractor".

Nowadays simulations can be easily performed easily by using a duplicate of the actual operator interface screen and a computer loaded with the plant control software. However, means (e.g. an auxiliary screen) are also needed such that all "digital" inputs from plant instrumentation (e.g. from flow sensors on chemical dilution water lines that are in the plant to ensure that concentrated chemical pumps can never start unless there is an adequate flow of dilution water) can be simulated by operation of software "switches" and similarly all "analogue" type inputs (e.g. from level transmitters in tanks) can likewise be simulated by manually controllable auxiliary input signals.

A second screen that can have the plant mimic diagram displayed at all times during the simulation may be desirable to assist in monitoring the outputs from the control system.

- *Electrical switchboards and motor control centers*: Similarly, in a new major project, a single supplier for all such items may be preferred. In a retrofit situation, the owner may have specific requirements as regards the supplier.
- *Site installation*: Nowadays there is a general preference for minimizing site installation time and costs through maximizing pre-assembly and delivering the plant to site mounted on a skid or number of skids. Both IX and membrane plants lend themselves to this approach. The specification will need to contain details of any physical constraints on the size of skids.

If there is a main contractor for an entire project, it would be normal for that contractor to perform the final installation of the skids and any piping or electrical interconnections between them, otherwise such work can be included in the plant suppliers' scope of work.

4.2.6.2   *Work by others*
Common items listed are supply of feed water, disposal of wastewater from the plant, handling of treated water, and supply of power. Other items could be provision of civil and building works as above, installation of the plant, supply of compressed air etc. If the control system software etc. is being provided separately then the simulation will require the software designer to demonstrate the control system software to both the plant contractor and the end-user. In such case the simulation would be an item listed under "work by others" in the plant specification. There could also be items specific to the innovation, e.g. an experimental resin or membranes may be supplied separately.

4.2.6.3   *Terminal points*
These are where "work by the contractor" and "work by others" interface. In a water treatment plant, normally most of them are either flanges in piping or electrical terminals in switchboards or control panels. The items above should be functionally detailed as comprehensively as possible in the specification. The more detailed the specification the less the likelihood of contractual disputes over "who provides what."

4.2.6.4   *Plant layout*
In principle only the general area in which the plant is to be located should be given in the specification and layout details left to the bidder/contractor. However, if space is limited, or supply of services to the plant is a significant issue, as either or both may be the case especially in a retrofit, layout may need to be specified in more detail.

4.2.6.5   *Information to be provided by the contractor*
This should be a comprehensive list of all information, including drawings, that is required by the end-user, either to facilitate work by others or to enable the end-user to operate and maintain the finished plant. Time deadlines for supply should be stated for all items that are relevant for construction and commissioning. Financial penalties for late supply of certain items may be included.
Examples include:

- In most cases, one of the first items required from the contractor is a detailed program, usually in a format specified by the end-user or main contractor and detailed in the specification, showing the periods for detailed design, manufacture, assembly, delivery, site work and commissioning of all major components of the plant and identifying the critical path and program floats.
- Where the civil works are being provided separately, plant layout and other information to enable the design and construction of those works is usually required very early in the contract period. Even where the civil works are being provided by the plant contractor, the end-user is likely to require information on, for example, what drainage facilities he must provide for final disposal of wastewater from the plant.

- Piping and instrumentation (P & I) drawings.
- Operating sequence charts. The great majority of IX plants use regenerable resins and these charts, together with the P & Is, enable the operating configuration of the plant (i.e. which valves are open, which pumps are running etc.) at each step of the regeneration and any other sequences such as start-up, to be traced. At least to some extent, information on process interlocks, short term delay timers etc. can be included in the charts. Most membrane plants also include sequential operations, either as part of normal service or at startup and/or shutdown.
- Where the control system software is being provided by others the plant contractor will need to provide fully detailed information of what is needed for the plant to be operated and controlled. The P & I diagram and operating sequence charts will form part of this information. The information should in particular detail what conditions will cause trips and what conditions will simply raise an alarm. This author recommends a general preference for "alarm only" unless the condition represents an immediate danger to the physical integrity of the plant, e.g. "water tank extra high level" would normally be alarm only but "water tank extra low level" should trip pumps drawing from the tank.

The format of this information may be detailed in the specification.

4.2.6.6 The feed water to the plant

In this author's experience, specifying the feed water, especially to an IX plant, is too often done poorly by Specification authors. Whilst the analysis of some waters, directly drawn from large underground aquifers, can be essentially unchanged for years or even decades, the analysis of most other waters will, to a greater or lesser extent, fluctuate over time. Where the application is treatment of process water, fluctuations, possibly severe, due to upsets in the upstream process must be considered in the plant conceptual design and specification.

It is poor practice just to examine what analyses of the feed water are available, take the maximum values found for each individual analyte, put them together as an analysis and call that the "design feed water". Even if the author makes adjustments to ensure that cations and anions balance, the resultant analysis may be one that may never be approached in practice and could unnecessarily result in higher plant cost. For example, in many areas of the world river waters can have very high organic and maybe silica contents during major wet season floods, but during such times the levels of dissolved inorganic salts is likely to be much lower than during the annual dry season or during droughts. So, for such waters, an analysis produced from combining highest individual analyses may well be unrealistic.

In many cases the approach of specifying a "design feed water", determined on the basis of such water being a time-weighted average or the feed water of highest frequency is satisfactory. The plant would be optimized to treat this water and likewise plant operating costs be calculated on the basis of this water being treated at all times. This could be supplemented by a "worst feed water" analysis (occasionally more than one) upon which the plant would still have to be able to produce its design output and treated water quality although operating costs could be higher than design.

In most IX plants the regeneration time is essentially fixed once the plant is built and there may be at best only a limited ability to increase its hydraulic throughput. Therefore the shorter the service run length at design conditions, the less the ability of the plant to maintain its design output should the ionic load on the resin increase. The risk of such an increase occurring will vary considerably but is clearly higher the less analytical data on the feed water is available. Similar considerations apply to MF and UF plants in respect of backwash times and solids loading. For NF and, especially, RO temperature of the feed water is critical. In some locations, inclusion and direct specification of feed water warming facilities in the plant may be clear-cut but in others it will not and if the decision is to be left to the bidder/contractor both seasonal feed water temperature and anticipated plant demand data as comprehensive as possible will be required to provide data for the bidders decision.

4.2.7   Guarantees

In all cases, the Specification author should ensure that guarantee requirements are reasonable. This is of particular importance in applications like those covered in this book where essentially complete removal of low or trace levels of highly dangerous contaminants is required. Even if the bidder does not initially object, in the case of a dispute the courts are very likely to disregard clauses that the court considers to be unfair or unreasonable to the contractor. Bearing this in mind, the Specification author may need to decide whether the guarantee can be absolute (i.e. applies at all times and is independent of feed water quality), or, for example, is more appropriately expressed as a time averaged required percentage removal within a specified range of feed water contaminant levels.

The accuracy of measurement of residual contaminant level in the treated water can occasionally become an issue that the Specification author must address at the specification writing stage.

Resin or membrane life may be a guarantee issue for any plant operating at elevated temperature or under significant oxidizing or fouling conditions. Especially in process water treatment applications, short term extreme conditions may be experienced until the main plant providing the feed water to the water treatment plant is fully commissioned and in normal operation. Under such circumstances, it may be appropriate to have conditions in the specification/contract that effectively allow the initial resin charge or membranes to be sacrificial and guarantees not to be applied until new replacements have been installed.

4.2.8   Evaluating bids

Generally the Specification author will have the primary responsibility for technical evaluation of bids and, in principle, should recommend to management that the lowest priced bid that conforms to the specification should be accepted. If operating or other costs vary substantially between bidders, "lowest priced" should mean lowest evaluated price when such costs are also taken into account.

This author considers the "departures from the specification" schedule to be the most important and his practice was to examine that schedule from all bids first. It can be reasonably expected that such examination, together with the knowledge of bid prices, should enable in relatively quick time the production of a "short list" of bids for detailed evaluation.

4.2.9   Contract administration

Once a bid is accepted and a contract formed, the Specification author may administer the contract through the subsequent phases of detailed design, plant manufacture, delivery to site, installation and commissioning until final handover to the operating personnel. He would also be responsible for arranging the completion of all "work by others" in accordance with the program.

However, in many cases, especially where the new water treatment plant is part of a major project, it is possible that the Specification author will hand over day-to-day administration of the contract etc. to a contract engineer or administrator. Especially where the water treatment plant is innovative, it is unlikely that such person would have the experience or knowledge of the specification author. Therefore it is important that the Specification author establish a sound working relationship with the engineer or administrator, especially during the time that the detailed design of the plant is being undertaken by the contractor.

This author typically requested the right to view, check, comment upon and recommend for approval:

- All drawings and documentation that concerned how the plant works. The principal items in this category are P & I diagrams, operating sequence charts and control system detailed description and/or process logic drawings.
- Any changes to the "process chemical design" of the plant proposed by the contractor.
- General assembly drawings.

- In an IX plant, drawings of the IX vessels and internals, plus other items like regenerant tanks and degasser towers.

Later in the contract period this author would typically

- Attend the control system simulation and assess the suitability of the "as simulated" system for the commissioning of the plant. After the simulation, further changes to the software would not be permitted unless and until proven necessary during plant commissioning.
- View and comment upon the operating instructions.

Except as may be specifically requested, there should not be any need for the Specification author to have substantial involvement during the manufacturing, delivery and site installation phases of the contract.

4.2.10   *Commissioning and performance evaluation*

This author believes that in principle even if the water treatment plant is new and innovative, in any sort of retrofit situation the operating staff who will run the plant long term are the best placed to supervise commissioning on a day-to-day basis and to gather the data for evaluation of the performance of the plant. In new projects that belief has to be qualified by "… assuming they are available". The water treatment plant operating staff may well have concurrent setup or commissioning responsibilities and ask for help from the specification author. This author has found that commissioning is a very valuable "experience builder" to be used in writing subsequent specifications.

4.2.11   *Evaluation of a specification*

Finally, at the completion of a project, a Specification author should be prepared to critically evaluate his own specification. Two questions to be asked are:

- Does the final plant perform in accordance with the reasonable requirements of the end-user and operate in a user-friendly manner? If either answer is "no" then the author should ask himself the question "could the reason for the operating problems have been reasonably foreseen when the specification was being prepared?".
- Were there any contract variations that were caused by deficiencies in the specification?

The aim of this chapter is to help future authors of specifications for water treatment plant using innovative technologies to be able to answer "yes" to the first question and "none" to the second question.

Part II
Innovative materials for removal
of arsenic and chromium

CHAPTER 5

Separation of arsenic from waters using inorganic adsorbents

Emür Henden & Tülin Deniz Çiftçi

5.1 INTRODUCTION

The stability and dominance of arsenic species depend on the pH of the solution. Under atmospheric or a more oxidizing environment, the predominant species is As(V), which, in the pH range of 6–9 exists predominantly as oxyanions, namely, $H_2AsO_4^-$ or $HAsO_4^{2-}$ (DeMarco et al., 2003; Hering, 1996; Mohan and Pittman, 2007) ($pK_1 = 2.2$, $pK_2 = 7.1$, $pK_3 = 11.5$). The ionic form of pentavalent As is relatively easy to remove (Korngold et al., 2001; Wang et al., 2009). Under reducing conditions, As(III) is predominant and exists predominantly as H_3AsO_3 ($pK_1 = 9.22$, $pK_2 = 12.3$, $pK_3 = 13.4$) at pH below 9.0. Under strong reducing conditions, arsine and elemental As are also formed but only rarely in the natural environment (Sharma and Sohn, 2009). Arsenic cannot be easily destroyed and can only be converted into different forms or transformed into insoluble compounds in combination with other elements, such as iron (Choong et al., 2007).

5.1.1 Arsenic removal techniques from water

Choice of technique for the removal of As depends on the composition of As contaminated water. Several methods such as oxidation, co-precipitation, membrane filtration, ion exchange, coagulation/filtration, and adsorption have been commonly used for the removal of As from waters.

Oxidation/precipitation method: The oxidation/precipitation method is a simple process, easily applied to large volumes of water. However, toxic by-products can occur and further removal treatments are then needed. The solid obtained after the precipitation method can be removed easily by sedimentation and filtration.

Membrane filtration method: The advantages of membrane filtration are: no toxic waste is produced and high removal efficiency is achieved for As(V), whereas the disadvantages are: high cost, pre-treatment step and high electrical energy are needed and also low removal efficiency is achieved for As(III).

Ion-exchange method: The effect of pH variation and the concentration of As in the liquid phase are not critical for the As removal efficiency by using the ion-exchange method. As(V) removal efficiency is relatively high, but for the removal of As(III) pre-oxidation is needed. Other ions may interfere. The removal efficiency of As at high concentrations is low.

Coagulation/filtration method: Coagulation/filtration is a simple and cost effective method for the removal of As. As(V) removal efficiency is high. It is applicable for large volume waters. A pre-oxidation step is needed for the removal of As(III) and a high amount of coagulant dose is required. After the coagulation, a filtration step is needed and toxic sludge occurs after the treatment.

Adsorption method: Adsorption is a cheap and highly effective method for the removal of both As(III) and As(V). No other chemicals are required. Some competitive ions, especially silicate and phosphate may interfere. Regeneration is needed (Litter et al., 2010; Mondal et al., 2013). At present most widely used As removal techniques on the small and large scale are based on or use, at some stages, adsorption processes.

5.2 ADSORBENTS USED

Adsorption is an effective and economic method for As removal. Also the other advantages of the method, such as high adsorption capacity, simple operation, easy handling, regeneration, low maintenance make it powerful. Many inorganic, organic and biological adsorption materials are used for the removal of As from waters. Especially iron and iron compounds such as granular ferric hydroxide (GFH), nano zero-valent iron, natural iron ores, iron(III) oxide (or hydroxides) (hematite, goethite, akagaenit, etc.) are used for this purpose. Sometimes, adsorbing compounds are coated onto support materials such as sand (Hsu *et al.*, 2008), zeolite (Jeon *et al.*, 2009), pumice (Reddy and Turner, 2007), silicagel (Çiftçi *et al.*, 2011), cement (Kundu and Gupta, 2007), pottery (Dong *et al.*, 2009), alumina (Kuriakose *et al.*, 2004), resin (Matsunaga *et al.*, 1996) as the support material for the removal of As. The usability of the adsorbent is increased when a support material is used, because it then becomes suitable for column separation (Çiftçi *et al.*, 2011).

While selecting the adsorbent, the concentration and the species of As, pH, competitive ions such as silicate, phosphate, sulfate, chloride and bicarbonate have to be taken into consideration. Adsorbents used for the removal of As have been given in review articles (Mohan and Pittman, 2007; USEPA/AWWA, 2005; Yadanaparthi *et al.*, 2009). As shown in Table 5.1, iron, aluminum, manganese, titanium, zirconium and copper-based materials have been most commonly used. When the capacity of the adsorbent is high, the amount of toxic residue will be a minimum. The other advantage of the adsorption method is that some of the adsorbents can be used several times because of their regeneration ability. Due to their high surface area, nanomaterials have high adsorption capacity and, therefore, are promising adsorbents.

Some of the adsorbents obtained from different materials and chemicals are described below. Nano materials are also described under the related sections.

5.2.1 *Iron-based adsorbents*

Most of the adsorbents used in recent years for the removal of As from waters are iron based. Some of the ferric hydroxide species such as iron (III) oxyhydroxide, hydroxide, goethite, hematite, iron (III) oxide coated materials have appeared to be promising. In rural areas, removal of As with naturally occurring iron may be one of the suitable methods because As contaminated waters probably contain iron at high concentration. If iron is present in As contaminated water, the As can be precipitated with iron and result in lowering of the concentration of As. In many parts of Bengal at community level, iron removal units (IRU) have been in use. The principle of IRU is sand filtration for removal of iron after aeration (Sharma *et al.*, 2014). Sharma *et al.* (2003) reported that if the conditions, such as an appropriate Fe:As ratio and the reaction time, are suitably selected, high levels of As concentrations can be decreased. Berg *et al.* (2006) reported that to achieve lower than $50\,\mu g\,L^{-1}$ As concentration, a 50 and higher Fe:As molar ratio was required.

Granular ferric hydroxide (GFH) (Banerjee *et al.*, 2008) has been frequently used as adsorbent for the removal of As(III) and As(V). Rate data was well fitted to the pseudo first-order kinetic model. The adsorption of both As(III) and As(V) by GFH was an endothermic process. The removal efficiencies were about 100% for both species. The adsorption capacity was calculated as $1.1\,mg\,g^{-1}$ for an equilibrium concentration of $10\,\mu g\,L^{-1}$ As(V).

Zhang and Itoh (2005) synthesized the adsorbent iron oxide-loaded melted slag (IOLMS), by loading iron (III) oxide on melted slag from a municipal solid waste incinerator. The adsorbent was effective for both As(V) and As(III) removal and its removal capabilities were 2.5 and 3 times of those of FeOOH, respectively. Similarly, for the removal of As(III) and As(V), Hsu *et al.* (2008) used iron-oxide coated sand (IOCS). They have reported that IOCS could be considered as a feasible and economical adsorbent for As removal.

Hematite-containing iron ores were tested by Zhang *et al.* (2004) for the removal of As(V). The adsorption capacity of the adsorbent was low ($0.4\,mg\,g^{-1}$). Silicate and phosphate decreased the adsorption efficiency while sulfate and chloride have a positive effect. They have minimized the strong negative effect of silicate by studies at pH around 5.

Çiftçi *et al.* (2011) synthesized and used ferric hydroxide supported on silicagel for the removal of both As(III) and As(V). Ferric hydroxide precipitation was realized at various pH values. The highest As removal capacity and removal efficiency were obtained by using the adsorbent obtained at relatively low pH 6.0. The initial pH of As solution did not significantly affect the removal efficiencies for As(III) and As(V) in the pH range of 3.1–9.7. The batch capacities were 16.2 mg As(III) g^{-1} and 17.7 mg As(V) g^{-1} for initial As concentration of 100 mg L^{-1}. It was shown, by the As speciation analysis at trace level, that As(III) is adsorbed onto ferric hydroxide partly without oxidation to As(V). The adsorbent was used successfully for the removal of As(III) and As(V) from drinking water, geothermal water and mineral water. The residual concentration of As was below the drinking water limit (10 µg L^{-1}) after the adsorption for both As(III) and As(V) species.

Nano-sized materials are effective as adsorbents and have high capacity for As removal. However, their application may be limited due to their tiny particle size. Therefore, Zhu *et al.* (2009) supported the nanoscale zero-valent iron onto activated carbon (NZVI/AC) by impregnating carbon with ferrous sulfate followed by reduction with NaBH$_4$. Maximum adsorption capacity of the adsorbent was 18.2 and 12.0 mg g^{-1}, for arsenite and arsenate, respectively. Phosphate and silicate decreased the removal of both arsenite and arsenate. NZVI/AC was regenerated by elution with 0.1 M NaOH and the adsorption capacity remained at the same level after eight cycles of sorption – desorption.

Xu *et al.* (2013) have synthesized superparamagnetic mesoporous ferrite nanocrystal clusters (MnFe$_2$O$_4$NCs) by a facile modified solvothermal-based method and used it for the removal of As(III). The adsorbent has high adsorption capacity (27.27 mg g^{-1}) because of its small constituent nanocrystal size and uniform mesoporous structure. Regeneration of the adsorbent was conducted with NaOH solutions at different concentrations and the results indicated good reusability of the adsorbent. The adsorption mechanism of As(III) on MnFe$_2$O$_4$ NCs has been reported to be through the formation of a surface complex by formation of As-O-M (M=Fe or Mn) linkages.

Türk and Alp (2014) synthesized Fe-hydrotalcite supported magnetite nanoparticle (M-FeHT) by a co-precipitation method. From the TEM images, it was observed that nanoparticles dispersed in Fe-hydrotalcite at an average diameter of 50 nm. Maximum adsorption of As was observed at pH around 9. The correlation coefficients of the Langmuir isotherm (0.9887 and 0.9998) were higher than the Freundlich isotherm (0.9395 and 0.9988) for both As(III) and As(V). The adsorption capacities were 0.121 and 1.28 mg g^{-1} for As(III) and As(V), respectively. Kinetic studies showed that a rapid adsorption of As was followed by a slower adsorption. The kinetic data was in good agreement with the pseudo-second order model for both species of As studied. The residual concentration of As was below the drinking water limit (10 µg L^{-1}) after the adsorption with M-FeHT.

Another adsorbent was synthesized by coating nano Fe$_3$O$_4$ onto the outer surface of polystyrene beads (Jiang *et al.*, 2012). The capacity of the adsorbent (139.3 mg g^{-1}) was about 78% greater than those of bulky Fe$_3$O$_4$ (78.4 mg g^{-1}). The authors reported that after the coating of Fe$_3$O$_4$ upon polystyrene beads, the particles were dispersed better in the liquid phase. Without a capacity loss, the adsorbent was effectively regenerated with 0.1 M NaOH solution. The adsorbent could be separated from the solution by employing a low magnetic field. The disadvantage of the adsorbent was that the Fe$_3$O$_4$ loading of the adsorbent might be slowly lost after repeated regeneration. Phosphate and silicate greatly competed with As(V), while chloride and nitrate did not decrease the adsorption efficiency.

Nanoscale zero-valent iron was supported on montmorillonite by reducing Fe^{3+} with BH$_4$ (Mt-nZVI) (Bhowmick *et al.*, 2014). The supporting surface areas were reported as 36.97 and 14.85 m^2 g^{-1} for Mt-nZVI and Mt, respectively. The maximum adsorption capacities of Mt-nZVI were 59.5 and 45.5 mg g^{-1} for As(III) and As(V), respectively, that are higher than that of both Mt and nZVI. The adsorption of both As(III) and As(V) was favorable and the type of the adsorption was indicated as chemisorption. From the XPS studies, the authors stated that, for the adsorption of As(V) onto Mt-nZVI, the adsorbed species were still in the As(V) form. However, As(III) was completely oxidized to As(V) by the Mt-nZVI surface. The adsorption efficiency of both As(III) and As(V) decreased by increasing the phosphate concentration, while nitrate, sulfate and bicarbonate did not significantly affect the adsorption efficiency. Regeneration studies showed that about 50–60% of adsorbed As was desorbed by 0.1 M NaOH in 4 h. Although the removal efficiencies were about 100% after 5 cycles, desorption efficiencies decreased to 30–40%.

5.2.2   Aluminum-based adsorbents

Alumina was also used frequently for As removal in the literature. Mesoporous alumina was synthesized by combining the three-block copolymer as a template with aluminum hydroxide sol at room temperature followed by isolation and calcining (Han et al., 2013). The equilibrium data was well fitted to the Langmuir isotherm and the maximum adsorption capacity was calculated as $36.6 \, mg \, g^{-1}$. The adsorption process of As(V) by the adsorbent was determined as physical. The presence of silicate, phosphate and fluoride ions caused a sharp fall in removal effectiveness, while the presence of nitrate and sulfate ions slightly decreased the efficiency.

Activated alumina was also used for the removal of As(III) by Singh and Pant (2004). Kinetics reveals that uptake of As(III) was very fast and the removal efficiency was strongly dependent on pH and temperature. The adsorption was exothermic. Maximum adsorption capacity was indicated as $2.29 \, mg \, g^{-1}$ from the Langmuir model.

5.2.3   Manganese-based adsorbents

Generally, the removal of As(V) is easier than the removal of As(III). Also it is known that As(III) species are efficiently oxidized to As(V) by the manganese-based adsorbents. Therefore, manganese-based adsorbents are used for both the oxidation and the removal of As. Yang et al. (2007) have compared the column systems packed with manganese coated sand (MCS) and iron coated sand (ICS). They have reported that the column system packed with an equal ratio of manganese coated sand and iron coated sand was identified as the best system due to a promising oxidation efficiency of As(III) to As(V) by MCS and adsorption of As(V) by both MCS and ICS. While the ions chloride, nitrate, perchlorate and sulfate did not significantly affect the adsorption efficiency, phosphate greatly interfered with the As(V) adsorption.

Zhang and Sun (2013) developed the adsorbent multifunctional micro-/nano-structured MnO_2 spheres for the removal of As. The BET surface area of the adsorbent was $162.54 \, m^2 \, g^{-1}$, whereas it was $32.0 \, m^2 \, g^{-1}$ for bulk MnO_2. The pore size distribution curve showed that the pore diameter was about 65 nm. They have reported that As(III) was effectively oxidized to As(V) and then its adsorption followed. Carbonate, sulfate and phosphate strongly affected the removal efficiency. The maximum As adsorption capacity was calculated from the Langmuir isotherm as $14.5 \, mg \, g^{-1}$. NaOH solution at a concentration of 1 M desorbed only 56% of adsorbed As(V); however, the adsorption ability of the regenerated adsorbent did not change. The adsorption was pH dependent, the capacity decreased with the increase of pH.

5.2.4   Zirconium-based adsorbents

Zirconium oxide is not soluble in water; it is an inert and non-toxic inorganic material that adsorbs As species. A hydro-thermal co-precipitation reaction was used for the modification of hydrated zirconium oxide ($ZrO(OH)_2$) nanoparticles with graphite oxide (GO), and the adsorbent was used for the simultaneous removal of As(III) and As(V) from drinking water (Luo et al., 2013). The monolayer adsorption capacities calculated from the Langmuir model were 95.15 and $84.89 \, mg \, g^{-1}$ for As(III) and As(V), respectively, which are 3.54 and 4.64 times that of $ZrO(OH)_2$ nanoparticles. The adsorbent showed good anti-interference ability for co-existing anions nitrate, bicarbonate, sulfate, chloride and fluoride, however, phosphate interfered with the adsorption of As(III) and As(V). $2 \, M \, HNO_3$ was used for the regeneration of the adsorbent and it was stable within five cycles.

Cui *et al.* (2012) synthesized a novel adsorbent, amorphous zirconium oxide nanoparticles (am-ZrO$_2$) using a hydrothermal process. The BET specific surface area of the adsorbent was 327.1 m^2 g^{-1}. The adsorbent was effective for removing As below 10 μg L^{-1}. The kinetic data was best fitted to the pseudo-second order model. The studies showed that the adsorption of As onto the adsorbent involved chemisorption. Phosphate ion strongly effected the removal efficiency of both As(III) and As(V), while the ions chloride, sulfate and fluoride showed no or a slight effect. At pH 7, the maximum adsorption capacities of the adsorbent were calculated as 83.2 and 32.5 mg g^{-1} for As(III) and As(V), respectively.

Highly porous nanostructured ZrO$_2$ spheres (Cui *et al.*, 2013) were used for the removal of both As(III) and As(V). The adsorbent has good properties such as non-toxicity, stability and resistant to acid or alkali. The adsorption of both species onto the adsorbent obeyed the Freundlich model. The isotherm parameter indicated a favorable adsorption. Over 90% of adsorbed As species were desorbed using 0.2 M NaOH and the regenerated adsorbent could be reused.

5.2.5 *Titanium-based adsorbents*

For the treatment of organic and inorganic compounds, numerous studies have been conducted for the photocatalytic reactions with TiO$_2$ suspensions. Bang *et al.* (2005) evaluated granular TiO$_2$ for the removal of both As(III) and As(V). The particle size of the adsorbent was between 30–100 mesh and the surface area was 250.7 m^2 g^{-1}. The removal efficiency of As(V) decreased from 99 to 70% when the pH increased from 7.3 to 10.3. Oppositely, the removal efficiency of As(III) increased from 84 to 95% when the pH increased from 4.6 to 8.5. At pH higher than 8.5, the removal efficiency of As(III) decreased. The point of zero charge pH of TiO$_2$ was reported as about 5.9–6.0. The decrease of the removal efficiency could be explained by the surface pH. At pH> pZC (point of zero charge), the adsorption of As(III) and As(V) (both in negative oxidation states) indicates a pZC surface complexation, rather than electrostatic interactions. The maximum adsorption capacities were calculated from the Langmuir isotherm as 32.4 and 41.4 mg g^{-1} for As(III) and As(V), respectively. At neutral pH 7.0, silicate did not significantly affect the adsorption capacity of the adsorbent for both As(III) and As(V). However, phosphate decreased the adsorption capacity of the adsorbent for As(V) by about 8%. The spent adsorbent was tested with 0.1 M acetic acid for the leachability of As. The concentration of As in the leachate was lower than the EPA regulatory level of As.

Pena *et al.* (2005) evaluated the effectiveness of nanocrystalline titanium dioxide (TiO$_2$) for the removal of As(V) and As(III) and photocatalytic oxidation of As(III). The competing anions (phosphate, silicate, and carbonate) had a moderate effect on the adsorption capacities of the TiO$_2$ for As(III) and As(V) at pH 7. The adsorption kinetics was described by a pseudo-second-order equation. The nanocrystalline TiO$_2$ is an effective adsorbent for As(V) and As(III) and an efficient photocatalyst.

5.2.6 *Copper-based adsorbents*

Manju *et al.* (1998) reported that, for the effective removal of As(III) from aqueous solutions, copper-impregnated coconut husk carbon could be used as an adsorbent. The adsorption was affected by the temperature. The adsorption capacity was increased by raising the temperature from 30–60°C. The process was endothermic and the maximum uptake occurred at pH 12.0. Thermodynamic parameters showed that the adsorption data fitted the Langmuir model. Copper (II) oxide nanoparticles (Goswami *et al.*, 2012) were synthesized and examined for the adsorption of As(III). The adsorbent was characterized by using TEM, BET, XRD, and FTIR. The shape of the adsorbent was cylindrical, the surface area was 52.11 m^2 g^{-1}, and the point of zero charge pH of the adsorbent was 6.8. The removal efficiency of As(III) decreased by decreasing the pH. When the pH decreased from 7 to 3.6, the removal efficiency decreased from about 98 to 75%. Sulfate and phosphate had a negative effect for the removal of As(III) by the adsorbent. The nature of the adsorption was endothermic and the Langmuir model fitted better than the Freundlich model.

The adsorption capacity of the adsorbent was $1.09 \, mg \, g^{-1}$ for As(III). Desorption efficiency was low (25%) with 0.5 M NaOH and about 4% decrease was observed in the adsorption efficiency after the regeneration.

5.2.7   Multiple oxide nanoparticles

Sometimes two or more metal/metal oxide containing mixtures have been used as adsorbent for the removal of As(III) and As(V).

Nanoscale nickel/nickel borides appear to be promising materials for As removal from drinking water because of their large active surface area and high As adsorption capacity (Henden et al., 2011).

Sabbatini et al. (2009) synthesized nanoscale iron oxide particles and deposited them on porous alumina tubes to develop tubular ceramic adsorbers. The As(V) removal capacity of the adsorbent was $3.16 \, mg \, g^{-1}$. The sorption efficiency remained intact during over eight adsorption-desorption cycles and there was no iron dissolution during the basic washing. However, the regeneration efficiency was low.

Basu and Ghosh (2013) studied the adsorption kinetics of nano-structured iron (III) – cerium (IV) mixed oxide (NICMO). The adsorbents synthesized were agglomerated crystalline nano-particles of dimension 10–20 nm. They have reported that the empirical composition of the adsorbent was $FeCe_{1.1}O_{7.6}$. Suppression of As(III) removal by the major groundwater occurring ions (chloride, silicate, sulfate, bicarbonate and phosphate) was less than occurred with As(V) removal, so that this adsorbent was suggested for the treatment of high As content groundwater where As(III) was the dominant species.

Graphite oxide (GO) was modified with Fe_3O_4 and MnO_2 nanoparticles (Fe_3O_4-RGO-MnO_2) (Luo et al., 2012). The adsorbent was magnetic. The adsorption of both As(III) and As(V) by the adsorbent was found to be favorable. The experimental data fitted the Langmuir model well. As compared with magnetite-reduced graphene oxide composites (Chandra et al., 2010), the adsorption capacity was higher. The adsorbent removed As below the drinking water limit ($10 \, \mu g \, L^{-1}$). In the pH range 2–10, while the adsorption capacity was not pH dependent significantly for As(III), the adsorption capacity decreased (about 25%) by pH increase for As(V). At pH higher than 10, the adsorption capacities decreased sharply for both As(III) and As(V). The maximum adsorption capacities of the adsorbent were 14.04 and $12.22 \, mg \, g^{-1}$ for As(III) and As(V), respectively.

The adsorbent Zr-doped akaganeite (Sun et al., 2013) was synthesized at different Zr:Fe ratios by co-precipitation. The adsorption of both As(III) and As(V) better obeyed the Freundlich model than the Langmuir model. The maximum adsorption capacities were calculated from the Langmuir isotherm as 120 and $60 \, mg \, g^{-1}$ for As(III) and As(V), respectively. XPS studies were carried out for the determination of oxidation states of the adsorbed As onto the adsorbent. The results showed that the As species kept their original oxidation states during or after the adsorption. At high concentrations of As, the main adsorption sites were –OH groups in the tunnel structure of the adsorbent.

Malana et al. (2011) synthesized nano-aluminum doped manganese copper ferrite polymer composite with an average crystallite size 13 nm. TGA study showed that the adsorbent remained stable over the studied temperature range from 298 to 1073 K. The experimental data fitted well to the Freundlich model. The adsorption process of As(III) was favorable and dominated by chemical forces. Kinetic studies showed that both internal and external mass transfers had taken place. The maximum adsorption capacity calculated from the Langmuir model was low, $0.053 \, mg \, g^{-1}$ for As(III).

A Fe_2O_3:$MnO_2 = 3$:1 nanoparticles mixture was synthesized by Andjelkovic et al. (2014) using mechanical – chemical treatment in a planetary ball mill. The point of zero charge pH of the adsorbent was 5.6. Removal efficiency decreased from 80 to 30% and 90 to 60% for As(III) and As(V), respectively, by increasing the pH from 4 to 10. Chloride had no significant effect for both As(III) and As(V) removal efficiency, while sulfate affected it slightly. Phosphate decreased the adsorption efficiency by about 10–15% and 12–18% for As(III) and As(V), respectively. The adsorption obeyed the pseudo-second order model. The maximum adsorption capacities calculated from the Langmuir isotherm were 2.89 and $3.84 \, mg \, g^{-1}$ for As(III) and As(V), respectively.

Table 5.1. Comparison of the adsorbents for As(III) and As(V) removal.

Adsorbent	Initial concentration [mg L⁻¹] As(III)	As(V)	Removal efficiency [%] As(III)	As(V)	pH	Adsorbent dose [g L⁻¹]	Adsorption capacity [mg g⁻¹] As(III)	As(V)	Isotherm	Kinetic	Water ions that suppress arsenic sorption	Potential for removing As below WHO limit	Adsorption (endothermic-exothermic)	Regeneration reagent	The point of zero charge pH$_{PZC}$	Ref.
Fe(VI)/Al(III) salts	0.5	*	99.7	*	6.5	*	*	*	*	*	Phosphate, silicate, bicarbonate	+	*	*	*	Jain et al. (2009)
Amorphe Iron(III) phosphate	0.5–100	0.5–100	*	*	*	1.0	21	10	*	*	*	*	*	*	3.7	Lenoble et al. (2005)
Crystalline-iron(III) phosphate	0.5–100	0.5–100	*	*	*	1.0	16	9	*	*	*	*	*	*	3.1	Lenoble et al. (2005)
Zero-valent iron	0.2	*	~100	*	7.0	2.0	*	*	*	Pseudo-first order	Borate and organic matter > nitrate and phosphate > chloride, carbonate and manganese	+	*	*	*	Biterna et al. (2010)
MnFe$_2$O$_4$ NCs	0.5–40	*	*	*	6.9	0.2 and 1.0	>27.3	*	Freundlich	Pseudo-second order	Chloride ~ silicate > sulfate > bicarbonate > phosphate	*	*	NaOH	*	Xu et al. (2013)
Nano FeCe$_{1.1}$O$_{7.6}$	4.8	4.5	*	*	7.0	2.0	2.42	2.11	*	Pseudo-second order	*	+	*	*	7.13	Basu and Gosh (2013)
Granular ferric hydroxide	0.1	0.1	90–97	95–99	6.5/7.5	0.25	*	3.13–4.57	Freundlich	Pseudo-first order	*	+	Endothermic	*	7.8	Banerjee et al. (2008)
Natural iron ores	*	0–30	*	*	4.5–6.5	1.0 and 5.0	*	0.4	Langmuir	Pseudo-first order	Silicate and phosphate	+	*	*	*	Zhang et al. (2004)

(continued)

Table 5.1. Continued.

Adsorbent	Initial concentration [mg L⁻¹] As(III)	As(V)	Removal efficiency [%] As(III)	As(V)	pH	Adsorbent dose [g L⁻¹]	Adsorption capacity [mg g⁻¹] As(III)	As(V)	Isotherm	Kinetic	Water ions that suppress arsenic sorption	Potential for removing As below WHO limit	Adsorption (endothermic-exothermic)	Regeneration reagent	The point of zero charge pH$_{PZC}$	Ref.
NZVI/AC	2.0	2.0	~100	~100	6.5	1.0	18.2	12.0	Freundlich + Langmuir		Phosphate and silicate	+	*	NaOH	7.4	Zhu et al. (2009)
GO-ZrO(OH)₂	2.0– 80	2.0– 80	*	*	7.0	0.5	95.18	84.89	Langmuir	Pseudo-second order	Phosphate	+	*	HNO₃	7.13	Luo et al. (2013)
δ-FeOOH	*	0–20	*	*	7.0	0.25	*	37.3	Langmuir-Redlich Peterson	Pseudo-second order	Nitrate	*	*	*	8.4	Faria et al. (2014)
IOCS	0.3	0.01– 0.5	99.5	99.5	5–8	4–24	0.0126	0.0221	Langmuir	*	*	+	*	*	7.0	Hsu et al. (2008)
Activated alumina	0.5– 1.5	*	96.2	*	7.6	1–13	*	*	Freundlich + Langmuir + DR	Pseudo-first order	*	*	Exothermic	*	*	Singh and Pant (2004)
Granular TiO₂	0.4– 80	0.4– 80	*	*	7.0	0.3–1.0	41.4	32.4	Langmuir	*	*	*	*	*	*	Bang et al. (2005)
Fe/glass composite	*	*	0.1– 1.0	97	6.0	100	*	*	*	*	*	–	*	*	*	Wang et al. (2009)
Iron and manganese coated sand	*	0.5– 10	*	>90	4.5	20	*	22	*	*	Phosphate	*	*	*	6.6	Yang et al. (2007)
MGNCs	*	0.5– 1.0	*	~100	*	1.0	*	2.3	Langmuir + Sips	*	*	*	*	*	*	Lee et al. (2014)
Fe₂O₃: MnO₂ = 3:1	0.2– 2.0	0.2– 2.0	*	*	4.0–7.0	0.5	2.89	3.84	Freundlich + Langmuir	Pseudo-second order	Sulfate and phosphate	+	*	*	5.6	Andjelkovic et al. (2014)
Leonardite	1.0– 80	1.0– 80	*	~100	3.0– 10.0	2.0	4.46	8.40	Freundlich + Langmuir	*	Phosphate Sulfate > nitrate > chloride	+	*	*	*	Chammui et al. (2014)
Copper(II) oxide nanoparticles	0.5– 1.0	*	~100	*	7.0	1.0	1.086	*	Langmuir	Pseudo-second order	Phosphate and sulfate	+	Endothermic	NaOH	*	Goswami et al. (2012)

Adsorbent															Reference
MION-tea magnetic iron oxide NP	0.2–4.0	*	*	7.0	1.0	188.7	153.8	Langmuir	Pseudo-second order	sulfate > phosphate > chloride > nitrate	*	Endothermic	NaOH	*	Lunge et al (2014)
Magnetic binary oxide particles (MBOP)	0–2.0	*	~99	7.0	1.0	16.94	*	Freundlich + Langmuir	Pseudo-first order	Chloride and phosphate	+	Exothermic	NaOH	*	Dhoble et al (2011)
Iron oxide hydroxide nanoflower (IOH)	0–1.0	*	*	7.28	1.0	0.475	*	Redlich-Peterson > Langmuir > Freundlich > Temkin	Pseudo-second order	Phosphate	+	Exothermic	NaOH	3.38	Raul et al (2014)
Aluminum doped manganese copper ferrite	0.02–0.2	*	*	6.0	3.3	0.053	*	Freundlich	Pseudo-second order	*	*	*		*	Malana et al (2011)
polymer Zr doped akaganeite	10	10	*	7.0	0.1	120	60	Freundlich	*	*	*	*	*	5.1	Sun et al (2013)
Fe-hydrotalcite	0.1–1.0	*	*	9.0	5.0	0.121	1.281	Langmuir	Pseudo-second order	*	+	*	*	*	Türk and Alp (2014)
Fe$_3$O$_4$-reduced graphite oxide-MnO$_2$ nanocomposites	0.01–10	~100	~100	2.0–10.0	0.5	14.04	12.22	Langmuir	Pseudo-second order	*	+	*	*	*	Luo et al (2012)
Montmorillonite supported nanoscale zero-valent iron (Mt-nZVI)	5.0	~100	~100	4.0–8.0	1.0	59.9	45.5	Freundlich + Langmuir	Pseudo-second order	Phosphate	+	*	NaOH	8.2	Bhowmick et al (2014)
Spherical polystyrene-supported nano-Fe$_3$O$_4$	*	1–50	*	6.0	0.5	*	139.3	Langmuir	Pseudo-first order	Phosphate and silicate	*	*	NaOH	*	Jiang et al (2012)
Ferric hydroxide supported on silicagel (FHSS)	0.1–1.0	~100	~100	3.1–9.7	5.0	16.2	17.7	*	*	*	+	*	*	*	Çiftçi et al (2011)

5.2.8   Comparison of adsorbents

The adsorbents used for the removal of As were compared in Table 5.1. Daus *et al.* (2004) reported comparison of five different adsorbents for the removal of arsenate and arsenite from water. The adsorbents were: granulated iron hydroxide (GIH), zirconium loaded activated carbon (Zr-AC), activated carbon (AC), Absorptionsmittel 3 (AM3) (a trade-named sorption medium), zero-valent iron (Fe(0)). The sorption kinetics of arsenate onto the materials followed the sequence Zr-AC>>GIH=AM3>Fe(0)>AC. A different sequence was obtained for arsenite (AC>>Zr-AC=AM3=GIH=Fe(0)). The removal of arsenite by using GIH, AM3 and Zr-Ac was slower and less efficient than for arsenate. While the removal rate of arsenate by AC was the slowest, the removal rate of arsenite by AC was the fastest. The difference was attributed to either a forced oxidation reaction or to strong binding of the non-charged arsenite molecules to the carbon atoms. The experiments showed that in spite of the relatively slow kinetics, the most promising adsorbent was GIH among these adsorbents. For the removal of arsenate, zirconyl activated carbon was efficient for fast removal.

As shown in Table 5.1, most of the adsorbents are iron-based. In recent years, the use of the adsorbents at nanoscale has increased, mainly because of their higher capacities and faster kinetics. In As removal, the most serious interference is caused by phosphate, which suppresses particularly As(III) removal seriously with most of the adsorbents. Most of the adsorbents in Table 5.1 seem to have potential for removing As below the WHO limit ($10 \mu g L^{-1}$) for drinking water (WHO, 2011).

REFERENCES

Andjelkovic, I., Nesic, J., Stankovic, D., Manojlovic, D., Pavlovic, M.B., Jovalekic, C. & Roglic, G. (2014) Investigation of sorbents synthesised by mechanical-chemical reaction for sorption of As(III) and As(V) from aqueous medium. *Clean Technologies and Environmental Policy*, 16 (2), 395–403.

Banerjee, K., Amy, G.L., Prevost, M., Nour, S., Jekel, M., Gallagher, P.M. & Blumenschein, C.D. (2008) Kinetic and thermodynamic aspects of adsorption of arsenic onto granular ferric hydroxide (GFH). *Water Research*, 42 (13), 3371–3378.

Bang, S., Patel, M., Lippincott, L. & Meng, X. (2005) Removal of arsenic from groundwater by granular titanium dioxide adsorbent. *Chemosphere*, 60 (3), 389–397.

Basu, T. & Ghosh, U.C. (2013) Nano-structured iron(III)-cerium(IV) mixed oxide: synthesis, characterization and arsenic sorption kinetics in the presence of co-existing ions aiming to apply for high arsenic groundwater treatment. *Applied Surface Science*, 283, 471–481.

Berg, M., Luzi, S., Trang, P.T.K., Viet, P.H., Giger, W. & Stüben, D. (2006) Arsenic removal from groundwater by household sand filters: comparative field study, model calculations, and health benefits. *Environmental Science & Technology*, 40 (17), 5567–5573.

Bhowmick, S., Chakraborty, S., Mondal, P., Renterghem, W.V., Berghe, S.V.D., Roman-Ross, G., Chatterjee, D. & Iglesias, M. (2014) Montmorillonite-supported nanoscale zero-valent iron for removal of arsenic from aqueous solution: kinetics and mechanism. *Chemical Engineering Journal*, 243, 14–23.

Biterna, M., Antonoglou, L., Lazou, E. & Voutsa, D. (2010) Arsenite removal from waters by zero valent iron: batch and column tests. *Chemosphere*, 78 (1), 7–12.

Chammui, Y., Sooksamiti, P., Naksata, W., Thiansem, S. & Arqueropanyo, O. (2014) Removal of arsenic from aqueous solution by adsorption on leonardite. *Chemical Engineering Journal*, 240, 202–210.

Chandra, V., Park, J., Chun, Y., Lee, J.W., Hwang, I.C. & Kim, K.S. (2010) Water-dispersible magnetite-reduced graphene oxide composites for arsenic removal. *ACS Nano*, 4, 3979–3986.

Choong, T.S.Y., Chuah, T.G., Robiah, Y., Gregory Koay, F.L. & Azni, I. (2007) Arsenic toxicity, health hazards and removal techniques from water: an overview. *Desalination*, 217, 139–166.

Çiftçi, T.D., Yayayürük, O. & Henden, E. (2011) Study of arsenic(III) and arsenic(V) removal from waters using ferric hydroxide supported on silica gel prepared at low pH. *Environmental Technology*, 32, 341–351.

Cui, H., Li, Q., Gao, S. & Shang, J.K. (2012) Strong adsorption of arsenic species by amorphous zirconium oxide nanoparticles. *Journal of Industrial and Engineering Chemistry*, 18, 1418–1427.

Cui, H., Su, Y., Li, Q., Gao, S. & Shang, J.K. (2013) Exceptional arsenic (III,V) removal performance of highly porous, nanostructured ZrO2 spheres for fixed bed reactors and the full-scale system modeling. *Water Research*, 47, 6258–6268.

Daus, B., Wennrich, R. & Weiss, H. (2004) Sorption materials for arsenic removal from water: a comparative study. *Water Research*, 38, 2948–2954.

DeMarco, M.J., SenGupta, A.K. & Greenleaf, J.E. (2003) Arsenic removal using a polymeric/inorganic hybrid sorbent. *Water Research*, 37, 164–176.

Dhoble, R.M., Lunge, S., Bhole, A.G. & Rayalu, S. (2011) Magnetic binary oxide particles (MBOP): a promising adsorbent for removal of As (III) in water. *Water Research*, 45, 4769–4781.

Dong, L., Zinin, P., Cowen, J.P. & Ming, L.C. (2009) Iron coated pottery granules for arsenic removal from drinking water. *Journal of Hazardous Materials*, 168, 626–632.

Faria, M.C.S., Rosemberg, R.S., Bomfeti, C.A., Monteiro, D.S., Barbosa, F., Oliveira, L.C.A., Rodriguez, M., Pereira, M.C. & Rodrigues, J.L. (2014) Arsenic removal from contaminated water by ultrafine δ-FeOOH adsorbents. *Chemical Engineering Journal*, 237, 47–54.

Goswami, A., Raul, P.K. & Purkait, M.K. (2012) Arsenic adsorption using copper (II) oxide nanoparticles. *Chemical Engineering Research and Design*, 90, 1387–1396.

Han, C., Li, H., Pu, H., Yu, H., Deng, L., Huang, S. & Luo, Y. (2013) Synthesis and characterization of mesoporous alumina and their performances for removing arsenic(V). *Chemical Engineering Journal*, 217, 1–9.

Henden, E., İşlek, Y., Kavas, M., Aksuner N., Yayayürük, O., Çiftçi, T.D. & İlktaç, R. (2011) A study of mechanism of nickel interferences in hydride generation atomic absorption spectrometric determination of arsenic and antimony. *Spectrochimica Acta* Part B, 66 (11-12), 793–798.

Hering, J.G. (1996) Risk assessment for arsenic in drinking water, limits to achievable risk levels. *Journal of Hazardous Materials*, 45, 175–184.

Hsu, J.C., Lin, C.J., Liao, C.H. & Chen, S.T. (2008) Removal of As(V) and As(III) by reclaimed iron-oxide coated sands. *Journal of Hazardous Materials*, 153, 817–826.

Jain, A., Sharma, V.K. & Mbuya, O.S. (2009) Removal of arsenite by Fe(VI), Fe(VI)/Fe(III), and Fe(VI)/Al(III) salts: effect of pH and anions. *Journal of Hazardous Materials*, 169, 339–344.

Jeon, C.S., Baek, K., Park, J.K., Oh, Y.K. & Lee, S.D. (2009) Adsorption characteristics of As(V) on iron coated zeolite. *Journal of Hazardous Materials*, 163, 804–808.

Jiang, W., Chen, X., Niu, Y. & Pan, B. (2012) Spherical polystyrene-supported nano-Fe$_3$O$_4$ of high capacity and low-field separation for arsenate removal from water. *Journal of Hazardous Materials*, 243, 319–325.

Korngold, E., Belayev, N. & Aronov, L. (2001) Removal of arsenic from drinking water by anion exchangers. *Desalination*, 141, 81–84.

Kundu, S. & Gupta, A.K. (2007) As(III) removal aqueous medium in fixed bed using iron oxide-coated cement (IOCC): experimental and modeling studies. *Chemical Engineering Journal*, 129, 123–131.

Kuriakose, S., Singh, T.S. & Pant, K.K. (2004) Adsorption of As(III) from aqueous solution onto iron oxide impregnated activated alumina. *Water Quality Research Journal of Canada*, 39, 258–266.

Lee, S.H., Cha, J., Sim, K. & Lee, J.K. (2014) Efficient removal of arsenic with magnetic nanoclusters. *Bulletin of the Korean Chemical Society*, 35 (2), 605–609.

Lenoble, V., Laclautre, C., Deluchat, V., Serpaud, B. & Bollinger, J.C. (2005) Arsenic removal by adsorption on iron(III) phosphate. *Journal of Hazardous Materials*, 123, 262–268.

Litter, M.I., Morgada, M.E. & Bundschuh, J. (2010) Possible treatments for arsenic removal in Latin American waters for human consumption. *Environmental Pollution*, 158, 1105–1118.

Lunge, S., Singh, S. & Sinha, A. (2014) Magnetic iron oxide (Fe$_3$O$_4$) nanoparticles from tea waste for arsenic removal. *Journal of Magnetism and Magnetic Materials*, 356, 21–31.

Luo, X., Wang, C., Luo, S., Dong, R., Tu, X. & Zeng, G. (2012) Adsorption of As (III) and As (V) from water using magnetite Fe$_3$O$_4$-reduced graphite oxide-MnO$_2$ nanocomposites. *Chemical Engineering Journal*, 187, 45–52.

Luo, X., Wang, C., Wang, L., Deng, F., Luo, S., Tu, X. & Au, C. (2013) Nanocomposites of graphene oxide-hydrated zirconium oxide for simultaneous removal of As(III) and As(V) from water. *Chemical Engineering Journal*, 220, 98–106.

Malana, M.A., Qureshi, R.B. & Ashiq, M.N. (2011) Adsorption studies of arsenic on nano aluminium doped manganese copper ferrite polymer (MA, VA, AA) composite: kinetics and mechanism. *Chemical Engineering Journal*, 172, 721–727.

Manju, G.N., Raji, C. & Anirudhan, T.S. (1998) Evaluation of coconut husk carbon for the removal of arsenic from water. *Water Research*, 32, 3062–3070.

Matsunaga, H., Yokoyama, T., Eldridge, R.J. & Bolto, B.A. (1996) Adsorption characteristics of arsenic(III) and arsenic(V) on iron(III)-loaded chelating resin having lysine-N$^{\alpha}$, N$^{\alpha}$-diacetic acid moitery. *Reactive Polymers*, 29, 167–174.

Mohan, D. & Pittman, C.U., Jr. (2007) Arsenic removal from water/wastewater using adsorbents – a critical review. *Journal of Hazardous Materials*, 142, 1–53.

Mondal, P., Bhowmick, S., Chatterjee, D., Figoli, A. & Bruggen, B.V.D. (2013) Remediation of inorganic arsenic in groundwater for safe water supply: a critical assessment of technological solutions. *Chemosphere*, 92, 157–170.

Pena, M.E., Korfiatis, G.P., Patel, M., Lippincott, L. & Meng, X. (2005) Adsorption of As(V) and As(III) by nanocrystalline titanium dioxide. *Water Research*, 39, 2327–2337.

Raul, P.K., Devi, R.R., Umlong, I.M., Thakur, A.J., Banerjee, S. & Veer, V. (2014) Iron oxide hydroxide nanoflower assisted removal of arsenic from water. *Materials Research Bulletin*, 49, 360–368.

Reddy, P.G. & Turner, C.D. (2007) Removal of arsenic from water with oxidized metal coated pumice. United States Patent Application Publication US 2007/0017871 A1.

Sabbatini, P., Rossi, F., Thern, G., Marajofsky, A. & Fidalgo de Cortalezzi, M.M. (2009) Iron oxide adsorbers for arsenic removal: a low cost treatment for rural areas and mobile applications. *Desalination*, 248, 184–192.

Sharma, A.K., Tjell, J.C. & Mosbaek, H. (2003) Removal of arsenic using naturally occurring iron. *Journal de Physique IV*, 107, 1223–1226.

Sharma, A.K., Tjell, J.C., Sloth, J.J. & Holm, P.E. (2014) Review of arsenic contamination, exposure through water and food and low cost mitigation options for rural areas. *Applied Geochemistry*, 41, 11–33.

Sharma, V.K. & Sohn, M. (2009) Aquatic arsenic: toxicity, speciation, transformations, and remediation. *Environment International*, 35, 743–759.

Singh, T.S. & Pant, K.K. (2004) Equilibrium, kinetics and thermodynamic studies for adsorption of As(III) on activated alumina. *Separation and Purification Technology*, 36, 139–147.

Sun, X., Hu, C., Hu, X., Qu, J. & Yang, M. (2013) Characterization and adsorption performance of Zr-doped akaganéite for efficient arsenic removal. *Journal of Chemical Technology and Biotechnology*, 88, 629–635.

Türk, T. & Alp, İ. (2014) Arsenic removal from aqueous solutions with Fe-hydrotalcite supported magnetite nanoparticle. *Journal of Industrial and Engineering Chemistry*, 20, 732–738.

USEPA/AWWA (2005) Adsorbent treatment technologies for arsenic removal. United States Environmental Protection Agency and American Water Works Association Research Foundation, Washington, DC and Denver, CO.

Wang, Y., Zhu, K., Wang, F. & Yanagisawa, K. (2009) Novel Fe/glass composite adsorbent for As(V) removal. *Journal of Environmental Sciences* (China), 21, 434–439.

WHO (2011) Guidelines for drinking water quality. 4th edition. World Health Organisation, Geneva, Switzerland. pp. 315–318.

Xu, W.H., Wang, L., Wang, J., Sheng, G.P., Liu, J.H., Yu, H.Q. & Huang, X.J. (2013) Superparamagnetic mesoporous ferrite nanocrystal clusters for efficient removal of arsenite from water. *Crystal Engineering Communications*, 15, 7895–7903.

Yadanaparthi, S.K.R., Graybill, D. & Wandruszka, R.V. (2009) Adsorbents for the removal of arsenic, cadmium, and lead from contaminated waters. *Journal of Hazardous Materials*, 171, 1–15.

Yang, J.K., Song, K.H., Kim, B.K., Hong, S.C., Cho, D.E. & Chang, Y.Y. (2007) Arsenic removal by iron and manganese coated sand. *Water Science and Technology*, 56, 161–169.

Zhang, F.S. & Itoh, H. (2005) Iron oxide-loaded slag for arsenic removal from aqueous system. *Chemosphere*, 60, 319–325.

Zhang, T. & Sun, D.D. (2013) Removal of arsenic from water using multifunctional micro-/nano-structured MnO_2 spheres and microfiltration. *Chemical Engineering Journal*, 225, 271–279.

Zhang, W., Singh, P., Paling, E. & Delides, S. (2004) Arsenic removal from contaminated water by natural iron ores. *Minerals Engineering*, 17, 517–524.

Zhu, H., Jia, Y., Wu, X. & Wang, H. (2009) Removal of arsenic from water by supported nano zero-valent iron on activated carbon. *Journal of Hazardous Materials*, 172, 1591–1596.

CHAPTER 6

Arsenic sorption on mono and binary metal oxide doped natural solids

Esra Bilgin Simsek & Ulker Beker

6.1 INTRODUCTION

Due to the increased awareness of the health risks of arsenic (As), significant effort has been devoted to As remediation from contaminated water. Several studies have demonstrated that As removal can be achieved by various techniques (Jain and Singh, 2012; Sharma *et al.*, 2013). Among them adsorption has been found to be an alternative option for the treatment of As-rich groundwater, which is especially well-accepted by the rural people of under-developed countries like India and Bangladesh, because of simple operation and requirement of less space (Gupta *et al.*, 2010). A wide range of low-cost adsorbents have been reported for As removal (Mohan and Pittman, 2007; Yadanaparthi *et al.*, 2009). Novel adsorbents such as iron-, alumina-, manganese-, titanium-oxide, as well as their composite oxides, have been developed to remove As present in water/wastewater (Bilgin Simsek *et al.*, 2013). Among these, iron oxides have received the greatest attention because of their promising binding affinity toward As and other toxic species naturally present in water, their availability, low cost and nontoxicity to the environment (Ahamed *et al.*, 2009; Mohan and Pittman, 2007). Especially, magnetic forms of these oxides can be separated from solution by using an external magnetic field.

There have been several researches reported about the usage of iron oxides combined with different materials and combinations of these oxides have high potential for water remediation due to their properties such as affinity and chemical stability. There have been several studies reporting that the mixed oxides of the metals showed better efficiency than many of the single metal oxides (Giles *et al.*, 2011; Hong *et al.*, 2010; Masue *et al.*, 2007; Silva *et al.*, 2010; Zhang *et al.*, 2005). Gupta *et al.* (2010) prepared manganese associated hydrous iron (III) oxide by heat treatment and they reported that the material could be used for the treatment of groundwater contaminated with high As. Zhang *et al.* (2003) synthesized Fe-Ce bimetal oxide powder by the coprecipitation method and found that the new adsorbent demonstrated a much better performance than activated alumina. Zhang *et al.* (2010); prepared bimetal oxide magnetic nanomaterials ($MnFe_2O_4$ and $CoFe_2O_4$) and examined As(V) and As(III) adsorption onto them. They found that the maximum adsorption capacities of on $MnFe_2O_4$ and $CoFe_2O_4$ were higher than on the referenced Fe_3O_4 and stated that might be caused by the increase of the surface hydroxyl groups.

On the other hand, distribution of As is handled by iron and aluminum oxides in most oxidized environments. The combined adsorption efficiency of these oxides is important to investigate since iron oxides are generally incorporated with aluminum in significant proportions (Cornell and Schwertmann, 2003). Masue *et al.* (2007) studied As adsorption/desorption behavior on Fe-Al hydroxide. They observed that when the Fe:Al molar ratio was 4:1 the bimetal hydroxide adsorbent gained almost equal As(V) adsorption capacity to Fe-hydroxide, yet the As(III) adsorption capacity on Fe-hydroxide was higher than that on the Fe-Al hydroxide. Ramesh *et al.* (2007) examined the As(III), As(V) and dimethylarsinate (DMA) adsorption onto polymeric Al/Fe modified montmorillonite. They found that the DMA adsorption capacity of adsorbent was higher compared to the literature and the adsorbent could be used in more than five cycles of sorption-desorption without any significant change of adsorption capacity ($<5\%$).

However, utilization of these oxides can bring difficulties in separation during treatment processes due to their fine powdered forms require a sedimentation or filtration process resulted in added cost and low mechanical resistance (Habuda-Stanic *et al.*, 2008; Li *et al.*, 2011). To overcome this problem, the effective method is to support or coat particles into/onto a porous matrix of larger size. The widely used carriers reported by researches are activated carbons (Chen *et al.*, 2007), polymers (Pan *et al.*, 2010), sands (Thirunavukkarasu *et al.*, 2002), clay minerals (Jiang and Zeng, 2003). In addition to these materials, natural zeolites have been recognized as alternative supporting materials, because of their mechanical and thermal properties, capability of exchanging cations and significant worldwide occurrence (Jiménez-Cedillo *et al.*, 2011; Šiljeg *et al.*, 2009; Wang and Peng, 2010).

Therefore, this study reports the synthesis and characterization of iron and aluminum oxides and their mixed forms doped with natural zeolite and the kinetic sorption behavior of As(V) onto these materials for removal from water. A suggestion of the sorption mechanism is also made.

6.2   MATERIALS AND METHODS

6.2.1   *Preparation of mono and binary metal oxides doped natural zeolite*

The metal oxides doped zeolites were prepared by a coprecipitation method as described in our previous work (Bilgin Simsek *et al.*, 2013): natural zeolite used in the present study was obtained from the Gördes-Manisa region of Turkey. The raw material was first conditioned with NaCl solution in order obtain the Na-form (ZNa) and then was treated with $FeCl_2$ and $AlCl_3$ solutions and a mixture of $AlCl_3$ and $FeCl_3$ solution. NaOH was added dropwise to each suspension in order to dope the metal oxides onto the surface of zeolite. The resultant samples which were modified with $FeCl_2$, $AlCl_3$ and a mixture of $AlCl_3$ and $FeCl_3$ solutions were denoted as ZFe, ZAl and ZAlFe, respectively. The synthesis route of ZFe sample is given in Figure 6.1.

6.2.2   *Characterization*

The configuration and morphology of samples and the elements distributed on the surface of the mixed oxides were examined by scanning electron microscopy (SEM) with an EDX (Philips XL30 SFEG). XRD measurement was performed on Philips Panalytical X'Pert Pro X-ray diffractometer using Cu Kα radiation.

6.2.3   *Adsorption kinetics*

Kinetics is able to predict the rate at which pollutant is removed from aqueous solutions in order to design suitable sorption treatment plants. Therefore, the study of adsorption kinetics in wastewater

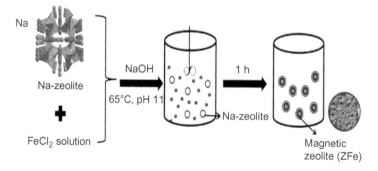

Figure 6.1.   Synthesis route of ZFe sample.

treatment is important as it provides helpful insights into the reaction pathways and also into the mechanism of sorption reactions (Ho and McKay, 1999). For kinetic adsorption experiments, 5 mg L^{-1} As(V) solution and defined amounts of ZFe, ZAl and ZAlFe adsorbents were contacted in a polyethylene bottle and the experiments were carried out at 25, 45 and 65°C in order to investigate the effect of temperature on adsorption rate and to calculate activation energy.

In order to analyze the controlling mechanism of the adsorption process, kinetic models were used to test the experimental data. The As(V) adsorption rates onto metal oxide immobilized zeolites were determined with four different kinetic models, i.e., the pseudo-first and pseudo-second order, the intraparticle diffusion and Elovich model.

In order to distinguish a kinetics equation based on concentration of solution and adsorption capacity of solid, Lagergren's first order rate equation has been established and called *pseudo-first order* (Ho, 2004):

$$\frac{dq_t}{dt} = k_1 \cdot (q_e - q_t) \tag{6.1}$$

where q_e and q_t are the amounts of As(V) adsorbed at equilibrium [mg g^{-1}] and at time t, respectively, and k_1 is the rate constant of pseudo-first order sorption (min^{-1}). From the after boundary conditions, $t=0$ to $t=t$ and $q_t=0$ to $q_t=q_t$, the integrated form of Equation (6.1):

$$\log(q_e - q_t) = \log(q_e) - \left(\frac{k_1}{2.303}\right)t \tag{6.2}$$

The slopes and intercepts of plots of $\log(q_e - q_t)$ versus t are used to determine the pseudo first-order rate constant k_1 and q_e.

Ho and McKay (1999) stated that the sorption capacity is proportional to the number of active sites occupied on the sorbent; then the kinetic rate law of *pseudo-second order* can be rewritten as:

$$\frac{dq_t}{dt} = k_2 \cdot (q_e - q_t)^2 \tag{6.3}$$

where k_2 is the rate constant of pseudo-second order sorption [g mg^{-1} min^{-1}]. After integration and applying boundary conditions ($t=0$ to $t=t$, $q_t=0$ to $q_t=q_t$), the integrated and linear form is:

$$\frac{t}{q_t} = \frac{1}{k_2 q_e^2} + \frac{1}{q_e}t \tag{6.4}$$

The values of q_e and k_2 are obtained from the slope and intercept of the straight line obtained by plotting t/q_t against t. Initial adsorption rate h [mg g^{-1} min^{-1}] can be calculated from the data of pseudo-second order kinetic model according to:

$$h = k_2 q_e^2 \tag{6.5}$$

The pseudo-second order model assumed that the rate-determining step might be a chemical sorption involving valence forces through sharing or exchange of electrons between adsorbent and adsorbate (Chen *et al.*, 2008; Ho, 2006).

The adsorption process can be controlled either by one or more steps: first one is bulk diffusion, the second is external mass transfer resistance and the third is intraparticle mass transfer resistance (Igwe and Abia, 2007). Generally, first and last steps are rapid, while pore and surface diffusion can be controlling factors of the overall process. Veličković *et al.* (2012) emphasized that the rate of surface complexation of As species is a fast process, generally in the order of milliseconds, regardless of the formation of monodentate or bidentate surface complexes. Weber and Morris (1963) proposed a parabolic equation namely *intraparticle diffusion model* in order to figure out the rate determining step. The intraparticle diffusion rate constant k_{id} [mg g^{-1} min$^{-0.5}$] is calculated by linearization of the curve $q_t = f(t^{0.5})$ as:

$$q_t = k_{id}t^{0.5} \tag{6.6}$$

The intercept of the plot of q_t versus $t^{0.5}$ gives an estimate of boundary layer thickness, the larger the value of the intercept, the greater the boundary layer diffusion effect is. If the regression of q_t versus $t^{0.5}$ is linear and passes through the origin, then intraparticle diffusion is the sole rate-limiting step (Weber and Morris, 1963; Bayramoglu et al., 2009).

The relationship (Equation (6.7)) – derived from the Fick's laws – was used to determine the intragranular diffusion coefficient D [m^2 min^{-1}]:

$$F(t) = \left[1 - \exp\left(-\frac{\pi^2 D^2 t}{r_0^2} \right) \right]^{1/2} \tag{6.7}$$

$F(t)$ represents the ratio $(C_i - C_t):(C_i - C_e)$ where C_e and C_i are the equilibrium and initial concentrations, respectively. C_t is the concentration at t time and r_0 is the equivalent spherical radius of particles (Hajjaji et al., 2001).

The Elovich equation is expressed as:

$$\frac{dq}{dt} = \alpha e^{-\beta q_t} \tag{6.8}$$

where, α is the initial sorption rate [mg g^{-1} min^{-1}]; β is surface activation energy for chemical sorption [g mg^{-1}]. The equation can be simplified by assuming $\alpha\beta_t \gg t$ and applying the boundary conditions ($q_t = 0$ at $t = 0$ and $q_t = q_t$ at $t = t$):

$$q_t = \frac{1}{\beta}\ln(\alpha\beta) + \frac{1}{\beta}\ln(t) \tag{6.9}$$

If a plot of q_t versus $\ln(t)$ yields a linear relationship, the sorption process fits the Elovich equation (Sparks, 1999).

6.2.4   Isotherm models

As(V) adsorption data were correlated with the theoretical isotherm models of Langmuir, Freundlich and Dubinin-Radushkevich:

- Langmuir isotherm model is based on the assumption that adsorption is limited to a mono-layer and the adsorbent surface is conceived as an energetically equivalent adsorption site for localized adsorption (Langmuir, 1918). The linearized model is expressed as:

$$\frac{1}{q_e} = \frac{1}{QbC_e} + \frac{1}{Q} \tag{6.10}$$

  where q_e is the adsorbed amount at equilibrium [mg g^{-1}], C_e is the equilibrium concentration of the adsorbate [mg L^{-1}], Q is the Langmuir monolayer sorption capacity [mg g^{-1}] and b is the Langmuir equilibrium constant [L mg^{-1}] related to the energy of adsorption and affinity of the adsorbent.

- Freundlich model is an empirical equation based on multilayer adsorption. The linearized form of model can be written as:

$$\ln q_e = \ln K_F + (1/n)\ln C_e \tag{6.11}$$

  where K_F and n are indicative isotherm parameters which can be determined by the linear plot of $\ln q_e$ versus $\ln C_e$.

- Dubinin-Radushkevich (DR) isotherm is generally applied to find the adsorption mechanism and the model assumes a heterogeneous surface. The isotherm is expressed by (Dubinin, 1960):

$$q_e = q_m \exp(-\beta\varepsilon^2) \tag{6.12}$$

$$\varepsilon = RT \ln(1 + 1/C_e) \tag{6.13}$$

where q_m is the maximum adsorption capacity [mg g^{-1}]; β is a constant which is related to adsorption energy. The DR isotherm has advantages like predicting quite fairly the experimental data over a wide concentration range and including the effect of temperature (Inglezakis and Poulopoulos, 2006).

6.2.5 *Desorption experiments*

Regeneration is an important feature for evaluating the adsorbents for their repeatability in use and also for environmental and economic reasons. In order to test the feasibility of metal oxide doped zeolites after As(V) adsorption for regeneration, the adsorbents were subjected to desorption tests. Suitable amounts of ZFe, ZAl and ZAlFe samples were added to bottles containing 20 mg L^{-1} As(V) concentration in order to have a final surface coverage of As(V) of about 95% (as determined by adsorption isotherms) and kept to react for 24 h at 25°C. Preliminary desorption experiments were carried out by varying percentages of NaOH solution (2, 5, 10 and 15% NaOH). Then, regeneration was tested by conducting adsorption/desorption cycles with a suitable desorption agent. Desorption efficiency (Equation (6.14)) is calculated for each cycle as a percentage of the amount of metal desorbed per unit mass of desorbent q_d [mg g^{-1}], to the amount of metal adsorbed onto the mineral per unit mass of adsorbent, q_a [mg g^{-1}] at equilibrium:

$$[\%] \text{ Desorption efficiency} = \frac{q_d}{q_a} \times 100 \tag{6.14}$$

6.3   RESULTS AND DISCUSSION

6.3.1 *Characterization studies*

As can be seen in Figure 6.2, clinoptilolite was the major crystalline phase detected in the XRD patterns of all samples. The peaks at $2\theta =$ 9.875°, 10.04°, 11.19°, 12.34°, 13.047°, 16.04°, 22.342°, 23.92°, 27.04°, 29.88°, 16.907°, 30.054°, and 33.575° were found to be in good agreement with the data of clinoptilolite (Joint Committee on Powder Diffraction Standards – JCPDS-39-1383). For ZFe sample, the peaks at $2\theta =$ 30°, 35°, 42° and 57° are common for magnetite and maghemite (JCPDS# 19-629, JCPDS# 39-1346) indicating magnetic properties of the sample. New peaks ($2\theta =$ 37.02°, 38.98°, 42.88°, 68.3° and 89.2°) were observed for the ZAlFe sample indicating the possibility of goethite or lepidocrocite phases, especially the peak at $2\theta = 41°$ is common for goethite. However, there was no significant peak detected for Al oxide-modified zeolite to indicate the presence of non-crystalline aluminum hydroxide.

The chemical composition of modified zeolites obtained by using EDX analysis is given in Table 6.1. Silicon and oxygen are major constituents along with aluminum, sodium, iron and potassium elements. The analysis indicated that the surface of ZAlFe and ZFe included 16.82% Fe and 17.31% Fe, respectively. It is evident from the EDX analysis that the iron percentage on the surface of ZAlFe (7.92% Fe) and ZFe (2.52% Fe) was reduced after As(V) adsorption. The aluminum content of the ZAl sample was found to be higher (16.99% Al) when compared the other samples. Furthermore, As content of modified zeolites after adsorption revealed that As(V) ions were adsorbed on the surface of adsorbents (2.14% As for ZAlFe, 0.83% As for ZFe and 0.42% As for ZAl).

6.3.2 *Adsorption kinetics*

Effect of contact time for modified zeolites was investigated and the results were shown in Figure 6.3. The removal percentages in the first eight hours for ZAl, ZFe and ZAlFe samples were calculated as 65, 74 and 81%, respectively. At the end of 24 h, 95% of As(V) was removed by the ZAlFe adsorbent, while that observed for ZFe was 83%. The ZAl sample, however, removed only 80% of As(V) after 24 h.

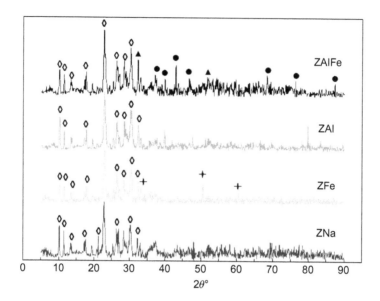

Figure 6.2.   XRD patterns of modified zeolites (◇: clinoptilolite, •: goethite, ▲: lepidocrite, +: magnetite or maghemite).

Table 6.1.   EDX surface analysis (before and after As(V) adsorption).

	Sample	Element [%, w/w]						
		O	Na	Al	Si	K	Fe	As
Before sorption	ZNa	37.73	3.38	8.46	43.19	0.98	1.13	–
	ZAl	32.39	3.57	16.99	40.06	0.8	2.08	–
	ZFe	27.57	6.06	7.25	38.72	0.63	17.31	–
	ZAlFe	34.68	5.45	11.8	38.23	1.08	16.82	–
After sorption	ZNa	45.26	4.12	8.23	40.02	1.11	1.11	0.15
	ZAl	38.87	3.67	13.88	38.62	1.15	1.68	0.42
	ZFe	42.68	3.87	8.26	40.92	1.34	2.52	0.83
	ZAlFe	45.48	4.94	10.01	30.32	0.5	7.92	2.14

The kinetic values of constants at different temperatures and their correlation coefficients for ZFe are summarized in the Table 6.2. It was observed that the q_e values estimated by the Lagergen model differ substantially from those measured experimentally, suggesting that the adsorption is not a pseudo-first order reaction. The pseudo-second order rate equation showed a correlation coefficient (R^2) of 0.998 for As(V) while that of the first order was 0.970 indicating the process can be better described by the second order reaction kinetics. Furthermore, the initial sorption rate (h) increases from 0.072 mg g^{-1} min^{-1} at 25°C to 0.282 mg g^{-1} min^{-1} at 65°C.

The results demonstrate that the adsorption process is endothermic. This can be attributed to the increased diffusion rate of the adsorbate molecules across the external boundary layer and within the pores.

Judging from the regression coefficients, the kinetic data is satisfactorily correlated by the pseudo-second order model (Table 6.3). The calculated adsorption capacities, q_e, are closer to those determined by experiments, q_e for the pseudo-second order model. The model capacities increased from 1.395 mg g^{-1} at 25°C to 1.494 mg g^{-1} at 65°C due to the favorable effect of

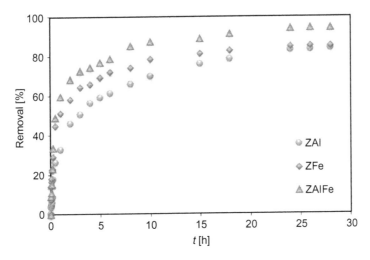

Figure 6.3.   Effect of contact time on As(V) adsorption on modified zeolites.

Table 6.2.   Comparison of kinetic models for As(V) adsorption onto ZFe.

Temperature	25°C	45°C	65°C
q_e [mg g^{-1}]	2.716	2.839	3.183
Pseudo-first order			
q_e [mg g^{-1}]	1.742	1.533	1.083
k_1 [min^{-1}]	0.007	0.006	0.002
R^2	0.970	0.967	0.565
Pseudo-second order			
q_e [mg g^{-1}]	2.728	2.835	2.960
k_2 [g mg^{-1} min^{-1}]	0.009	0.011	0.032
h [mg g^{-1} min^{-1}]	0.072	0.090	0.282
R^2	0.998	0.997	0.997
Intraparticle diffusion			
k_{id-1} [mg g^{-1} min$^{-0.5}$]	0.102	0.084	0.086
R^2	0.914	0.902	0.860
k_{id-2} [mg g^{-1} min$^{-0.5}$]	0.019	0.019	0.005
R^2	0.990	0.995	0.974
D [nm^2 min^{-1}]	0.867	1.230	2.023
Elovich			
β [g mg^{-1}]	2.759	3.246	3.357
α [mg g^{-1} min^{-1}]	0.469	1.811	9.319
R^2	0.988	0.990	0.934

temperature on adsorption capacity. Moreover, the diffusion coefficient of the ZAl samples was lower ($D_{ZAl} = 0.218$ nm^2 min^{-1}) when compared with the ZFe samples ($D_{ZFe} = 0.867$ nm^2 min^{-1}) indicating that the adsorption rate of iron oxide particles was higher than the aluminum ones.

In Figure 6.4b, the plot of the pseudo-second order kinetics was linear with good correlation coefficients (>0.99, Table 6.4) indicating the applicability of the model for ZAlFe samples. Moreover, it is important to mention that the parameter h for ZAlFe was found to be 10 times greater ($h_{ZAlFe} = 0.137$ mg g^{-1} min^{-1}) than that for ZAl ($h_{ZAl} = 0.012$ mg g^{-1} min^{-1}) and ZFe

Table 6.3.   Comparison of kinetic models for As(V) adsorption onto ZAl.

Temperature	25°C	45°C	65°C
q_e [mg g^{-1}]	1.338	1.401	1.473
Pseudo-first order			
q_e [mg g^{-1}]	1.076	1.051	1.058
k_1 [min^{-1}]	0.006	0.005	0.007
R^2	0.959	0.966	0.932
Pseudo-second order			
q_e [mg g^{-1}]	1.395	1.393	1.494
k_2 [g mg^{-1} min^{-1}]	0.006	0.009	0.011
h [mg g^{-1} min^{-1}]	0.012	0.018	0.026
R^2	0.991	0.993	0.996
Intraparticle diffusion			
k_{id-1} [mg g^{-1} min$^{-0.5}$]	0.075	0.050	0.061
R^2	0.972	0.977	0.983
k_{id-2} [mg g^{-1} min$^{-0.5}$]	0.001	0.001	0.013
R^2	0.083	0.995	0.872
D [nm^2 min^{-1}]	0.218	1.015	1.060
Elovich			
β [g mg^{-1}]	5.112	5.409	5.082
α [mg g^{-1} min^{-1}]	0.075	0.110	0.165
R^2	0.970	0.961	0.932

Table 6.4.   Comparison of kinetic models for As(V) adsorption onto ZAlFe.

Temperature	25°C	45°C	65°C
q_e [mg g^{-1}]	3.691	3.831	3.849
Pseudo-first order			
q_e [mg g^{-1}]	2.502	1.866	0.974
k_1 [min^{-1}]	0.009	0.006	0.006
R^2	0.979	0.817	0.855
Pseudo-second order			
q_e [mg g^{-1}]	3.700	3.780	3.900
k_2 [g mg^{-1} min^{-1}]	0.010	0.011	0.027
h [mg g^{-1} min^{-1}]	0.137	0.146	0.405
R^2	0.993	0.999	0.999
Intraparticle diffusion			
k_{id-1} [mg g^{-1} min$^{-0.5}$]	0.349	0.287	0.463
R^2	0.931	0.938	0.959
k_{id-2} [mg g^{-1} min$^{-0.5}$]	0.032	0.020	0.016
R^2	0.967	0.957	0.901
D [nm^2 min^{-1}]	1.376	2.972	4.510
Elovich			
β [g mg^{-1}]	1.829	1.816	2.996
α [mg g^{-1} min^{-1}]	0.421	0.662	0.495
R^2	0.972	0.962	0.955

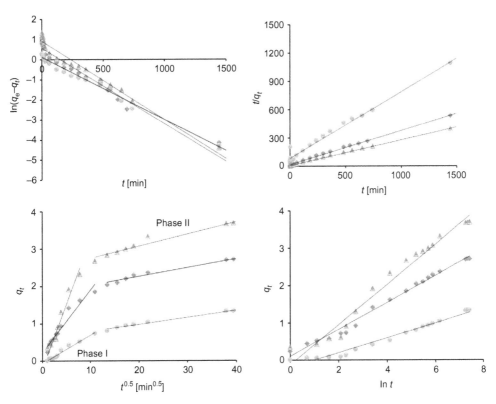

Figure 6.4. Kinetic modeling of As(V) adsorption onto modified zeolites (▲ ZAlFe, ◇ ZFe, ● ZAl)
(a) Pseudo-first order kinetics; (b) Pseudo-second order kinetics; (c) Intraparticle diffusion;
(d) Elovich model plots.

($h_{ZFe} = 0.072$ mg g^{-1} min^{-1}) owing to the synergistic influence of iron oxides associated with aluminum on the adsorption velocity of As(V).

Moreover, the plots of q_t versus $t^{0.5}$ indicate that the intraparticle diffusion model also describes well the adsorption processes (Fig. 6.4c). According to this model, if the adsorption of a solute is controlled by the intraparticle diffusion process, the plot gives a straight line. It is clear from the figure that two linear consecutive steps are observed: first linear portion (phase I) and second linear part (phase II). In phase I approximately 75% of As(V) was rapidly taken up by ZAlFe within one hour, this being attributable to the immediate utilization of the most readily available adsorbing sites on the adsorbent surface (Kumar *et al.*, 2009). The second phase indicates very slow diffusion of the adsorbate from the surface site into the inner pores. Velićković *et al.* (2012) stated that the first part is related to the instantaneous adsorbate bonding at the most readily available adsorbing sites, and the second is related to the transport of As species inside adsorbent micro pores. Hence, the initial portion of As(V) adsorption by ZFe, ZAl and ZAlFe might be governed by the initial intraparticle transport of As(V) controlled by a surface diffusion process and the later part controlled by pore diffusion. However, neither plots of intraparticle model passed through the origin (Fig. 6.3). Özcan *et al.* (2006) reported that this could be explained as if intraparticle diffusion was involved in the adsorption process; it was not the rate-controlling step.

Consequently, for modified adsorbents, As(V) adsorption kinetics fits the following order: pseudo-second order > intraparticle diffusion > Elovich > pseudo-first order. Therefore, it may be concluded that As(V) adsorption onto modified zeolites consists of chemical interactions due

Table 6.5.   The activation energy values of As(V)
adsorption onto modified zeolites.

Sample	E_a [kJ mol^{-1}]	R^2
ZFe	27.33	0.9812
ZAl	18.07	0.9989
ZAlFe	25.32	0.9928

to the fact that the pseudo-second order model suggests that the adsorption process involves the chemisorption mechanism. Analogous phenomena have also been observed in the literature for As(V) adsorption on various adsorbents (Bilici Baskan and Pala, 2011; Gupta and Ghosh, 2009; Hong et al., 2011; Jiménez-Cedillo et al., 2009).

6.3.3   Activation energy

The effect of temperature on As(V) adsorption was investigated using the linear plot of the Arrhenius equation (5.14) for the determination of activation energy (Moore and Pearson, 1981):

$$\ln k = \ln A_0 - \frac{E_a}{RT} \tag{6.15}$$

where k is the rate constant [g mg^{-1} min^{-1}], E_a is the Arrhenius activation energy [kJ mol^{-1}], R is the universal gas constant [8.314 J mol^{-1} K^{-1}], A_0 is the pre-exponential factor (frequency factor), and T is the absolute temperature [K]. The values of rate constants (k_2) from the pseudo-second-order model which fitted the adsorption system were used to calculate the activation energy of the sorption process.

The activation energies (E_a) of each sample were calculated from the slope of the plot of $\ln k_2$ vs. $1/T$ at initial As(V) concentration of 5 mg L^{-1}. The activation energy values were found to be 27.33, 18.07 and 25.32 kJ mol^{-1} for ZFe, ZAl and ZAlFe, respectively.

Sparks (1999) reported that E_a values for film diffusion are typically <25 kJ mol^{-1} while that for intraparticle diffusion processes are in the range of 21 to 42 kJ mol^{-1}. The obtained data (Table 6.5) thus indicated that the diffusion processes might be the rate-limiting step in As(V) adsorption by the three adsorbents.

6.3.4   Adsorption isotherms

The adsorption isotherm models were calculated at 5 mg L^{-1} initial As(V) concentration at ambient temperature. Figures 6.5–6.6 represent the experimental and theoretical As(V) adsorption isotherms of raw and modified zeolites.

According to the IUPAC classification, the adsorption isotherms for metal oxide doped zeolites resemble the typical Type-I isotherm, while the isotherm for ZNa samples was fitted to the Type-III isotherm indicating weak adsorbate–adsorbent interactions. Moreover, the coefficient of determination (R^2) values of ZNa were found to be too low due to the low adsorption capacity and the Freundlich, Langmuir and DR models were not fitted well.

The dimensionless separation factor (R_L) of the Langmuir model can be used to predict the affinity between the adsorbent and adsorbate (Avcı Tuna et al., 2013). R_L values of the ZFe, ZAl and ZAlFe samples were in the range of 0.001–0.002 showing As(V) adsorption was favorable.

As seen in Figures 6.5–6.6, the DR and Freundlich models were the best fitting isotherms for modified zeolites. The chi-square values (χ^2) of the DR and Freundlich isotherm models were equal to zero while they were found higher in the case of the Langmuir model.

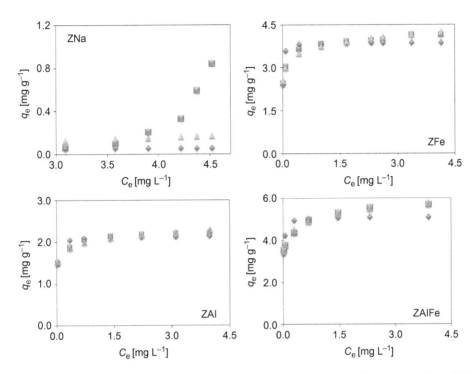

Figure 6.5. Experimental and theoretical adsorption isotherms (■ Experimental; ▲ Freundlich; ◆ Langmuir).

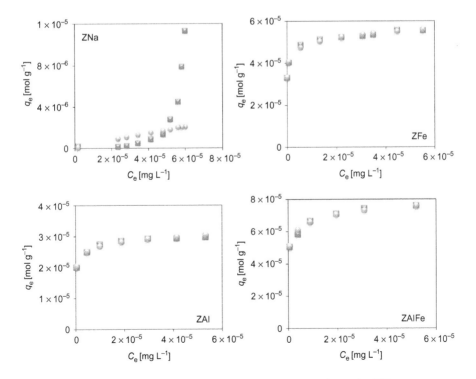

Figure 6.6. Experimental and theoretical adsorption isotherms (■ Experimental; ● DR).

Table 6.6.   Desorption efficiencies of similar adsorbents.

Adsorbent	Reagent	Desorption [%]	Reference(s)
Al-Zeolite	1% NaOH	99	Xu *et al.* (2002)
Al-Fe-Montmorillonite	5–15% NaCl	85 (10% NaCl)	Ramesh *et al.* (2007)
Al-Fe hydroxide	Na_2HPO_4	80–90 (pH 3 and 11)	Masue *et al.* (2007)
Mn-Fe oxide	0.4% NaOH	90	Zhang *et al.* (2010)
Gibbsite, $Al(OH)_x$	KH_2PO_4	48–56	Pigna *et al.* (2006)
Goethite, Ferrihydrite	KH_2PO_4	18–23	Pigna *et al.* (2006)
Fe-Si oxide	pH adjustment	73 (at pH 12)	Mahmood *et al.* (2012)
Magnetite	$MgCl_2$	24	Mamindy-Pajany *et al.* (2011)
Magnetite	Na_2HPO_4	37	
Goethite	$MgCl_2$	4	Mamindy-Pajany *et al.* (2011)
Goethite	Na_2HPO_4	34	
Fe-Montmorillonite	KOH	Max. pH 9.5	Luengo *et al.* (2011)
Al-Zeolite	10% NaOH	98.9	Present study
Fe-Zeolite	10% NaOH	98.6	Present study
Al-Fe-Zeolite	10% NaOH	96.8	Present study

Considering both R^2 (0.998) and χ^2 (0.0007) values, the DR isotherm model fitted the experimental data better than the Freundlich model indicating As(V) adsorption on metal oxide doped zeolites take places on multilayer.

6.3.5   Desorption experiments

As is very strongly adsorbed by oxide minerals and adsorption protects many natural environments from widespread toxicity problems. As desorbs from the oxide surfaces by increasing pH, especially above pH 8.5 (Smedley and Kinniburgh, 2002). Previous regeneration studies were carried out by similar adsorbents using HCl, NaCl, $MgCl_2$, KOH, phosphate (PO_4^{3-}) and NaOH (Table 6.6). Generally, the best regeneration percentages were obtained by using PO_4^{3-} and NaOH solutions. Henke (2009) stated that phosphates (PO_4^{3-}) and silicates (SiO_4^{4-}) have the same tetrahedral structure as As(V) and due to these similarities they could desorb As(V) from clay, iron, aluminum, and other sorbents over a wide range of pH values.

In the present study, NaOH was used to regenerate the spent adsorbents. In the first cycle, the effect of alkali strength was investigated (Fig. 6.7) and the best regeneration was obtained with 10% and 15% NaOH solutions. For ZFe samples, the desorption efficiency increased from 89.4% (2% NaOH) to 98.6% by using 10% NaOH solution. This increase could be due to the increase in concentration of the hydroxyl anions in the alkaline media which is responsible for anion exchange removal of As(V) from the loaded surface into the aqueous phase according to the following mechanism (Mahmood *et al.*, 2012):

$$FeH_2AsO_{4(s)} + OH^- \rightleftarrows FeOH_{(s)} + H_2AsO_{4(aq)}^- \qquad (6.16)$$

Dong *et al.* (2012) also found a similar increasing trend in the desorption with the increase in pH while studying As(V) desorption from nano zero-valent iron. As there were no significant differences in desorption efficiencies between 10 and 15% NaOH reagents, further adsorption/desorption tests were carried out using 10% NaOH solution.

As can be seen from Figure 6.8, regeneration percent fell across the cycles. The highest desorption efficiency was calculated for ZAl sorbent which has the lowest adsorption capacity. As(V) desorption rates were obtained as 98.9, 98.6 and 96.8% for ZAl, ZFe and ZAlFe, respectively. The regeneration percent of ZAl sorbent decreased to 74.5% after seven adsorption-desorption

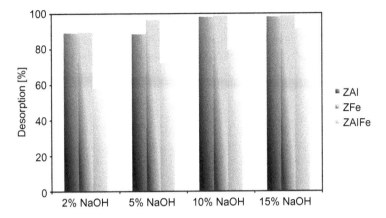

Figure 6.7.   The effect of alkali strength on As(V) desorption.

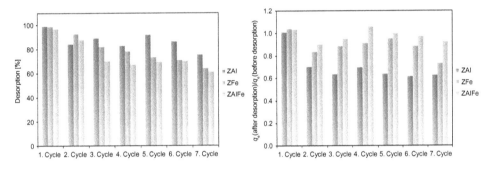

Figure 6.8.   Desorption efficiencies and adsorption capacity change for each cycle.

cycles while that was found as 60.9% for ZAlFe samples. These results demonstrate that As was strongly adsorbed onto the ZAlFe adsorbent, suggesting specific interactions on the surface.

When the same adsorbents after regeneration were used for As(V) adsorption (Fig. 6.8), the results indicated that binary metal oxide doped zeolite (ZAlFe) could be used for more than seven cycles of adsorption-desorption without any significant change of adsorption capacity ($<7\%$).

6.4   CONCLUSIONS

This study focused on adsorption of As(V) ions onto mono and binary metal oxide doped natural zeolite from aqueous solution. The adsorption capacities of ZAl, ZFe and ZAlFe were found to be 1.338, 2.716 and 3.691 mg g^{-1}, respectively. The rate of the adsorption is governed by a pseudo-second order rate equation. Further, the As(V) adsorption is controlled by a particle diffusion mechanism in a majority of cases. The initial sorption rates of ZAlFe were found to be 10 times greater than those for ZAl and ZFe owing to the synergistic influence of iron oxides associated with aluminum on the adsorption velocity. The heterogeneity of the adsorbent surface was confirmed, as the isotherm data fitted well to the DR and Freundlich models for modified samples while for Na-form zeolite the isotherm model type indicated weak adsorbate-adsorbent interactions. Regeneration results showed that binary metal oxide doped zeolite (ZAlFe) could be used for more than seven cycles of adsorption-desorption without any significant change of adsorption capacity.

Taking into consideration present findings, it can be concluded that iron and aluminum doped zeolites are viable adsorbents for As(V) removal from aqueous media. The present study demonstrates the theory that Al-Fe incorporation is beneficial for As(V) adsorption. In addition, the magnetic character of ZFe can provide an easy and rapid separation of the adsorbent from aqueous solution by a simple magnetic process. Furthermore, evaluation of naturally occurring zeolites with metal oxides in As remediation will also contribute to the economy of the world.

REFERENCES

Ahamed, S., Hussam, A. & Munir, A.K.M. (2009) Groundwater arsenic removal technologies based on sorbents: field applications and sustainability. In: Ahuja, S. (ed.) *Handbook of water purity and quality*. Academic Press, Waltham, MA. pp. 379–417.

Avcı Tuna, A.Ö., Özdemir, E., Bilgin Simsek, E. & Beker, U. (2013) Removal of As(V) from aqueous solution by activated carbon-based hybrid adsorbents: impact of experimental conditions. *Chemical Engineering Journal*, 223, 116–128.

Bayramoglu, G., Altintas, B. & Arica, M.Y. (2009) Adsorption kinetics and thermodynamic parameters of cationic dyes from aqueous solutions by using a new strong cation exchange resin. *Chemical Engineering Journal*, 152, 339–346.

Bilgin Simsek, E., Özdemir, E. & Beker, U. (2013) Zeolite supported mono- and bimetallic oxides: promising adsorbents for removal of As(V) in aqueous solutions. *Chemical Engineering Journal*, 220, 402–411.

Bilici Baskan, M. & Pala, A. (2011) Removal of arsenic from drinking water using modified natural zeolite. *Desalination*, 281, 396–403.

Chen, A.H., Liu, S.C. & Chen, C.Y. (2008) Comparative adsorption of Cu(II), Zn(II), and Pb(II) ions in aqueous solution on the crosslinked chitosan with epichlorohydrin. *Journal of Hazardous Materials*, 154, 184–191.

Chen, W., Parette, R., Zou, J., Cannon, F.S. & Dempsey, B.A. (2007) Arsenic removal by iron-modified activated carbon. *Water Research*, 41, 1851–1858.

Cornell, R.M. & Schwertmann, U. (2003) *The iron oxides: structure, properties, reactions, occurrence and uses*. Wiley, Weinheim, Germany.

Dong, H., Guan, X. & Lo, I.M.C. (2012) Fate of As(V)-treated nano zero-valent iron: determination of arsenic desorption potential under varying environmental conditions by phosphate extraction. *Water Research*, 46, 4071–4080.

Dubinin, M.M. (1960) The potential theory of adsorption of gases and vapors for adsorbents with energetically non-uniform surface. *Chemical Reviews*, 60, 235–266.

Giles, D.E., Mohapatra, M., Issa, T.M., Anand, S. & Singh, P. (2011) Iron and aluminium based adsorption strategies for removing arsenic from water. *Journal of Environmental Management*, 92, 3011–3022.

Gupta, K. & Ghosh, U.C. (2009) Arsenic removal using hydrous nanostructure iron(III)-titanium(IV) binary mixed oxide from aqueous solution. *Journal of Hazardous Materials*, 161, 884–892.

Gupta, K., Maity, A. & Ghosh, U.C. (2010) Manganese associated nanoparticles agglomerate of iron(III) oxide: synthesis, characterization and arsenic(III) sorption behavior with mechanism. *Journal of Hazardous Materials*, 184 (1–3), 832–842.

Habuda-Stanic, M., Kalajdzic, B., Kuleš, M. & Velic, N. (2008) Arsenite and arsenate sorption by hydrous ferric oxide/polymeric material. *Desalination*, 229, 1–9.

Hajjaji, M., Kacim, S., Alami, A., El Bouadili, A. & El Mountassir, M. (2001) Chemical and mineralogical characterization of a clay taken from the Moroccan Meseta and a study of the interaction between its fine fraction and methylene blue. *Applied Clay Science*, 20, 1–12.

Henke, K.R. (2009) Waste treatment and remediation technologies for arsenic. In: Henke, K.R. (ed.) *Arsenic environmental chemistry, health threats and waste treatment*. Wiley, New York, NY. pp. 351–415.

Ho, Y.S. (2004) Citation review of Lagergren kinetic rate equation on adsorption reaction. *Scientometrics*, 59, 171–177.

Ho, Y.S. (2006) Second-order kinetic model for the sorption of cadmium onto tree fern: a comparison of linear and non-linear methods. *Water Research*, 40, 119–125.

Ho, Y.S. & McKay, G. (1999) Pseudo-second order model for sorption processes. *Process Biochemistry*, 34, 451–465.

Hong, H.J., Farooq, W., Yang, J.S. & Yang, J.W. (2010) Preparation and evaluation of Fe & Al binary oxide for arsenic removal: comparative study with single metal oxides. *Separation Science and Technology*, 45, 1975–1981.

Hong, H.J., Yang, J.S., Kim, B.K. & Yang, J.W. (2011) Arsenic removal behavior by Fe-Al binary oxide: thermodynamic and kinetic study. *Separation Science and Technology*, 46 (16), 2531–2538.

Igwe, J.C. & Abia, A.A. (2007) Adsorption kinetics and intraparticulate diffusivities for bioremediation of Co(II), Fe(II) and Cu(II) ions from waste water using modified and unmodified maize cob. *International Journal of Physical Sciences*, 2 (5), 119–127.

Inglezakis, V.J. & Poulopoulos, S.G. (2006) Adsorption and ion exchange. Chapter 4 in: Inglezakis, V.J. & Poulopoulos, S.G. (eds.) *Adsorption, ion exchange and catalysis; design of operations and environmental applications*. Elsevier Science Publishers, Amsterdam, The Netherlands. p. 269.

Jain, J.K. & Singh, R.D. (2012) Review: technological options for the removal of arsenic with special reference to South East Asia. *Journal of Environmental Management*, 107, 1–18.

Jiang, J.Q. & Zeng, Z. (2003) Comparison of modified montmorillonite adsorbents. Part II: The effects of the type of raw clays and modification conditions on the adsorption performance. *Chemosphere*, 53, 53–62.

Jiménez-Cedillo, M.J., Olguín, M.T. & Fall, Ch. (2009) Adsorption kinetic of arsenates as water pollutant on iron, manganese and iron-manganese-modified clinoptilolite-rich tuffs. *Journal of Hazardous Materials*, 163, 939–945.

Jiménez-Cedillo, M.J., Olguín, M.T., Fall, Ch. & Colin, A. (2011) Adsorption capacity of iron- or iron-manganese-modified zeolite-rich tuffs for As(III) and As(V) water pollutants. *Applied Clay Science*, 54, 206–216.

Kumar, E., Bhatnagar, A., Ji, M., Jung, W., Lee, S.-H., Kim, S.-J., Lee, G., Song, H., Choi, J.-Y., Yang, J.S. & Jeon, B.H. (2009) Defluoridation from aqueous solutions by granular ferric hydroxide (GFH). *Water Research*, 43, 490–498.

Langmuir, I. (1918) The adsorption of gases on plane surfaces of glass, mica and platinum. *Journal of the American Chemical Society*, 40, 1361–1403.

Li, Z., Jean, J.S., Jiang, W.T., Chang, P.H. & Chen, C.J. (2011) Removal of arsenic from water using Fe-exchanged natural zeolite. *Journal of Hazardous Materials*, 187, 318–323.

Luengo, C., Puccia, V. & Avena, M. (2011) Arsenate adsorption and desorption kinetics on a Fe(III)-modified montmorillonite. *Journal of Hazardous Materials*, 186, 1713–1719.

Mahmood, T., Din, S.U., Naeem, A., Mustafa, S., Waseem, M. & Hamayun, M. (2002) Adsorption of arsenate from aqueous solution on binary mixed oxide of iron and silicon. *Chemical Engineering Journal*, 192, 90–98.

Mamindy-Pajany, Y., Hurel, C., Marmier, N. & Roméo, M. (2011) Arsenic (V) adsorption from aqueous solution onto goethite, hematite, magnetite and zero-valent iron: effects of pH, concentration and reversibility. *Desalination*, 281, 93–99.

Masue, Y., Loeppert, R.H. & Kramer, T.A. (2007) Arsenate and arsenite adsorption and desorption behavior on coprecipitated aluminum: iron hydroxides. *Environmental Science & Technology*, 41, 837–842.

Mohan, D. & Pittman, C.U., Jr. (2007) Arsenic removal from water/wastewater using adsorbent – a critical review. *Journal of Hazardous Materials*, 142, 1–53.

Moore, J.W. & Pearson, R.G. (1981) *Kinetics and mechanisms*. 3rd edition. John Wiley & Sons, New York, NY.

Özcan, A., Öncü, E.M. & Özcan, A.S. (2006) Kinetics, isotherm and thermodynamic studies of adsorption of Acid Blue 193 from aqueous solutions onto natural sepiolite. *Colloids and Surfaces A*, 277, 90–97.

Pan, B.J., Qiu, H., Pan, B.C., Nie, G.Z., Xiao, L.L., Lv, L., Zhang, W.M., Zhang, Q.X. & Zheng, S.R. (2010) Highly efficient removal of heavy metals by polymer-supported nanosized hydrated Fe (III) oxides: behavior and XPS study. *Water Research*, 44, 815–824.

Pigna, M., Krishnamurti, G.S.R. & Violante, A. (2006) Kinetics of arsenate sorption-desorption from metal oxides: effect of residence time. *Soil Science Society of American Journal*, 70, 2017–2027.

Ramesh, A., Hasegawa, H., Maki, T. & Ueda, K. (2007) Adsorption of inorganic and organic arsenic from aqueous solutions by polymeric Al/Fe modified montmorillonite. *Separation and Purification Technology*, 56, 90–100.

Sharma, A.K., Tjell, C.J., Sloth, J.J. & Holm, P.E. (2013) Review of arsenic contamination, exposure through water and food and low cost mitigation options for rural areas. *Applied Geochemistry*, 41, 11–33.

Šiljeg, M., Cerjan Stefanovic, Š., Mazajb, M., Novak Tušar, N., Arćon, I., Kovac, J., Margeta, K., Kaucic, V. & Zabukovec Logar, N. (2009) Structure investigation of As(III)- and As(V)-species bound to Fe-modified clinoptilolite tuffs. *Microporous and Mesoporous Materials*, 118, 408–415.

Silva, J., Mello, J.W.V., Gasparon, M., Abrahão, W.A.P., Ciminelli, V.S.T. & Jong, T. (2010) The role of Al-goethites on arsenate mobility. *Water Research*, 44, 5684–5692.

Smedley, P.L. & Kinniburgh, D.G. (2002) A review of the source, behaviour and distribution of arsenic in natural waters. *Applied Geochemistry*, 17, 517–568.

Sparks, D.L., (1999) Kinetics of sorption/release reactions at the soil mineral/water interface. In: Sparks, D.L. (ed.) *Soil physical chemistry*. 2nd edition. CRC Press, Boca Raton, FL. pp. 135–191.

Thirunavukkarasu, O.S., Viraraghavan, T., Subramanian, K.S. & Tanjore, S. (2002) Organic arsenic removal from drinking water. *Urban Water*, 4 (4), 415–421.

Velićković, Z., Vuković, G.D., Marinković, A.D., Moldovan, M.S., Perić-Grujić, A.A., Uskoković, P.S. & Ristić, M.D. (2012) Adsorption of arsenate on iron(III) oxide coated ethylenediamine functionalized multiwall carbon nanotubes. *Chemical Engineering Journal*, 181–182, 174–181.

Wang, S. & Peng, Y. (2010) Natural zeolites as effective adsorbents in water and wastewater treatment. *Chemical Engineering Journal*, 156, 11–24.

Weber, J.W.J. & Morris, J.C. (1963) Kinetics of adsorption on carbon from solution. *Journal of the Sanitary Engineering Division*, 89 (2), 31–60.

Xu, Y., Nakajima, T. & Ohki, T. (2002) Adsorption and removal of arsenic(V) from drinking water by aluminum-loaded Shirasu-zeolite. *Journal of Hazardous Materials*, B92, 275–287.

Yadanaparthi, S.K.R., Graybill, D. & Wandruszka, R. (2009) Adsorbents for the removal of arsenic, cadmium, and lead from contaminated waters. *Journal of Hazardous Materials*, 171, 1–15.

Zhang, S., Niu, H., Cai, Y., Zhao, X. & Shi, Y. (2010) Arsenite and arsenate adsorption on coprecipitated bimetal oxide magnetic nanomaterials: $MnFe_2O_4$ and $CoFe_2O_4$. *Chemical Engineering Journal*, 158, 599–607.

Zhang, Y., Yang, M. & Don, X.M. (2005) Arsenate adsorption on a Fe-Ce bimetal oxide adsorbent: role of surface properties. *Environmental Science & Technology*, 39, 7246–7253.

CHAPTER 7

Low-cost adsorbents for arsenic separation from wastewaters

Evgenia Iakovleva, Marjatta Louhi-Kultanen & Mika Sillanpää

7.1 INTRODUCTION

Arsenic is a naturally occurring metalloid, widely distributed in rocks and soils, natural waters, air, and in small amounts in all living organisms. In the Earth's crust, arsenic (As) is rarely found as a pure metal, but often as a component in sulfur-containing minerals, the most common of which is arsenopyrite. Commercially, As is produced as arsenic trioxide or as a pure metal. It is used in pesticides, wood preservatives, and metal alloys production. Leading As producers in the world are China, Chile, Morocco and Peru. In addition to production from arsenopyrite and other minerals, As is also produced as a by-product of the copper and gold mining industries during treatment of sulfur-containing ores.

7.1.1 Arsenic removal mechanisms

To find the most efficient adsorbent for As removal, it is necessary to know its chemical properties and the mechanism of removal. Arsenites dominate in reducing aqueous environments, and arsenates in oxidation environments. Arsenites are less active kinetically. Therefore, for their removal, initial preoxidation to arsenates with ozone, chlorine or hydrogen peroxide is usually used. Various strains of bacteria accelerate the oxidation of arsenite to arsenate and transform inorganic As to alkyl- and methyl-arsenic compounds.

The behavior of As in aqueous solutions depends on its oxidation state, as determined by studies (Lizama *et al.*, 2011; Nguyen *et al.*, 2009). Arsenate ions are easily fixed with components, such as clay particles, phosphate gels, humus and calcites. Aluminum and iron hydroxide groups on the adsorbent surface are extremely effective for binding arsenates. Figure 7.1 shows the interaction mechanism of aluminum hydroxide with arsenate. Arsenic bound with iron or aluminum oxides can be released easily through hydrolysis by lowering the redox potential. However, desorption of As from clay and calcite adsorbents is difficult.

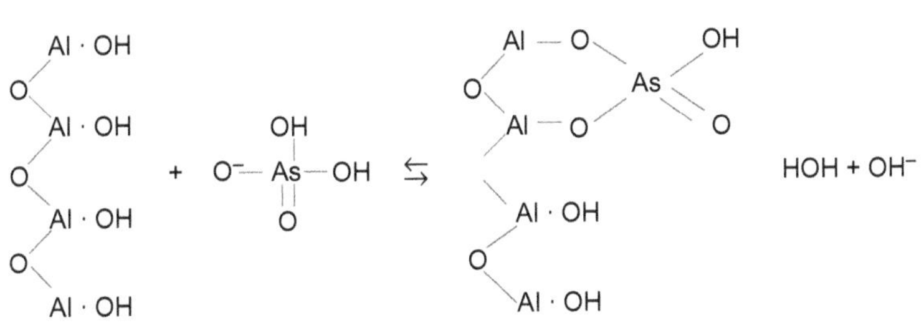

Figure 7.1. The interaction mechanism of aluminum hydroxide and As(V).

7.2   METHODS OF ARSENIC EXTRACTION

To eliminate health risks, a thorough treatment of As containing wastewaters is necessary. There are several methods of wastewater treatment aimed at removing As, such as:

- Precipitative processes
  - ○ Coagulation/filtration
  - ○ Coagulation assisted microfiltration
  - ○ Enhanced coagulation
  - ○ Lime softening
- Ion exchange
- Photo-oxidation
- Membrane processes
  - ○ microfiltration
  - ○ ultrafiltration
  - ○ nanofiltration
  - ○ reverse osmosis
  - ○ electrodialysis reversal
- Adsorptive process
  - ○ Fe/Mn oxidation
  - ○ activated alumina
  - ○ iron oxide coated sand
  - ○ sulfur-modified iron
  - ○ granular ferric hydroxide
  - ○ iron filling

Adsorption is the cheapest, easiest and most environmentally friendly process. Recently, the use of nanoparticles for As removal from water has become popular. Nowadays it is a promising technology allowing purifying water from As and other pollutants. On the other hand, the synthesis of nanoparticles is still expensive. Below, the main groups of low-cost adsorbents for wastewater purification from As are discussed, and a description of their pros and cons and a comparative analysis of these adsorbents are presented.

7.3   LOW-COST ADSORBENTS AND THEIR EFFICIENCY

In addition to affordable price, in order to be applied commercially adsorbents must meet other requirements. They need to be selective to As, be able to extract as without precipitation, and there should be an inexpensive way to regenerate them for reuse. Cheap adsorbents, such as by-products of agriculture and industries, various strains of bacteria, chitosanes, zeolites, and nano-particles produced by non-expensive methods, are discussed in this chapter.

7.3.1   *Biosorbents*

Agricultural by-products: Agricultural by-products, such as rice polish, fish scale, chicken fat, charcoal or coconut fiber, may be good alternatives for commercial adsorbents, because they are cheap and are produced in large quantities year after year. Their main advantage is that for use as adsorbents they require minimal pretreatment. For example, rice polish was subjected to double washing with water, crushing and sieving to $<178\,\mu m$ before use for As adsorption (Ranjan *et al.*, 2009). Initial As concentration was $1000\,\mu g\,L^{-1}$, adsorbent concentration $20\,g\,L^{-1}$. Less than 60 min for As(III) and 40 min for As(V) were needed for the equilibrium to be established, and removal efficiency of As(III) and As(V) was 95 and 97%, respectively. The sorption capacity for As(III) was found to be $138.88\,\mu g\,g^{-1}$ ($20°C$, $pH = 7.0$), for As(V) – $147.05\,\mu g\,g^{-1}$ ($20°C$, $pH = 4.0$). These data allow us to conclude that rice polish might be a good biosorbent for wastewater purification and can be recommended for successful As removal in developing countries (Ranjan *et al.*, 2009).

Fish scale, chicken fat, coconut fiber and charcoal were tested as adsorbents for As(III) and As(V) removal (Rahaman *et al.*, 2008). It appeared that only fish scale effectively extracted both As(V) and As(III) forms with removals of 90 and 82%, respectively. On the other hand, each of four adsorbents removed As(III) with different efficiency, observed rates being 82, 55, 50 and 23% with fish scale, coconut fiber, charcoal and chicken fat, respectively. At the same time, As(V) separation from liquid was only successful with fish scale and chicken fat. The maximum adsorption capacities of fish scale, estimated from the Langmuir adsorption model, are 26.67 and 24.75 $\mu g\,g^{-1}$ for As(V) and As(III), respectively. The optimum pH value for removing both As(III) and As(V) was determined as 4.0. Fish scales demonstrated high removal efficiencies in the range of initial As concentrations of 200–1000 $\mu g\,L^{-1}$. Although this concentration range is often prevalent in groundwater, fish scale can be tested as a potential adsorbent for As(V) and As(III) separation from wastewaters (Rahaman *et al.*, 2008).

Agricultural by-products are potentially prospective adsorbents for As removal. These materials are produced in immense amounts around the world, and hence sufficient quantities of adsorbents could be available. However, for each potential adsorbent, further detailed study of their capacity and properties is needed.

7.3.2   *Bacteria*

As mentioned above, arsenites are thermodynamically less active than arsenates. Bacteria are able to oxidize As(III) to As(V), enabling its further removal from the solution. In a review by Wang and Chen (2009), different classes of bacteria are observed for As removal from wastewaters. Some strains showed good adsorption properties, for example, *Penicillium canescens* and *Penicillium purpurogenum* have a biosorption capacity 26.4 and 35.6 mg g^{-1}, respectively. So they can be considered promising adsorbents for As removal.

In a study by Srivastava *et al.* (2011), it was observed that the adsorption of positive metal ions increases with the increasing pH of a solution, because more functional groups for metal ion binding are available due to deprotonation, resulting in high adsorption at higher pH values.

The use of bacteria might be a viable alternative technology for the removal of As(III) and As(V) from groundwaters (Kao *et al.*, 2013). The technology is based on bacterial oxidation and further removal of As(V). The study allows us to conclude that oxidation of As(III)-contaminated groundwater by a native isolated bacterium, followed by As(V) removal using bacterial biomass, is potentially effective and cost-efficient technology for treatment of As(III)-contaminated water.

Obviously the method of As removal with various strains of bacteria is environmentally non-hazardous and low-cost. Nevertheless, further research investigating the potential impact of used bacteria on the environment is needed.

7.3.3   *Industrial by-products*

The management of solid and liquid wastes is a global problem of modern society. Waste management is a process aimed at reducing the effect of wastes on human health and the environment. The mining industry produces large amounts of wastes, and the issue of waste treatment arises constantly. On the other hand, wastewaters require more effective and low-cost treatment methods. One of the most advantageous ideas for modern industries is waste-free production, where wastes are used for wastewater treatment at the same factories they are generated in. In a review by Iakovleva and Sillanpää (2013), various industrial wastes for purification of mining wastewaters from different toxic elements, including As, are described in detail. The most effective industrial wastes are considered in this chapter.

7.3.3.1   *Red mud*

Red mud or red sludge is highly alkaline with a pH ranging from 10 to 13 as solid waste from the Bayer process for alumina production. Red mud is composed of a mixture of metallic oxides, mainly iron oxide (up to 60%). The annual generation of sludge in the world is 82 million tons. Red mud is a problem, as it cannot be disposed of easily. Ongoing research aims at finding alternative uses for the sludge, for example, as an agent for increasing the alkalinity of wastewater, as an adsorbent for specific toxic elements, or as raw material for new geopolymer-like compounds, etc. (Hajjaji *et al.*, 2013).

Ferrous-based red mud sludge can be used as a low-cost and effective material for As removal by Fe-As co-precipitation and adsorption. Phosphate can greatly reduce the As removal efficiency while the presence of carbonate had no significant effect, as was observed in a study by Li, Y. *et al.* (2012).

Red mud, which is a waste from bauxite processing, has been explored as an adsorbent for As removal from wastewaters. Soner *et al.* (2000) have also confirmed the influence of pH on As(III) and As(V) removal. Tests showed that As(III) removal was successful at pH 9.5, whereas pH ranges from 1.1 to 3.2 promoted As(V) removal. The adsorption capacity of red mud was 4.32 and 5.07 μmol g^{-1} for As(III) and As(V), respectively. The initial concentration of As was 133.5 μmol g^{-1}, and 20 g L^{-1} of adsorbent was used. The removal rates of As(III) and As(V) with 100 g L^{-1} of adsorbent were approximately 90 and 99%, respectively.

A study by Bertocchi *et al.* (2006) presents an investigation on the use of red mud for As removal from a disused mine tailings dam. The results of this study show that red mud performs best at low pH. The batch and columns tests demonstrate effective properties of red mud for As removal.

Many researchers have noted that modified red mud has the best adsorption capacity for As removal, as discussed below. Arsenate removal from a synthetic solution with chemically modified red mud (Bauxol) and activated Bauxsol coated sand by batch and column methods are observed in Genç-Fuhrman *et al.* (2005). The batch experiments indicate that adsorption equilibrium is reached in about 4 h, however slow adsorption will continue for at least 21 days. The column method was also applied for 21 days. The column method shows a better adsorption capacity than the batch one. Although modified fly ash also showed good adsorption properties, the disadvantage of this material is the length of time needed to remove arsenate from the solution to the threshold limit value (TLV). This fact increases the cost of the use of fly ash, unless it is used for a passive method of wastewater treatment.

The comparative analysis of As adsorption with unmodified and modified with FeCl$_3$ mud shows that modified one has a much higher adsorption capacity than unmodified material and its capacity was 68.5 mg g^{-1} (Zhang *et al.*, 2008). The adsorption capacity of the modified adsorbent measured after regeneration with NaOH shows that the material restores itself fully in the regeneration process.

Red mud unmodified and muds modified with phosphogypsum were studied as adsorbents for As bioavailability reduction (Lopes *et al.*, 2013). Modified red mud adsorbed As 3.5 times better than unmodified mud with adsorption capacities of 3333 and 150 mg kg^{-1}, respectively. The presence of Ca^{2+} increased the efficiency of As adsorption, leading to the formation of ternary complexes [Ca$_3$(AsO$_4$)$_2 \cdot$ 4H$_2$O]. Concurrently with studies involving the effect of Ca^{2+}, Cl$^-$ and HCO$_3^-$ on arsenate removal, the authors concluded that competition between HCO$_3^-$ and arsenate took place. A mixture including 75% red mud and 25% phosphogypsum showed the best adsorption capacity for As removal.

Red mud exhibits better sorption capacity than other by-products, in particular fly ash, probably because it contains larger amounts of Fe and Al oxides and hydroxides, which are generally used for wastewater treatment from As and other pollutants (Soner *et al.*, 2000).

Red mud, as a waste product, economically, is a very attractive material for adsorption of As. The important factor in removing various As forms is pH. For example, As(V) is removed at a narrow pH range, approximately from 1 to 3.5. At a pH of 4, a sharp decline in As(V) adsorption was observed. The advantage of red mud is the fact that it may be reused in certain metallurgical processes which use red mud as a source of iron. However, the difficulties in solid-liquid separation diminish the exploitability of red mud as an adsorbent. On the other hand, the liquid phase of red mud constituting a weak alkaline aluminate solution may be used for As removal by coagulation.

7.3.3.2 *Fly ash*

Fly ash generated at a coal-fired power station was investigated for remediating contaminated mine sites of As and other toxic elements (Bertocchi *et al.*, 2006). Tests were carried out by batch and column methods. Column tests showed that fly ash significantly reduced the amounts of As, Cd, Cu, Pb and Zn in wastewaters. These tests were conducted in comparison with red mud as adsorbent material. Red mud is more effective than fly ash, apparently due to the presence of Fe/Al oxides and hydroxides alongside with a larger surface area and stability in a neutral and acidic solution. These properties make red mud one of the best candidates for wastewater treatment materials.

The activation of fly ash and its further use as an adsorbent for As removal was studied by Li, Y. *et al.* (2009). In their study, the examined As concentration was 50 mg L^{-1}, corresponding to the average concentration in industrial wastewater, and the concentration of modified fly ash was 40 g L^{-1}. For modification, the authors used fly ash with high iron content and activated it with 1 M HCl and NaOH solutions. Modification allowed the surface area of the adsorbent to be increased by up to 22 times and particle size to be increased. The increase of the particle size facilitates the separation of the liquid and solid phases after the adsorption process. It seems that modified fly ash has a variety of advantages over traditional iron-oxide adsorbents for As removal, such as:

- avoiding the addition of an extra iron source thus reducing the cost of As removal;
- increasing the particle size and making it easy to separate the end product from an aqueous system after adsorption activity;
- changing the surface structure of powder particles into porous structure, thus increasing the surface area and promoting As removal capacity;
- easy preparation since all the chemical reactions occur under ordinary conditions.

However, the modification process requires large amounts of purified water: (i) for the preliminary washing of the adsorbent, with the object of removing alkali and alkaline earth elements from the material so as to reduce the consumption of hydrochloric acid in the process of adsorbent synthesis; (ii) for the final washing of the prepared adsorbent. The use of large quantities of purified water is a serious disadvantage of this method, unless it would be possible to reuse this water in production cycles at a particular plant (Li, Y. *et al.*, 2009).

Another method of fly ash modification has been studied by Balsamo *et al.* (2010). Fly ash modified with HCl was compared to unmodified fly ash for arsenate removal from synthetic wastewater. The results show that the modified fly ash has a better adsorption capacity. This can probably be explained by oxidation caused by the HCl solution on the adsorbent's surface. The authors specified that the material should be thoroughly washed prior to use, because otherwise As was released from the adsorbent (about 0.05 mg g^{-1}) during the batch test of leaching fly ash with suprapure water. This can be a serious obstacle for the use of fly ash as an adsorbent.

Modification of fly ash with magnesia and manganese for arsenate removal has been observed by Li, Q. *et al.* (2012). Modification with magnesia was found to be better in every aspect, such as equilibrium time, optimum amount of adsorbent, removal capacity, and interlocking with competing anions (carbonate and dihydric phosphate).

In contrast to red mud, the As removal capacity of fly ash increases with increasing pH of the solution. This material can be an alternative low-cost adsorbent for As removal from wastewaters; however, some principal limitations must be taken into account. Firstly, in some cases, desorption of As from the material can occur, because fly ash generated as a waste at thermal power plants is a potentially significant anthropogenic source of As (Pandey *et al.*, 2011). Secondly, small particles of fly ash are difficult to separate from liquids before the wastewater treatment process. The surface area and particle size need to be increased, and this can be accomplished by modification.

In contrary to red mud, the adsorption ability of fly ash increases with increasing pH for As removal, which makes the material attractive in cases when red mud cannot be used. Some principal limitations thus are needed to be taken into account. First, possibilities of desorption of

As and second, small particles of fly ash making difficult solid-liquid separation before treatment process. Modification is able not only to increase the surface area of material but also to increase particle size.

7.3.3.3   Pyrite

Waste pyrite was tested as adsorbent for As adsorption by (Bulut *et al.*, 2014). The adsorption conditions were studied and tested in process water using dried ground pyrite. Removal of As was effective and running at 99%. FTIR analysis showed characteristic spectral bands indicating that As was chemically adsorbed by the pyrite surface through formation of scorodite. High As uptake capacity and cost-effectiveness make waste pyrite a potentially attractive material for As removal.

7.3.3.4   Hydroxide, oxides and oxide complexes

A detailed study of As removal using granular ferric hydroxide (GFH) in a controlled batch system was conducted by Banerjee *et al.* (2008). It was determined that temperature and pH have an important role in the kinetic mechanism of As adsorption. The authors mentioned that with an increase in temperature, the uptake rates of As(III) and As(V) also increased. As(V) exhibits greater removal rates than As(III) at lower pH levels. As(V) is best removed in a protonated liquid form; however, the capacity for adsorption of As(III) is greater when neutral surface sites are predominant.

The removal process of As(III) is usually more difficult compared to that of As(V). A study by Wu *et al.* (2013) demonstrates, however, that by-products of Fe-removal plants have high effectiveness for As(III) removal from synthetic wastewater with an initial concentration of As of 1–$120\,mg\,L^{-1}$. The maximum adsorption amount of As(III) was $59.7\,mg\,g^{-1}$. The uptake of As(III) by Fe(III) hydroxide arises from the ligand exchanges between As and sulfate.

Similar results were obtained by Manna and Ghosh (2007). Hydrated stannic oxide showed a better adsorption capacity for As(III) removal in comparison to As(V). These experiments were conducted for drinking water purification, and adsorption volumes were 15.85 and $4.3\,mg\,g^{-1}$ for As(III) and As(V), respectively (see Table 7.1).

Some authors (Ardau *et al.*, 2013, Meng *et al.*, 2002) have studied the influence of phosphate, silicate, carbonate, and bicarbonate ions for As(III) and As(V) removal with iron hydroxides from synthetic wastewater. According to the results, phosphate and silicate ions could significantly reduce the removal of As(V) at high concentrations of these anions, and the removal of As(III) could be substantially reduced even low anion concentrations. Other metal ions had moderate effect on the As removal.

Bimetal hydroxides seem to be good adsorption materials for metals and As removal. The cost of these adsorbents is much lower than that of commercial oxide adsorbents. For example, a low-cost adsorbent-based on iron and cerium was synthesized and studied by Zhang *et al.* (2010a; 2010b). Iron-cerium bimetal hydroxide was prepared by co-precipitation and contained 80% of Fe and 20% of Ce oxides. Adsorption from synthetic wastewater with initial concentration of As(V) $1\,mg\,L^{-1}$ was studied both in batch and column. The material exhibits adsorption characteristics similar to those of commercial adsorbents. However, larger size particles can more easily be separated from the liquid phase. A desorption study showed that 89% of As can be released, after which the adsorbent can be reused.

Relatively limited information is available regarding the impact of temperature on the adsorption kinetics and equilibrium capacities of granular ferric hydroxide (GFH) for As(V) and As(III) removal from aqueous solutions.

Alongside with hydroxides, Fe oxides as well as Fe and other metal oxides have been used for extraction of As from wastewaters.

Another candidate for a low cost adsorbent is Fe(III) oxide processed with slag from a municipal solid waste incinerator (Zhang and Itoh, 2005). Removal of both As(III) and As(V) was relatively low – approximately 70%. But the fact that the adsorbent is cheap, environmentally acceptable and potentially effective makes further investigation of this adsorbent preparation technique quite promising.

Table 7.1. Data of arsenic adsorption with different adsorbents from wastewaters.

Sorbent	Concentration of As	pH	T [°C]	Adsorbent dose	Adsorption capacity	Removal [%]	References
Rice polish	1000 μg L⁻¹	As(III) 7 / As(V) 4	20	20 g L⁻¹	138.88 μg g⁻¹ / 147.05 μg g⁻¹	95 / 97	Ranjan et al. (2009)
Fish scale	200–1000 μg L⁻¹	4	20		24.75 μg g⁻¹ (As(III)) / 26.67 μg g⁻¹ (As(V))	82 / 90	Rahaman et al. (2008)
Penicillium canescens					26.4 μg g⁻¹		Wang and Chen (2009)
Penicillium purpurogenum					35.6 μg g⁻¹		Wang and Chen (2009)
Red mud	133.5 μmol g⁻¹	As(III) 9.5 / As(V) / 1.1–3.2	20	20 g L⁻¹	4.32 μmol g⁻¹ / 5.07 μmol g⁻¹	90 / 99	Soner et al. (2000)
Bauxsol-coated sand	10 mg L⁻¹	2.6	20	10 g L⁻¹		68	Bertocchi et al. (2006)
Activated Bauxsol-coated sand	10 mg L⁻¹	4.5 / 7 / 7	20	5 g L⁻¹	3.32 mg g⁻¹ / 1.64 mg g⁻¹ / 2.14 mg g⁻¹		Genç-Fuhrman et al. (2005)
As(V)	100–1300 μmol L⁻¹	5.5	24	0.3 g L⁻¹	250–3000 mg kg⁻¹	75–100	Lopes et al. (2013)
As(III)	133.5 μmol L⁻¹	3.2 / 9.5	24	20 g L⁻¹	5.07 μmol g⁻¹ / 4.32 μmol g⁻¹	95 / 90	Soner et al. (2000)
Fly ash	10 g L⁻¹	7.5	20	10 g L⁻¹	0.05 g g⁻¹	98	Balsamo et al. (2010)
	500 μg L⁻¹	8	25	2 g L⁻¹	190 μg g⁻¹	99	Li, Q. et al. (2012)
Pyrite	10 mg L⁻¹	5		5 g L⁻¹		99	Bulut et al. (2014)
Fe(III) hydroxide, oxides and oxide complexes	100 μg L⁻¹	6.5	20	2.5 mg L⁻¹		99	Banerjee et al. (2008)
As(III)	5 g L⁻¹	7	25	0.6 g L⁻¹	59.7 mg g⁻¹		Wu et al. (2013)
As(V)	1 mg L⁻¹	7	27		4.30 mg g⁻¹	95	Manna and Ghosh (2007)
As(III)					15.85 mg g⁻¹	80	
As(V)	10 mg L⁻¹	5	25		18.2 mg g⁻¹	80	Zhang et al. (2010a; 2010b)
As(V)	150 mg L⁻¹	2.5	20	4 g L⁻¹	10 mg g⁻¹	100	Zhang and Itoh (2005)
As(III)					35 mg g⁻¹	75	
As(V)	50 g L⁻¹	3		2.5 g L⁻¹	38 mg g⁻¹	94	Repo et al. (2012)
As(III)		7			55 mg g⁻¹	99	
As(V)	25 mg L⁻¹	10		4 g L⁻¹		95	Ahn et al. (2003)
As(III)		9				90	
As(V)	10 μM	7	25	20 g L⁻¹		100	Rahman et al. (2013)

(continued)

Table 7.1. Continued.

Sorbent	Concentration of As	pH	T [°C]	Adsorbent dose	Adsorption capacity	Removal [%]	References
Clay minerals and silicates As(V), As(III)	10 mg L^{-1}	2	25	10 g L^{-1}	3.39 mg g^{-1}	95	Oh et al. (2012)
Calcite	50 μM	8.3	25	0.5 g L^{-1}	20 mg g^{-1}	83	Alexandratos et al. (2007)
		4	25		1.73 mg g^{-1}		Markovski et al. (2014)
CaCO$_3$ + αFeOOH					21.0 mg g^{-1}	90	
CaCO$_3$ + αMnO$_2$					10.36 mg g^{-1}	90	
CaCO$_3$ + αFeOOH + αMnO$_2$					42.0 mg g^{-1}		
Zeolites							
As(V)	750 mg L^{-1}	7	25	1 g L^{-1}		99	Medina et al. (2010)
As(V)	2 mM	6	20	0.01 M	40 μmol g^{-1}	75	Shukla et al. (2013)
As(V)	2 mg L^{-1}	10	60	100 g L^{-1}	0.007 mg g^{-1}		Jiménez-Cedillo et al. (2011; 2013)
As(III)							
As(V)	5 mg L^{-1}	5		1 g L^{-1}	0.026 mg g^{-1}		Simsek et al. (2013)
2Na		5			1.50 mg g^{-1}		
2Na-Al		3			3.86 mg g^{-1}		
2Na-AlFe(III)		3			3.02 mg g^{-1}		
2Na-Fe(III)							
Chitosan							
As(V)	400 mg L^{-1}	6	20	0.5 g L^{-1}	600 mg g^{-1}		Gérente et al. (2010)
As(V)	3000 μg L^{-1}	3	24	0.5 g L^{-1}	10000 μg g^{-1}		Kwok (2009), Kwok et al. (2014)
Nanoparticles							
As(V)	14.6 mg L^{-1}	7	–	6 g L^{-1}	82.7 mg g^{-1}		Zhang et al. (2013)
As(III)					122.3 mg g^{-1}		
Al-cryo (Al nanoparticles)		2	20		20 mg g^{-1}	100	Önnby et al. (2012)
MIP-cryo (polyacrylanide cryogel)		4			17 mg g^{-1}	92	

Table 7.2.   Composition [%] and properties of adsorbent form V.G. Khlopin Radium
Institute and Tampere University of Technology, respectively.

Al	1.7
Ca	14.4
Fe	7.2
K	0.3
S	17.6
Si	0.2
Specific surface area [$m^2 g^{-1}$]	62.5
Adsorption capacity [$cm^3 g^{-1}$]	0.13

Unmodified and modified at 25°C lepidocrocite was used for As removal in a study by Repo *et al.* (2012). The adsorption capacity for As(III) was 55 mg L^{-1} and for As(V) 38 mg L^{-1}, respectively. Taking into account that As(III) was removed by the unmodified (cheap) adsorbent quite effectively and the optimal pH was 7, it can be concluded that these adsorbents are practically applicable for As removal in the conditions studied.

According to a comparison analysis, steel mill waste materials are more effective for As separation from wastewaters than zero-valent Fe (Ahn *et al.*, 2003). This efficiency may be due to the presence of calcium, which promotes formation of calcium As compounds, this being more effective than adsorption by iron oxide. It is suggested that As may also be removed by adsorption onto iron oxides if the pH is lowered to near-neutral. The high separation rates of As (about 98%) allow to confirm that steel wastes as adsorbents can be placed in situ as permeable reactive barriers to control the subsurface release of leachate from tailing containment systems.

Iron-rich sand from household filters has also been tested for As removal from contaminated water. It showed good adsorption properties for As removal. The proposed process is a cost-effective scheme which includes the option of recycling the washing solvent in addition to the decontamination of the spent As sludge (Rahman *et al.*, 2013).

In another study, Fe-Mn binary-oxide was applied to treat As contaminated wastewater. The removal efficiency was about 99% for both As forms. Furthermore, poly-aluminum chloride (PACl) was added as a coagulant for the solid-liquid separation. The binary oxide with PACl showed high effectiveness, low cost, safety and easy processing for As separation (Wu *et al.*, 2013).

Promising initial results in the study of an industrial by-product, containing about 7% of iron oxides, 14% of Ca and 18% of S, as a potential adsorbent for As(III) and As(V) extraction in mining wastewater treatment were obtained (Iakovleva *et al.*, 2013). The composition of the adsorbent was studied with X-ray fluorescent analysis. The surface of the adsorbent material is multilayer and loose, providing a relatively large surface area (62.5 $m^2 g^{-1}$) and porosity (Table 7.2). The microstructure of the adsorbent was examined using a scanning electron microscope (SEM) (Figure 7.2). Non-modified adsorbent for As removal from synthetic mining process water with 10% NH_4 was used in the initial experiments.

Maximum adsorption capacities of the adsorbent were 0.18 and 0.13 $cm^3 g^{-1}$ for As(III) and As(V), respectively. Removal of As was effective: approximately 98% for As(III) and 93% for As(V). Further experiments are needed to study the influence of interfering components on As removal.

Iron oxides and hydroxides as well as iron bi- and polyoxides with other metals exhibit high affinity for As and can be effectively used for the removal of both As forms from contaminated waters under normal conditions.

S4800 30.0kV 7.8mm x5.00k 2012-05-28   10.0um

Figure 7.2.   SEM image of adsorbent microstructure.

7.3.4   Clay minerals and silicates

Steel-making slag, which contains about 36% Fe and 35% Ca, is a potential adsorbent for the extraction of both As forms. Slag removed 95–100% of the As from a solution with initial pH 2 (Oh *et al.*, 2012). One of the main advantages of the material is that only a small amount of toxic elements were leached from the slag during the pretreatment stage.

The As uptake mechanism of calcite has also been investigated (Alexandratos *et al.*, 2007). Arsenate ions show affinity for calcite surfaces. The optimal pH for the solution was found to be 8.3 (Table 7.1). These results are similar to those of arsenate adsorption by other minerals and phosphate adsorption by calcite. The results indicated that As(V) interacted strongly with the calcite surface, similarly to often-cited phosphate, and uptake could occur via both adsorption and coprecipitation. Calcite may be considered effective for partial removal of dissolved arsenate from aquatic and soil systems.

Three different methods of calcite modification with goethite (α-FeOOH), α-MnO$_2$, and mixture of goethite and α-MnO$_2$, were observed by Markovski *et al.* (2014). The modified adsorbents were compared with unmodified highly porous calcium carbonate. The highest adsorption capacity was accomplished by a series of composite adsorbents based on solvothermal synthesis of highly porous calcite and subsequent precipitation of goethite, α-MnO$_2$ and goethite/α-MnO$_2$. Their maximum adsorption capacities were 21 and 42 mg g^{-1}, correspondingly, while the unmodified and modified calcites by MnO$_2$ only exhibited adsorption capacities of 1.7 and 10 mg g^{-1}, respectively (Table 7.1). Application of ultrasound had a large impact on improving adsorption performance, whereas modification gave the best results with an optimal goethite and hybrid system goethite/α-MnO$_2$ loading.

Factors influencing arsenate sorption on calcites has been studied widely (Sø *et al.*, 2008). Sorption increases with decreasing alkalinity, indicating a competition over sorption sites between arsenate and (bi)carbonate. The pH also affects the sorption behavior, likely in response to changes in arsenate speciation or protonation/deprotonation of the adsorbing arsenate ion. Finally, sorption is influenced by the ionic strength, possibly due to electrostatic effects. Arsenites, unlike arsenates, are not sorbed effectively on calcites.

Calcites can be used for As(V) removal from wastewaters and AMD treatment, but they do not seem applicable for As(III). They can also be used as substrates for treatment of abandoned mines. To use them effectively, process conditions need to be selected carefully in order to avoid possible precipitation during the adsorption process.

7.3.5   *Zeolites*

Zeolites produced from highly reactive fly ash were studied as adsorbents for As(V) ($740\,\mu g\,L^{-1}$) removal from an aqueous solution of $Na_2HAsO_4\cdot7H_2O$ by Medina *et al.* (2010). Modified zeolite showed good adsorption capacity for As (99%). The authors proposed a method of surface area modification with KOH and NaOH to promote the adsorption of anionic species, and this could provide an advanced novel material for selective As and other anions removal.

However, Shukla *et al.* (2013) confirm that unmodified zeolite produced from fly ash does not have a good adsorption capacity for As. They proposed modification of zeolite with $Fe(NO_3)_3$. Iron enrichment of artificial zeolite changed the material's physical and chemical properties and allowed good adsorption properties for purifying liquids of As.

Shukla *et al.* (2013) also showed that modified natural zeolite has high adsorption ability for As(III) and As(V) removal from synthetic solutions.

A method where natural zeolites are modified with a NaCl solution and then with $FeCl_3$ and $MnCl_2$ accept to receive two types of adsorbents (zeolite/Fe and zeolite/Mn) with high adsorption capacities for As(III) and As(V) with an initial concentration of $2\,g\,L^{-1}$. The presence of Mn on the zeolite structure significantly enhances As removal (Jiménez-Cedillo *et al.*, 2011; 2013).

Experiments involving the addition of Fe and Al oxides to the structure of natural clinopteolite have also been conducted (Simsek *et al.*, 2013). The obtained adsorbents exhibited enhanced arsenate uptake capacities compared to the unmodified material. Maximum adsorption capacity was observed at pH 5.

Zeolites have a good capacity for As removal and should be examined in greater detail and extent for this field of application.

7.3.6   *Chitosan*

Chitosan is a linear polysaccharide produced by deacetylation of chitin from crabs and shrimp crustacean shells. Chitin is cheap and affordable; for example, in 2009–2010, 80,000 metric tons of chitosan and chitin were produced only in India as by-products of crab and fish processing. Being of biological origin, chitosan is easily decomposed, which makes it environmentally very attractive.

Some studies have considered chitosan an effective adsorbent for high concentrations of As in aqueous solutions (Gérente *et al.*, 2010; Kwok, 2009; Kwok *et al.*, 2014).

Kwok and McKay (2009) investigated removal of As with an initial concentration of up to $10\,mg\,L^{-1}$ using chitosan as an adsorbent. Adsorption equilibrium was reached after 30 min. Thereafter, slow increasing of the solution pH with simultaneous partial desorption of As was observed.

Chitosans have not been widely investigated for As removal from wastewater. The easy recycling of these materials is one of their presumed advantages compared to other materials.

7.3.7   *Nanoparticles*

In modern practice, nanoparticles and nanostructures have generated great interest in the scientific society. They are very effective in many areas, including wastewater treatment. Nevertheless, their practical application is still expensive. However, nowadays several low-cost methods for the production of nanoparticle-based adsorbents have been developed.

To obtain a highly efficient and low-cost adsorbent for As removal from water, a novel nano-structured Fe-Cu binary oxide (Cu:Fe molar ratio – 1:2) was synthesized via a facile co-precipitation method (Zhang *et al.*, 2013). The results indicated that the material had excellent performance in removing both As(V) and As(III) from water, and the maximal adsorption capacities for As(V) and As(III) were 82.7 and 122.3 $mg\,g^{-1}$ at pH 7.0, respectively. The values are favorable, compared to those reported in literature using other adsorbents. The coexisting sulfate and carbonate had no significant effect on As removal. However, the presence of phosphate, especially in high concentrations, obviously inhibited As removal. Moreover, the Fe-Cu binary oxide could be easily regenerated using NaOH solution, and subsequently reused.

Three varieties of adsorbents with nano- and micro-structures were studied for removal of As(V) by Önnby *et al.* (2012). The investigated materials include aluminum nanoparticles (Alu-NPs, <50 nm) incorporated in amine rich cryogels (Alu-cryo), and molecular imprinted polymers (<38 μm) in polyacrylamide cryogels (MIP-cryo). Both of these composites showed good adsorption properties; however, aluminum nanoparticles proved about three times more effective for As(V) removal than polyacrylamide cryogels. These materials also adsorbed well in competitive tests carried out in the presence of copper and zinc ions. The production method of aluminum nanoparticles observed in this work is inexpensive and can be applied commercially (Önnby *et al.*, 2012).

Maghemite nanoparticles, which are non-expensive adsorbents, were tested by Tuutijärvi *et al.* (2012) for arsenate removal in binary solutions containing various ions, such as sulfate, nitrate, phosphate and silicate. Phosphate and silicate ions were found to have a great effect on arsenate removal, while sulfate and nitrate ions affected removal only moderately.

7.4   CONCLUSION

In conclusion, it can be said that certain cheap adsorbents can be used to remove As from aqueous solutions with high efficiency. For example, calcite can be used for separation of As(V), and iron oxides and hydroxides have the greatest affinity for As(III). Adsorbents such as zeolites, chitosans and red mud are almost equally good for removal of both forms of As, provided the provisional modifications of adsorbents and adsorption conditions (pH, temperature, interfering components) are appropriate. Bacteria, in turn, are initially used for converting As(III) into As(V), followed by removal from the solutions. Each considered adsorbent and their optimal conditions of use require further studies.

ACKNOWLEDGEMENTS

The authors are grateful to the Finnish Funding Agency for Technology and Innovation (TEKES) for financial support. The authors thank Dr. Victoria Vergizova for her thoughtful and constructive review of this chapter.

REFERENCES

Ahn, J.S., Chon, C., Moon, H. & Kim, K. (2003) Arsenic removal using steel manufacturing byproducts as permeable reactive materials in mine tailing containment systems. *Water Research*, 37, 2478–2488.

Alexandratos, V.G., Elzinga, E.J. & Reeder, R.J. (2007) Arsenate uptake by calcite: macroscopic and spectroscopic characterization of adsorption and incorporation mechanisms. *Geochimica et Cosmochimica Acta*, 71, 4172–4187.

Ardau, C., Frau, F. & Lattanzi, P. (2013) New data on arsenic sorption properties of Zn-Al sulphate layered double hydroxides: influence of competition with other anions. *Applied Clay Science*, 80–81, 1–9.

Balsamo, M., Di Natale, F., Erto A., Lancia, A., Momtagnaro, F. & Santoro, L. (2010) Arsenate removal from synthetic wastewater by adsorption onto fly ash. *Desalination*, 263, 58–63.

Banerjee, K., Amy, G.L., Prevost, M., Nour, S., Jekel, M., Gallagher, P.M. & Blumenschein, C.D. (2008) Kinetic and thermodynamic aspects of adsorption of arsenic onto granular ferric hydroxide (GFH). *Water Research*, 42, 3371–3378.

Bertocchi, A.F., Ghiani, M., Peretti, R. & Zucca, A. (2006) Red mud and fly ash for remediation of mine sites contaminated with As, Cd, Cu, Pb and Zn. *Journal of Hazardous Materials*, 134, 112–119.

Bulut, G., Yenial, Ü., Emiroglu, E. & Sirkeci, A.A. (2014) Arsenic removal from aqueous solution using pyrite. *Journal of Cleaner Production*, 84, 526–532.

Genc-Fuhrman, H., Bregnhoj, H. & Mcconchie, D. (2005) Arsenate removal from water using sand-red mud columns. *Water Research*, 39, 2944–2954.

Gerente, C., Andres, Y., Mckay, G. & Le Clorec, P. (2010) Removal of arsenic(V) onto chitosan: from sorption mechanism explanation to dynamic water treatment process. *Chemical Engineering Journal*, 158, 593–598.

Iakovleva, E. & Sillanpää, M. (2013) The use of low-cost adsorbents for wastewater purification in mining industries. *Environmental Science and Pollution Research*, 20 (11), 7878–7899.

Iakovleva, E., Sitarz, M., Mäkilä, E. & Sillänpää, M. (2013) Solid wastes for wastewaters purification. *9th Fennoscandian Exploration and Mining Conference, 29–31 October 2013, Levi, Lapland, Finland.*

Jimenez-Cedillo, M.J., Olguin, M.T., Fall, C. & Colin-Cruz, A. (2011) Adsorption capacity of iron- or iron-manganese-modified zeolite-rich tuffs for As(III) and As(V) water pollutants. *Applied Clay Science*, 54, 206–216.

Jimenez-Cedillo, M.J., Olguin, M.T., Fall, C. & Colin-Cruz, A. (2013) As(III) and As(V) sorption on iron-modified non-pyrolyzed and pyrolyzed biomass from *Petroselinum crispum* (parsley). *Journal of Environmental Management*, 117, 242–252.

Hajjaji, W., Andrejkovicova, S., Zanelli, C., Alshaaer, M., Dondi, M., Labrinch, J.A. & Rocha, F. (2013) Composition and technological properties of geopolymers based on metakaolin and red mud. *Materials & Design*, 52, 648–654.

Kao A., Chu, Y., Hsu, F. & Liao, V.H. (2013) Removal of arsenic from groundwater by using a native isolated arsenite-oxidizing bacterium. *Journal of Contaminant Hydrology*, 155, 1–8.

Kwok, K.C.M. (2009) Novel model development for sorption of arsenate on chitosan. *Chemical Engineering Journal*, 151, 122–133.

Kwok, K.C.M., Koong, L.F., Chen, G. & Mckay, G. (2014) Mechanism of arsenic removal using chitosan and nanochitosan. *Journal of Colloid and Interface Science*, 416, 1–10.

Li, Q., Xu, X., Cui, H., Pang, J., Wei, Z., Sun, Z. & Zhai, J. (2012) Comparison of two adsorbents for the removal of pentavalent arsenic from aqueous solutions. *Journal of Environmental Management*, 98, 98–106.

Li, Y., Zhang, F. & Xiu, F. (2009) Arsenic (V) removal from aqueous system using adsorbent developed from a high iron-containing fly ash. *Science of the Total Environment*, 407, 5780–5786.

Li, Y., Wang, J., Peng, X., Ni, F. & Luan, Z. (2012) Evaluation of arsenic immobilization in red mud by CO_2 or waste acid acidification combined ferrous (Fe^{2+}) treatment. *Journal of Hazardous Materials*, 199–200, 43–50.

Lizama, K.A., Fletcher, T.D. & Sun, G. (2011) Removal process for arsenic in constructed wetlands. *Chemosphere*, 84, 1032–1043.

Lopes, G., Guilherme, L.R.G., Costa, E.T.S., Curi, N. & Penha, H.G.V. (2013) Increasing arsenic sorption on red mud by phosphogypsum addition. *Journal of Hazardous Materials*, 262, 1196–1203.

Manna, B. & Ghosh, U.C. (2007) Adsorption of arsenic from aqueous solution on synthetic hydrous stannic oxide. *Journal of Hazardous Materials*, 144, 522–531.

Markovski, J.S., Dokic, V., Milosavljevic, M., Mitric, M., Peric-Grujic, A.A., Onjia, A.E. & Marinkovic, A.D. (2014) Ultrasonic assisted arsenate adsorption on solvothermally synthesized calcite modified by goethite, α-MnO_2 and goethite/α-MnO_2. *Ultrasonics Sonochemistry*, 21, 790–801.

Medina, A., Gamero, P., Almanza, J.M., Vargas, A., Montoya, A., Vargas, G. & Izquierdo, M. (2010) Fly ash from a Mexican mineral coal. II. Source of W zeolite and its effectiveness in arsenic (V) adsorption. *Journal of Hazardous Materials*, 181, 91–104.

Meng, X., Korfiatis, G.P., Bang, S. & Bang, K.W. (2002) Combined effects of anions on arsenic removal by iron hydroxides. *Toxicology Letters*, 133, 103–111.

Nguyen, C.M., Bang, S., Cho, J. & Kim, K.-W. (2009) Performance and mechanism of arsenic removal from water by a nanofiltration membrane. *Desalination*, 245, 82–94.

Oh, C., Rhee, S., Oh, M. & Park, J. (2012) Removal characteristics of As (III) and As (V) from acidic aqueous solution by steel making slag. *Journal of Hazardous Materials*, 213–214, 147–155.

Önnby, L., Pakade, V., Mattiasson, B. & Kirsebom, H. (2012) Polymer composite adsorbents using particles of molecularly imprinted polymers or aluminum oxide nanoparticles for treatment of arsenic contaminated waters. *Water Research*, 46, 4111–4120.

Pandey, V.C., Singh, J.S., Singh, R.P., Singh, N. & Yunus, M. (2011) Arsenic hazards in coal fly ash and its fate in Indian scenario. *Resources, Conservation and Recycling*, 55, 819–835.

Rahaman, M.S., Basu, A. & Islam, M.R. (2008) The removal of As(III) and As(V) from aqueous solutions by waste materials. *Bioresource Technology*, 99, 2815–2823.

Rahman, I.M.M., Begum, Z.A., Sawai, H., Maki, T. & Hasegawa, H. (2013) Decontamination of spent iron-oxide coated sand from filters used in arsenic removal. *Chemosphere*, 92, 196–200.

Ranjan, D., Talat, M. & Hasan, S.H. (2009) Biosorption of arsenic from aqueous solution using agricultural residue "rice polish". *Journal of Hazardous Materials*, 166, 1050–1059.

Repo, E., Mäkinen, M., Rengaraj, S., Natarajan, G., Bhatnagar, A. & Sillanpää, M. (2012) Lepidocrocite and its heat-treated forms as effective arsenic adsorbents in aqueous medium. *Chemical Engineering Journal*, 180, 159–169.

Shukla, E.A., Johan, E., Henmi, T. & Matsue, N. (2013) Arsenate adsorption on iron modified artificial zeolite made from coal fly ash. *Procedia Environmental Science*, 17, 279–284.

Simsek, E.B., Özdemir, E. & Beker, U. (2013) Zeolite supported mono- and bimetallic oxides: promising adsorbents for removal of As(V) in aqueous solutions. *Chemical Engineering Journal*, 220, 402–411.

So, H.U., Postma, D., Jakobsen, R. & Larsen, F. (2008) Sorption and desorption of arsenate and arsenite on calcite. *Geochimica et Cosmochimica Acta*, 72, 5871–5884.

Soner, A.H., Altundogan, S., Tumen, F. & Bildik, M. (2000) Arsenic removal from aqueous solutions by adsorption on red mud. *Waste Management*, 20, 761–767.

Srivastava, P.K., Vaish, A., Dwivedi, S., Chakrabarty, D., Singh, N. & Tripathi, R.D. (2011) Biological removal of arsenic pollution by soil fungi. *Science of the Total Environment*, 409, 2430–2442.

Tuutijärvi, T., Repo, E., Vahala, R., Sillanpää, M. & Chen, G. (2012) Effect of competing anions on arsenate adsorption onto maghemite nanoparticles. *Chinese Journal of Chemical Engineering*, 20, 505–514.

Wang, J. & Chen, C. (2009) Biosorbents for heavy metals removal and their future. *Biotechnology Advances*, 27, 195–226.

Wu, K., Liu, R., Li, T., Liu, H., Peng, J. & Qu, J. (2013) Removal of arsenic(III) from aqueous solution using a low-cost by-product in Fe-removal plants – Fe-based backwashing sludge. *Chemical Engineering Journal*, 226, 393–401.

Zhang, F. & Itoh, H. (2005) Iron oxide-loaded slag for arsenic removal from aqueous system. *Chemosphere*, 60, 319–325.

Zhang, G., Ren, Z., Zhang, X. & Chen, J. (2013) Nanostructured iron(III)-copper(II) binary oxide: a novel adsorbent for enhanced arsenic removal from aqueous solutions. *Water Research*, 47, 4022–4031.

Zhang, S., Liu, C., Luan, Z., Peng, X., Ren, H. & Wang, J. (2008) Arsenate removal from aqueous solutions using modified red mud. *Journal of Hazardous Materials*, 152, 486–492.

Zhang, Y., Dou, X., Yang, M., He, H., Jing, C. & Wu, Z. (2010a) Removal of arsenate from water by using an Fe-Ce oxide adsorbent: effects of coexistent fluoride and phosphate. *Journal of Hazardous Materials*, 179, 208–214.

Zhang, Y., Dou, X., Zhao, B., Yang, M., Takayama, T. & Kato, S. (2010b) Removal of arsenic by a granular Fe-Ce oxide adsorbent: fabrication conditions and performance. *Chemical Engineering Journal*, 162, 164–170.

CHAPTER 8

Polymeric sorbents for selective chromium removal

Luděk Jelínek, Helena Parschová & Michal Němeček

8.1 INTRODUCTION

While the problem of water contamination with chromium (Cr) is sometimes narrowed only to the problem of contamination with hexavalent Cr species, the trivalent species are also important. The main reason why attention is focused on hexavalent chromium (Cr(VI)) is its toxicity and suspected carcinogenicity (Bagchi *et al.*, 2002). Trivalent chromium (Cr(III)) was for a long time considered to be an essential element for mammals, however, recent studies dispute this role (Di Bona *et al.*, 2011). Nevertheless, it is still used as a dietary supplement and its toxicity was found to be very low (Staniek *et al.*, 2011).

Chromium-containing wastewaters usually arise from metal plating (Tenório and Espinosa, 2001) and tanning (Balasubramanian and Pugalenthi, 1999) industries. Another source of Cr contamination is processing of Cr ores (Geelhoed *et al.*, 2003). Wastewaters from the tanning industry contain 100–500 mg L^{-1} of trivalent Cr which is accompanied by comparable concentrations of Ca, Mg and Na cations (Petruzzelli *et al.*, 1996).

Speciation of Cr in waters depends mainly on its oxidation state and the pH. With some simplification, Cr(III), is present in the cationic form, while Cr(VI), is present in the form of oxyanions. The speciation will be discussed in detail later. The high charge of the Cr^{3+} cation allows its separation by means of ion exchange. Similarly, chromates, which are analogous to sulfates, can be removed by anion exchange. The advantage of ion exchange, compared to the commonly used precipitation, is the attainment of lower concentrations of Cr in the treated water. In the case of alkaline precipitation, the residual Cr concentration in real solutions can be at the mg L^{-1} level (Almeida and Boaventura, 1998). Also, during the ion-exchange column regeneration, solutions containing high concentrations of Cr are obtained. Depending on the selectivity of the process, Cr species can be separated from accompanying ions, further improving the reusability of the Cr solution.

8.2 TRIVALENT CHROMIUM

8.2.1 *Speciation*

The dominant form of trivalent Cr in water at neutral pH is its electro-neutral hydroxo complex [Cr(OH)$_3$]^0 while at lower pHs of 5–6 it is the [Cr(OH)]$^{2+}$ (Icopini and Long, 2002; Richard and Bourg, 1991). Only at low pHs (<4) is hydrated Cr^{3+} cation predominant (Fukushima *et al.*, 1995). However, in nature, trivalent Cr is commonly bound to humic substances, forming anionic complexes (Fukushima *et al.*, 1995; Icopini and Long, 2002). In a similar manner, electro-neutral or anionic chelates with carboxylic acids, such as nitrilotriacetic acid (NTA), ethylenediaminetetraacetic acid (EDTA) and citric acid are formed (Kornev and Mikryukova, 2004). These complexes usually exhibit color described as violet or purple (Cherney *et al.*, 1954; Hamm, 1953). Carboxylic acids are used in the electroplating industry (Vinokurov *et al.*, 2005) and such complexes can therefore be found in industrial wastewaters.

As trivalent Cr can be present in solution in a variety of species of cationic, anionic or electro-neutral nature, it is necessary to choose suitable sorbent and working conditions to achieve its removal.

8.2.2   Ion exchangers

8.2.2.1   Strong acid cation exchangers

Strong acid cation exchangers (SAC) would be the logical choice for the removal of Cr cations from aqueous solution. The selectivity of strong acid cation exchangers is based only on electrostatic interactions. The selectivity coefficient of the Cr^{3+} cation is therefore close to that of Ca^{2+} (Bonner and Smith, 1957; Bonner et al., 1958). SAC are suitable for removal of target metal from solutions, which do not contain an excess of accompanying cations (such as Ca^{2+}, Mg^{2+}, Na^+) that would compete with Cr. This is especially true in the case of polishing and reuse of rinse water in the electroplating industry. In such cases, demineralized water is often used and there are therefore only trace amounts of accompanying cations.

The strong acid nature of SAC enables operation at low pHs, so acidic waste waters can be treated directly. The optimum pH for Cr(III) sorption onto the commercial SAC Lewatit S 100 was found to be 3.5 (Gode and Pehlivan 2006), which corresponds to the prevalence of the free Cr^{3+} cation. In another article, which deals with sorption of Cr(III) onto SAC Indion 790, removal of Cr(III) was constant in the pH range of 0.5 to 3.5, with a steep decrease of sorption efficiency observed above pH 3.5 (Sahu et al., 2009).

8.2.2.2   Weak acid cation exchangers

Compared to strong acid cation exchangers, weak acid cation exchangers (WAC) show an ability to form complexes with transition metal cations. The resulting bond is then a combination of electro-static and donor-acceptor interactions (Snukiškis et al., 2000). WAC prefer Cr(III) over alkaline and alkaline earth metals, which are commonly found in water. However, a disadvantage comes from the weak acid nature of WAC, which prevents them from operating at lower (roughly < 4) pHs.

Removal of Cr(III) using SAC was tested on solutions similar to that of tannery wastewater. It was capable of Cr removal but the regeneration was difficult. A novel method of oxidative stripping of Cr(III) species by their oxidation with alkaline hydrogen peroxide was found promising (Kocaoba and Akcin, 2004).

An interesting example of utilization of a weak acid cation exchanger is the use of a typical mixed bed of WAC Purolite S 106 and the weak base anion exchanger (WBA) Duolite A7 for removal of Cr from tannery waste water. Also, in this case, the regeneration of weak acid cation exchanger loaded with Cr(III) was proven to be difficult. The WBA resin acted as an acidity buffer and showed a synergistic effect on the column performance both in the sorption and regeneration runs (Petruzzelli et al., 1996).

8.2.2.3   Chelating resins

The most common chelating resin having iminodiacetic acid (IDA) as a functional group behaves similarly to carboxylic resins. Chelating resins are also limited by the dissociation of the functional group, which is more complicated. The first carboxylic group dissociates at about pH 4, the second carboxylic group dissociates at pH 7.4 and at an alkaline pH (12.3) the nitrogen in the IDA group deprotonates (Cohen and Heitner-Wirguin, 1969).

The optimum pH for removal of Cr(III) by the IDA chelating resin Chelex 100 was found to be 4.5 and the capacity of the resin calculated from the Langmuir isotherm was $0.29\,mmol\,g^{-1}$ of resin (Gode and Pehlivan, 2006) The optimum pH corresponds with the dissociation of the functional group and the speciation of Cr(III).

The same group of authors also investigated a chelating sorbent based on sporopollenin biopolymer functionalized with bis-diaminoethyl glyoxim. However, its sorption capacity was about two orders of magnitude lower than that of carboxylic acid functionalized sporopollerin (Gode and Pehlivan, 2007).

A composite sorbent based on polyvinyl alcohol, citric acid and chitosan was tested for Cr(III) removal. The optimum pH for sorption was found to be 6. The composite was regenerated with $0.5 \, mol \, L^{-1}$ sulfuric acid and was reused (Zuo and Balasubramanian, 2013).

8.3   CHROMATES

8.3.1   *Speciation*

Chromates are similar to other oxyanions in group 6 of the periodic table of elements. Similarly to molybdates and tungstates, chromates can polymerate at lower pHs and higher metal concentrations. The principal species existing in solution are chromates (CrO_4^{2-}), bichromates ($HCrO_4^-$) and dichromates ($Cr_2O_7^{2-}$). In a $0.145 \, mol \, kg^{-1}$ solution of Cr(VI), chromates prevail at pH higher than 6, while at lower pHs, dichromates prevail. In diluted acidic solutions, bichromates prevail (Hoffmann *et al.*, 2001). There are of course other species that can be encountered in industrial wastewaters.

In solution with a high excess of chlorides and at higher temperatures, CrO_3Cl^- species are present. At a temperature of $150°C$ in a solution containing $0.01 \, mol \, L^{-1}$ Cr(VI) and $5 \, mol \, L^{-1}$ of NaCl, CrO_3Cl^- are the prevailing species between pH 3 and 6 (Palmer *et al.*, 1987). Such species were also reported in hydrochloric acid together with the $CrSO_7^-$ species that were found in sulfuric acid (Haight *et al.*, 1964).

8.3.2   *Ion exchangers*

8.3.2.1   *Anion exchangers*
Anion exchangers, both strongly and weakly basic, are commercially available with a variety of functional groups and matrices (gel, macroporous, acrylate, etc.). Strong base anion exchangers (SBA) interact with the anions only electrostatically and are able to operate in the whole pH range encountered in wastewaters.

SBA having quaternary ammonium functional groups derived from trimethylamine are called type I, while SBA having functional groups derived from 2-(methylamino)ethanol are called type II. Type II SBA resin Purolite A 310E was found to be more effective than type I resin Purolite A 600 for the removal of low levels ($30 \, \mu g \, L^{-1}$) of Cr(VI) from water of mineralization $560 \, mg \, L^{-1}$ of TDS and pH 7.3 (Bahowick *et al.*, 2007).

Chromate removal can be a very interesting application for nitrate-selective SBA. They are similar to the type I but contain longer aliphatic chains (ethyl, propyl, or butyl) on the quaternary ammonium functional group. In a comparison of chromate sorption from tap water containing $3 \, mg \, L^{-1}$ of Cr(VI), nitrate-selective SBA Relite A 490 with type I and type II anion exchangers, the nitrate-selective resin clearly outperforms both types. Sulfates showed a stronger effect on chromate sorption compared to chlorides. Regeneration was carried out using a reduction step with sodium bisulfite followed by regeneration with hydrochloric or sulfuric acid. An accelerated stability test, carried out at $50°C$, showed decrease of the resin capacity loaded with a concentration of Cr(VI) in the range of $10–23 \, g \, L^{-1}$ to about 65% of the original value (Korngold *et al.*, 2003).

Another type of suitable strong base anion exchanger is a quaternized poly(4-vinylpyridine). These perform well and offer better resistance to oxidation than conventional poly(styrene-co-divinylbenzene)-based resins. The most important feature of quaternized poly(4-vinylpyridine) is its selectivity.

Weak base anion exchangers (WBA), which can also be also used for removal of chromates, are produced with a variety of functional groups and some of them can contain several amino groups to increase their capacity. Such ion exchangers, having multiple N donors in functional groups, can form strong chelates with transition metals, such as copper, and can be utilized for their removal (Hudson and Matejka 1989). An example can be the weak base anion exchanger Purolite A 830 with diethylenetriamine (DETA) functional groups. This ion exchanger has a nitrogen content of 16.88% (Wolowicz and Hubicki 2012). Their main disadvantage is caused by the necessity of protonation of the functional groups, which narrows the operational pH range. On the other hand, this problem facilitates their regeneration.

Lewatit MP 62, which is a weak base anion exchanger with tertiary amine functional groups, was compared with Lewatit M 610, which is a strong base anion exchanger (type II). Both anion exchangers were capable of Cr(IV) removal within the studied pH range 3–5.5. Lewatit M 610 showed stronger affinity towards chromates (Gode and Pehlivan, 2005). Unfortunately, sorption efficiency in a higher pH region, where the degree of protonation of the weak base anion exchanger would play a role, was not studied. It should be noted, that some weak base anion exchangers contain a proportion of strong base (quaternary ammonium) to increase swelling.

This is the case with Lewatit MP 64. It was compared to strong base anion exchanger (type I) Lewatit M 500, a high capacity weak anion exchanger with DETA groups Purolite A 830 and Purolite S 110 with N-methylglucamine. The breakthrough capacity of a model solution containing 5 mg L^{-1} of Cr(VI) was compared. As in the previous case, strong base anion exchanger Lewatit M 500 showed the highest affinity (breakthrough capacity), followed by Purolite A 830, Lewatit MP 64 and Purolite S 110 (which will be discussed in Section 8.3.2.2).

It is interesting that Purolite A 830 did not outperform the strong base anion exchanger. Adding to its high capacity, it also obviously formed a very stable complex with the DETA functional groups. This is shown by the regeneration attempt with 1 mol L^{-1} of NaOH followed by HCl, which yielded only 30% (Nemecek *et al.*, 2013).

The dominance of strong base anion exchangers is shown also in another study featuring Lewatit MP 64 and Lewatit MP 500, which is a macroporous variant of Lewatit M 500. Lewatit MP 500 exhibited 6 as the optimal pH for sorption. In the case of Lewatit MP 64 it was 5 (Pehlivan and Cetin, 2009).

A natural polymer that can be classified as a weak base anion exchanger, chitosan, was also tested for chromate removal. This polysaccharide, prepared by de-acetylation of chitin, is a close relative of N-methylglucamine sorbents, which will be discussed later.

Cross-linked chitosan was compared with modified cross-linked chitosan as a sorbent for chromates. Conditioning (protonating) of cross-linked chitosan with hydrochloric acid was proven to be an important step in greatly increasing its sorption capacity. Modification of chitosan with chloracetic acid to create carboxylated chitosan slightly increased its sorption ability. Substantial improvement of the sorption ability of chitosan was achieved by converting the carboxylic groups of carboxylated chitosan to amide with ethylenediamine, introducing another amino group to the structure of chitosan (Kousalya *et al.*, 2010).

In a recent study, chitosan was compared to activated carbon and carbon nanotubes. Its performance was close to that of activated carbon and it outperformed both single and multi-wall carbon nanotubes. After the sorption only about 2–3% of residual Cr was reduced to Cr(III) (Jung *et al.*, 2013).

8.3.2.2 *N-methylglucamine*

The N-methylglucamine (syn. methylamino-glucitol) functional group is commonly used for the removal of borates from aqueous solutions, where it exhibits great selectivity (Parschová *et al.*, 2007). It can, however be used for removal of metal oxyanions as well (Matejka *et al.*, 2001; 2004; Schilde and Uhlemann, 1993).

The selective sorbent Wofatit MK 51 having the N-methylglucitol functional group was tested for removal of various oxyanions of metalloids and metals (B, Al, Ga, Ge, Pb, V, Cr, Mo). With respect to chromates, the authors state, that chromates "... *show no tendency to form complexes with polyols* ..." (Schilde and Uhlemann, 1993).

The article, however, states important findings about color change during the sorption and desorption, which it ascribes to the redox reaction. Unfortunately, no mechanism of such a reaction is given (Schilde and Uhlemann, 1993). Also, a recent study states color change from yellow to green of chromate during sorption on sorbent Purolite S 110, which has the *N*-methylglucitol functional group. Also, the recovery ratio for the sequential regeneration with 1 mol L^{-1} of NaOH followed by HCl, was found to be 87% (Nemecek *et al.*, 2013).

8.3.2.3 *Polypyrrole*
Organic/organic composite, nanofibers with core-shell structure based on PAN and PPy, were tested for chromate removal in the batch process. After electro-spinning of PAN nanofibers, PPy monomer was polymerized on them. This material showed sorption properties similar to other PPy-based materials. The amount of Cr(VI) species adsorbed onto the composite decreased with increasing pH in the range of pH 2 to 5. Reduction of the Cr(VI) species to Cr(III) by the PPy was observed (Wang *et al.*, 2013). It should be noted, that in contrast to other sorbents such as *N*-methylglucamine, oxidation of PPy can be easily reversed. It can be electrochemically reduced/oxidized repeatedly (Weidlich *et al.*, 2001).

In the above mentioned study, the effect of accompanying ions was tested by additions of NaCl and Cu^{2+} and Ni^{2+} ions. In the concentration range 0 to 0.1 mol L^{-1} NaCl, the effect of increasing ionic strength was only mild, decreasing the adsorbed amount to the 88.3% of the amount adsorbed in the absence of accompanying ions. Cu^{2+} ions (and to a lesser extent also Ni^{2+} ions) were taken up by the PPy, decreasing the amount of Cr(VI) adsorbed. Due to the limited chemical stability of the nanofibers, NaOH solution of rather low concentration of 0.01 mol L^{-1} was found to be best for the regeneration of the material. The reuse of the material was tested 5 times. The material retained 80% of its original capacity (Wang *et al.*, 2013).

Polypyrrole (PPy) was also tested in the form of composites with inorganic materials. The sorption efficiency of a magnetic nanocomposite of PPy-coated Fe$_3$O$_4$ decreased almost linearly in the range of pH 2 to 11. In the regeneration of the composite with 0.5 mol L^{-1} NaOH solution only 14% of the adsorbed Cr(VI) was recovered. Unfortunately, this finding is not elaborated and is in contrast to the proposed sorption mechanism, which is simple ion exchange of chlorides in PPy$^+$Cl$^-$ for the Cr(VI) anionic species (Bhaumik *et al.*, 2011).

Another study deals with PPy-coated palygorskite. The findings are similar the study dealing with Fe$_3$O$_4$ composite. The ion-exchange mechanism was in this case corroborated by the presence of Cl$^-$ ions in the equilibrium solution (Yao *et al.*, 2012).

8.3.2.4 *Bi-functional chelating exchangers*
An interesting composite material is one based on silica having thiol and ethylene diamine triacetate as a functional groups. The thiol group acts as reducing agent while the ethylene diamine triacetate acts as chelating moiety for the Cr(III) species. It has a big advantage over thiol-functionalized silica. Interestingly, the optimal pH for chromate sorption is about 2.5 (Zaitseva *et al.*, 2013). It can be connected to the proposed oxidation mechanism of –SH moieties leading to strongly acidic –SO$_3$H functional groups. Similarly to other chelating cation exchangers, it was difficult to regenerate the sorbent. Moreover, even if the regeneration is successful, the regenerated sorbent will have part of the thiol moieties oxidized.

Another bi-functional polymer derived from cellulose-containing natural materials, containing both carboxylic and quaternary ammonium groups was tested for simultaneous removal of chromates and cupric cation. Citric acid, choline and dimethyloldihydroxyethylene urea were attached to the cellulose backbone. The capacity of materials derived from soybean hulls, corn husks and sugarcane bagasse was lower than that of poly(styrene-*co*-divinylbenzene)-based ion exchangers but comparable or better than the cellulose-based ion exchangers (Whatman QA-52 and CM-52). The authors state that the price of the modified cellulose-containing waste is 2.16 US$ kg^{-1} and therefore lower than that of commercial ion exchangers (Marshall and Wartelle 2006).

8.4 CONCLUSIONS

Selection of a suitable ion exchanger for each of the stable oxidations states of Cr represents a compromise between selectivity, capacity, reusability, price and other factors. When dealing with waters containing low levels of other ions both Cr(III) and Cr(VI) species can be easily removed using commercially available strong acid and strong base ion exchangers. They offer high capacity, stability, easy regeneration and reasonable price.

In the presence of other ions, which can compete for the ion-exchange sites, we have to look for selective ion exchangers. The negative side of high selectivity is the stability of complexes of our target metal with functional groups of the ion exchanger or sorbent. It can result in difficult or impossible regeneration. The resulting inability to reuse the sorbent increases the price of treated water.

There is no ideal sorbent for Cr(III) or Cr(VI) in every water matrix, but there are plenty of ingenious ways we can choose from for the removal of Cr species.

REFERENCES

Almeida, M.A.F. & Boaventura, R.A.R. (1998) Chromium precipitation from tanning spent liquors using industrial alkaline residues: a comparative study. *Waste Management*, 17, 201–209.

Bagchi, D., Stohs, S.J., Downs, B.W., Bagchi, M. & Preuss, H.G. (2002) Cytotoxicity and oxidative mechanisms of different forms of chromium. *Toxicology*, 180, 5–22.

Bahowick, S., Dobie, D. & Kumamoto, G. (2007) Ion-exchange resin for removing hexavalent chromium from ground water at treatment facility c: data on removal capacity, regeneration efficiency, and operation. Lawrence Livermore National Laboratory, Livermore, CA.

Balasubramanian, S. & Pugalenthi, V. (1999) Determination of total chromium in tannery waste water by inductively coupled plasma-atomic emission spectrometry, flame atomic absorption spectrometry and UV-visible spectrophotometric methods. *Talanta*, 50, 457–467.

Bhaumik, M., Maity, A., Srinivasu, V.V. & Onyango, M.S. (2011) Enhanced removal of Cr(VI) from aqueous solution using polypyrrole/Fe_3O_4 magnetic nanocomposite. *Journal of Hazardous Materials*, 190, 381–390.

Bonner, O.D., Jumper, C.F. & Rogers, O.C. (1958) Some cation-exchange equilibria on dowex 50 at 25°. *Journal of Physical Chemistry*, 62, 250–253.

Bonner, O.D. & Smith, L.L. (1957) A selectivity scale for some divalent cations on Dowex 50. *Journal of Physical Chemistry*, 61, 326–329.

Cherney, P.J., Crafts, B., Hagermoser, H.H., Boule, A.J., Harbin, R. & Zak, B. (1954) Determination of ethylenediaminetetraacetic acid as chromium complex. *Analytical Chemistry*, 26, 1806–1809.

Cohen, R. & Heitner-Wirguin, C. (1969) Copper species sorbed in ion-exchangers. II. The chelating iminodiacetate resin. *Inorganica Chimica Acta*, 3, 647–650.

Di Bona, K., Love, S., Rhodes, N., Mcadory, D., Sinha, S., Kern, N., Kent, J., Strickland, J., Wilson, A., Beaird, J., Ramage, J., Rasco, J. & Vincent, J. (2011) Chromium is not an essential trace element for mammals: effects of a "low-chromium" diet. *Journal of Biological Inorganic Chemistry*, 16, 381–390.

Fukushima, M., Nakayasu, K., Tanaka, S. & Nakamura, H. (1995) Chromium(III) binding abilities of humic acids. *Analytica Chimica Acta*, 317, 195–206.

Geelhoed, J.S., Meeussen, J.C.L., Roe, M.J., Hillier, S., Thomas, R.P., Farmer, J.G. & Paterson, E. (2003) Chromium remediation or release? Effect of iron(II) sulfate addition on chromium(VI) leaching from columns of chromite ore processing residue. *Environmental Science & Technology*, 37, 3206–3213.

Gode, F. & Pehlivan, E. (2005) Removal of Cr(VI) from aqueous solution by two Lewatit-anion exchange resins. *Journal of Hazardous Materials*, 119, 175–182.

Gode, F. & Pehlivan, E. (2006) Removal of chromium(III) from aqueous solutions using Lewatit S 100: the effect of pH, time, metal concentration and temperature. *Journal of Hazardous Materials*, 136, 330–337.

Gode, F. & Pehlivan, E. (2007) Sorption of Cr(III) onto chelating b-DAEG-sporopollenin and CEP-sporopollenin resins. *Bioresource Technology*, 98, 904–911.

Haight, G.P., Richardson, D.C. & Coburn, N.H. (1964) A spectrophotometric study of equilibria involving mononuclear chromium(VI) species in solutions of various acids. *Inorganic Chemistry*, 3, 1777–1780.

Hamm, R.E. (1953) Complex ions of chromium. IV. The ethylenediaminetetraacetic acid complex with chromium(III) 1. *Journal of American Chemical Society*, 75, 5670–5672.

Hoffmann, M.M., Darab, J.G. & Fulton, J.L. (2001) An infrared and X-ray absorption study of the equilibria and structures of chromate, bichromate, and dichromate in ambient aqueous solutions. *Journal of Physical Chemistry A*, 105, 1772–1782.

Hudson, M.J. & Matejka, Z. (1989) Extraction of copper by selective ion exchangers with pendent ethyleneimine groups – investigation of active states. *Separation Science and Technology*, 24, 1417–1426.

Icopini, G.A. & Long, D.T. (2002) Speciation of aqueous chromium by use of solid-phase extractions in the field. *Environmental Science & Technology*, 36, 2994–2999.

Jung, C., Heo, J., Han, J., Her, N., Lee, S.-J., Oh, J., Ryu, J. & Yoon, Y. (2013) Hexavalent chromium removal by various adsorbents: powdered activated carbon, chitosan, and single/multi-walled carbon nanotubes. *Separation and Purification Technology*, 106, 63–71.

Kocaoba, S. & Akcin, G. (2004) Chromium(III) removal from wastewaters by a weakly acidic resin containing carboxylic groups. *Adsorption Science & Technology*, 22, 401–410.

Kornev, V.I. & Mikryukova, G.A. (2004) Coordination compounds of chromium(III) with different complexones and citric acid in aqueous solutions. *Russian Journal of Coordination Chemistry*, 30, 895–899.

Korngold, E., Belayev, N. & Aronov, L. (2003) Removal of chromates from drinking water by anion exchangers. *Separation and Purification Technology*, 33, 179–187.

Kousalya, G.N., Rajiv Gandhi, M. & Meenakshi, S. (2010) Sorption of chromium(VI) using modified forms of chitosan beads. *International Journal of Biological Macromolecules*, 47, 308–315.

Marshall, W.E. & Wartelle, L.H. (2006) Chromate (CrO_4^{2-}) and copper (Cu^{2+}) adsorption by dual-functional ion exchange resins made from agricultural by-products. *Water Research*, 40, 2541–2548.

Matejka, Z., Parschova, H., Ruszova, P., Jelinek, L., Benes, M. & Hruby, M. (2001) Selective uptake and separation of (Mo, V, W, Ge)-oxoanions by synthetic sorbents having polyol-moieties and by polysaccharide-based biosorbents. *Abstracts of Papers, 222nd ACS National Meeting, 26–28 August 2001, Chicago, IL, United States*. IEC-004.

Matejka, Z., Parschova, H., Ruszova, P., Jelinek, L., Houserova, P., Mistova, E., Benes, M. & Hruby, M. (2004) Selective uptake and separation of oxoanions of molybdenum, vanadium, tungsten, and germanium by synthetic sorbents having polyol moieties and polysaccharide-based biosorbents. *Fundamentals and Applications of Anion Separations, Proceedings American Chemical Society National Meeting*. pp. 249–261.

Neagu, V. & Mikhalovsky, S. (2007) Removal of hexavalent chromium by new quaternized crosslinked poly(4-vinylpyridines). *Journal of Hazardous Materials*, 183, 533–540.

Nemecek, M., Parschova, H. & Slapakova, P. (2013) Sorption of CrVI ions from aqueous solutions using anion exchange resins. *Ion Exchange Letters*, 6, 8–11.

Palmer, D., Wesolowski, D. & Mesmer, R.E. (1987) A potentiometric investigation of the hydrolysis of chromate (vi) ion in NaCl media to 175°C. *Journal of Solution Chemistry*, 16, 443–463.

Parschová, H., Mištová, E., Matějka, Z., Jelínek, L., Kabay, N. & Kauppinen, P. (2007) Comparison of several polymeric sorbents for selective boron removal from reverse osmosis permeate. *Reactive and Functional Polymers*, 67, 1622–1627.

Pehlivan, E. & Cetin, S. (2009) Sorption of Cr(VI) ions on two Lewatit-anion exchange resins and their quantitative determination using UV-visible spectrophotometer. *Journal of Hazardous Materials*, 163, 448–453.

Petruzzelli, D., Tiravanti, G. & Passino, R. (1996) Cr(III)/A1(III)/Fe(III) ion binding on mixed bed ion exchangers. Synergistic effects of the resins behavior. *Reactive and Functional Polymers*, 31, 179–185.

Richard, F.C. & Bourg, A.C.M. (1991) Aqueous geochemistry of chromium: a review. *Water Research*, 25, 807–816.

Sahu, S.K., Meshram, P., Pandey, B.D., Kumar, V. & Mankhand, T.R. (2009) Removal of chromium(III) by cation exchange resin, Indion 790 for tannery waste treatment. *Hydrometallurgy*, 99, 170–174.

Schilde, U. & Uhlemann, E. (1993) Separation of several oxoanions with a special chelating resin containing methylamino-glucitol groups. *Reactive Polymers*, 20, 181–188.

Snukiškis, J., Gefeniene, A. & Kaušpediene, D. (2000) Cosorption of metal (Zn, Pb, Ni) cations and non-ionic surfactant (alkylmonoethers) in polyacrylic acid-functionalized cation-exchanger. *Reactive and Functional Polymers*, 46, 109–116.

Staniek, H., Krejpcio, Z., Iwanik, K., Szymusiak, H. & Wieczorek, D. (2011) Evaluation of the acute oral toxicity class of trinuclear chromium(III) glycinate complex in rat. *Biological Trace Element Research*, 143, 1564–1575.

Tenório, J.A.S. & Espinosa, D.C.R. (2001) Treatment of chromium plating process effluents with ion exchange resins. *Waste Management*, 21 (7), 637–642.

Vinokurov, E.G., Demidov, A.V. & Bondar, V.V. (2005) Physicochemical model for choosing complexes for chromium-plating solutions based on Cr(III) compounds. *Russian Journal of Coordination Chemistry*, 31, 14–18.

Wang, J., Pan, K., He, Q. & Cao, B. (2013) Polyacrylonitrile/polypyrrole core/shell nanofiber mat for the removal of hexavalent chromium from aqueous solution. *Journal of Hazardous Materials*, 244–245, 121–129.

Weidlich, C., Mangold, K.M. & Jüttner, K. (2001) Conducting polymers as ion-exchangers for water purification. *Electrochimica Acta*, 47, 741–745.

Wolowicz, A. & Hubicki, Z. (2012) Applicability of new acrylic, weakly basic anion exchanger Purolite A-830 of very high capacity in removal of palladium(II) chloro-complexes. *Industrial and Engineering Chemistry Research*, 51, 7223–7230.

Yao, C., Xu, Y., Kong, Y., Liu, W., Wang, W., Wang, Z., Wang, Y. & Ji, J. (2012) Polypyrrole/palygorskite nanocomposite: a new chromate collector. *Applied Clay Science*, 67–68, 32–35.

Zaitseva, N., Zaitsev, V. & Walcarius, A. (2013) Chromium(VI) removal via reduction-sorption on bi-functional silica adsorbents. *Journal of Hazardous Materials*, 250–251, 454–461.

Zuo, X. & Balasubramanian, R. (2013) Evaluation of a novel chitosan polymer-based adsorbent for the removal of chromium (iii) in aqueous solutions. *Carbohydrate Polymers*, 92, 2181–2186.

CHAPTER 9

Solvent-impregnated resins (SIRs) for Cr(VI) removal

Nalan Kabay & Marek Bryjak

9.1 INTRODUCTION

The mutagenic and carcinogenic effects of Cr(VI) on biological systems are well known (Güell *et al.*, 2008). It gives serious damage to lungs, kidney, liver and the gastric system (Al-Ohtman *et al.*, 2012; Sahu *et al.*, 2008). Therefore, the limit for discharge of Cr(VI) into the surface waters is defined as $0.1\,mg\,L^{-1}$ (Ozcan *et al.*, 2010). Contrary to Cr(VI) species or chromates, Cr(III) is less toxic and very insoluble at neutral to alkaline pH. Therefore, chemical precipitation is the most often preferred method for the treatment of wastewaters with a high Cr(VI) concentration. During the process, Cr(VI) is reduced to Cr(III) first, followed by the adjustment of pH which is needed to form insoluble $Cr(OH)_3$ as a precipitate (Lin and Huang, 2008; SenGupta, 2002). However, for the removal of Cr(VI) at trace levels, chemical precipitation is not very effective due to the low reduction efficiency of Cr(VI) to Cr(III) (Edwards, 1995; Sağ and Kutsal, 1995).

For relatively low concentrations of Cr(VI), ion exchange can be employed to remove Cr(VI) by commercially available anion-exchange resins (Edebali and Pehlivan, 2010; SenGupta and Clifford, 1986; SenGupta and Lim, 1988; Tenorio and Espinosa, 2001). Following ion-exchange separation, Cr(VI) can be enriched during the regeneration step and Cr(VI) in the regenerant solution can be easily converted to less harmful $Cr(OH)_3$ (SenGupta, 2002).

Cr(VI) removal by the liquid-liquid extraction method also was extensively studied in the literature. The commercially available extractants Alamine 336 and Aliquat 336 were highly effective for Cr(VI) extraction with a high degree of extraction (Senol, 2004). Other extractants such as tri-*n*-butyl phosphate (Sahu *et al.*, 2008), Cyanex 272 (Lanagan, 2003), terbutyl ammonium bromide (Venkateswaran and Palanivelu, 2004), Cyanex 923 (Agrawal *et al.*, 2008) were also employed for the separation of Cr(VI).

Liquid-liquid extraction is a very fast and very effective process for metal separations from aqueous solutions (Wionczyk *et al.*, 2006). This method has been used for many years for a variety of separations in hydrometallurgy (Prados and Gutierrez-Cervello, 2011). It is possible to separate and pre-concentrate metals from aqueous solutions in the presence of complexing anionic species by liquid-liquid extraction (Kumar *et al.*, 2010; Shukla and Rao, 2002). Separation is performed through contact with an immiscible organic phase to form salts or complex compounds giving a favorable distribution between the aqueous and organic phases. For a selective metal extraction, chelating ligands like diketones, oximes, amides, dithiocarbamates etc. have been used (Kumar *et al.*, 2010; Shukla and Rao, 2002). During the hydrometallurgical applications of liquid-liquid extraction, intimate contact of organic solvents with aqueous solution containing metal ions in extraction, scrubbing and stripping in the following steps results in loss of organic solvents due to entrainment and their solubility in aqueous solution. Therefore, the possible loss of hazardous and expensive extractants which are partly soluble in water is the major drawback of the liquid-liquid extraction process. In industrial scale operations, loss of organic solvent and hence environmental pollution, offensive odors and fire risk would be substantial, since large volumes of organic solvents will be used. The liquid-liquid extraction process is performed in complicated mixer-settlers and not effective for the separation of metals from dilute solutions (Nguyen *et al.*, 2013).

Adsorption and ion-exchange methods are effective for dilute metal solutions but they have lower capacity and lower selectivity toward target metal ions than the solvent extraction method. Therefore, their utilization is uneconomical and unacceptable for such purposes (Nguyen et al., 2013).

Chelating ion-exchange resins are produced by incorporating organic ligands as functional groups which are covalently attached to the polymer matrix in order to get a high selectivity for the target metal ion. These resins are highly selective but they are expensive and complete elution of metal from the loaded resin is not so easy. In addition, chelating ion-exchange resins generally exhibit slow kinetics that are primarily intraparticle diffusion, although they have a high affinity toward most of the metal cations. Ion-exchange kinetics for chelating ion-exchange resins could be improved by decreasing the degree of crosslinking, the hydrophobic nature of the polymer matrix, and the size of the resin beads. Poor kinetics is one of the major limitations for selectivity of chelating ion-exchange resins. The effective intraparticle diffusivities for metal ions within chelating ion-exchange resins are several orders of magnitude lower than they are in the solvent phase (SenGupta, 2002).

Solvent-impregnated resins (SIRs) can be considered as an alternative ion-exchange material by overcoming this drawback. They can be easily prepared with readily available liquid extractants, since covalent attachment of organic functional groups onto the polymer matrix is not required. SIRs can combine the unique features and process advantages of liquid-liquid extraction and ion-exchange in a single unit operation without the drawbacks of entrainment, emulsification, and difficulty in regeneration. Commercially available liquid-liquid extractants have been employed along with polymeric or inorganic adsorbents having a high porosity to prepare SIRs (Kabay et al., 2010; SenGupta, 2002; Warshawsky, 1981).

This chapter will review the past and recent studies on SIRs for metal ion separations, and focus especially on Cr(VI) extraction by SIRs.

9.2   PREPARATION OF SIRs FOR METAL ION SEPARATION

SIRs were developed in order to combine the synergistic-action of liquid-liquid extraction and ion-exchange methods in a single separation process. Warshawsky (1974) and Grinstead (1971) were the first scientists who introduced the term "solvent-impregnated resins (SIRs)" to the ion-exchange terminology and used these materials for metal ion separation. Warshawsky (1974) employed diphenylglyoximes to be impregnated on crosslinked polystyrene beads for nickel ion extraction with a selectivity factor greater than 100.

The preparation of SIRs is done by incorporating an extractant into a porous polymer matrix by a physical impregnation technique (Warshawsky, 1981). Several impregnation methods have been suggested in order to prepare SIRs. The two most common methods (dry and wet impregnations) are represented in Figure 9.1 (Zagorodni, 2007). According to Warshawsky and Berkovitz (1979), a "wet impregnation method" method was necessary with hydrophobic extractants while a "dry impregnation method" is preferred for hydrophilic extractants such as amines, ethers and ketones.

Schematic representation of a solute extraction through an SIR is shown in Figure 9.2. The organic extractant E filled up the pores of a polymer/inorganic particle. The solute S to be extracted from the aqueous solution is transported into the organic extractant phase during the extraction process. It forms a complex ES. As a result of complexation of the solute with the extractant in the pores, the extraction equilibrium shifts towards the organic phase. Thus, the solute extraction from the aqueous phase is completed (Burghoff et al., 2008).

Various hydrophobic/hydrophilic polymer adsorbents and liquid extractants such as amines, phosphoric, phosphinic, thiophosphinic acids, and esters, some solvating extractants (TBP, TOPO, CPMO, DMDBTDMA) etc. have been employed to prepare various SIRs (Kabay et al., 2010). During impregnation, the liquid extractants are retained within the pores of polymer adsorbents by Van der Waals-forces, hydrophobic, π-π interactions or capillary forces. Therefore, the loss of extractant could be less than that of liquid-liquid extraction (Burghoff et al., 2008). Since

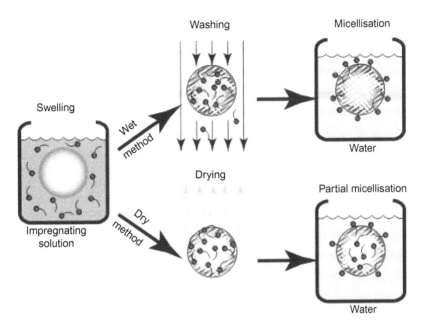

Figure 9.1. Most common methods for preparation of SIRs (Reproduced with permission from Zagorodni, 2007. Copyright of Elsevier).

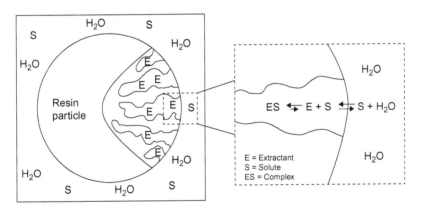

Figure 9.2. Extraction mechanism for SIRs (Burghoff *et al.*, 2008) (Reproduced with permission from Burghoff *et al.*, 2008. Copyright of Elsevier).

the pores of the polymer adsorbent are filled with the liquid extractant, almost the whole pore volume will be available for the extraction of solute. Therefore, the capacity of the SIRs will be quite high.

The inherent advantages of such hybrid materials are as follows (Kabay *et al.*, 2010):

- The availability of specific and selective extractants;
- A simple way of preparation;
- The well-known mechanisms of interactions between metal ions and extractants;
- Elimination of third phase formation observed in the case of liquid-liquid extraction;
- Possibility to handle unclear solutions;
- Making the liquid-solid separation process continuous.

SIRs have been widely used for chromatographic separations of toxic elements, extraction of rare and valuable metals, radioanalytical separations, purification of wet process phosphoric acid and other environmental applications. In a review paper published in 2010, preparation techniques and various applications of SIRs are explained in detail (Kabay *et al.*, 2010).

Interest in preparation and application of SIRs in metal ion removal has continued. The application of SIRs for separation of rare earths by polyethersulfone beads containing D2EHPA as extractant was reported by Yadav *et al.* (2013). Selective recovery of copper, cobalt and nickel from aqueous chloride media was investigated by using SIRs containing acidic organophosphorus extractants by Guo *et al.* (2012). Adhikari *et al.* (2011) used 2-ethylhexylphosphonic acid mono-2-ethylhexyl ester as the extractant for preparation of SIRs in order to remove zinc ions from an electrodeless nickel plating bath. The SIRs based on 1-phenyl-1,3-tetradecanedione (C11phDK)/tri-*n*-octylphosphine oxide (TOPO) was employed by Nishihama *et al.* (2011) for selective recovery of lithium ions from seawater. Guo *et al.* (2013) studied the selective recovery of valuable metal ions such as copper, cobalt and lithium from spent Li-ion batteries using SIRs.

Some alternative supports such as kapok fibers were also used for preparation of solvent-impregnated fibers (SIFs) and they were applied for recovery of a number of metal ions (Higa *et al.*, 2011). More research is needed to utilize solvent-impregnated fibers for metal separations. Preconcentration of metal ions by Donnan dialysis combining membrane with a cationic exchange textile and a SIR was investigated by Berdous and Akretche (2013).

Recently, some ionic liquids have been employed for the preparation of SIRs. Ionic liquids have some advantages such as high thermal stability, low solubility in water and low vapor pressure (Navarro *et al.*, 2012). Two methods are followed to prepare new types of SIRs using ionic liquids. One is the conventional method based on solvent evaporation following impregnation (Arias *et al.*, 2009; 2011; Blahusiak *et al.*, 2011; Gallardo *et al.*, 2008; Navarro *et al.*, 2010; Yang *et al.*, 2012). The other is immobilization of an ionic liquid into the biopolymer capsules (Campos *et al.*, 2008; Vincent *et al.*, 2008a; 2008b). Loh *et al.* (2013) used the solvent-impregnated agarose gel liquid phase microextraction for preconcentration of polyaromatic hydrocarbons prior to gas chromatography-mass spectrometry (GC-MS) analysis.

SIRs were also considered as promising materials for removal of some organic impurities at trace concentrations in wastewater streams containing pyridine and pyridine derivatives (Bokhove *et al.*, 2012; 2013).

Most recently, so called "*tunable aqueous polymer phase-impregnated resins*" based on a combination of the SIR principle and aqueous two-phase systems were introduced in order to avoid the drawbacks of aqueous two-phase extraction, such as long phase separation times due to persistent emulsification (Burghoff, 2013). These materials are employed for removal of proteins and other biomolecules from aqueous solutions. However, more studies should be performed to be able to understand the possible limitations and potential applications of these new materials.

The loss of extractant limits the use of SIRs in large scale applications due to the decrease in their sorption capacities and to possible environmental damage. Some papers were published on stabilization of SIRs by employing different strategies. One of them is based on removal of some part of the extractant attached weakly to the support. The other one is coating SIRs with a protective barrier (Kabay *et al.*, 2010; Trochimczuk *et al.*, 2004). Recently, there are some new approaches for giving a higher stability to SIRs. Hosseini *et al.* (2011) carried out nitration of benzene rings present in the backbone of the resin before the impregnation process. According to their findings, such treatment resulted in SIRs with a faster rate of adsorption, higher capacity and sorption rate than SIRs prepared in the ordinary fashion. Liu *et al.* (2013) proposed a coating method using a PVA-boric acid crosslinking technique to get SIRs with better stability.

Recently, microcapsules enclosing extractants without coating were also introduced as useful solid-phase extraction systems. The microcapsules could be single-core, multicore and matrix types. The solvent evaporation technique was employed to prepare single-core microcapsules. On the other hand, multi-core microcapsules include a chelating extractant dispersed in the polymer matrix as microdroplets. Alginates could be used as polymers to prepare such multi-core microcapsules for metal ion removal (Kobayashi *et al.*, 2010).

SIRs were also applied for removal of fermentation broth products having a tendency to form emulsions and affecting the fermentation due to phase toxicity (Van der Berg *et al.*, 2010). Thus, in-situ product recovery (ISPR) using capsules or SIRs was reported to be a useful technique for removing inhibiting products directly from fermentation broths by decreasing the total manufacturing cost as well (Van der Berg *et al.*, 2010).

Binary sorbents were developed by following the idea of mixed ionic extractants introduced in 1970s–1980s. Kuzmin *et al.* (2012) prepared binary sorbents containing a salt of a cation-exchange organic extractant (dialkyldithiophosphoric acid) in the matrix of a strongly basic anion exchanger for metal ion recovery. According to Kuzmin *et al.* (2012), there are three possibilities for the preparation of binary sorbents:

- Solid matrix-binary extractant;
- Anion-exchange resin-cation exchange extractant;
- Cation-exchange resin-anion exchange extractant.

It will be possible to extract cations and anions from the solution at the same time. However, there may be a possible change of the character of the interphase change during metal ion sorption on such sorbents due to the acid-base interaction only as a mode of immobilization of organic acids into a polymeric organic matrix (Kuzmin *et al.*, 2012). The same authors reported that binary sorbents prepared thus were quite stable in sorption-desorption processes. On the other hand, it was noted that the stability of these sorbents can be improved by using low-crosslinked anion exchangers to increase the large free volume and by using organic cation-exchange extractants with a molecular volume that is not large (Kuzmin *et al.*, 2012).

9.2.1   *Elimination of Cr(VI) by SIRs*

It was reported that the extractants containing amine groups such as tricaprylylmethylammonium chloride (or trioctylmethylammonium chloride) so-called "Aliquat 336" are effective for the removal of Cr(VI) by the solvent extraction method (Senol, 2004).

Some researchers published reports on the solvent extraction technique for the recovery of Cr(VI) from industrial effluents using Aliquat 336 based on their laboratory-scale work (Alonso, *et al.*, 1996; 1997; Galan, *et al.*, 1994; Salazar *et al.*, 1992). On the other hand, there has not been any practical application of the solvent extraction technique in the industry for separation of Cr(VI) from aqueous solution.

The SIRs prepared using Aliquat 336 as extractant were effective for removal of Cr(VI) from aqueous solution. They were prepared by a wet impregnation technique using both a hydrophobic styrene-divinylbenzene polymeric structure and a hydrophilic methacryclic-based polymer matrix. According to the batch-mode sorption result obtained, the SIRs containing Aliquat 336 exhibited a good performance for Cr(VI) removal from aqueous solution (Kabay *et al.*, 2003). Vincent and Guibal (2001) studied the influence of pH on the extraction of Cr(VI) by Aliquat 336 as a carrier in a hollow-fiber membrane module of chitosan. Since the existence of anionic species of Cr(VI) is highly pH dependent, this will also influence the removal efficiency and extraction mechanism of Cr(VI). A speciation diagram of Cr(VI) showing the relative distribution of various Cr(VI) species in an aqueous solution versus pH is depicted in Figure 9.3 (Saha *et al.*, 2004). The anion-exchange reactions between the chloride ion on the ion exchanger and the chromate species are shown in Figure 9.4 (Saha *et al.*, 2004).

As seen in Figure 9.5, the effect of Aliquat 336 employed during the impregnation of a polymeric adsorbent influenced the breakthrough capacity of the resulting SIRs (Kuşku, 2013). On the other hand, it was possible to regenerate these SIRs by using a NaOH-NaCl mixture (Kuşku, 2013).

SIRs prepared by using ionic liquid Cyphos IL 104 (trihexyl phosphonium bis 2,4,4-trimethylpentyl phosphinate) and Amberlite XAD-7 resin were also employed for removal of Cr(VI) (Yang *et al.*, 2012). It was reported that the optimum pH range for Cr(VI) extraction was 0–2 and the sorption capacity of SIRs decreased with an increase in pH.

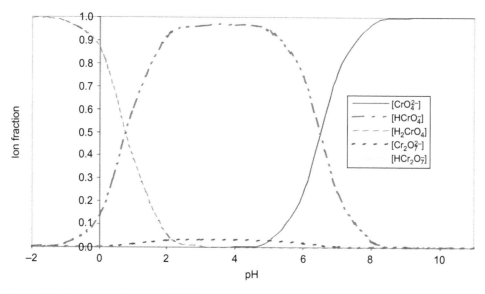

Figure 9.3. Speciation diagram of Cr(VI) with a total concentration of 0.001 M (Reproduced with permission from Saha *et al.*, 2004. Copyright of Elsevier).

Nguyen *et al.* (2013) studied the removal of Cr(VI) from acidic chloride media using an SIR prepared with Cyanex 923 (a mixture of four trialkyl phosphine oxide) and Amberlite XAD-7HP resin. The Cr loaded SIR was eluted effectively by 0.1 M NaOH solution and it was possible to reuse the SIR for three cycles with no loss in the capacity.

9.2.2   *Stabilization of SIRs selective for Cr(VI) removal*

Trochimczuk *et al.* (2004) stabilized SIRs which are prepared by Aliquat 336 as extractant by using the post-impregnation technique. For this, a hydrophilic polymer (polyvinyl alcohol) was used for coating Aliquat 336-impregnated SIRs and vinyl sulfone causing a chemical crosslinking. Thus, a protective barrier on the SIR prevented leaching of Aliquat 336 from its internal pore structure towards the external aqueous solution. Stabilization of SIRs is shown in Figure 9.6.

According to the results obtained, it was possible to prepare SIRs with a high stability by a post-impregnation technique. Due to the protective barrier around SIRs, kinetic performances of SIRs containing Aliquat 336 were significantly influenced by the extent of coating and the degree of crosslinking (Kabay *et al.*, 2004a).

The column sorption performance of coated and crosslinked SIRs containing Aliquat 336 for Cr(VI) remained almost the same during column-mode recycle tests. On the other hand, the nitrogen content of uncoated SIR decreased significantly after each cycle due to the leakage of Aliquat 336 from uncoated SIRs with time. Almost complete elution of Cr(VI) from stabilized SIRs was obtained with NaOH-NaCl mixture. It was also noted that pre-conditioning of stabilized SIRs with NaOH-NaCl mixture improved their sorption/elution performances (Kabay *et al.*, 2004b).

According to the literature, chloride ions did not influence the column sorption performance of SIRs for Cr(VI) extraction, although the influence of sulfate ions was similar for both Amberlite IRA 904 and SIRs impregnated with Aliquat 336 (Kabay *et al.*, 2005).

There are other water-soluble polymers that can form a hydrogel-layer on the surface of SIR particles. Some preliminary studies on such materials were conducted as part of a CHILTUR-POL2, 7FP MC Action, project. There, as the porous polymer Dowex-XUS-43594.00 resin (Dow Chem) was employed. The resin has *N*-methyl-D-glucamine ligands anchored to the matrix of poly(styrene-co-divinylbenzene).

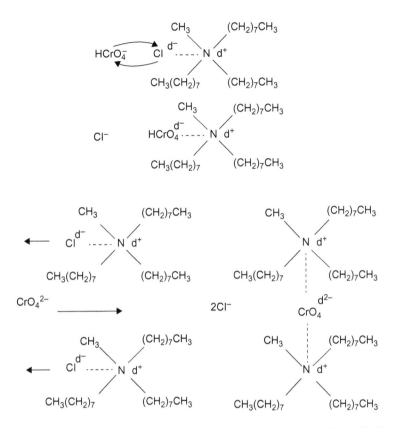

Figure 9.4.   Ion-exchange mechanisms between chromate ions and chloride ion on Aliquat 336 (Reproduced with permission from Saha *et al.*, 2004. Copyright of Elsevier).

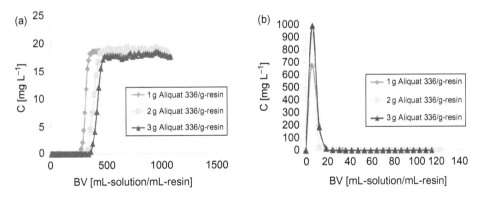

Figure 9.5.   (a) Effect of Aliquat 336 used to prepare SIR on breakthrough profiles, (b) Elution profiles of Cr(VI) (Adapted from Kuşku, 2013 with permission of the author).

The specifications by the resin manufacturer show the minimum ion-exchange capacity of the resin as $0.7\,eq\,L^{-1}$, porosity at the level of 50–55% and mean particle size of $550\,\mu m$. This resin was impregnated with Aliquat 336 ($1\,g\,g^{-1}$ resin), coated with poly(ethyleneimine) and crosslinked by glutaraldehyde. Application studies showed that the sorption capacity for Cr(VI) changed in the following sequence: Dowex-XUS << Dowex-XUS impregnated with Aliquat

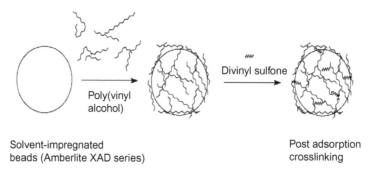

Solvent-impregnated
beads (Amberlite XAD series)

Post adsorption
crosslinking

Figure 9.6.   Stabilization of SIR by coating with PVA and chemical crosslinking (Reproduced with permission from Trochimczuk *et al.*, 2004. Copyright of Elsevier).

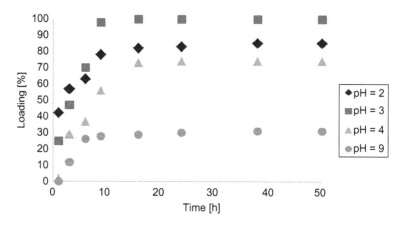

Figure 9.7.   Sorption of Cr(VI) on PEI/XAD4 sample.

336 < Aliquat 336 impregnated Dowex-XUS coated with PEI and crosslinked. The detailed kinetic studies on the mechanism that controls the sorption process revealed the following facts:

- When porous polymer was used as the sorbent, the critical feature was particle diffusion;
- When the pores were filled with Aliquat 336, the chemical reaction and reacted layer models described the process kinetics with the same precision as particle diffusion;
- In the case of use of PEI to coat particles the external hydrogel layer made the mechanisms the same. Hence, film diffusion and particle diffusion contributed to sorption to the same extent.

Sorbent stability tests showed that PEI coated SIR particles were stable during the sorption studies and their ion-exchange capacity did not change significantly with time. However, the question if the poly(ethyleneimine) layer stabilized the SIR resins or participated in the transport of Cr(VI) was still open. To verify it, XAD4 particles coated with PEI and crosslinked with glutaraldehyde were checked for the sorption of Cr(VI) from aqueous solutions. The amount of polymer used for coating changed from 0.5 to 2.0 g PEI g^{-1} XAD4. It was found that batch sorption of Cr(VI) was related to the amount of deposited PEI, the solution pH and time of the process. The obtained data were plotted in Figure 9.7.

According to the obtained results, PEI implemented a double function: as the extractant protector that did not allow the extractant to leak out to the external phase and as the specific sorbent that controlled sorption of Cr(VI) and transported it towards the core of particles.

Most recently, surface modification of SIR by the solvent-nonsolvent method was reported by Chen *et al.* (2013). According to Chen *et al.* (1992), a two-step solvent-nonsolvent method is based on the immobilizing of a surface-active extractant by phase segregation. During the solvent treatment step, the surface-active extractant is first sorbed by the polymer matrix swollen, while the phase segregation due to the existence of nonsolvent facilitates the immobilization of the extractant along with adsorption in the nonsolvent treatment step. Chen *et al.* (2013) studied the removal of Cr(VI) using Aliquat 336-impregnated resin prepared by the solvent-nonsolvent method.

9.3 CONCLUSIONS

SIRs prepared by using tricaprylylmethylammonium chloride (Aliquat 336) could be efficiently used for extraction of Cr(VI) from aqueous solution at trace concentrations. Since the leakage of extractant from SIRs with time is not acceptable for environmental reasons, the stability of SIRs needs to be improved. This could be achieved by coating SIRs with PVA or PEI followed by chemical crosslinking. Selection of proper water-soluble polymers as well as optimization of their degree of crosslinking should be performed for each particular system. It should be profitable to use polymers that facilitate transport of Cr(VI) and coating layers that do not limit the kinetics of sorption greatly.

ACKNOWLEDGEMENT

The authors acknowledge the grants from EC (FP7 CHILTURPOL2, Project no. 269153) and Ege University (Project No. EU-2012-BIL-026).

REFERENCES

Adhikari, C.R., Kumano, H. & Tanaka, M. (2011) Selective removal of zinc from an electroless nickel plating bath by solvent impregnated resin using 2-ethyhexylphosphonic acid mono-2-ethylhexyl ester as the extractant. *Solvent Extraction and Ion Exchange*, 29, 323–336.

Agrawal, A., Pal, C. & Sahu, K.K. (2008) Extractive removal of chromium (VI) from industrial waste solution. *Journal of Hazardous Materials*, 159, 458–464.

Al-Ohtman, Z.A., Ali, R. & Naushad, M. (2012) Hexavalent chromium removal from aqueous medium by activated carbon prepared from peanut shell: adsorption kinetics, equilibrium and thermodynamic studies. *Chemical Engineering Journal*, 184, 238–247.

Alonso, A.I., Irabien, A. & Ortiz, M.I. (1996) Nondispersive extraction of Cr(VI) with Aliquat 366: influence of carrier concentration. *Separation Science and Technology*, 31 (2), 271–282.

Alonso, A.I., Galan, B., Irabien, A. & Ortiz, I. (1997) Separation of Cr (VI) with Aliquat 336: chemical equilibrium modeling. *Separation Science and Technology*, 32 (9), 1543–1555.

Arias, A., Navarro, R., Saucedo, I., Gallardo, V., Martinez, M. & Guibal, E. (2009) Cadmium(II) recovery from HCl solutions using Amberlite XAD-7 impregnated with an ionic liquid (Cyphos IL-101). *Reactive and Functional Polymers*, 71, 1050–1070.

Arias, A., Saucedo, I., Navarro, R., Gallardo, V., Martinez, M. & Guibal, E. (2011) Cadmium (II) recovery from hydrochloric acid solutions using Amberlite XAD-7 impregnated with a tetraalklyl phosphonium ionic liquid. *Reactive and Functional Polymers*, 71, 1059–1070.

Berdous, D. & Akretche, D. (2013) Recovery and enrichment of heavy metals using new hybrid Donnan dialysis systems combining cation exchange textile and solvent impregnated resin. *Korean Journal of Chemical Engineering*, 30 (7), 1485–1492.

Bokhove, J., Schuur, B. & Haan, A.B. (2012) Solvent design for trace removal of pyridines from aqueous streams using solvent impregnated resins. *Separation and Purification Technology*, 98, 410–418.

Bokhove, J., Schuur, B. & Haan, A.B. (2013) Resin screening for the removal of pyridine-derivatives from waste-water by solvent impregnated resin technology. *Reactive and Functional Polymers*, 73, 595–605.

Blahusiak, M., Schlosser, S. & Martak, J. (2011) Extraction of butyric acid by a solvent impregnated resin containing ionic liquid. *Reactive and Functional Polymers*, 71, 736–744.

Burghoff, B. (2013) Tunable aqueous polymer phase impregnated resins: a combination of the extractant impregnated resin principle and aqueous two-phase systems. *Separation Science and Technology*, 48, 2159–2163.

Burghoff, B., Goetheer, E.L.V. & de Haan, A.B. (2008) Solvent impregnated resins for the removal of low concentration phenol from water. *Reactive and Functional Polymers*, 68, 1314–1324.

Campos, K., Vincent, T., Bunio, P., Trochimczuk, A. & Guibal, E. (2008) Gold recovery from HCl solutions using Cyphos IL-101 (a quaternary phosphonium ionic liquid) immobilized in biopolymer capsules. *Solvent Extraction and Ion Exchange*, 26, 570–601.

Chen, J.H. & Ruckenstein, E. (1992) Generation of porous polymer surface by solvent-nonsolvent treatment. *Journal of Applied Polymer Science*, 45, 377–382.

Chen, J.H., Hsu, K.C. & Chang, Y.M. (2013) Surface modification of hydrophobic resin with tricapryl-methylammonium chloride for the removal of trace hexavalent chromium. *Industrial and Engineering Chemistry Research*, 52, 11,685–11,694.

Edebali, S. & Pehlivan, E. (2010) Evaluation of Amberlite IRA96 and Dowex 1 X 8 ion exchange resins for the removal of Cr(VI) from aqueous solution. *Chemical Engineering Journal*, 161, 179–186.

Edwards, J.D. (1995) *Industrial wastewater treatment*. CRC Lewis Publishers, Boca Raton, FL.

Galan, B., Urtiaga, A.M., Alonso, A.I., Irabien, J.A. & Ortiz, M.I. (1994) Extraction of anions with Aliquat 336: chemical equilibrium modeling. *Industrial and Engineering Chemistry Research*, 1994, 33 (7), 1765–1770.

Gallardo, V., Navarro, R., Saucedo, I., Avila, M. & Guibal, E. (2008) Zinc(II) extraction from hydrochloric acid solutions using Amberlite XAD-7 impregnated with Cyphos IL 101 (tetrade-cyl(trihexyl)phosphonium chloride). *Separation Science and Technology*, 43, 2434–2459.

Grinstead, R.R. (1971) Report by the Dow Chemical Co., on contract No.14-12-808 to the Water Quality Office of the US Environmental Protection Administration.

Guo, F., Nishihama, S. & Yoshizuka, K. (2012) Selective recovery of copper, cobalt, and nickel from aqueous chloride media using solvent impregnated resins. *Solvent Extraction and Ion Exchange*, 30, 579–592.

Guo, F., Nishihama, S. & Yoshizuka, K. (2013) Selective recovery of valuable metals from spent Li-ion batteries using solvent-impregnated resins. *Environmental Technology*, 34, 1307–1317.

Güell, R., Antico, E., Salvado, V. & Fontas, C. (2008) Efficient hollow fiber supported liquid membrane system for the removal and preconcentration of Cr(VI) at trace levels. *Separation and Purification Technology*, 62, 389–398.

Higa, N., Nishihama, S. & Yoshizuka, K. (2011) Adsorption of Europium (III) by solvent impregnated kapok fibers containing 2-ethylhexyl phosphonic acid mono-2-ethylhexyl ester. *Solvent Extraction Research and Development, Japan*, 18, 187–192.

Hosseini, M.S. & Bandegharaei, A.H. (2011) Comparison of sorption behavior of Th(IV) and U(VI) on modified impregnated resin containing quinizarin with that conventional prepared impregnated resin. *Journal of Hazardous Materials*, 190, 755–765.

Kabay, N., Arda, M., Saha, B. & Streat, M. (2003) Removal of Cr(VI) by solvent impregnated resins (SIR) containing Aliquat 336. *Reactive and Functional Polymers*, 54, 103–115.

Kabay, N., Arda, M., Trochimczuk, A. & Streat, M. (2004a) Removal of chromate by solvent impregnated resins (SIRs) stabilized by coating and chemical crosslinking-I. Batch-mode sorption studies. *Reactive and Functional Polymers*, 59, 9–14.

Kabay, N., Arda, M., Trochimczuk, A. & Streat, M. (2004b) Removal of chromate by solvent impregnated resins (SIRs) stabilized by coating and chemical crosslinking-II. Column-mode sorption studies. *Reactive and Functional Polymers*, 59, 15–22.

Kabay, N., Solak, O., Arda, M., Topal, U., Yuksel, M., Trochimczuk, A. & Streat, M. (2005) Packed column study of the sorption of hexavalent chromium by novel solvent impregnated resins containing Aliquat 336: effect of chloride and sulfate ions. *Reactive and Functional Polymers*, 64, 75–82.

Kabay, N., Cortina, J.L., Trochimczuk, A. & Streat, M. (2010) Solvent-impregnated resins (SIRs)-methods of preparation and their applications. *Reactive and Functional Polymers*, 70, 484–496.

Kobayashi, T., Yoshimoto, M. & Nakao, K. (2010) Preparation and characterization of immobilized extractant in PVA gel beads for an efficient recovery of copper (II) in aqueous solution. *Industrial and Engineering Chemistry Research*, 49, 11,652–11,660.

Kumar, V., Sahu, S.K. & Pandey, B.D. (2010) Prospects for solvent extraction processes in the Indian context for the recovery of base metals. A review. *Hydrometallurgy*, 103, 45–53.

Kuşku, Ö. (2013) *Preparation of solvent impregnated resins from various polymeric adsorbents and their utilization for removal of Cr(VI) ions from aqueous solutions*. MS Thesis. Ege University, Izmir, Turkey.

Kuzmin, V.I., Kholkin, A.I. & Kuzmina, V.N. (2012) New sorbents for metal salts based on cation exchange extractants and anion exchange resins: their preparation and properties. *Solvent Extraction and Ion Exchange*, 30, 553–565.

Lanagan, M.D. & Ibana, D.C. (2003) The solvent extraction and stripping of chromium with Cyanex 272. *Minerals Engineering*, 16, 237–243.

Lin, Y.T. & Huang, C.P. (2008) Reduction of Cr(VI) by pyrite in dilute aqueous solutions. *Separation and Purification Technology*, 63, 191–200.

Liu, J., Gao, X., Liu, C., Zhang, S., Liu, X., Li, H., Liu, C. & Jin, R. (2013) Adsorption properties and mechanism for Fe(II) with solvent impregnated resins containing HEHEHP. *Hydrometallurgy*, 137, 140–147.

Loh, S.H., Sanagi, M.M., Wan, A.W.I. & Hasan, N.H. (2013) Solvent-impregnated agarose gel liquid phase microextraction of polycyclic aromatic hydrocarbons in water. *Journal of Chromatography A*, 1302, 14–19.

Navarro, R., Saucedo, I., Lira, M.A. & Guibal, E. (2010) Gold(III) recovery from HCl solutions using Amberlite XAD-7 impregnated with an ionic liquid (Cyphos IL-101). *Separation Science and Technology*, 45, 1950–1962.

Navarro, R., Saucedo, I., Gonzales, C. & Guibal, E. (2012) Amberlite XAD-7 impregnated with Cyphos IL-101 (tetraalkylphosphonium ionic liquid) for Pd(II) recovery from HCl solutions. *Chemical Engineering Journal*, 185–186, 226–235.

Nishihama, S., Onishi, K. & Yoshizuka, K. (2011) Selective recovery process of lithium from seawater using integrated ion exchange methods. *Solvent Extraction and Ion Exchange*, 29, 421–431.

Nguyen, N.V., Lee, J.C., Jeong, J. & Pandey, B.D. (2013) Enhancing the adsorption of chromium (VI) from the acidic chloride media using solvent impregnated resin (SIR). *Chemical Engineering Journal*, 219, 174–182.

Ozcan, S., Tor, A. & Aydın, M.E. (2010) Removal of Cr(VI) from aqueous solution by polysulfone microcapsules containing Cyanex 923 as extraction reagent. *Desalination*, 259, 179–186.

Prados, J.C. & Gutierrez-Cervello, G. (2011) *Water purification and management*. Springer, The Netherlands.

Sag, Y. & Kutsal, T. (1995) Biosorption of heavy metals by *Zoogloea ramigera*: use of adsorption isotherms and a comparison of biosorption characteristics. *Chemical Engineering Journal*, 60, 181–189.

Saha, B., Gill, R.J., Bailey, D.G., Kabay, N. & Arda, M. (2004) Sorption of Cr(VI) from aqueous solution by Amberlite XAD-7 resin impregnated with Aliquat 336. *Reactive and Functional Polymers*, 60, 223–244.

Sahu, S.K., Verma, V.K., Bagchi, D., Kumar, V. & Pandey, B.D. (2008) Recovery of chromium (VI) from electroplating effluent by solvent extraction with tri-*n*-butyl phosphate. *Indian Journal of Chemical Technology*, 15, 397–402.

Salazar, E., Ortiz, M.I., Urtiaga, A.M. & Irabien, J.A. (1992) Equilibrium and kinetics of chromium (VI) extraction with Aliquat 336. *Industrial and Engineering Chemistry Research*, 31 (6), 1516–1522.

Sengupta, A.K. (2002) Principles of heavy metals separation: an introduction. In: SenGupta, A.K. (ed.) *Environmental separation of heavy metals-engineering processes*. Lewis Publishers, Boca Raton, FL. pp. 1–14.

SenGupta, A.K. & Clifford, D. (1986) Important process variables in chromate ion exchange. *Environmental Science & Technology*, 20, 149–157.

SenGupta, A.K. & Lim, L. (1988) Modeling chromate ion exchange processes. *AIChE Journal*, 34 (12), 2019–2024.

Senol, A. (2004) Amine extraction of chromium (VI) from aqueous acidic solutions. *Separation and Purification Technology*, 36, 63–75.

Shukla, R. & Rao, G.N. (2002) Solvent extraction of metals with potassium-dihydro-bispyrazolyl-borate. *Talanta*, 57, 633–639.

Tenorio, J.A.S. & Espinosa, D.C.R. (2001) Treatment of chromium plating process effluents with ion exchange resin. *Waste Management*, 21, 637–642.

Trochimczuk, A.W., Kabay, N., Arda, M. & Streat, M. (2004) Stabilization of solvent impregnated resins (SIRs) by coating with water soluble polymers and chemical crosslinking. *Reactive and Functional Polymers*, 59, 1–7.

Van der Berg, C., Boon, F., Roelands, M., Bussmann, P., Goetheer, E., Verdoes, D. & Van der Wielen, L. (2010) Techno-economic evaluation of solvent impregnated particles in a bioreactor. *Separation and Purification Technology*, 74, 318–328.

Venkateswaran, P. & Palanivelu, K. (2004) Solvent extraction of hexavalent chromium with tetrabutyl ammonium bromide from aqueous solution. *Separation and Purification Technology*, 40, 279–284.

Vincent, T. & Guibal, E. (2001) Cr(VI) Extraction using Aliquat 336 in a hollow fiber module made of chitosan. *Industrial and Engineering Chemistry Research*, 40 (5), 1406–1411.

Vincent, T., Parodi, A. & Guibal, E. (2008a) Pt recovery using Cyphos IL-101 immobilized in biopolymer capsules. *Separation and Purification Technology*, 62, 470–479.

Vincent, T., Parodi, A. & Guibal, E. (2008b) Immobilization of Cyphos IL-101 in biopolymer capsules for the synthesis of Pd sorbents. *Reactive and Functional Polymers*, 68, 1159–1169.

Warshawsky, A. (1974) Polystyrene impregnated with β-diphenylglyoxime, a selective reagent for palladium. *Talanta*, 21, 624–626.

Warshawsky, A. (1981) Extraction with solvent-impregnated resins. In: Marinsky, J.A. & Marcus, Y. (eds.) *Ion exchange and solvent extraction.* Volume 8. Marcel Dekker Inc., New York, NY. pp. 229–310.

Warshawsky, A. & Berkovitz, H. (1979) Hydroxyoxime solvent impregnated resins for selective copper extraction. *Transactions of the Institution of Mining and Metallurgy Section*, C88, c36–c43.

Wionczyk, B., Apostoluk, W. & Charewicz, W.A. (2006) Solvent extraction of chromium (III) from spent tanning liquors with Aliquat 336. *Hydrometallurgy*, 82, 83–92.

Yadav, K.K., Singh, D.K., Anitha, M., Varshney, L. & Singh, H. (2013) Studies on separation of rare earths from aqueous media by polyethersulfone beads containing D2EHPA as extractant. *Separation and Purification Technology*, 118, 350–358.

Yang, X., Zhang, J., Guo, L., Zhang, Y. & Chen, J. (2012) Solvent impregnated resin prepared using ionic liquid Cyphos IL 104 for Cr(VI) removal. *Transactions of Nonferrous Metals Society of China*, 22, 3126–3130.

Zagorodni, A.A. (2007) *Ion exchange materials – properties and applications.* Elsevier, Amsterdam, The Netherlands.

CHAPTER 10

Arsenic and chromium removal using ion-exchange processes

Bernabé L. Rivas, Bruno F. Urbano & Cristian Campos

10.1 INTRODUCTION

Various treatment methods (conventional and advanced) have been proposed for the removal of arsenic and chromium pollutants from groundwater under both laboratory and field conditions. Commonly reported methods for sequestering As and Cr from contaminated water using chemical and physicochemical treatment methods include oxidation-reduction/precipitation technologies (Earle *et al.*, 2009; Li, S. *et al.*, 2012; Lizama *et al.*, 2011; Mondal *et al.*, 2013; Mukhopadhyay *et al.*, 2007; Ozturka *et al.*, 2012), coagulation/co-precipitation (Amrose *et al.*, 2013; Chen *et al.*, 2009; Guan *et al.*, 2011; Li, Y. *et al.*, 2012; Mukhopadhyay *et al.*, 2007), membrane technologies (Chang and Liu 2012; Rivas *et al.*, 2011; Sánchez and Rivas, 2011; Toledo *et al.*, 2013), sorption/ion-exchange technologies (Badruddoza *et al.*, 2013a; Khan and Baig, 2013; Miller *et al.*, 2000; Petruzzelli *et al.*, 1996; Tenório and Espinosa, 2001), and biochemical transformations (Kanmani *et al.*, 2012; Ozturka *et al.*, 2012; Singha *et al.*, 2012). These technologies are considered reference methods to assess the need for and feasibility of advanced methods.

The cost of applying common water treatment technologies is a critical problem. Many of these approaches are marginally cost effective or difficult to implement in developing countries (Bissen and Frimmel, 2003b; Malaviya and Singh, 2011). Thus, the need exists for a treatment strategy that is simple and robust, which addresses the local resources and constraints. Sorption operations, including adsorption and ion exchange, are a potential alternative water and wastewater treatment. This chapter focuses on using ionic exchange process (IEP) separation as a cost-effective technology for the removal of As and Cr. The following section introduces the technology and its mechanism of operation.

Specific oxides and organic compounds are known to convert Cr(III) back to Cr(VI), and recognition of this type of redox chemistry has resulted in the assessment that all Cr compounds are potentially carcinogenic. The carcinogenicity, toxicity and widespread contamination of Cr in populated areas have earned both species of this transition metal positions on the shortlist of US Environmental Protection Agency (USEPA) priority pollutants (Ellis *et al.*, 2002; WHO, 2004)

10.2 ARSENIC AND CHROMIUM REMOVAL USING ION EXCHANGE

10.2.1 *General aspect of the ion-exchange process*

Ion exchange is a sorption reaction in which an ion sorbs to the surface by removing another ion of the same valence or multiple ions of lower valence. In this manner, ion exchange never changes the surface charge (Sengupta and Kim, 1988). These materials act as a second phase when immersed in an aqueous solution, and the functional groups located near the interphase boundary (surface of the solids) are reachable for the ions dissolved in the aqueous phase. The processes by which the ions are sorbed onto the surface correspond to the adsorption process, but many of the ions are usually involved in the ion-exchange process. The property of ion exchange is a consequence of Donnan exclusion: "*when the resin is immersed in a medium in which it is insoluble, the counter-ions are mobile and can be exchanged for other counter-ions from the surrounding medium*"; ions of the same type of charge as the bound ions do not have free movement into and out of the sorbent (Sengupta and Kim, 1988; Wójcik *et al.*, 2011). The desirable properties for sorbents and ion-exchange media include (i) rapidly and effectively removing large amounts of ions, (ii) capability of being reused, (iii) high durability in water, and (iv) reasonable costs (Zagorodni, 2007).

The sorption isotherm is the most important measurement to evaluate the capacity of the ion-exchange materials. Studies have reported using batch experiments and/or continuous systems by monitoring the concentration of aqueous phase As and/or Cr in the absence and presence of the solid species in order to reproduce natural and/or anthropogenic conditions for remediation (Bissen and Frimmel, 2003a; Chiavola *et al.*, 2012; Sengupta and Kim, 1988; Zagorodni, 2007). Other important parameters include time-dependent ion exchange, which can be easily performed and yield important kinetic information on the binding/redox processes in the reaction vessel in laboratory and pilot treatments (Badruddoza *et al.*, 2013b; Chen *et al.*, 2009; Duranğlu *et al.*, 2012; Sengupta and Kim, 1988). These approaches represent a very powerful method to assess the affinity of As and/or Cr ions in response to the ion-exchange materials. In most cases, metal ion sorption obeys Langmuir and Freundlich sorption isotherms. However, the sorption capacity alone is not the only factor used to assess the potential of a sorbent. Emphasis should also be placed on the fast sorption kinetics, stability, and reusability of the sorbent (Slater and Crits, 1992; Slater and Wieck-Hansen, 1992; Slater *et al.*, 1992b). Thus, these criteria should be taken into consideration when selecting the sorbent material.

The selection of an appropriate material for As and Cr removal is dependent on the characteristics of the contaminated water, such as the total dissolved solids (TDS) content, pH, redox conditions, microbial activity, the presence of any organic or inorganic species that might interfere with treatment, and the concentration of the As and/or Cr species. Ion-exchange materials can be mineral or biological in origin, zeolites, industrial byproducts, agricultural wastes, biomass and polymeric materials. In a previous report, we discussed the speciation of the pollutants dominant in removal technology.

The advantages of the ion-exchange process include its relatively well-known process, natural or commercial available materials, well-defined capacity, and regeneration. The disadvantages are that a large amount of the exchanger must be replaced after the operation is achieved, the regeneration process produces a sludge disposal problem, and low selectivity is apparent, specifically in the case of As removal.

The use of polymeric materials or resins is a promising technique based on the adsorption/exchange of cations or anions on synthetic resins with essential characteristics of regeneration after the elution/release of ions (Inamuddin *et al.*, 2012). These materials are commonly prepared from styrene and various levels of the cross-linking agent 1,4-divinyl benzene, which controls the porosity of the particles and mechanical properties (Inamuddin *et al.*, 2012; Slater and Wieck-Hansen, 1992; Slater *et al.*, 1992a; 1992d). A schematic route for the selection of different commercially available resins for metal ions remediation is shown in Figure 10.1. These ionic polymers exhibit two types of ions, those that are bound within the structure and those that are oppositely charged counter-ions, which are free.

10.2.2   *Arsenic removal*

Among the diverse technologies for As removal, ion exchange is distinguished by its easy operation, commercially available operation and reuse. An example of materials that are able to remove As using ion exchange include the zero-valent iron (Gupta *et al.*, 2012; Mamindy-Pajany *et al.*, 2011), aluminum (oxy)(hydr)oxides (Han *et al.*, 2013; Patra *et al.*, 2012; Suresh Kumar *et al.*, 2013), hydrotalcites (Chetia *et al.*, 2012; Goh *et al.*, 2008; Jobbágy and Regazzoni, 2013), titanium dioxides (Guan *et al.*, 2012), and ion-exchange resins (Clifford and Ghurye, 2002). Ion-exchange resins have been widely studied in terms of As removal. These materials consist of a crosslinked polymer matrix bearing attached functional groups. Most of the functional groups for anion exchange are amine groups that are able to accept a proton or a quaternary ammonium groups (Zagorodni, 2007).

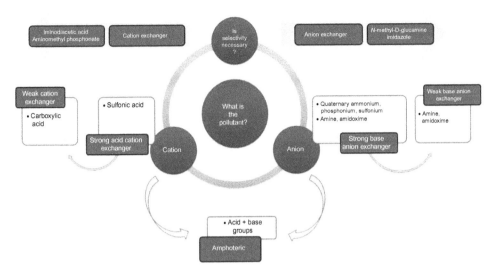

Figure 10.1.   Diagram for the selection of ion exchange resins.

Different studies have investigated the efficiency of As sorption using commercial ion-exchange resins, such as Amberlite IRA-400, Purolite A 505, Relite A 490, and Diaion WA 20/30 (Awual *et al.*, 2013; Chiavola *et al.*, 2012; Korngold *et al.*, 2001). Anion-exchange resins have a higher affinity towards divalent anions than towards monovalent anions, and the maximum sorption is reached between pH 6.0 and 9.0, where the concentration of divalent anions is maximal. In this pH range, the As(III) species are still neutral, and it is not possible to remove them using ion exchange, therefore, a pre-oxidation step is needed. However, the main limitations of these materials are their low affinity for As species in the presence of interference of anions, such as sulfate, phosphate and chloride. A proposed affinity series for anion-exchange resins is the following:

$$SO_4^{2-} > HAsO_4^{2-} > CO_3^{2-}, NO_3^- > Cl^- > H_2AsO_4^- > HCO_3^- \gg H_3AsO_3 \qquad (10.1)$$

This series clearly indicates that sulfate is the most interfering anion for As removal, the divalent arsenical has a higher affinity than the monovalent arsenical, and the species with the lowest affinity is As(III), which is due to its uncharged condition at neutral pH. In a multicomponent system, the solute-solute competition for the active site can originate due to a decrease in the As sorption; for example, in Bangladesh, water samples contain interfering concentrations as high as 0.3–3.0 mg L^{-1}, 6–28 mg L^{-1} and 50–671 mg L^{-1} of phosphate, silicates, and bicarbonate, respectively (Meng *et al.*, 2001). Specifically, in the case of sulfates, this anion exhibits the most marked effect on As sorption and ion-exchange process, which were not recommended for As removal when the sulfate concentration exceeded 120 mg L^{-1} (Ghurye *et al.*, 1999).

The USEPA has evaluated the cost and efficiency of different As removal technologies, and it has indicated that the current ion-exchange technologies do not satisfy the maximum concentration level of As in water (0.01 mg L^{-1}), suggesting several changes in ion-exchange processes (USEPA, 2002). Thus, the scientific community has addressed the efforts towards obtaining new polymeric materials with enhanced properties, namely, higher sorption capacities, selectivity, etc. These polymers consist of materials with novel ligand functionalities (Alexandratos, 2007; Donia *et al.*, 2011; Pustam and Alexandratos, 2010) or novel support matrices, such as hybrid, fibers and composites (Chandra *et al.*, 2010; Greenleaf *et al.*, 2006; Pan *et al.*, 2009; Urbano *et al.*, 2012b; Wang *et al.*, 2010).

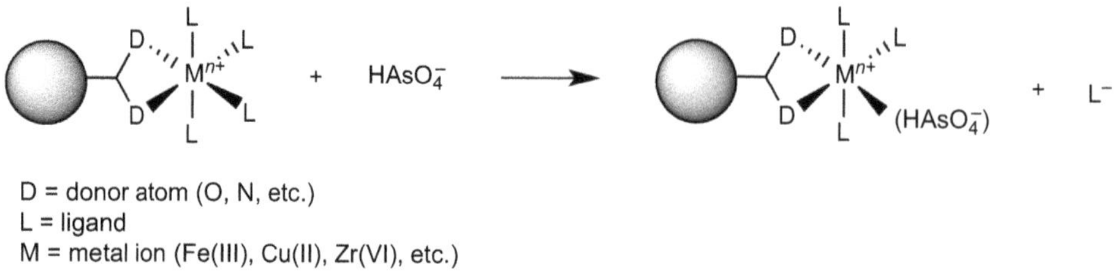

D = donor atom (O, N, etc.)
L = ligand
M = metal ion (Fe(III), Cu(II), Zr(VI), etc.)

Figure 10.2.   Scheme of the mechanism for arsenate uptake by ligand exchange process.

Regarding the novel ligand functionalities, the N-methyl-D-glucamine group has shown interesting selectivity properties for As removal. Ion-exchange resins bearing N-methyl-D-glucamine functional groups are available commercially and have been used for selective boron uptake (Kabay et al., 2007; Santander et al., 2013; Wolska and Bryjak, 2013). Unlike the strong base anion-exchange resins, which exhibit a higher sorption where divalent arsenate is the major species (alkaline media), the N-methyl-D-glucamine-based resin shows a greater sorption at acidic pHs and selectivity towards As in the presence of sulfate (Dambies et al., 2004). The effectiveness of N-methyl-D-glucamine for As sorption has been confirmed in previous studies (Sayin et al., 2010; Wei et al., 2011a; 2011b). Also, a resin of glycidylmethacrylate and N, N-methylene-bis-acrylamide, which was modified using tetraethylenepentamine was reported (Donia et al., 2011). The resin exhibited a fast As sorption and the highest sorption between pH 4 and 7. Based on the affinity of sulfur-based compound towards arsenite instead of arsenates, some approaches have been implemented to study the preparation of mercapto sorbent for the selective sorption of arsenite and further use as an analytical step for determining the concentration of arsenite and arsenate in the sample (Bennett et al., 2011; Boyaci et al., 2011).

Metal-loaded ligand exchange resins have received particular attention due to their potential to overcome several limitations of strong base anion-exchange resins, for example, the interference of sulfate (Dambies, 2004). In addition, the interaction between arsenate/arsenite with the metal is stronger, providing the possibility to enhance the sorption capacities and selectivity. Metal-loaded resins can be obtained using different polymer matrices, such as strong cation-exchange resins, macroporous polymers, chelating resins, or biosorbent (Ghimire et al., 2003; Muñoz et al., 2002). Diverse metal-cations have been immobilized onto polymer matrices and subsequently studied for As sorption, such as Fe(III) (Cumbal and Sengupta, 2005), Cu(II) (An et al., 2010), Zr(IV) (Awual et al., 2011), Y(III) (Shao et al., 2008) (Fig. 10.2). The requirement of a polymer matrix for loading with cations is the presence of suitable functional groups that are able to form stable complexes. Most functionalized polymers used are iminodiacetic acid resins, polyhydroxamic acid resins, and lysinediacetic resins.

Previously, copper-based ligand exchangers were obtained using ion-exchange resin matrices bearing nitrogen rich functional groups, such as iminodiacetic acid and pyridyl groups (Ramana and SenGupta, 1992). The study showed significantly higher selectivities in the metal cation exchanger for arsenate in the presence of sulfate. More recently, exchangers bearing Cu(II) ions were synthesized using functional XAD resins with two types of pyridinyl functional groups with 2 or 3 donor nitrogens per functional group. The resins were tested using an aqueous mixture of arsenate/sulfate solution, which resulted in the high sorption of sulfate, reaching separation factors As(V)/SO_4^{2-} as high as 10 (An et al., 2010). The selectivity of the Cu(II)-based ligand exchanger has been confirmed in other studies (An et al., 2005; Tao et al., 2011).

Other ligand exchangers that have exhibited effective performance are zirconium-based exchangers. These resins have the ability to retain both arsenite and arsenate ions and are selective in the presence of interfering anions (Suzuki et al., 2000; Zhu and Jyo, 2001). The direct removal of arsenite is of great interest because the uncharged condition of arsenite in a broad range of pHs requires a pre-oxidation step for arsenate.

In the case of using a novel polymer matrix for As removal, different approaches have been reported. For example, the affinity between metal cations and inorganic arsenicals has encouraged the use of metal oxide nanoparticles, such as oxides of Fe(III), Zr(IV), Ti(IV), and Al(III). These nanoparticles have been incorporated into functionalized polymer matrices, which provide suitable hydraulic properties for potential application (Sarkar *et al.*, 2011). Hydrated Fe(III) oxide (HFO) particles can selectively adsorb dissolved metalloids such as As oxyacids or oxyanions (Vatutsina *et al.*, 2007). Iron particles have been loaded onto different supports to evaluate their performance for As sorption, such as the case of polystyrene beads (Jiang *et al.*, 2012), macroporous composite gels (Savina *et al.*, 2011), chitosan (Cho *et al.*, 2012; Gang *et al.*, 2010; Horzum *et al.*, 2013), and glass fibers (Wang *et al.*, 2010). Moreover, montmorillonite polymer clay nanocomposites bearing *N*-methyl-glucamine function have been synthesized (Urbano *et al.*, 2012a; 2012b); the composite exhibits a high sorption capacity and selectivity in the presence of sulfate and phosphate.

10.2.3   *Chromium removal using ion exchange*

Among the physicochemical methods developed for Cr removal from waste water, ion-exchange processes have been widely used to remove heavy metals due to their many advantages, such as their high treatment capacity, high removal efficiency and fast kinetics, as previously described.

Cr(III) responds similarly to a positively charged ion when sorbing onto surfaces. Agricultural wastes have been applied as adsorbents for trivalent Cr remediation from water. The most commonly used agricultural by-products include coir pith (Parab *et al.*, 2006), husks (Eromosele *et al.*, 1996), cork powder (Machado *et al.*, 2002), rice bran (Oliveira *et al.*, 2005), santbush (Sawalha *et al.*, 2005), carrot residues (Nasernejad *et al.*, 2005), and tannin gel (Nakano *et al.*, 2001). The removal of Cr(III) from aqueous solutions using these materials follows the Langmuir model and removed above 80% of Cr.

Activated carbon adsorbents are widely used in the removal of heavy metal contaminants, in product purification, and the pollution control of Cr(III) was more effectively adsorbed using thermal treated activated carbons and zeolites (Covarrubias *et al.*, 2005; Dinesh and Charles 2006; Lyubchik *et al.*, 2004). NaA zeolite has previously been used for Cr(III) at neutral pH, whereas other research used 4A zeolite, which was synthesized by the dehydroxylation of low grade Kaolin (Basaldella *et al.*, 2007).

The uptake of Cr using synthetic polymeric resins has also shown effective results. There is an increasing need to identify cost-effective and more efficient adsorbents with a good sorption-IE capacity compared with other sorbents. Ion-exchange resins for Cr remediation exhibit different functional groups that are capable of retaining Cr(III) and Cr(VI) (Slater and Crits, 1992; Slater *et al.*, 1992c).

The most common Cr(III) exchangers are strongly acidic resins with sulfonic groups ($-SO_3H$) and weakly acidic resins with carboxylic acid groups ($-COOH$). Commercial cation-exchange resins with $-COOH$ functionality that are more commonly used include Purolite C 105 and C 106, while those used $-SO_3H$ functionality with the styrene matrix are Amberlite IR 120 and Amberlite 252 (Petruzzelli *et al.*, 1996). The ion exchanger Indion 790 is a resin with a good retention capacity for Cr(III) and is a macro-porous strongly acidic cation-exchange resin of the sulfonated polystyrene group (selective for the sorption of Cr(III) >92% in the pH range 0.5–3.5) (Sahu *et al.*, 2009). Other types of resins include Chelix 100, which is a resin for sulfate solutions and which possesses amine and carboxylic acids groups (Chanda and Rempel, 1997).

Other cation-exchange resins used in Cr(III) removal include Amberjet 1200H and 1500H, and Amberlist IRN97H, which demonstrate a remarkable increase in sorption capacity for Cr compared to other adsorbents. The adsorption process, which is pH dependent, shows a maximum removal of Cr in the pH range of 2–6 for an initial Cr concentration of 10 mg L^{-1} (Rengaraj *et al.*, 2003).

Other novel cation exchangers for liquid phase extraction are di(2-ethylhexyl)phosphoric acid (D2EHPA) and bis(2,4,4-trimethylpentyl)phosphinic acid (Cyanex®272). These exchangers have been studied for naphtha purifications (Pandey *et al.*, 1996). Finally, other materials include hybrid nanocomposites, such as epoxy-cross-linked poly ethylenimine gel-coated silica (PEI), with saturation capacities of 3.2 mM of Cr(III) g^{-1} of dry resin (Chanda and Rempel, 1997). The sorption of Cr(III) from its chloride solution is low; however, its removal can be enhanced significantly using NaCl, with more than 100% Cr(III) sorption from Cr(III) in low metal concentrations and 0.5 M NaCl. Other important properties of this material include the high regeneration capacity and kinetics, with the $t_{1/2}$ value in 4 N HCl being only 25 s in Cr(III) elution.

Cr(VI) removal using ion-exchange is an efficient technology recommended by the United States Environmental Protection Agency (USEPA) and has been used successfully by many water utilities to produce drinking water with Cr levels below the USEPA's Maximum Contaminant Level (MCL) of 0.1 mg L^{-1} (Pakzadeh and Batista, 2011). The uptake of Cr by the IEP is reversible and has a good potential for the removal/recovery of Cr(VI) from aqueous solutions. This process is pH dependent, and the maximum removal of Cr is within the pH range of 2–6 (Petruzzelli *et al.*, 1996, Tenório and Espinosa 2001)

Cost-effective adsorbents such as dolomite (Albadarin *et al.*, 2012), modified cellulose (Kumar *et al.*, 2012) and coir pith (Suksabye *et al.*, 2008), clinopyrrhotite (Lu *et al.*, 2006), tea waste (Malkoc and Nuhoglu, 2006), inorganic solid as SiO$_2$ (Buerge and Hug, 1999), TiO$_2$ (Deng and Stone, 1996b), Al$_2$O$_3$ (Deng and Stone, 1996a) and zeolites (Kumar *et al.*, 2012) are known for their good adsorptive capacities. Natural zeolites and clays have gained significant interest, mainly due to their valuable properties as ion exchangers. Clays are characterized by their excellent cation-exchange capacity and good surface area for the effective adsorption of heavy metals and water treatment (Jiang and Ashekuzzaman, 2012; Zhou, 2010). Montmorillonite is typical 2:1 smectite clay, which can be converted to its sodium form via treatment with sodium chloride. Furthermore, the sodium ion in the interlayer can be substituted with organoammonium cations (Jiang and Ashekuzzaman, 2012). Recently, organo-modified montmorillonite clay together with a novel nanocomposite resin has shown excellent potential in the adsorption of As(V) and Cr(VI) (Urbano *et al.*, 2012a). Common surface-specific approaches for studying Cr interaction with geosorbents are generally limited by relatively low sensitivities. Consequently, most surface studies in the available literature have been performed using Cr concentrations that exceed those found in most contaminated and uncontaminated soils by several orders of magnitude.

For Cr(VI) remediation, the resin used is strongly basic (quaternary ammonium anion-exchange resin) with a styrene-divinylbenzene copolymer gel matrix and is capable of reducing Cr(VI) concentrations to below the detection limit (Wójcik *et al.*, 2011). Cr(VI) responds similarly to an anion (such as CrO$_4^{2-}$ and HCrO$_4^-$), and thus, the sorption of Cr(VI) decreases with increasing pH. At low pH values, the surfaces will be neutral or positively charged, resulting in a charge attraction. As the pH increases, the surfaces are deprotonated, which increases the attraction between Cr(III) and the surface. Thus, the sorption is enhanced as the pH increases (Slater *et al.*, 1992d).

Remediation using IE has been investigated in batch adsorption processes and performed as a function of time, adsorbent dosage, pH and temperature to evaluate the adsorption capacity of resins. The IE process, which is pH dependent, indicates the maximum removal of Cr(VI) in the pH range of 1–5 for an initial concentration of 100 mg L^{-1} Cr(VI) (Pehlivan and Cetin, 2009; Petruzzelli *et al.*, 1996; Tenório and Espinosa, 2001).

Edebali and Pehlivan (2010) report the use of Amberlite IRA96 and Dowex 1×8 resins and found that the optimum pH for Cr(VI) adsorption was 3.0. In addition, more than 93% removal was achieved under optimal conditions. The maximum adsorption capacities were 0.46 and 0.54 mmol g^{-1} of Amberlite IRA96 and Dowex 1 × 8 resin for Cr(VI) ion, respectively.

New quaternized crosslinked poly(4-vinylpyridine) prepared using nucleophilic substitution reactions of 4-vinylpyridine: 1,4-divinylbenzene copolymers of gel and porous structure with halogenated compounds, such as benzyl chloride and 2-chloracetone, were used to remove Cr(VI) from an aqueous solution (Neagu and Mikhalovsky, 2010). Sorption studies were performed to determine the effect the competitive adsorption chromate/sulfate, and these results revealed the selectivity of the pyridine adsorbents towards Cr ions due to the formation of a sandwich arrangement with the Cr anion and functional groups attached to the quaternary nitrogen atom. Competition by other anions (namely, SO_4^{2-}, NO_3^-, and Cl^-) was not a problem in most applications because Cr(VI) has a higher affinity for all polymeric anion exchangers (Neagu and Mikhalovsky, 2010). Increasing the ratio of competing ion concentration to the Cr concentration may appreciably change the Cr selectivity.

In other studies, a synthetic Dowex 2-×4 ion-exchange resin was employed to investigate the uptake of Cr(VI) from real plating waste water, which achieved 100% removal (Madhavi *et al.*, 2013).

Finally, the advantage of these materials was its capacity for regeneration. This regeneration was typically accomplished using NaOH and alkaline brine (Slater and Crits, 1992). Cr(VI) in the regeneration effluent was either disposed in a concentrated form or recovered for reuse. Janin *et al.* (2009) studied elution using NH_4OH (4 M) and H_2SO_4 (10%), which were efficient elution solutions for Dowex M4195 and Amberlite IR120 resins, respectively, because they obtained 81% of the Cr recovered from the column studies.

The Cr remediation in wastewater containing a mixture of Cr(III) and Cr(VI) is difficult. In that case, the use of mixed bed resin was reported by Pramanik *et al.* (2007). The resin was a polystyrene 1,4-divinylbenzene copolymer resin synthesized and coupled with diazotized anthranilic acid. The resulting polymer contained azophenolcarboxylate and was evaluated in terms of the sorption capacity for Cr(III) and Cr(VI). The recoveries were $96.9 \pm 2.9\%$ and $96.2 \pm 2.1\%$ for Cr(III) and Cr(VI), respectively, at the 95% confidence level.

Other technologies have been reported, for example, Kabay *et al.* (2003) prepared a solvent-impregnated resin (SIR) with aliquot 336, which was used for the batch removal of Cr(VI). It was reported that the sorption capacity of SIR increases with an increasing impregnation ratio. Another synthetic ion-exchange resin, Ambersep 132 was also prepared to recover chromic acid from a synthetic plating solution (Lin and Kiang, 2003). The anion-exchange resins extracted Cr(VI) using an ion-pairing mechanism. IRA-900, a poly styrene matrix – divinyl benzene cross-linked with trimethyl quaternary amine functionality – was a strongly basic exchanger, whereas IRA-458 consisted of an acrylic matrix that contained a quaternary amine, which is a weak basic resin with lower chromate selectivity (Sengupta and Kim, 1988). These resins also removed other anions along with Cr(VI).

Surfactant-modified sodium montmorillonite and tannin-immobilized activated clay (Brum *et al.*, 2010; Li, W. *et al.*, 2012) have been utilized for the adsorption of hexavalent Cr using specific long chain amines, such as dodecylamine (Kumar *et al.*, 2012), cellulose-clay (Kumar *et al.*, 2012) and chitosan-clay (Kumar *et al.*, 2012; Pandey and Mishra, 2011) composites have a very good adsorption capacity for Cr. The simultaneous adsorption of phenol and hexavalent Cr on natural red clay modified with hexadecyltrimethyl ammonium bromide (HDTMA) is another method that has been recently reported (Plaska *et al.*, 2012). Other techniques used for the removal of Cr(VI) include extraction with trialkylamines. Some studies have found that the chromic acids, similar to other inorganic acids, are extracted with amine solutions in the presence of other interfering ions (Deptula, 1968; Riedel, 1985; Senol, 2004). One difficulty with this remediation is the use of high quantities of organic solvent to extract the amine-Cr(VI) ions. Moreover, the improvement in the extraction using long-chain amine-modified clay materials has also been explored to its fullest potential in the remediation of heavy metals. Kumar *et al.* (2012) reported the interaction mechanism of hexavalent Cr with dodecylamine-modified sodium montmorillonite and its potential towards the remediation of tannery wastewater.

ACKNOWLEDGMENTS

The authors are grateful for grants from FONDECYT (Grant No 1110079), REDOC (MINEDUC Project UCO1202 at U. de Concepción), PIA (Grant Anillo ACT 130) and acknowledge the CHILTURPOL2 grant (MCA, 7FP Project no. 269153). B. F. Urbano thanks to Fondecyt Initiation (Grant No 11121291).

REFERENCES

Albadarin, A.B., Mangwandi, C., Al-Muhtaseb, A.H., Walker, G.M., Allen, S.J. & Ahmad, M.A.M. (2012) Kinetic and thermodynamics of chromium ions adsorption onto low-cost dolomite adsorbent. *Chemical Engineering Journal*, 179, 193–202.

Alexandratos, S.D. (2007) New polymer-supported ion-complexing agents: design, preparation and metal ion affinities of immobilized ligands. *Journal of Hazardous Materials*, A139, 467–470.

Amrose, S., Gadgil, A., Srinivasan, V., Kowolik, K., Muller, M., Jessica Huang, J. & Kostecki, R. (2013) Arsenic removal from groundwater using iron electrocoagulation: effect of charge dosage rate. *Journal of Environmental Science and Health*, Part A, 48, 1019–1030.

An, B., Steinwinder, T.R. & Zhao, D. (2005) Selective removal of arsenate from drinking water using a polymeric ligand exchanger. *Water Research*, 39, 4993–5004.

An, B., Fu, Z., Xiong, Z., Zhao, D. & Sengupta, A.K. (2010) Synthesis and characterization of a new class of polymeric ligand exchangers for selective removal of arsenate from drinking water. *Reactive and Functional Polymers*, 70, 497–507.

Awual, M.R., El-Safty, S.A. & Jyo, A. (2011) Removal of trace arsenic(V) and phosphate from water by a highly selective ligand exchange adsorbent. *Journal of Environmental Sciences (China)*, 23, 1947–1954.

Awual, M.R., Hossain, M.A., Shenashen, M.A., Yaita, T., Suzuki, S. & Jyo, A. (2013) Evaluating of arsenic(V) removal from water by weak-base anion exchange adsorbents. *Environmental Science and Pollution Research*, 20, 421–430.

Badruddoza, A.Z.M., Shawon, Z.B.Z., Rahman, M.T., Hao, K.W., Hidajat, K. & Uddin, M.S. (2013) Ionically modified magnetic nanomaterials for arsenic and chromium removal from water. *Chemical Engineering Journal*, 225, 607–615.

Basaldella, E.I., Vázquez, P.G., Iucolano, F. & Caputo, D. (2007) Chromium removal from water using LTA zeolites: effect of pH. *Journal of Colloid and Interface Science*, 313, 574–578.

Bennett, W.W., Teasdale, P.R., Panther, J.G., Welsh, D.T. & Jolley, D.F. (2011) Speciation of dissolved inorganic arsenic by diffusive gradients in thin films: selective binding of As III by 3-mercaptopropyl-functionalized silica gel. *Analytical Chemistry*, 83, 8293–8299.

Bissen, M. & Frimmel, F.H. (2003a) Arsenic – a review. Part I: Occurrence, toxicity, speciation, mobility. *Acta Hydrochimica et Hydrobiologica*, 31, 9–18.

Bissen, M. & Frimmel, F.H. (2003b) Arsenic – a review. Part II: Oxidation of arsenic and its removal in water treatment. *Acta Hydrochimica et Hydrobiologica*, 31, 97–107.

Boyaci, E. (2011) Synthesis, characterization and application of a novel mercapto- and amine-bifunctionalized silica for speciation/sorption of inorganic arsenic prior to inductively coupled plasma mass spectrometric determination. *Talanta*, 85, 1517–1525.

Brum, M.C., Capitaneo, J.L. & Oliveira, J.F. (2010) Removal of hexavalent chromium from water by adsorption onto surfactant modified montmorillonite. *Minerals Engineering*, 23, 270–272.

Buerge, I.J. & Hug, S.J. (1999) Influence of mineral surfaces on chromium(VI) reduction by iron(II). *Environmental Science & Technology*, 3, 4285–4291.

Chanda, M. & Rempel, G.L. (1997) Chromium(III) removal by epoxy-cross-linked poly(ethylenimine) used as gel-coat on silica. 1. Sorption characteristics. *Industrial and Engineering Chemistry Research*, 36, 2184–2189.

Chandra, V., Park, J., Chun, Y., Lee, J.W., Hwang, I.C. & Kim, K.S. (2010) Water-dispersible magnetite-reduced graphene oxide composites for arsenic removal. *ACS Nano*, 4, 3979–3986.

Chang, F.F. & Liu, W.J. (2012) Arsenate removal using a combination treatment of precipitation and nanofiltration. *Water Science and Technology*, 65, 296–302.

Chen, Q., Luo, Z., Hills, C., Xue, G. & Tyrer, M. (2009) Precipitation of heavy metals from wastewater using simulated flue gas: Sequent additions of fly ash, lime and carbon dioxide. *Water Research*, 43, 2605–2614.

Chetia, M., Goswamee, R.L., Banerjee, S., Chatterjee, S., Singh, L., Srivastava, R.B. & Sarma, H.P. (2012) Arsenic removal from water using calcined Mg-Al layered double hydroxide. *Clean Technologies and Environmental Policy*, 14, 21–27.

Chiavola, A., D'Amato, E. & Baciocchi, R. (2012) Ion exchange treatment of groundwater contaminated by arsenic in the presence of sulphate. Breakthrough experiments and modeling. *Water, Air, & Soil Pollution*, 223, 2373–2386.

Cho, D.W., Jeon, B.H., Chon, C.M., Kim, Y., Schwartz, F.W., Lee, E.S. & Song, H. (2012) A novel chitosan/clay/magnetite composite for adsorption of Cu(II) and As(V). *Chemical Engineering Journal*, 200–202, 654–662.

Clifford, D.A. & Ghurye, G.L. (2002) Metal-oxide adsorption, ion exchange, and coagulation-microfiltration for arsenic removal from water. In: Frankenberger, W.T. (ed.) *Environmental chemistry of arsenic*. Marcel Dekker Inc., New York, NY. p. 217.

Covarrubias, C., Arriagada, R., Yáñez, J., García, R., Angélica, M., Barros, S.D., Arroyo, P. & Sousa-Aguiar, E.F. (2005) Removal of chromium(III) from tannery effluents, using a system of packed columns of zeolite and activated carbon. *Journal of Chemical Technology and Biotechnology*, 80, 899–908.

Cumbal, L. & Sengupta, A.K. (2005) Arsenic removal using polymer-supported hydrated iron(III) oxide nanoparticles: role of Donnan membrane effect. *Environmental Science & Technology*, 39, 6508–6515.

Dambies, L. (2004) Existing and prospective sorption technologies for the removal of arsenic in water. *Separation Science and Technology*, 39, 603–627.

Dambies, L., Salinaro, R. & Alexandratos, S. (2004) Immobilized *N*-methyl-D-glucamine as an arsenate-selective resin. *Environmental Science & Technology*, 38, 6139–6146.

Deng, B. & Stone, A.T. (1996a) Surface-catalyzed chromium(VI) reduction: reactivity comparisons of different organic reductants and different oxide surfaces. *Environmental Science & Technology*, 30, 2484–2494.

Deng, B. & Stone, A.T. (1996b) Surface-catalyzed chromium(VI) reduction: the TiO_2-CrVI-mandelic acid system. *Environmental Science & Technology*, 30, 463–472.

Deptula, C. (1968) Extraction of chromium(VI) from sulphuric acid solutions by means of tri-*n*-octylamine. *Journal of Inorganic and Nuclear Chemistry*, 30, 1309–1316.

Dinesh, M. & Charles, P.J. (2006) Activated carbons and low cost adsorbents for remediation of tri- and hexavalent chromium from water. *Journal of Hazardous Materials*, B137, 762–811.

Donia, A.M., Atia, A.A. & Mabrouk, D.H. (2011) Fast kinetic and efficient removal of As(V) from aqueous solution using anion exchange resins. *Journal of Hazardous Materials*, 191, 1–7.

Durangñlu, D., Trochimczukb, A.W. & Bekera, U. (2012) Kinetics and thermodynamics of hexavalent chromium adsorption onto activated carbon derived from acrylonitrile-divinylbenzene copolymer. *Chemical Engineering Journal*, 187, 193–202.

Earle, S.T., Alcock, B.E., Lowry, J.P. & Breslin, C.B. (2009) Remediation of chromium(VI) at polypyrrole-coated titanium. *Journal of Applied Electrochemistry*, 39, 1251–1257.

Edebali, S. & Pehlivan, E. (2010) Evaluation of Amberlite IRA96 and Dowex 1×8 ion-exchange resins for the removal of Cr(VI) from aqueous solution. *Chemical Engineering Journal*, 161, 161–166.

Ellis, A.S., Johnson, T.M. & Bullen, T.D. (2002) Chromium isotopes and the fate of hexavalent chromium in the environment. *Science*, 295, 2060.

Eromosele, I.C., Eromosele, C.O., Orisakiya, J.O. & Okufi, S. (1996) Binding of chromium and copper ions from aqueous solutions by shea butter (*Butyrospermum parkii*) seed husks. *Bioresource Technology*, 58, 25–29.

Gang, D.D., Deng, B. & Lin, L. (2010) As(III) removal using an iron-impregnated chitosan sorbent. *Journal of Hazardous Materials*, 182, 156–161.

Ghimire, K.N., Inoue, K., Yamaguchi, H., Makino, K. & Miyajima, T. (2003) Adsorptive separation of arsenate and arsenite anions from aqueous medium by using orange waste. *Water Research*, 37, 4945–4953.

Ghurye, G.L., Clifford, D.A. & Tripp, A.R. (1999) Combined arsenic and nitrate removal by ion exchange. *Journal of the American Water Works Association*, 91, 85–96.

Goh, K.H., Lim, T.T. & Dong, Z. (2008) Application of layered double hydroxides for removal of oxyanions: a review. *Water Research*, 42, 1343–1368.

Greenleaf, J.E., Lin, J.C. & Sengupta, A.K. (2006) Two novel applications of ion exchange fibers: arsenic removal and chemical-free softening of hard water. *Environmental Progress*, 25, 300–311.

Guan, X., Dong, H., Ma, J. & Lo, I.M.C. (2011) Simultaneous removal of chromium and arsenate from contaminated groundwater by ferrous sulfate: batch uptake behavior. *Journal of Environmental Sciences (China)*, 23, 372–380.

Guan, X., Du, J., Meng, X., Sun, Y., Sun, B. & Hu, Q. (2012) Application of titanium dioxide in arsenic removal from water: a review. *Journal of Hazardous Materials*, 215–216, 1–16.

Gupta, A., Yunus, M. & Sankararamakrishnan, N. (2012) Zerovalent iron encapsulated chitosan nanospheres – A novel adsorbent for the removal of total inorganic arsenic from aqueous systems. *Chemosphere*, 86, 150–155.

Han, C., Pu, H., Li, H., Deng, L., Huang, S., He, S. & Luo, Y. (2013) The optimization of As(V) removal over mesoporous alumina by using response surface methodology and adsorption mechanism. *Journal of Hazardous Materials*, 254–255, 301–309.

Horzum, N., Demir, M.M., Nairat, M. & Shahwan, T. (2013) Chitosan fiber-supported zero-valent iron nanoparticles as a novel sorbent for sequestration of inorganic arsenic. *RSC Advances*, 3, 7828–7837.

Inamuddin, M., Luqman, M. & Park, J.-S. (2012) Structure, synthesis and general properties of ion exchangers. In: Inamuddin, M. & Luqman, M. (eds.): *Ion-exchange technology. I. Theory and materials*. Springer, The Netherlands. pp. 211–232.

Janin, A., Blais, J.-F., Mercier, G. & Drogui, P. (2009) Selective recovery of Cr and Cu in leachate from chromated copper arsenate treated wood using chelating and acidic ion exchange resins. *Journal of Hazardous Materials*, 169, 1099–1105.

Jiang, J.Q. & Ashekuzzaman, S.M. (2012) Development of novel inorganic adsorbent for water treatment. *Current Opinion in Chemical Engineering*, 1, 1–9.

Jiang, W., Chen, X., Niu, Y. & Pan, B. (2012) Spherical polystyrene-supported nano-Fe_3O_4 of high capacity and low-field separation for arsenate removal from water. *Journal of Hazardous Materials*, 243, 319–325.

Jobbágy, M. & Regazzoni, A.E. (2013) Complexation at the edges of hydrotalcite: the cases of arsenate and chromate. *Journal of Colloid and Interface Science*, 393, 314–318.

Kabay, N., Arda, M., Saha, B. & Streat, M. (2003) Removal of Cr(VI) by solvent impregnated resins (SIR) containing Aliquat 336. *Reactive and Functional Polymers*, 54, 103–115.

Kabay, N., Sarp, S., Yuksel, M., Arar, O. & Bryjak, M. (2007) Removal of boron from seawater by selective ion exchange resins. *Reactive and Functional Polymers*, 67, 1643–1650.

Kanmani, P., Aravind, J. & Preston, D. (2012) Remediation of chromium contaminants using bacteria. *International Journal of Environmental Science & Technology*, 9, 183–193.

Khan, A.A. & Baig, U. (2013) Preparation of new polymethylmethacrylate-silica gel anion exchange composite fibers and its application in making membrane electrode for the determination of As(V). *Desalination*, 319, 10–17.

Korngold, E., Belayev, N. & Aronov, L. (2001) Removal of arsenic from drinking water by anion exchangers. *Desalination*, 141, 81–84.

Kumar, A.S.K., Kalidhasan, S., Rajesh, V. & Rajesh, N. (2012) Application of cellulose-clay composite biosorbent toward the effective adsorption and removal of chromium from industrial wastewater. *Industrial and Engineering Chemistry Research*, 51, 58–69.

Li, S., Li, T., Li, G., Li, F. & Guo, S. (2012) Enhanced electrokinetic remediation of chromium contaminated soil using approaching anodes. *Frontiers of Environmental Science & Engineering*, 6, 869–874.

Li, W., Tang, Y., Zeng, Y., Tong, Z., Liang, D. & Cui, W. (2012) Adsorption behavior of Cr(VI) ions on tannin-immobilized activated clay. *Chemical Engineering Journal*, 193–194, 88–95.

Li, Y., Wang, J., Su, Y., Luan, Z. & Liu, J. (2012) Coagulation of arsenic adsorbed ferrihydrite with the use of polyaluminium chloride (PAC) or polyferric sulfate (PFS). *Desalination and Water Treatment*, 49, 157–164.

Lin, S.H. & Kiang, C.D. (2003) Chromic acid recovery from waste acid solution by an ion exchange process: equilibrium and column ion exchange modeling. *Chemical Engineering Journal*, 92, 193–199.

Lizama, A.K., Fletcher, T.D. & Sun, G. (2011) Removal processes for arsenic in constructed wetlands. *Chemosphere*, 84, 1032–1043.

Lu, A., Zhong, S., Chen, J., Shi, J., Tang, J. & Lu, X. (2006) Removal of Cr(VI) and Cr(III) from aqueous solutions and industrial wastewaters by natural clino-pyrrhotite. *Environmental Science & Technology*, 40, 3064–3069.

Lyubchik, S.I., Lyubchik, A.I., Galushko, O.L., Tikhonova, L.P., Vital, J., Onseca, I.M. & Lyubchik, S.B. (2004) Kinetics and thermodynamics of the Cr(III) adsorption on the activated carbon from co-mingled wastes. *Colloids and Surfaces A Physicochemical Engineering Aspects*, 242, 151–158.

Machado, R., Carvalho, J.R. & Correia, M.J.N. (2002) Removal of trivalent chromium(III) from solution by biosorption in cork powder. *Journal of Chemical Technology and Biotechnology*, 77, 1340–1348.

Madhavi, V., Bhaskar Reddy, A.V., Reddy, K.G., Madhavi, G. & Prasad, T.N.K.V. (2013) An overview on research trends in remediation of chromium. *Research Journal of Recent Science*, 2, 71–83.

Malaviya, P. & Singh, A. (2011) Physicochemical technologies for remediation of chromium-containing waters and wastewaters. *Critical Reviews in Environmental Science & Technology*, 41, 1111–1172.

Malkoc, E. & Nuhoglu, Y. (2006) Fixed bed studies for the sorption of chromium(VI) onto tea factory waste. *Chemical Engineering Science*, 61, 4363–4372.

Mamindy-Pajany, Y., Hurel, C., Marmier, N. & Roméo, M. (2011) Arsenic (V) adsorption from aqueous solution onto goethite, hematite, magnetite and zero-valent iron: effects of pH, concentration and reversibility. *Desalination*, 281, 93–99.

Meng, X., Korfiatis, G.P., Christodoulatos, C. & Bang, S. (2001) Treatment of arsenic in Bangladesh well water using a household co-precipitation and filtration system. *Water Research*, 35, 2805–2810.

Miller, G.P., Norman, D.I. & Frisch, P.L. (2000) A comment on arsenic species separation using ion exchange. *Water Research*, 34, 1397–1400.

Mondal, P., Bhowmick, S., Chatterjee, D., Figoli, A. & Van der Bruggen, B. (2013) Remediation of inorganic arsenic in groundwater for safe water supply: a critical assessment of technological solutions. *Chemosphere*, 92, 157–170.

Mukhopadhyay, B., Sundquist, J. & Schmitz, R.J. (2007) Removal of Cr(VI) from Cr-contaminated groundwater through electrochemical addition of Fe(II). *Journal of Environmental Management*, 82, 66–76.

Muñoz, J.A., Gonzalo, A. & Valiente, M. (2002) Arsenic adsorption by Fe(III)-loaded open-celled cellulose sponge. Thermodynamic and selectivity aspects. *Environmental Science & Technology*, 36, 3405–3411.

Nakano, Y., Takeshita, K. & Tsutsumi, T. (2001) Adsorption mechanism of hexavalent chromium by redox within condensed-tannin gel. *Water Research*, 35, 496–500.

Nasernejad, B., Zadeh, T.E., Pour, B.B., Bygi, M.E. & Zamani, A. (2005) Camparison for biosorption modeling of heavy metals (Cr(III), Cu(II), Zn(II)) adsorption from wastewater by carrot residues. *Process Biochemistry*, 40, 1319–1322.

Neagu, V. & Mikhalovsky, S. (2010) Removal of hexavalent chromium by new quaternized crosslinked poly(4-vinylpyridines). *Journal of Hazardous Materials*, 183, 533–540.

Oliveira, E.A., Montanher, S.F., Andrade, A.D., Nabrega, J.A. & Rollemberg, M.C. (2005) Equilibrium studies for the sorption of chromium and nickel from aqueous solutions using raw rice bran. *Process Biochemistry*, 40, 3485–3490.

Ozturka, S., Kayab, T., Aslimc, B. & Tan, S. (2012) Removal and reduction of chromium by *Pseudomonas* spp. and their correlation to rhamnolipid production. *Journal of Hazardous Materials*, 231–232, 64–69.

Pakzadeh, B. & Batista, J.R. (2011) Chromium removal from ion-exchange waste brines with calcium polysulfide. *Water Research*, 45, 3055–3064.

Pan, B., Zhang, W., Lv, L., Zhang, Q. & Zheng, S. (2009) Development of polymeric and polymer-based hybrid adsorbents for pollutants removal from waters. *Chemical Engineering Journal*, 151, 19–29.

Pandey, B.D., Cote, G. & Bauer, D. (1996) Extraction of chromium (III) from spent tanning baths. *Hydrometallurgy*, 40, 343–357.

Pandey, S. & Mishra, S.B. (2011) Organic-inorganic hybrid of chitosan/organoclay bionanocomposites for hexavalent chromium uptake. *Journal of Colloid and Interface Science*, 361, 509–520.

Parab, H., Joshi, S., Shenoy, N., Lali, A., Sarma, U.S. & Sudersanan, M. (2006) Determination of kinetic and equilibrium parameters of the batch adsorption of Co(II), Cr(III) and Ni(II) onto coir pith. *Process Biochemistry*, 41, 609–615.

Patra, A.K., Dutta, A. & Bhaumik, A. (2012) Self-assembled mesoporous γ-Al_2O_3 spherical nanoparticles and their efficiency for the removal of arsenic from water. *Journal of Hazardous Materials*, 201–202, 170–177.

Pehlivan, E. & Cetin, S. (2009) Sorption of Cr(VI) ions on two Lewatit-anion exchange resins and their quantitative determination using UV-visible spectrophotometer. *Journal of Hazardous Materials*, 163, 448–453.

Petruzzelli, D., Passino, R. & Tiravantit, G. (1996) Ion exchange process for chromium removal and recovery from tannery wastes. *Industrial and Engineering Chemistry Research*, 34, 2612–2617.

Płaska, A.G., Majdan, M., Pikus, S. & Sternik, D. (2012) Simultaneous adsorption of chromium(VI) and phenol on natural red clay modified by HDTMA. *Chemical Engineering Journal*, 179, 140–150.

Pramanik, S., Dey, S. & Chattopadhyay, P. (2007) A new chelating resin containing azophenolcarboxylate functionality: synthesis, characterization and application to chromium speciation in wastewater. *Analytica Chimica Acta*, 584, 469–476.

Pustam, A.N. & Alexandratos, S.D. (2010) Engineering selectivity into polymer-supported reagents for transition metal ion complex formation. *Reactive and Functional Polymers*, 70, 545–554.

Ramana, A. & Sengupta, A.K. (1992) Removing selenium(IV) and arsenic(V) oxyanions with tailored chelating polymers. *Journal of Environmental Engineering*, 118, 755–775.

Rengaraj, S., Joo, C.K., Kim, Y. & Yi, J. (2003) Kinetics of removal of chromium from water and electronic process wastewater by ion exchange resins: 1200H, 1500H and IRN97H. *Journal of Hazardous Materials*, B102, 257–275.

Riedel, G.F. (1985) The relationship between chromium(VI) uptake, sulfate uptake, and chromium(VI) toxicity in the estuarine diatom *Thalassiosira pseudonana*. *Aquatic Toxicology*, 7, 191–204.

Rivas, B.L., Pereira, E.D., Palencia, M. & Sánchez, J. (2011) Water-soluble functional polymers in conjunction with membranes to remove pollutant ions from aqueous solutions. *Progess in Polymer Science*, 36, 294–322.

Sahu, S.K., Meshram, P., Pandey, B.D., Kumar, V. & Mankhand, T.R. (2009) Removal of chromium(III) by cation exchange resin, Indion 790 for tannery waste treatment. *Hydrometallurgy*, 99, 170–174.

Sánchez, J. & Rivas, B.L. (2011) Cationic hydrophilic polymers coupled to ultrafiltration membranes to remove chromium (VI) from aqueous solution. *Desalination*, 279, 338–343.

Santander, P., Rivas, B.L., Urbano, B.F., Yilmaz İpek, İ., Özkula, G., Arda, M., Yüksel, M., Bryjak, M., Kozlecki, T. & Kabay, N. (2013) Removal of boron from geothermal water by a novel boron selective resin. *Desalination*, 310, 102–108.

Sarkar, S., Chatterjee, P.K., Cumbal, L.H. & SenGupta, A.K. (2011) Hybrid ion exchanger supported nanocomposites: sorption and sensing for environmental applications. *Chemical Engineering Journal*, 166, 923–931.

Savina, I.N., English, C.J., Whitby, R.L.D., Zheng, Y., Leistner, A., Mikhalovsky, S.V. & Cundy, A.B. (2011) High efficiency removal of dissolved As(III) using iron nanoparticle-embedded macroporous polymer composites. *Journal of Hazardous Materials*, 192, 1002–1008.

Sawalha, M.F., Gardea-Torresdey, J.L., Parsons, J.G., Saupe, G. & Peralta-Videa, J.R. (2005) Determination of adsorption and speciation of chromium species by saltbush (*Atriplex canescens*) biomass using a combination of XAS and ICP-OES. *Microchemical Journal*, 81, 122–132.

Sayin, S., Ozcan, F. & Yilmaz, M. (2010) Synthesis and evaluation of chromate and arsenate anions extraction ability of a N methylglucamine derivative of calix[4]arene immobilized onto magnetic nanoparticles. *Journal of Hazardous Materials*, 178, 312–319.

Sengupta, A.K. & Kim, L. (1988) Modeling chromate ion-exchange processes. *American Institute of Chemical Engineers Journal*, 34, 2019–2029.

Senol, A. (2004) Amine extraction of chromium(VI) from aqueous acidic solutions. *Separation and Purification Technology*, 36, 63–75.

Shao, W., Li, X., Cao, Q., Luo, F., Li, J. & Du, Y. (2008) Adsorption of arsenate and arsenite anions from aqueous medium by using metal(III)-loaded amberlite resins. *Hydrometallurgy*, 91, 138–143.

Singha, S., Sarkar, U., Mondal, S. & Saha, S. (2012) Transient behavior of a packed column of *Eichhornia crassipes* stem for the removal of hexavalent chromium. *Desalination*, 297, 48–58.

Slater, M.J. & Crits, G.J. (1992) Improvements in regeneration, resin separations, resin residuals removal in condensate polishing. In: Slater, M.J. (ed.) *Ion exchange advances*. Published for SCI by Elsevier Applied Science, London, UK, New York, NY, pp. 81–88.

Slater, M.J. & Wieck-Hansen, K. (1992) Leachables from cation resins in demineralization and polishing plants and their significance in water/steam circuits. In: Slater, M.J. (ed.) *Ion exchange advances*. Published for SCI by Elsevier Applied Science, London, UK, New York, NY, pp. 128–135.

Slater, M.J., Belfer, S., Binman, S. & Korngold, E. (1992a) Preparation of chelating hollow fibers based on chlorosulfonated polyethylene and its uses for metal extraction. In: Slater, M.J. (ed.) *Ion exchange advances*. Published for SCI by Elsevier Applied Science, London, UK, New York, NY, pp. 17–24.

Slater, M.J., Cowan, C.J., Cox, M., Croll, B.T., Holden, P., Joseph, J.B., Rees, A.J. & Squires, R.C. (1992b) Development of exchange, a continuous ion exchange process using powdered resins and cross-flow filtration. In: Slater, M.J. (ed.) *Ion exchange advances*. Published for SCI by Elsevier Applied Science, London, UK, New York, NY, pp. 49–56.

Slater, M.J., Dale, J. & Irving, J. (1992c) Comparison of strong base resin types. In: Slater, M.J. (ed.) *Ion exchange advances*. Published for SCI by Elsevier Applied Science, London, UK, New York, NY, pp. 33–40.

Slater, M.J., Eldridge, R.J. & Vickers, S. (1992d) Preparation and chromate selectivity of weakly basic ion exchangers based on macroporous polyacrylonitrile. In: Slater, M.J. (ed.) *Ion exchange advances*. Published for SCI by Elsevier Applied Science, London, UK, New York, NY, pp. 25–32.

Suksabye, P., Thiravetyan, P. & Nakbanpote, W. (2008) Column study of chromium(VI) adsorption from electroplating industry by coconut coir pith. *Journal of Hazardous Materials*, 160, 56–62.

Sullivan, C., Tyrer, M., Cheeseman, C.R. & Graham, N.J.D. (2010) Disposal of water treatment wastes containing arsenic – a review. *Science of the Total Environment*, 408, 1770–1778.

Suresh Kumar, P., Onnby, L. & Kirsebom, H. (2013) Arsenite adsorption on cryogels embedded with iron-aluminium double hydrous oxides: possible polishing step for smelting wastewater? *Journal of Hazardous Materials*, 250–251, 469–476.

Suzuki, T.M., Bomani, J.O., Matsunaga, H. & Yokoyama, T. (2000) Preparation of porous resin loaded with crystalline hydrous zirconium oxide and its application to the removal of arsenic. *Reactive and Functional Polymers*, 43, 165–172.

Tao, W., Li, A., Long, C., Fan, Z. & Wang, W. (2011) Preparation, characterization and application of a copper (II)-bound polymeric ligand exchanger for selective removal of arsenate from water. *Journal of Hazardous Materials*, 193, 149–155.

Tenório, J.A.S. & Espinosa, D.C.R. (2001) Treatment of chromium plating process effluents with ion exchange resins. *Waste Management*, 21, 637–642.

Toledo, L., Rivas, B.L., Urbano, B.F. & Sánchez, J. (2013) Novel *N*-methyl-D-glucamine-based water-soluble polymer and its potential application in the removal of arsenic. *Separation and Purification Technology*, 103, 1–7.

USEPA (2002) Proven alternatives for aboveground treatment of arsenic in groundwater. Environmental Protection Agency, Washington, DC.

Urbano, B.F., Rivas, B.L., Martinez, F. & Alexandratos, S.D. (2012a) Equilibrium and kinetic study of arsenic sorption by water-insoluble nanocomposite resin of poly[*N*-(4-vinylbenzyl)-*N*-methyl-D-glucamine]-montmorillonite. *Chemical Engineering Journal*, 193–194, 21–30.

Urbano, B.F., Rivas, B.L., Martinez, F. & Alexandratos, S.D. (2012b) Water-insoluble polymer-clay nano-composite ion exchange resin based on *N*-methyl-D-glucamine ligand groups for arsenic removal. *Reactive and Functional Polymers*, 72, 642–649.

Vatutsina, O.M., Soldatov, V.S., Sokolova, V.I., Johann, J., Bissen, M. & Weissenbacher, A. (2007) A new hybrid (polymer/inorganic) fibrous sorbent for arsenic removal from drinking water. *Reactive and Functional Polymers*, 67, 184–201.

Wang, J., Li, X., Ince, J.S., Yue, Z. & Economy, J. (2010) Iron oxide-coated on glass fibers for arsenic removal. *Separation Science and Technology*, 45, 1058–1065.

Wei, Y.-T., Zheng, Y.-M. & Chen, J.P. (2011a) Uptake of methylated arsenic by a polymeric adsorbent: process performance and adsorption chemistry. *Water Research*, 45, 2290–2296.

Wei, Y.-T., Zheng, Y.-M. & Paul Chen, J. (2011b) Enhanced adsorption of arsenate onto a natural polymer-based sorbent by surface atom transfer radical polymerization. *Journal of Colloid and Interface Science*, 356, 234–239.

WHO (2004) Guidelines for drinking-water quality. World Health Organization, Geneva, Switzerland.

Wójcik, G., Neagu, V. & Bunia, I. (2011) Sorption studies of chromium(VI) onto new ion exchanger with tertiary amine, quaternary ammonium and ketone groups. *Journal of Hazardous Materials*, 190, 544–552.

Wolska, J. & Bryjak, M. (2013) Methods for boron removal from aqueous solutions – a review. *Desalination*, 310, 18–24.

Zagorodni, A.A. (2007) *Ion exchange materials properties and applications*. Elsevier BV, Amsterdam, The Netherlands.

Zhou, C.H. (2010) Emerging trends and challenges in synthetic clay-based materials and layered double hydroxides. *Applied Clay Science*, 48, 1–4.

Zhu, X. & Jyo, A. (2001) Removal of arsenic(V) by zirconium(IV)-loaded phosphoric acid chelating resin. *Separation Science and Technology*, 36, 3175–3189.

CHAPTER 11

Water-soluble polymers in conjunction with membranes to remove arsenic and chromium

Bernabé L. Rivas, Julio Sánchez & Leandro Toledo

11.1 INTRODUCTION

Chromium (Cr) and arsenic (As) are highly toxic chemical species associated with serious environmental pollution and several diseases.

Chromium species are mainly present in wastewater from different industries, such as metal plating, paints and pigments, leather tanning, textile dyeing, printing inks, and in additives for wood preservation (Lakshmipathiraj *et al.*, 2008; USEPA, 1998; Von Burg and Liu, 1993). In contrast, high As concentrations in the environment originate from natural sources as well as human activities, including waste chemicals, the smelting of As-bearing minerals, the burning of fossil fuels, and the application of As compounds in many products (Bissen and Frimmel, 2003).

In aqueous solution, Cr and As species are present mainly as oxyanions depending on the pH. Chromium species exist mainly with hexavalent and trivalent oxidation states (Cr(VI) and Cr(III), respectively) (Lakshmipathiraj *et al.*, 2008). Arsenic species present in water are mainly arsenate, As(V), and arsenite, As(III) (Bissen and Frimmel, 2003; Cullen and Reimer, 1989; Ferguson and Gavis, 1972).

The World Health Organization (WHO) recommends a maximum Cr(VI) concentration of $50\,\mu g\,L^{-1}$ and As concentration of $10\,\mu g\,L^{-1}$ in drinking water (Mertz, 1993; USEPA, 2001; WHO, 1993; 2001).

Several alternative methods have been reported for the removal of Cr and As, including chemical precipitation/coagulation, ion exchange, membrane filtration, adsorption, and biological processes (Baran *et al.*, 2006; Bundschuh *et al.*, 2010; Gupta *et al.*, 2001; Johnston and Heijnen, 2001; Kabay *et al.*, 2010; Kozlowski and Walkowiak, 2002; Mohan *et al.*, 2005). However, many of these processes are not widely used due to several disadvantages, such as incomplete toxic species removal, requirements for expensive equipment and monitoring systems, or the generation of toxic sludge or other waste products that require disposal (Baran *et al.*, 2006). The complete extraction of these toxic species from drinking water, wastewater, and industrial effluent to reach acceptable levels represents a true challenge.

Recent investigations show the possibility of removing Cr and As oxyanion species using a hybrid method of membrane separation called polymer enhanced ultrafiltration (Aroua *et al.*, 2007; Cañizares *et al.*, 2008; Geckeler and Volchek, 1996; Korus and Loska 2009; Rivas *et al.*, 2006; 2007; 2009; 2011; Sánchez and Rivas, 2010; 2011a; 2011b). This method is also known as liquid-phase polymer-based retention (LPR). The LPR technique involves the simultaneous use of ultrafiltration membranes and functional water-soluble polymers (WSPs) to separate and concentrate low molecular weight species (LMWS) in aqueous solution, e.g. the metal ions. These low molecular weight species interact with the functional groups of water-soluble polymers to form new polymer–LMWS species with a size larger than the pore diameter of the membrane and are therefore retained (Rivas *et al.*, 2003; Spivakov *et al.*, 1985).

In LPR, two types of experiments can be identified: (i) a washing method, which is an elution method based on continuous diafiltration by the addition of solvent at constant volume and (ii) an enrichment method, which is a concentration method based on continuous diafiltration by the addition of solvent and LMWS at a constant volume. This method is used to determine the maximum retention capacity of the water-soluble polymer.

In the present chapter, the removal of chromate and arsenate oxyanions is analyzed by the LPR technique using water-soluble polymers through the washing and enrichment methods. The extracting agents studied are polymers containing quaternary ammonium salts and a polymer based on *N*-methyl-D-glucamine. These polymers were synthesized and fractionated through membranes with different cut-offs.

Using the washing method, the removal of Cr(VI) and As(V) was performed at different pHs and with WSPs with molecular weights above 50,000 g mol^{-1}. The blank experiment without WSPs was performed previously at the same pHs. The interference of other ions, such as NaCl and Na$_2$SO$_4$ in solution, using the washing method at constant ionic strength, was analyzed for anion removal. The effect of counter-ions and the optimum polymer:anion molar ratio for efficient separation were also analyzed.

Through enrichment experiments, the maximum retention capacity (*MRC*) was determined for Cr(VI) and As(V). The FTIR spectra of the polymer and polymer-chromate were compared before and after the *MRC*.

Finally, the retention-elution process of arsenate was investigated by changing the pH of the solution from basic to acidic in two cycles.

11.2   PREPARATION OF WATER-SOLUBLE POLYMERS

All WSPs were prepared by free-radical polymerization. Approximately five grams of each monomer and 1 mol-% ammonium persulfate (APS) used as an initiator were dissolved in 40–60 mL of water under an inert atmosphere. The reaction mixture was kept at 70°C under N$_2$ (g) for 24 hours (Sánchez and Rivas, 2011a). The following monomers were used for the free-radical polymerization: [3-(acryloylamino) propyl] trimethylammonium chloride (ClAPTA), (ar-vinyl benzyl)trimethylammonium chloride solution (ClVBTA), [2-(acryloyloxy) ethyl]trimethylammonium chloride solution (ClAETA), and [2-(acryloyloxy)ethyl] trimethylammonium methyl sulfate solution (SAETA). All these monomers were purchased commercially.

N-methyl-D-glucamine-based WSP was prepared in two stages: first, the vinyl macromonomer glycidyl methacrylate–*N*-methyl-D-glucamine (GMA-NMG) was synthesized from their respective precursors, and then it was polymerized, as described before (Toledo *et al.*, 2013; Bıçak *et al.*, 2000).

All polymers were dissolved in water and purified by ultrafiltration membranes of poly(ethersulfone) with different molar mass cut-offs (MMCO) (10,000, 30,000, 50,000, and 100,000 Da). The maximum yield (95%) was obtained with the fraction above 100,000 g mol^{-1} (Korus and Loska, 2009; Sánchez and Rivas, 2011a). The WSP structures with quaternary ammonium groups are shown in Figure 11.1, and the WSP structure with *N*-methyl-D-glucamine is shown in Figure 11.2.

11.3   LPR TECHNIQUE

In this study, two different modes of LPR were used to remove Cr(VI) and As(V) oxyanions. The first method was the washing method (Fig. 11.3a), which is a batch-like procedure wherein washing is performed with water at constant pH. Before conducting ultrafiltration, the pH of the solution was adjusted between 3 and 9 by adding 10^{-1} M HNO$_3$ or NaOH in separate experiments. The pH was determined by a pH meter (H. Jürgen and Co.). A solution of 300 mg L^{-1} K$_2$Cr$_2$O$_7$ or Na$_2$HAsO$_4$·7H$_2$O was used. The mixture polymer-anion was stirred at room temperature and then placed in the ultrafiltration cell. The solution was introduced to the ultrafiltration cell and washed with reservoir water at the same pH. The ultrafiltration was performed under a total pressure of 1 bar (0.1 MPa) using an ultrafiltration membrane with a molecular weight cut-off (MWCO) of

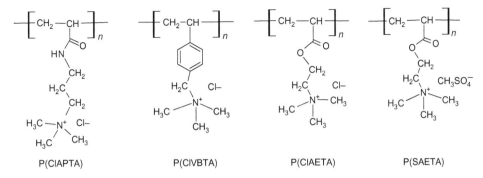

P(ClAPTA) P(ClVBTA) P(ClAETA) P(SAETA)

Figure 11.1. Structures of WSP: poly[3-(acryloylamine) propyl] trimethyl ammoniumchloride, P(ClAPTA), poly(ar-vinyl benzyl) trimethylammonium chloride, P(ClVBTA), poly[2-(acryloyloxy) ethyl] trimethylammonium chloride, P(ClAETA), and poly[2-(acryloyloxy) ethyl] trimethylammonium methyl sulfate, P(SAETA).

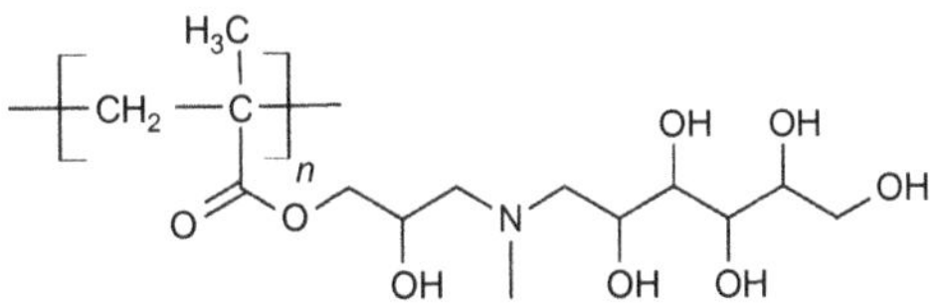

Figure 11.2. Structure of poly(glycidyl methacrylate-N-methyl-D-glucamine), P(GMA-NMG).

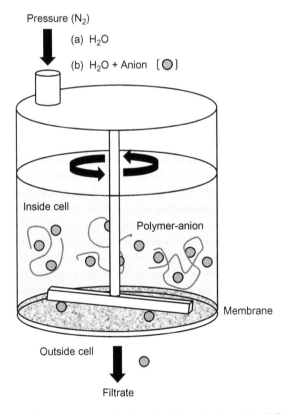

Figure 11.3. Procedure of oxyanion removal using the LPR technique. The different experiments: (a) washing method and (b) enrichment method.

10,000 Da. The total cell volume was kept constant during the filtration process. In the case of the pH studies, all experiments were performed with a solution of polymer and Cr(VI) or As(V) (20:1 polymer:anion mole ratio) using 30 mg L^{-1} of Cr(VI) or As(V). The Cr and As concentrations were measured in the filtrate by atomic absorption spectrometry (AAS). The results of the Cr(VI) and As(V) uptake are systematically presented as the percentage of the retention R [%] versus the filtration factor Z (volume of filtrate/volume of the cell).

The second ultrafiltration mode is the enrichment method, analogous to a column method (see Fig. 11.3b). A solution containing the toxic oxyanions to be separated was passed from the reservoir through the ultrafiltration cell containing a WSP solution. Both cell and reservoir solutions could be adjusted to the same values of pH and ionic strength.

The enrichment method, to determine the *MRC* of WSPs was performed by passing a solution of 2.5×10^{-3} mol L^{-1} of Cr(VI) (131 mg L^{-1}) from the reservoir to a cell containing 2.5×10^{-4} mol of WSPs and 150 mL of filtrate. In the case of As, 4×10^{-3} mol L^{-1} of As(V) solution and 8×10^{-4} mol of WSP were used, and 300 mL of total filtrate was collected.

In the retention-elution process, the enrichment method and washing method were alternately used. In both cases, a blank experiment (in the absence of the WSPs) was included to evaluate the interaction of the membrane with the toxic anions.

11.4   REMOVAL OF Cr(VI) AND As(V) BY LPR WASHING METHOD

Two values need to be defined to determine the retention capacity of Cr(VI) or As(V) oxyanions from solution: (i) retention (R) and (ii) filtration factor (Z).

Retention is the fraction of toxic oxyanions remaining in the cell according to the expression:

$$R = [\text{Anion}_{\text{cell}}]/[\text{Anion}_{\text{init}}] \qquad (11.1)$$

where [Anion$_{\text{cell}}$] corresponds to the absolute quantity of oxyanions that are retained in the cell and [Anion$_{\text{init}}$] is the absolute quantity of oxyanions at the start of the experiment.

The filtration factor (Z) is the ratio between the total volume of permeates (V_p) and the volume of retentate (V_{cell}):

$$Z = V_p/V_{\text{cell}} \qquad (11.2)$$

Depending on the experimental data, a graph (retention profile) in which R is represented as a function of Z can be drawn.

11.4.1   *Effect of pH on Cr(VI) and As(V) removal*

Chromium (VI) and As(V) normally exist in the anionic form in aqueous environments. Chromium is mainly present in water as $Cr_2O_7^{2-}$, $HCrO_4^-$, or CrO_4^{2-}, depending on the pH and concentration. At pH values below 1, the predominant species is chromic acid (H_2CrO_4). In acidic media with a pH value of 2–4, Cr(VI) exists mostly in the form of dichromate ($Cr_2O_7^{2-}$) ions. At pH between 4 and 6, $Cr_2O_7^{2-}$ and $HCrO_4^-$ ions exist in equilibrium, and under alkaline conditions (pH 8), it exists predominantly as the chromate anion (CrO_4^{2-}) (Ansari and Delavar, 2010; Korus and Loska, 2010). As(V) species coexist in an aqueous media as oxyanions according to the pH: $H_2AsO_4^-$, $HAsO_4^{2-}$ and AsO_4^{3-}; pK_{a1}: 2.2; pK_{a2}: 7.0 and pK_{a3}: 11.5, respectively (Rivas *et al.*, 2007).

The removal experiments were performed as a function of pH by the washing method. The pH was previously adjusted to different pHs in both the LPR cell and the water reservoir. The resulting polymer:anion solution (20:1 mole ratio) was carried out with the LPR-technique by the washing method. Under these experimental conditions, at pH 9, the retention of Cr(VI) (Fig. 11.4) and As(V) (Fig. 11.5) was at a maximum for the WSPs with a quaternary ammonium group (Sánchez and Rivas, 2011a; Rivas *et al.*, 2011). The P(GMA–NMG) showed increased uptake capacity for As(V) at slightly acidic pH (Fig. 11.6).

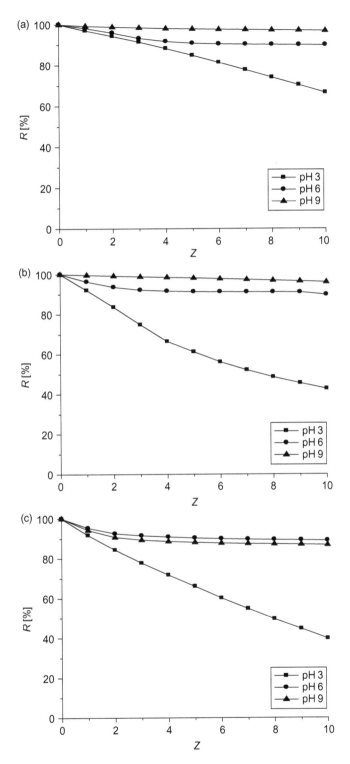

Figure 11.4. Retention profile of Cr(VI) using (a) P(ClVBTA), (b) P(ClAPTA), and (c) P(SAETA) at different pHs, with 2×10^{-4} mol absolute polymer and 1×10^{-5} mol absolute Cr(VI) (reproduced with permission from Sanchez and Rivas 2012, copyright of Wiley).

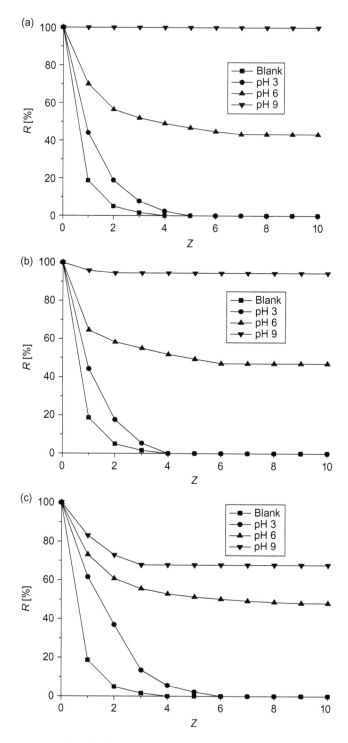

Figure 11.5. Retention profile of As(V) using (a) P(ClVBTA), (b) P(ClAPTA), and (c) P(SAETA) at different pHs, with 2×10^{-4} mol absolute polymer and 1×10^{-5} mol absolute As(V) (reproduced with permission from Sanchez and Rivas 2012, copyright of Wiley).

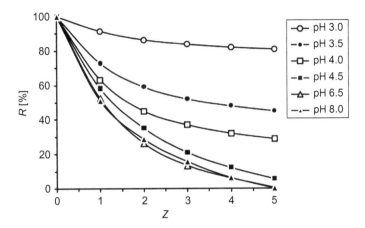

Figure 11.6.   Retention profile of As(V) using P(GMA-NMG) (reproduced with permission from Sanchez and Rivas 2012, copyright of Wiley).

The retention capacity of these oxyanions by the WSPs is mainly due to the presence of a positively charged group. In the case of P(ClVBTA), P(ClAPTA), and P(SAETA), the charged group corresponded to quaternary ammonium. For P(GMA-NMG), the protonation of nitrogen in the amino tertiary group at acidic pH produces a positive charge on the polymer (Toledo *et al.*, 2013).

The polarity of the functional group is assumed to be a parameter to control the selectivity of the ion exchange. The interactions are produced mainly through anion exchange between the counter-ions of the polymer and the Cr or As anion species, preferably with divalent charge, for WSPs with quaternary ammonium groups. This can be corroborated by the higher retention capacity of the polymers at pH 9 because divalent CrO_4^{2-} or $HAsO_4^{2-}$ species are predominant. The Cr and As removal decreased at pH 6 because $Cr_2O_7^{2-}$ and $HCrO_4^-$ or $HAsO_4^{2-}$ and $H_2AsO_4^-$ ions exist in equilibrium. The retention of Cr and As was lower at pH 3 compared at basic pH in both cases. However, the removal of Cr(VI) by the LPR technique using the washing method with WSPs was suitable over a wide range of pHs (Sánchez and Rivas, 2011a). WSP with N-methyl-D-glucamine interacted preferentially with the monovalent species, which are found in higher concentrations at pH 3 (Fig. 11.6). When considering the pK_a of the tertiary amine in the polymer ($pK_a = 6.2$), it is possible to deduce that the amount of protonated amine in the polymer determines the percentage retention of the species $H_2AsO_4^-$ obtained, assuming that there is ion exchange between the two ionic species. However, it was not possible to confirm that the retention ability was purely due to the amine function, as it is also known that the presence of polyols can improve interactions with oxyanions (Smith *et al.*, 2005).

On the other hand, we measured the Cr recovery in blank experiments, without WSPs, to determine the influence of the ultrafiltration membrane for the Cr retention. To compare with the WSP, the same experimental conditions were used: 5.71×10^{-6} mol absolute Cr(VI) in 10 mL of solution and 1 bar (0.1 MPa) of pressure. The results showed that the membrane interacts with Cr oxyanions, preferentially at basic pH (Fig. 11.7), for a wide range of Z. However, these interactions were weak because, at $Z = 10$, the removal of Cr was less than 10% at pH 9 (Sánchez and Rivas, 2011a). We can consider the depreciable influence of the membrane for Cr removal in our experimental conditions.

11.4.2   *Effect of polymer counter-ions on Cr(VI) and As(V) removal*

The results demonstrate the influence of the counter-ions of the WSP. Polymers with chloride exchanger groups, such as P(ClAPTA), P(ClAETA), and P(ClVBTA), show a higher ability to

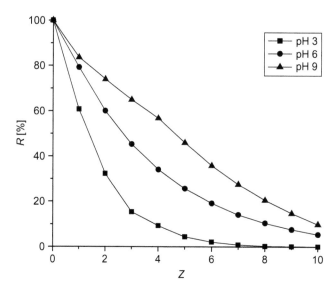

Figure 11.7.   Blank removal experiments without cationic hydrophilic polymers at different pHs using 1×10^{-5} mol absolute Cr(VI) in 10 mL of solution and 1 bar (0.1 MPa) of pressure (reproduced with permission from Sanchez and Rivas 2012, copyright of Wiley).

remove Cr(VI) or As(V) ions than polymers that contain methyl sulfate as the anion exchanger group, P(SAETA), under the same conditions. Polymers with chloride exchanger groups show the highest capacity to remove these oxyanions at basic pHs. These results can be attributed to the easier release of the chloride anions in comparison with the methyl sulfate anions, which are associated with the quaternary ammonium groups (see Figs 11.4 and 11.5).

These results confirm that the WSPs with chloride counter-ions used in conjunction with ultrafiltration membranes might be a useful technique for Cr or As removal from contaminated solutions.

11.4.3   *Optimum polymer:anion molar ratio*

The removal of Cr(VI) and As(V) was optimized by changing the polymer:anion ratio. The influence of the concentration of polymers on Cr and As removal was studied using the washing method.

Different polymer:Cr(VI) molar ratios, such as 40:1, 20:1, 10:1, and 5:1 were prepared at pH 9. The results of the retention R [%] of Cr(VI) with a filtration factor of $Z = 10$ are shown in Table 11.1. The retention capacity was limited by the polymer concentration when 5.71×10^{-6} mol absolute Cr(VI) (30 mg L^{-1}) was used. The results indicate an optimum molar ratio of 10:1 (polymer:Cr(VI)) for all the polymers to reach the maximum Cr(VI) removal. The maximum removal in these molar ratio conditions was 100% for P(ClAPTA), 99% for P(ClVBTA), and 88% for P(SAETA).

The influence of the concentration of the polymers P(ClAETA) and P(ClVBTA) on As removal was studied using the washing method. Different polymer:As(V) molar ratios (31:1, 20:1, 10:1, 6:1, and 3:1) were prepared at pH 9. The results of the retention R [%] of As(V) with a filtration factor of $Z = 10$, P(ClVBTA) and P(ClAETA) by the washing method are presented in Table 11.2. The retention capacity of WSP was determined when a range of 10–84 mg L^{-1} As was used.

The results indicate an optimum ratio of 20:1 (polymer:arsenate) for the complete removal of arsenate. This ratio occurred over a range of two orders of magnitude (2×10^{-4} and 7×10^{-5} mol) of polymer. This finding is important from the point of view of application due to the high

Table 11.1. Behavior of different molar ratios of polymer:Cr(VI) in the removal of chromate using P(ClVBTA), P(ClAPTA), and P(SAETA) at pH 9 (reproduced with permission from Sanchez and Rivas 2012, copyright of Wiley).

Molar ratio Polymer:Cr(VI)	Moles of polymer	Moles of Cr(VI)	P(ClVBTA) R [%]	P(ClAPTA) R [%]	P(SAETA) R [%]
(40:1)	2.28×10^{-4}	5.71×10^{-6}	96	94	74
(20:1)	1.14×10^{-4}	5.71×10^{-6}	97	96	87
(10:1)	5.71×10^{-5}	5.71×10^{-6}	99	100	88
(5:1)	2.86×10^{-5}	5.71×10^{-6}	96	93	69

Table 11.2. Behavior of different molar ratios of polymer:As(V) in the removal of arsenate using P(ClVBTA) and P(ClAETA) at pH 9 (reproduced with permission from Sanchez and Rivas 2012, copyright of Wiley).

Molar ratio Polymer:As(V)	Moles of polymer	Moles of As(V)	P(ClVBTA) R [%]	P(ClAETA) R [%]
(31:1)	7×10^{-5}	2.25×10^{-4}	70	84
(20:1)	7×10^{-5}	3.45×10^{-6}	10	100
(20:1)	2×10^{-4}	1.00×10^{-5}	100	100
(10:1)	7×10^{-5}	6.90×10^{-6}	88	59
(6:1)	7×10^{-5}	1.12×10^{-5}	77	60
(3:1)	7×10^{-5}	2.25×10^{-5}	54	14

Table 11.3. Behavior of different molar ratio polymer:As(V) in the removal of arsenate using P(GMA-NMG) at pH 3 (reproduced with permission from Sanchez and Rivas 2012, copyright of Wiley).

Molar ratio Polymer:As(V)	Moles of polymer	Moles of As(V)	P(GMA-NMG) R [%]
(70:1)	5.6×10^{-4}	8.0×10^{-6}	82
(50:1)	4.0×10^{-4}	8.0×10^{-6}	48
(20:1)	1.6×10^{-4}	8.0×10^{-6}	32
(1:1)	8.0×10^{-6}	8.0×10^{-6}	6

efficiency of the polymer with respect to the recovery of the toxic species, even at high concentrations.

At pH 3, the removal of As(V) using different concentrations of P(GMA-NMG) (see Table 11.3) was carried out. The results showed that the anion uptake capacity increases with an increasing amount of polymer in the feed solution. The polymer:arsenate molar ratio of 70:1 was found to be optimal for achieving 80% removal. This finding may be attributed to the higher availability of active groups for the ionic species to remove.

11.4.4 *Competitive effect of other monovalent and divalent anions on arsenate retention at constant ionic strength*

WSPs present the highest retention of arsenate species by the LPR technique when no other anions are present in the solution.

To determine the influence of other anions, different experiments in the presence of divalent and monovalent anions, such as a sulfate and chloride, were performed using different concentrations of these salts at basic pH. In this study, we used the washing method at constant ionic strength, adding to both the reservoir and the ultrafiltration cell different concentrations of NaCl and Na_2SO_4 in separate experiments with a P(ClAETA):As(V) mole ratio of 20:1 and a P(GMA-NMG):As(V) mole ratio of 70:1 inside the ultrafiltration cell.

The arsenate retention decreased with the increasing salt concentration and the increased charge of the added anions. The decrease in the retention was due to the presence of the added salts declining in the following order $Na_2SO_4 > NaCl$.

According to the literature (Berdal *et al.*, 2010), the order of interference in As retention is trivalent ions > divalent ions > monovalent ions. The effect of added electrolytes on As binding to the WSPs can be understood as being due to the competition between arsenate and other anions for the binding sites on the polymer. The affinity of anions to bind onto the polymer is similar to the behavior observed in the ion-exchange resin containing ammonium groups when removing As by ion-exchange processes (Sánchez and Rivas, 2011b). Another way to explain the effect is that the electrical double layer is compressed around the polymer as the ionic strength increases, thus reducing the polymer's electric potential. Divalent anions produce a greater reduction in As retention than monovalent anions because divalent anions bind more strongly to the polymer's charged sites and compress the electric double layer around the polymer more effectively than monovalent anions (William, 1992). It is reasonable that sulfate and chloride anions present different interferences toward arsenate retention. The results prove the adsorption of the interfering ions at the same active sites on the polymer, especially in the case of sulfate, which, like arsenate, has a tetrahedral structure and a divalent charge at basic pH.

For P(ClAETA), the results showed that As retention decreased from 96 to 20% at $Z=10$ when 1×10^{-3} mol L^{-1} of sodium sulfate was added. Moreover, arsenate retention dropped to zero when the sulfate ion concentration increased to 5×10^{-3} mol L^{-1} (Fig. 11.8). On the other hand, the competition between arsenate and monovalent chloride was lower than between sulfate and arsenate. In a separate experiment, when the minimum chloride concentration was added, corresponding to 1×10^{-3} mol L^{-1}, the retention capacity of arsenate decreased from 96 to 55% at $Z=10$ (Fig. 11.8). This behavior shows that when the concentration of chloride increased, it blocked the polymer active sites and the retention of arsenate gradually decreased. These results proved that when the ionic strength increases, the retention capacity of the polymer decreases due to the competition between ions in solution. This behavior depends directly of the type and charge of ion interfering. Even at a low concentration, interfering ions block and diminish the removal capability of the WSP.

For P(GMA-NMG), the results showed that As retention decreased from 82.4% to 72.0%, 62.3%, and 51.9% when the chloride concentration was 48 mg L^{-1}, 280 mg L^{-1}, and 560 mg L^{-1}, respectively (Fig. 11.9a). The decrease of retention in the presence of sulfate ions did not vary significantly with the increasing concentration (Fig. 11.9b). When sulfate ions were present at low concentrations (38 mg L^{-1}), the retention decreased to 63.1%. Figure 11.9c shows the retention capability of P(GMA-NMG) in the presence of both interferent ions. Although the retention of As decreased under these conditions, the polymer was still capable of retaining As, attaining 43.0% retention. This fact shows that there is a degree of selectivity for As(V), which can be attributed to the presence of polyol groups.

11.5   MAXIMUM RETENTION CAPACITY BY THE ENRICHMENT METHOD

The maximum retention capacity of Cr or As ions by WSPs was determined through the enrichment method. The maximum retention capacity (*MRC*) is defined as:

$$MRC = (MV)/Pm \tag{11.3}$$

where *Pm* is the amount of polymer [g], *M* is initial concentration of anions [mg L^{-1}], and *V* is the volume of the filtrate (volume set) containing the toxic anion [mL] that passes through the membrane. Assuming a quantitative retention, the enrichment factor (*E*) is a measurement of the binding capacity of the polymer and is determined as follows:

$$E = [P(MRC)]/M \tag{11.4}$$

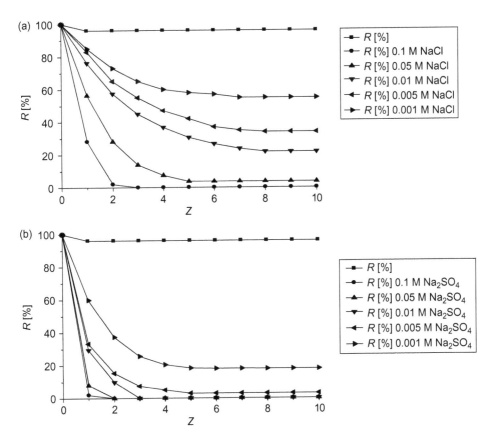

Figure 11.8. Retention profile of As(V) by P(ClAETA) in the presence of different concentrations of (a) NaCl and (b) Na$_2$SO$_4$ in both the reservoir and the ultrafiltration cell at pH 8, using a molar ratio of 20:1 polymer:As(V) (3.2 × 10^{-4} mol:1.6 × 10^{-5} mol) (reproduced with permission from Sanchez and Rivas 2012, copyright of Wiley).

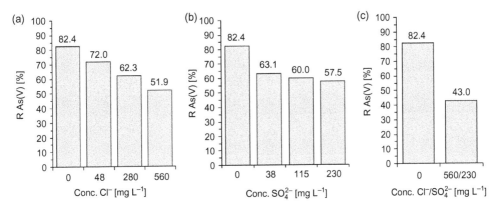

Figure 11.9. Retention profile of As(V) by P(GMA-NMG) in presence of different concentrations of (a) NaCl, (b) Na$_2$SO$_4$, and (c) both interferents in the reservoir and ultrafiltration cell at pH 3, using a molar ratio of 70:1 polymer:As(V) (reproduced with permission from Sanchez and Rivas 2012, copyright of Wiley).

Table 11.4. Maximum retention capacities of Cr(VI) and enrichment factors of cationic water-soluble polymers (reproduced with permission from Sanchez and Rivas 2012, copyright of Wiley).

Polymer	Maximum retention capacity (MRC) [mg Cr(VI)/g polymer]	Enrichment factor (E)
P(ClVBTA)	164	7.5
P(ClAPTA)	152	6.5
P(SAETA)	90	5.0

Table 11.5. Maximum retention capacities of As(V) and enrichment factors of water-soluble polymers (reproduced with permission from Sanchez and Rivas 2012, copyright of Wiley).

Polymer	Maximum retention capacity (MRC) [mg As(V)/g polymer]	Enrichment factor (E)
P(ClAPTA)	380	7.5
P(ClDDA)	369	9.4
P(SAETA)	79	2.5

where P is the concentration of the polymer [g L^{-1}]. As the anion-polymer interactions are processes in equilibrium, a lower slope in the rate of increase of the anion concentration in the filtrate is observed. From the difference in the slopes, the amount of anions bound to the polymer and free in solution, as well as the maximum retention capacity, can be calculated.

The MRC and E for all the polymers are summarized in Table 11.4 for chromate and in Table 11.5 for arsenate.

The highest Cr retention capacity was found for polymers with counter-ions of Cl$^-$, such as P(ClVBTA) (Fig. 11.10, curve ●) and P(ClAPTA) (Fig. 11.10, curve ▲), compared to P(SAETA) (Fig. 11.10, curve ▼), which contains CH$_3$OSO$_3^-$. The nature of the counter-ion was a more important factor for the maximum retention of Cr(VI) ions.

The interaction between the polymer and Cr(VI) was not purely electrostatic, presumably because of the formation of a coordination bond between a partially movable functional group on the polymeric network and another on the oppositely charged Cr anion. This pairing may be explained by the water structure induced by ion pairing, where the larger and more polarizable ions disrupt the local water structure and associate more easily with a given quaternary ammonium ion (Barron and Fritz, 1984; Sánchez and Rivas, 2011a).

The FTIR spectra of P(ClVBTA) before and after the maximum retention capacity of the polymer with Cr(VI) are shown in Figure 11.11. In the high region of the spectra, only the vibrations of a functional group corresponding to P(ClVBTA) could be identified. Some modifications were observed in the spectra in the 700 to 1700 cm^{-1} range. The band intensity at 1641 cm^{-1}, corresponding to the C=C stretching vibration was taken as a reference. Following the addition of Cr(VI), a new band appeared at 1380 cm^{-1} from the chromate groups corresponding to ν (Cr=O). The band at 891 cm^{-1} was assigned to the ν (Cr–O) stretching vibration (Miller and Wilkins, 1952). The band intensity at 1380 cm^{-1} indicates that there is an interaction between the polymer and the Cr oxyanions.

11.5.1   Desorbing of arsenate: the retention-elution process

To study the arsenate retention-elution process, the enrichment method and the washing method were alternately used. In these experiments, P(ClAETA) and P(SAETA), which differ only in their counter-ions, were studied. The first step of the experiment was the saturation of the polymers through the enrichment method using the conditions previously described: the enrichment method was performed at pH 8, using 8×10^{-4} mol of polymer in the ultrafiltration cell (20 mL) and adding

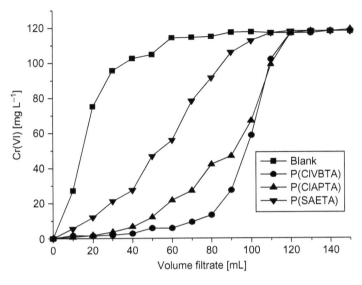

Figure 11.10. Maximum retention capacity of Cr(VI) using P(ClVBTA), P(ClAPTA), and P(SAETA) as extracting agents at pH 9. Mole ratio of 2.5×10^{-4} mol of polymer and solution of Cr(VI) 2.5×10^{-3} M. The blank (■) is the experiment without polymer at pH 9 (reproduced with permission from Sanchez and Rivas 2012, copyright of Wiley).

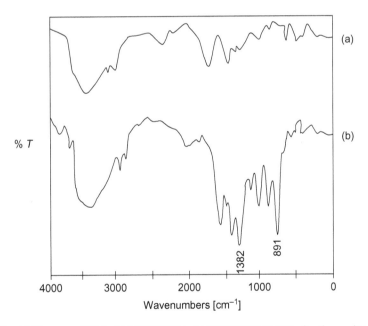

Figure 11.11. FTIR spectra (KBr) of (a) P(ClVBTA), (b) P(ClVBTA)-Cr(VI) after the maximum retention capacity at pH 9.

a solution of 4×10^{-3} M As(V) to the reservoir. After reaching saturation, the polymer:As(V) solution was washed in the ultrafiltration cell with reservoir water buffered at pH 3, in a similar manner as the washing method. It was assumed that the polymer activity can be recovered in the strongly acidic media and that this did not significantly affect the polymer's active sites

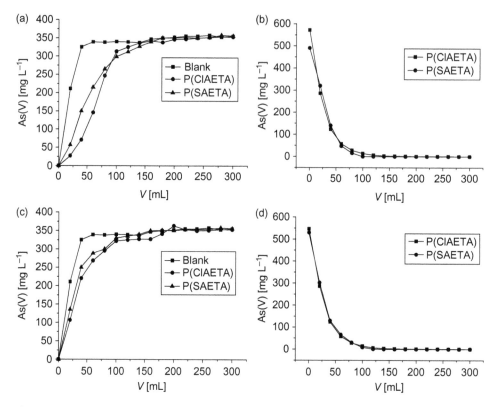

Figure 11.12.   Retention-elution process of arsenate ions using P(ClAETA) and P(SAETA). (a) First charge process of polymers through the enrichment method at pH 8 and (b) first discharge process of polymers using the washing method at pH 3 with 1×10^{-1} M HCl. (c) Recharge of polymers through the enrichment method at pH 8 and (d) second discharge process of polymers using the washing method at pH 3 with 1×10^{1} M HCl (reproduced with permission from Sanchez and Rivas 2012, copyright of Wiley).

because acidic pH was used in the radical polymerization. The same retention-elution process was repeated twice for each polymer to determine the capacity of arsenate delivery and to regenerate the extracting ability of the WSP.

Figure 11.12 shows the retention-elution behavior for both polymers. Figure 11.12a presents the enrichment process reaching the same *MRC* obtained previously for both polymers at pH 8. The values of *MRC* were 165 mg g^{-1} for P(ClAETA) and 79 mg g^{-1} for P(SAETA), and the total filtrate volume was 300 mL.

After the charge process, the discharge process was initiated by changing the pH from basic to acidic using a buffered solution of 1×10^{-1} M HCl. Figure 11.12b presents the discharge process of the arsenate ions from both polymers when the polymer-arsenate was in contact with the acid solution (pH 3) from the reservoir. The first discharge of arsenate was effective and was conducted almost entirely in the first 100 mL of solution when a higher ion arsenate concentration was discharged from P(ClAETA) compared with P(SAETA) at the same volume. Both polymers discharged the total amount of arsenate at 300 mL of filtrate.

Figure 11.12c shows that the second charge process did not improve the polymers' maximum retention capacity compared with the first charge process. P(ClAETA) lost the capacity to remove arsenate, and P(SAETA) was only slightly better at the same conditions. The values of *MRC* were 83 mg g^{-1} for P(ClAETA) and 47 mg g^{-1} for P(SAETA), and the total filtrate volume was 300 mL.

This result is most likely due to the presence of more species in the solution when the pH was adjusted from basic to acidic in the discharge process and from acidic to basic in the second charge process. Finally, the second discharge process (Fig. 11.12d) showed almost the same behavior in both polymers, releasing most of the arsenate ions into the filtrate in the first 100 mL in a similar manner.

11.6   CONCLUSIONS

Liquid-phase polymer-based retention (LPR) has proved to be a convenient method to significantly retain arsenate or chromate anions from solution using polymers with quaternary ammonium and *N*-methyl-D-glucamine groups.

The study of the pH using the washing method showed the highest retention capacity of the polymers with quaternary ammonium groups at basic pH for both toxic anions. In contrast, P(GMA-NMG) showed an optimum removal at pH 3. This fact shows the importance of the type of functional groups present in the polymer structure.

The polymer P(SAETA), containing bulky counter-ions ($CH_3OSO_3^-$) that are more hydrophobic than Cl^- ions, showed lower retention capacity. Thus, the nature of the anionic exchanger groups appears to be an important factor in the retention.

The study of the polymer:anion ratio showed the optimum molar ratio was 10:1 and 20:1 for efficient Cr and As removal, respectively, whereas for the P(GMA-NMG) and As(V), an optimum molar ratio of 70:1 was achieved.

The decrease in the retention ability of the cationic polymer is most likely due to an increase in the solution's ionic strength following the addition of Na_2SO_4, higher than that NaCl, which induced a change in polarization.

The enrichment method shows the maximum retention capacity for Cr(VI) and As(V) anions between 79 to 165 mg anion retained/g polymer at pH 9.

The interaction between the polymer and Cr(VI) was not purely electrostatic. The FTIR spectra of P(ClVBTA) before and after the maximum retention capacity of the polymer show differences due to the presence of Cr(VI).

The retention-elution process shows that it was possible to perform the discharge process of the arsenate ions from the polymers when the polymer-arsenate was in contact with the acid solution from the reservoir.

In the future, this experiment should be repeated several times to determine at what point it is possible to use the same polymer in the retention-elution process.

ACKNOWLEDGMENTS

The authors are grateful for grants from FONDECYT (Grant No 1110079), REDOC (MINEDUC Project UCO1202 at U. of Concepción), and PIA (Grant Anillo ACT 130) and acknowledge the CHILTURPOL2 grant (MCA, 7FP Project no. 269153). Julio Sánchez thanks FONDECYT (No 11140324) and CIPA.

REFERENCES

Ansari, R. & Delavar, A.F. (2010) Removal of Cr(VI) ions from aqueous solutions using poly 3-methyl thiophene conducting electroactive polymers. *Journal of Polymers and the Environment*, 18, 202–207.

Aroua, M.K., Zuki, F.M. & Sulaiman, N.M. (2007) Removal of chromium ions from aqueous solutions by polymer-enhanced ultrafiltration. *Journal of Hazardous Materials*, 147, 752–758.

Baran, A., Bıçak, E., Baysal, S.H. & Onal, S. (2006) Comparative studies on the adsorption of Cr(VI) ions on to various sorbents. *Bioresource Technology*, 98, 661–665.

Barron, R. & Fritz, J. (1984) Effect of functional group structure on the selectivity of low-capacity anion exchangers for monovalent anions. *Journal of Chromatography*, 284, 13–25.

Berdal, A., Verrie, D. & Zaganiaris, E., Removal of arsenic from potable water by ion exchange resins. In: Greig, J.A. (ed.) *Ion exchange at the millennium, Proceedings of IEX*. Imperial College Press, London, UK. p. 101.

Bıçak, N., Ozbelge, H.O., Yilmaz, L. & Senkal, B.F. (2000) Crosslinked polymer gels for boron extraction derived from *N*-glucidol-*N*-methyl-2-hydroxypropyl methacrylate. *Macromolecular Chemistry and Physics*, 201, 577–584.

Bissen, M. & Frimmel, F.H. (2003) Arsenic – a review. Part I: Occurrence, toxicity, speciation, mobility. *Acta Hydrochimica et Hydrobiologica*, 31, 9–18.

Bundschuh, J., Litter, M., Ciminelli, V.S.T., Morgada, M.E., Cornejo, L., Garrido Hoyos, S., Hoinkis, J., Alarcon-Herrera, M.T., Armienta, M.A. & Bhattacharya, P. (2010) Emerging mitigation needs and sustainable options for solving the arsenic problems of rural and isolated urban areas in Latin America – a critical analysis. *Water Research*, 44, 5828–5845.

Cañizares, P., Perez, A., Llanos, J. & Rubio, G. (2008) Preliminary design and optimisation of a PEUF process for Cr(VI) removal. *Desalination*, 223, 229–237.

Cullen, R.W. & Reimer, K. (1989) Arsenic speciation in the environment. *Chemical Reviews*, 89, 713–764.

Ferguson, J.F. & Gavis, J. (1972) A review of the arsenic cycle in natural waters. *Water Research*, 6, 1259–1274.

Geckeler, K.E. & Volchek, K. (1996) Removal of hazardous substances from water using ultrafiltration in conjunction with soluble polymers. *Environmental Science & Technology*, 30, 725–734.

Gupta, V.K., Shrivastava, A.K. & Jain, N. (2001) Biosorption of chromium (VI) from aqueous solutions by green algae *Spirogyra* species. *Water Research*, 35, 4079–4085.

Johnston, R. & Heijnen, H. (2001) Safe water technology for arsenic removal. In: Ahmed, M.F., Ali, M.A. & Adeel, Z. (eds.) *Technologies for arsenic removal from drinking water*. Bangladesh University of Engineering and Technology, Dhaka. Bangladesh. pp. 1–22.

Kabay, N., Bundschuh, J., Hendry, B., Bryjak, M., Yoshizuka, K., Bhattacharya, P. & Anaç, S. (eds.) (2010) *The global arsenic problem: challenges for safe water production*. CRC Press, Boca Raton, FL.

Korus, I. & Loska, K. (2009) Removal of Cr(III) and Cr(VI) ions from aqueous solutions by means of polyelectrolyte-enhanced ultrafiltration. *Desalination*, 247, 390–395.

Kozlowski, C.A. & Walkowiak, W. (2002) Removal of chromium (VI) from aqueous solutions by polymer inclusion membranes. *Water Research*, 36, 4870–4876.

Lakshmipathiraj, P., Bhaskar Raju, G., Raviatul Basariya, M., Parvathy, S. & Prabhakar, S. (2008) Removal of Cr (VI) by electrochemical reduction. *Separation and Purification Technology*, 60, 96–102.

Mertz, W. (1993) Chromium in human nutrition: a review. *Journal of Nutrition*, 123, 626–633.

Miller, F.A. & Wilkins, C.H. (1952) Infrared spectra and characteristic frequencies of inorganic ions. *Analytical Chemistry*, 24, 1253–1294.

Mohan, D., Singh, K.P. & Singh, K.V. (2005) Removal of hexavalent chromium from aqueous solution using low-cost activated carbons derived from agricultural waste materials and activated carbon fabric cloth. *Industrial and Engineering Chemistry Research*, 44, 1027–1042.

Rivas, B.L. & Aguirre, M.C. (2009) Water-soluble polymers: optimization of arsenate species retention by ultrafiltration. *Journal of Applied Polymer Science*, 112, 2327–2333.

Rivas, B.L., Pereira, E. & Moreno-Villoslada, I. (2003) Water-soluble polymer-metal ion interactions. *Progress in Polymer Science*, 28, 173–208.

Rivas, B.L., Aguirre, M.C. & Pereira, E. (2006) Retention properties of arsenate anions of water-soluble polymers by a liquid-phase polymer-based retention technique. *Journal of Applied Polymer Science*, 102, 2677–2684.

Rivas, B.L., Aguirre, M.C. & Pereira, E. (2007a) Cationic water-soluble polymers with the ability to remove arsenate through an ultrafiltration technique. *Journal of Applied Polymer Science*, 106, 89–94.

Rivas, B.L., Aguirre, M.C., Pereira, E., Moutet, J.C. & Saint Aman, E. (2007b) Capability of cationic water-soluble polymers in conjunction with ultrafiltration membranes to remove arsenate ions. *Polymer Engineering Science*, 47, 1256–1261.

Rivas, B.L., Pereira, E., Palencia, M. & Sánchez, J. (2011) Water-soluble functional polymers in conjunction with membranes to remove pollutant ions from aqueous solutions. *Progress in Polymer Science*, 36, 294–322.

Sánchez, J. & Rivas, B.L. (2010) Arsenic extraction from aqueous solution: electrochemical oxidation combined with ultrafiltration membranes and water-soluble polymers. *Chemical Engineering Journal*, 165, 625–632.

Sánchez, J. & Rivas, B.L. (2011a) Cationic hydrophilic polymers coupled to ultrafiltration membranes to remove chromium (VI) from aqueous solution. *Desalination*, 279, 338–343.

Sánchez, J. & Rivas, B.L. (2011b) Arsenate retention from aqueous solution by hydrophilic polymers through ultrafiltration membranes. *Desalination*, 270, 57–63.

Sánchez, J. & Rivas, B.L. (2012) Liquid phase polymer based retention of chromate and arsenate oxy-anions. *Macromolecular Symposia*, 317–318, 123–136.

Smith, B.F., Robison, T.W., Carlson, B.J., Labouriau, A., Khalsa, G.R.K., Schroeder, N.C., Jarvinen, G.D., Lubeck, C.R., Folkert, S.L. & Aguino, D.I. (2005) Boric acid recovery using polymer filtration: studies with alkyl monool, diol, and triol containing polyethylenimines. *Journal of Applied Polymer Science*, 97, 1590–1604.

Spivakov, B.Y., Geckeler, K. & Bayer, E. (1985) Liquid-phase polymer-based retention – the separation of metals by ultrafiltration on polychelatogens. *Nature*, 315, 313–315.

Toledo, L., Rivas, B.L., Urbano, B.F. & Sánchez, J. (2013) Novel *N*-methyl-D-glucamine-based water-soluble polymer and its potential application in the removal of arsenic. *Separation and Purification Technology*, 103, 1–7.

USEPA (1998) *Toxicological review of hexavalent chromium.* Environmental Protection Agency, Washington, DC.

USEPA (2001) National primary drinking water regulations; arsenic and clarifications to compliance and new source contaminants monitoring. *Federal Register*, 60 (14), 6976–7066, Environmental Protection Agency, Washington, DC.

Von Burg, R. & Liu, D. (1993) Chromium and hexavalent chromium. *Journal of Applied Toxicology*, 13, 225–230.

WHO (1993) Guidelines for drinking-water quality. World Health Organization, Geneva, Switzerland.

WHO (2001) Arsenic Compounds, Environmental Health Criteria 224. World Health Organization, Geneva, Switzerland.

William, R.A. (1992) *Colloid and surface engineering: applications in the process industries.* Butterworth-Heinemann, Oxford, UK. p. 11.

CHAPTER 12

Solid waste materials for arsenic and chromium removal

Izabela Polowczyk, Tomasz Koźlecki, Justyna Ulatowska & Anna Bastrzyk

12.1 INTRODUCTION

A variety of methods and materials have been developed for removal of heavy metals and metal-loids from wastewater and synthetic wastewater (Fu and Wang, 2011). Treatment processes include among others chemical precipitation, ion exchange, adsorption, and mixed coprecipitation/adsorption techniques (Bailey et al., 1999). All methods have advantages and disadvantages, however adsorption is well studied, since it is an efficient and versatile one for As and Cr removal (Owlad et al., 2009). Comparing the adsorbent materials, the crucial parameter is expense, which varies for individual sorbent relative to local availability and required pretreatment. Many efforts have been made to produce cheap adsorbents and replace commercially available activated carbons and other costly techniques of wastewater treatment (Wan Ngah and Hanafiah, 2008). Generally, an adsorbent can be considered as low-cost if requires little processing, is available in excess in nature or is a waste material or industrial by-product (Bailey et al., 1999).

The industrial wastes are generally inorganic materials, however biomass, a by-product of the fermentation industry, or biosolids, a by-product of sewage and wastewater treatment plants, are composed of organic matter. The agricultural by-products are lignocellulosic biomass (Sud et al., 2008).

The most often used for heavy metal removal from water and wastewater are the following industrial waste: fly ash, blast furnace slag, red mud, waste sludge, carbon slurry, and fermentation biomass. The agricultural by-products, such as sawdust, bark, coir pith, straw, bagasse, olive industry waste, cobs, seeds, peat, leaves, husks, bran, shells and others are also commonly applied (Wan Ngah and Hanafiah, 2008). Activated carbons prepared from solid waste can also be regarded as cheap, if the adsorptive properties achieved compensate for the required processing expenses (Bailey et al., 1999). They also have been used for removal of As and Cr, however, they were of no concern in this work.

In this chapter we want to present the application of solid waste materials as sorbents of As and Cr ions. Solid waste materials were divided into two groups, industrial wastes and agricultural by-products, according to the origin. The adsorbents considered were characterized by composition, physico-chemical properties, metal uptake equilibrium and kinetic results. The possible mechanisms of As and Cr removal provided in the literature are presented. Additionally, in two tables, separately for As and Cr, over 100 exemplary adsorbents are listed, with the maximum reported uptakes, optimal conditions and adsorption equilibrium and kinetic models studied.

12.2 SOLID INDUSTRIAL BY-PRODUCTS

Industrial wastes and by-products are produced abundantly and are materials normally posing disposal and environmental problems. Occasionally, they may be used as construction materials or backfill and some of them have significant heating power or recycling potential. The composition of these materials depends on their origin and very often they are a mixture of more or less hazardous components. Safe disposal or reuse of such materials is a big concern. Industrial waste is one of the potential low-cost adsorbents for heavy metal removal (Babel and Kurniawam, 2003). For many years these materials have been investigated as potential adsorbents for water and wastewater purification, as a substitute for commercial costly adsorbents such as resins or activated carbons (Wang and Wu, 2006). What is more, very often low-cost adsorbents exhibit a sorption capacity bigger or comparable with commercially available adsorbents (Demirbas, 2008), especially when subjected to appropriate modifications.

12.2.1   Fly ash

Fly ash is a common by-product of coal incineration. This material consists mostly of silica and alumina oxide particles. However, the particular components in fly ash can be different, depending on combustion conditions and the nature of the fuel (Vassilev and Vassileva, 2005). Literature data revealed that fly ash can be employed as a sorbent to remove heavy metals from aqueous solution (Bertrocchi *et al.*, 2006; Wang and Wu, 2006).

A class C (high calcium oxide content) raw fly ash from thermal power plant was used as a sorbent for As(V) removal by Diamadopoulos *et al.* (1993). The specific surface area was found to be $0.8 \, \text{m}^2 \, \text{g}^{-1}$. The results showed that the adsorption of As(V) decreased with an increase in pH of the suspension. The best removal of As(V) was obtained at pH 4 and the calculated maximum capacity was $27.8 \, \text{mg} \, \text{g}^{-1}$. Equilibrium was reached after 72 hours. The adsorption data fitted well by both Freundlich and Langmuir isotherm models.

Li *et al.* (2009) presented studies on class F (high silica and alumina content) raw fly ash from thermal power plant as a sorbent of As(V). The BET surface area of raw fly ash was $6.23 \, \text{m}^2 \, \text{g}^{-1}$. It was shown that the adsorption density was $11.2 \, \text{mg} \, \text{g}^{-1}$ at pH 2.5, initial concentration of $50 \, \text{mg} \, \text{L}^{-1}$, dosage of sorbent $2 \, \text{g} \, \text{L}^{-1}$ and room temperature. The authors developed also a new route for utilization of fly ash. They prepared a sorbent high iron containing class F fly ash containing a high amount of amorphous FeOOH on the surface of the adsorbent. This new sorbent is characterized by high surface area, $140.07 \, \text{m}^2 \, \text{g}^{-1}$ and a more porous structure than raw fly ash. Effective removal of As(V) was realized in the pH range of 1–8.0. The best removal was observed at pH 2.5, initial As concentration of $50 \, \text{mg} \, \text{L}^{-1}$, adsorbent dosage of $2 \, \text{g} \, \text{L}^{-1}$, contact time 15 h and 25°C. In the same conditions, it was shown that $40 \, \text{g} \, \text{L}^{-1}$ of modified fly ash is sufficient to remove the total amount of As(V) ($50 \, \text{mg} \, \text{L}^{-1}$). The adsorption data fitted the Langmuir model and the calculated value of maximum capacity was $19.46 \, \text{mg} \, \text{g}^{-1}$. Kinetic data showed that equilibrium was reached after 8 hours of contact.

Balsamo *et al.* (2010) also used class F coal fly ash for As(V) removal from both distilled and mineral water. The BET surface area was found to be $19.0 \, \text{m}^2 \, \text{g}^{-1}$. The shape of adsorption isotherms for both investigated waters remained almost unchanged in mineral water because of the releasing of anions in distilled water from raw fly ash. The maximum adsorption was achieved for two ranges of pH, 2–4 and 10–12. The adsorption capacity in these ranges of pH and at $10 \, \text{mg} \, \text{L}^{-1}$ of As(V) was about $0.5 \, \text{mg} \, \text{g}^{-1}$. In order to increase the adsorption capacity for As(V), the authors developed acid-treated class F fly ash as a sorbent. The samples were prepared by immersing of raw fly ash in solutions of HCl at different concentrations, ranging from 0.01 to 10 M. It was observed that As was more effectively removed from water when more acidic solutions were applied. Taking into account the cost of the process, the authors used 1 M HCl-treated fly ash. The experimental data showed that the maximum capacity for the modified sorbent was reached at pH 2–4 and it was about $1.1 \, \text{mg} \, \text{g}^{-1}$ at $10 \, \text{mg} \, \text{L}^{-1}$ of As(V), 1g of sorbent per 0.1 L of solution and 20°C. This modification did not change the BET surface area but led to surface oxidation, which enhanced the removal of As(V) and shifted the pH$_{\text{PZC}}$ to pH 4.0. Moreover, treatment of fly ash with HCl solution resulted in leaching of ions from fly ash, especially calcium, which was also released to water when the raw material was used.

Removal of As(III) using fly ash was investigated by Polowczyk *et al.* (2010, 2011) and Ulatowska *et al.* (2014). In their studies, fly ash taken from different power plants and fluidal boilers, the Turów power plant (unit No. 4 and 6) and the Zgierz power plant, were used. Polowczyk *et al.* (2010a) used class C fly ash from the Turów brown coal power plant (unit No. 6) as a sorbent for removal of As(III) from synthetic water and wastewater. The BET surface area of the sorbent was $5.40\,m^2\,g^{-1}$. The adsorption data fitted the Langmuir isotherm adsorption model well, and the maximum adsorption capacity was calculated to be $32.87\,mg\,g^{-1}$ at pH 11–12.5, total initial concentration of $1000\,mg\,L^{-1}$, adsorbent dosage of $20\,g\,L^{-1}$, and 25°C. The kinetics data followed the pseudo-second-order model. It was observed that this type of fly ash can be used to remove heavy metals from industrial wastewater.

Fly ash is a fine-particle material which is very difficult to remove from aqueous solution, and there is a need to use an additional step to separate it from solution. In many branches of industry powdered materials are used in a compressed form of granules, pellets and briquettes. Agglomerates of class C and class F fly ash with addition of blast-furnace cement as a binder were used in removal of As(III) from water in batch and column studies (Polowczyk *et al.*, 2007). The equilibrium experimental data were fitted by the Freundlich isotherm, while kinetic data using both the Elovich and Lagergren models. Polowczyk *et al.* (2010b) presented the possibility of use of granulated class C fly ash from a brown coal-burning power plant (unit No. 4) as a cheap, irreversible adsorbent for removal of As(III) from concentrated solution. After granulation and hardening of agglomerates, all the calcium oxide disappeared and the new phases appeared; ettringite ($3CaO \cdot Al_2O_3 \cdot 3CaSO_4 \cdot 32H_2O$) and a C-S-H phase (calcium silicate hydrate). The experiments were performed in batch and column tests. The BET surface area of the agglomerates was found to be $8.50\,m^2\,g^{-1}$. The adsorption data followed the Freundlich isotherm equation and the pseudo-second order kinetic model. The calculated maximum adsorption capacity was 2.17 and $2.23\,mg\,g^{-1}$ in column and batch tests, respectively. The conditions were as follows; sorbent dose of $250\,g\,L^{-1}$, total As concentration $1000\,mg\,L^{-1}$, pH 11–12 and 25°C. Use of non-agglomerated class C fly ash from a brown coal-burning power plant (unit No. 4) for removal of As(III) was also investigated (Polowczyk *et al.*, 2011). The BET surface area of the adsorbent was $6.50\,m^2\,g^{-1}$. For the non-granulated adsorbent, the adsorption data followed the Langmuir isotherm model and the pseudo-second-order kinetic equation. The maximum adsorption capacity was $36.65\,mg\,g^{-1}$. It was shown that the purification of wastewater is possible by this type of fly ash.

Ulatowska *et al.* (2014) applied raw and granulated class C fly ash from a brown coal and biomass-burning power plant. The BET surface of raw and granulated sorbent was found to be 17.30 and $25.9\,m^2\,g^{-1}$, respectively. During the agglomeration a new phase C-S-H phase was created. It was found that adsorption of As(III) onto powder and agglomerated fly ash was not much affected over a range of pH (5–12). The experimental maximum removal of As was observed at 25 and $30\,g\,L^{-1}$ of adsorbent for powder and agglomerated fly ash, and it was 13.5 and $5.7\,mg\,g^{-1}$, respectively. The authors noticed that adsorption studies showed that the data well fitted the Langmuir and Freundlich isotherm models for non-agglomerated and agglomerated fly ash. The kinetics data followed the pseudo-second order model. Adsorption of As(III) on both forms of fly ash turned out to be an endothermic and spontaneous process. The maximum adsorption capacities on FA are collected in Table 12.1.

In the removal of Cr(III) and Cr(VI), fly ash of different origins has been used, and has been investigated by many research groups. Two different types of fly ash from coal-burning thermal power plants were studied as potential adsorbents for Cr(VI) removal from aqueous solution (Bayat, 2002). One of the adsorbents was found to be a class F while the second was class C fly ash. The specific surface area was found to be 0.12 and $0.34\,m^2\,g^{-1}$ for class F and class C FA, respectively. The particle size analysis showed less fine particles ($-75\,\mu m$) in class F FA, 15%, compared with class C FA, 75%. The adsorption equilibrium for Cr(VI) was established within 2 h at 20°C. The removal of Cr(VI) was affected by pH and higher at pH 4.0 for class C FA (25.5%) and pH 3.0 for class F FA (32%) and decreased with increasing pH. The adsorption of Cr(VI) on both types of fly ash was described well by neither the Langmuir nor Freundlich isotherm models. Comparing the adsorption capacities of both fly ashes (0.70 and $0.88\,mg\,g^{-1}$ for class-C and class-F fly ash, respectively) it was found that they were nearly three times less than that of activated carbon ($2.75\,mg\,g^{-1}$) for the removal of Cr(VI).

Table 12.1. Various arsenic adsorbents from literature: origin, process conditions, adsorption capacities and equilibrium and kinetics models applied.

Adsorbent	Adsorbent origin	Element	pH	T [°C]	Adsorption capacity [mg g^{-1}]	Isotherm model	Kinetic model	Reference
Industrial waste								
Red mud	Aluminum plant	As(III) As(V)	9.5 3.2	25	0.66 0.51	L	PFO	Altundoğan et al. (2000)
Activated red mud	Aluminum plant	As(III)	7.2	25–70	0.88 (25°C) 0.66 (40°C) 0.59 (55°C) 0.34 (70°C)	L	PFO	Altundoğan et al. (2002)
		As(V)	3.5		0.94 (25°C) 1.12 (40°C) 1.28 (55°C) 1.33 (70°C)			
Activated red mud	Aluminum refinery	As(V)	6.3	23	1.08	L	N/A	Genç et al. (2003)
Activated red mud	Aluminum refinery	As(V) As(III)	4.5 6.8	23	7.60 0.54	L	PFO	Genç-Fuhrman et al. (2004a; 2004b)
FeCl$_3$-modified red mud	Aluminum corporation	As(V)	6.0 7.0 9.0	20	68.5 50.6 23.2	L	PSO	Zhang et al. (2008)
Red mud with phosphogypsum	Local source	As(V)	3.5	24	3.3	L	N/A	Lopes et al. (2013)
Fe$_2$O$_3$ NP obtained from red mud	Aluminum refinery	As(V)	2.5	25	29.6	F	PSO	Akin et al. (2012)
Pisolite	Manganese oxide mining	As(V)	6.5	N/A	1.29 (batch test) 1.5 (column test)	N/A	N/A	Pereira et al. (2007)
Activated pisolite	Manganese oxide mining	As(V)	6.5	N/A	3.17 (batch test) 3.5 (column test)	N/A	N/A	Pereira et al. (2007)
Steel making slag: ECD,OGS, BOFS*	Steel mill	As(III) As(V)	11–12	r.t.	<1.25	N/A	N/A	Ahn et al. (2003)
BFS*		As(III) As(V)	6–10		0.50 (pH 6.0) 0.25 (pH 10) 0.90 (pH 6.0) 0.15 (pH 10)			

Material	Source	As species	pH	Temp (°C)	Capacity	Mechanism	Model	Reference
Steel making slag / FeOOH-loaded steel slag		As(V) As(V) As(III)	2.5	20	7.0 78.5 30.1	N/A	N/A	Zhang and Itoh (2005)
Iron slag waste	Steel production plant	As(V) As(III)	7.0	20	4.04 0.82	L	PSO	Jovanović et al. (2011)
BFS*	Local steel mill	As(III) As(V)	7	N/A	0.82 4.04	L	PSO	Lekić et al. (2013)
Modified-BFS*		As(III) As(V)			0.70 2.79			
Class C-FA*	Thermal power plant	As(V)	4.0 7.0 10	20	27.80 10.0 7.0	L, F	N/A	Diamadopoulos et al. (1993)
Fly ash	Power plant	As(V)	10.6	N/A	0.75	N/A	N/A	Bertocchi et al. (2006)
Class C-FA*	Power plant	As(III)	11.0–12.5	25	32.87	L	PSO	Polowczyk et al. (2010a)
Agglomerated class C-FA*	Power plant	As(III)	11–12.0	25	2.17 (column test) 2.2 (batch test)	F	PSO	Polowczyk et al. (2010b)
Class C-FA*	Power plant	As(III)	12.9	25	36.65	L	PSO	Polowczyk et al. (2011)
Class C-FA*	Power plant	As(III)	10–11.5	25	13.5	L	PSO	Ulatowska et al. (2014)
Agglomerated class C-FA*			8.5–10.7		5.7	F		
Class F-FA* high iron oxide class F-FA	Power plant	As(V)	2.5	25	11.2 19.46	L	N/A	Li et al. (2009)
Class F-FA* acid-treated class F-FA	Power plant	As(V)	2–4	20	0.50 1.10	N/A	N/A	Balsamo et al. (2010)
Bagasse FA*	Sugar factory	As(III) As(V)	2.0–12.0	20–30	21.51 (20°C) 20.92 (25°C) 19.90 (30°C) 19.76 (20°C) 20.74 (25°C) 18.87 (30°C)	L, F, TI	PSO	Ali et al. (2014)
Modified bagasse FA*	Sugarcane juice shop	As(III) As(V)	6.0	25	0.0177 (batch test) 0.085 (column test) 0.0176 (batch test) 0.084 (column test)	N/A	N/A	Roy et al. (2013)
Laterite by-product	Local quarries	As(III) As(V)	7.0	N/A	0.30 0.92	L, F	PSO	Glocheux et al. (2013)

(continued)

Table 12.1. Continued.

Adsorbent	Adsorbent origin	Element	pH	T [°C]	Adsorption capacity [mg g⁻¹]	Isotherm model	Kinetic model	Reference
Fe(III)/Cr(III) hydroxide waste	Petrochemical industries	As(V)	4.0	32	11.02	L, F	PFO	Namasivayam and Senthikumar, (1998)
Biochars:	Local source	As(V)	N/A			F	PSO	Agrafioti et al. (2014)
BC-RH*			9.0		0.0026			
BC-SS*			7.0		0.0042			
BC-SW*			9.0		0.0035			
Agricultural by-products								
Rice husks	Rice milling industry	As(V)			0.013	N/A	N/A	Amin et al. (2006)
		As(III)			0.016			
Rice polish	Local rice mill	As(V)	4.0	20	0.147	L, F, DR	PSO	Ranjan et al. (2009)
		As(III)	7.0		0.139			
Rice husk		As(III)	6.0	30	224.4	F	PSO	Kamsonlian et al. (2012a)
Modif. rice husk	Local source	As(V)	7.5	25	18.0	L	N/A	Lee et al. (1999)
Modif. rice husk	Local source	As(III)	7.0	22	0.77	L	N/A	Chaudhuri and Mohammed (2012)
Banana peel	Fruit waste	As(III)	7.0	35	0.84	F	PSO	Kamsonlian et al. (2012b)
Fe(III) modified spruce sawdust	Local source	As(V)	8.0		9.26	L	PSO	Urik et al. (2009)
Pine sawdust	Local wood mill	As(V)	4.0		4.40	L-F	PSO	López-Leal et al. (2012)
			7.0		2.60			
FeCl-SW*			4.0		12.80			
			7.0		5.90			
FeNit-SW*			4.0		6.00			
			7.0		4.60			
Fe(III)-load orange waste	Local source	As(III)	9.0	30	68.1	N/A	N/A	Ghimire et al. (2003)
		As(V)	3.0					
Zr(IV)-load orange waste	Local source	As(V)	3.0	30	88.0 (batch test)	L	PSO	Biswas et al. (2008a)
					36.3 (column test)			
		As(III)	10.0		130.0 (batch test)			
					35.1 (column test)			
La(III)-load orange waste	Local source	As(III)	10.0	30	43.0 (batch test)	L	PSO	Biswas et al. (2008b)
					10.5 (column test)			
		As(V)	6.5		42.0 (batch test)			
					14.2 (column test)			
Activated white chicken feathers	Poultry industry	As(III)	2.0	25	13.0	L	N/A	Teixeria and Ciminelli (2005)

*FA – fly ash; Class C-FA – high calcium oxide content; Class F-FA – high silica and alumina content; FeCl-SW – ferric chloride modified sawdust; FeNit-SW – ferric nitrate modified sawdust; BC-RH – biochar from rice husk; BC-SS – biochar from sewage; BC-SW – biochar from solid waste; BFS – blast furnace slag from steel production; ECD – evaporation cooler dust; OGS – oxygen gas sludge; BOFS – basic oxygen furnace slag.

The potential of aluminum and iron-impregnated fly ash (IFA) for the removal of hexavalent Cr from aqueous solutions was investigated (Banerjee *et al.*, 2004). The FTIR analysis showed the raw adsorbent was a class F fly ash with a high content of SiO_2 (52%) and Al_2O_3 (27%), a minor contribution of Fe_2O_3 (about 6%) and loss of ignition of about 13%. Adsorption efficiency was found to be dependent on the initial concentration and the equilibrium data fitted the Langmuir adsorption model well. The maximum adsorption capacities of FA, Al-impregnated FA, and Fe-impregnated FA for Cr(VI) were found to be 1.38, 1.82, and 1.67 mg g^{-1}, respectively. The equilibrium was established in 30 min. Rate constants were evaluated in terms of a first-order kinetic model. With increase in temperature the adsorption of Cr(VI) by FA, Al-IFA, and Fe-IFA increased at concentration of 5 mg L^{-1} and pH of 2.0. The calculated thermodynamic parameters confirmed the endothermic nature of the adsorption process for both raw and metal-impregnated fly ash. The adsorption was found to be pH-dependent and there was an initial increase up to pH 2.0 followed by a continuous decrease. The positive effect of Al and Fe impregnation on adsorption was probably caused by newly developed active sites (Al or Fe hydroxide layers) formed due to the electrostatic interaction between $Al(OH)_3$, $Fe(OH)_3$, and SiO_2 on IFA.

The capability of zeolites synthesized from different types of fly ash (ZFAs) to remove Cr(III) from aqueous solutions was investigated in a batch mode (Wu *et al.*, 2008). The zeolites were prepared by boiling fly ash of different coal origins and chemical composition with 2.0M NaOH with reflux for 48 h. It was found that the main chemical composition of the ZFAs and their raw fly ash changed and the SiO_2 content decreased significantly. No other constituent changes appeared in the chemical composition. Both the ZFAs and their raw fly ash consisted mainly of Si and Al components, followed by Ca and Fe components, while the MgO content was very low. The effect of pH on Cr(III) removal and the adsorption isotherms at a constant pH 4 were investigated using three representative ZFAs with low, medium and high-calcium contents. The results showed that zeolites from FA had greater ability than raw fly ash to remove Cr(III). As can be seen from Table 12.2, hydroxysodalite produced from a high-Ca fly ash demonstrated higher sorptive capacity for Cr(III) than the NaP1 zeolite from medium- and low-Ca fly ashes. It was concluded that ZFAs and high-Ca fly ash might be promising materials for the Cr(III) removal from wastewater.

The application of fly ash porous pellets as adsorbent of Cr(III) ions from synthetic aqueous solution was studied by Papandreou *et al.* (2011). Experiments were carried out using porous pellets developed from class C fly ash derived from a coal-burning power plant. The raw material, with mean particle diameter of about 75 μm, was formed into spherical agglomerates after wetting and tumbling in a laboratory scale granulator. Kinetic studies indicated that sorption followed a pseudo-second-order model. The equilibrium adsorption data for Cr fitted the Langmuir isotherm equation. The adsorption capacity of fly ash pellets reached 22.88 mg g^{-1} at 25°C and pH 2.0. The results showed that the adsorption rate of Cr(III) was substantially increased with temperature (25–60°C), indicating the endothermic phenomenon of chemisorption. The stabilization of metal saturated pellets in a concrete structure was studied. Leaching tests showed excellent heavy metals stabilization in concrete blocks.

The fly ashes obtained from the agricultural by-products are also discussed in this paragraph, since their composition after the burning process is similar to those produced in thermal power plants. The fly ash from the combustion of poultry litter was assessed as an adsorbent for Cr(III) from aqueous solution by Kelleher *et al.* (2002). The material was taken from the second cyclone of a fluidized-bed combustor. The BET surface area of adsorbent was found to be 11 m^2 g^{-1}, and the pore size distribution indicated a mesoporous arrangement. It was found that uptake attained equilibrium between 15 and 30 minutes. Adsorption data was best described by the Langmuir isotherm model. The maximum adsorption at pH 5.3 was 53, 61 and 106.5 mg g^{-1} at 20, 30 and 40°C, respectively. The negative value of the Gibbs free energy change indicated the feasibility of the process and the spontaneous nature of adsorption. The positive value of enthalpy change indicated the adsorption as an endothermic process. The positive entropy change reflects an affinity of Cr(III) to the adsorbent surface. Kinetic studies suggested that adsorption followed the pseudo-second order model. Film diffusion limited the adsorption rate at low initial Cr concentrations, while pore diffusion became more important at higher initial concentrations.

Table 12.2. Various chromium adsorbents from literature: origin, process conditions, adsorption capacities and equilibrium and kinetics models applied.

Adsorbent	Adsorbent origin	Element	pH	T [°C]	Adsorption capacity [mg g^{-1}]	Isotherm model	Kinetic model	Reference
Industrial waste								
Clarified sludge	Steel industry	Cr(VI)	3.0	30	26.31	L	PFO, PSO	Bhattacharya et al. (2008)
Biogas residual slurry	Biogas plant	Cr(VI)	2.0	30	5.87	L, F	PFO	Namasivayam and Yamuna (1995)
Carbon slurry	Ammonia plant	Cr(VI)	2.5	30–60	24.05 (30°C) 25.15 (45°C) 25.64 (60°C)	L	PFO	Singh and Tiwari (1997)
Activated carbon slurry	Fertilizer industry	Cr(VI)	2.0	30–50	15.24 (30°C) 6.14 (40°C) 5.08 (50°C)	L, F	PSO	Gupta et al. (2010)
Distillery sludge	Distillery unit	Cr(VI)	3.0	N/A	5.7	L, F	PFO	Selvaraj et al. (2003)
Wine processing waste sludge	Wine processing factory	Cr(III)	4.0	20–50	10.46 (20°C) 13.45 (30°C) 15.40 (40°C) 16.36 (50°C)	L, F	PFO, IPD	Li et al. (2004)
Mucor meihi biomass	Fermentation industry	Cr(III)	4.0	N/A	59.8	L	N/A	Tobin and Roux (1998)
Dead fungal biomass of *Aspergillus niger*	Fermentation processes	Cr(VI)	2.0	22–28	20.0	N/A	PFO	Park et al. (2005)
Blast furnace sludge	Steel mill	Cr(III)	N/A	20–80	9.55 (20°C) 9.46 (40°C) 12.06 (60°C) 16.05 (80°C)	L, F	N/A	López-Delgado et al. (1998)
Activated red mud	Aluminum industry	Cr(VI)	5.2	30	30.74	L, F	N/A	Pradhan et al. (1999)
Activated red mud	Aluminum industry	Cr(VI)	2.0	30–50	22.67 (30°C) 21.58 (40°C) 21.06 (50°C)	L, F	PFO	Gupta et al. (2001)
Spent activated clay	Edible oil refinery company	Cr(VI)	2.0	4–40	0.743 (4°C) 0.756 (14°C) 0.957 (24°C) 1.422 (40°C)	L	PFO	Weng et al. (2008)

Adsorbent	Source	Metal			Capacity	Isotherm	Kinetics	Reference
Foundry sand	Iron foundry industry	Cr(VI)	2.5	25–55	1.99 (25°C) 2.40 (35°C) 2.50 (45°C) 3.14 (55°C)	L	PSO	Campos et al. (2013)
Class C-FA*	Coal power plant	Cr(VI)	4.0	20	0.70	L, F	N/A	Bayat (2002)
Class F-FA*			3.0		0.88			
Class F-FA*	Coal power plant	Cr(III)	6.8	20–25	2.65	L	N/A	Wang et al. (2004)
Fly ash Al-impregn. FA FA Fe-impregn. FA	Coal power plant	Cr(VI)	2.0	30	1.38 1.82 1.67	L	PFO	Banerjee et al. (2004)
Coal fly ash Zeolites from FA (ZFA*)	Coal power plant	Cr(III)	3.4	20	1.8 l-Ca FA 9.0 m-Ca FA 32.2 h-Ca FA 25.2 l-Ca ZFA 39.4 m-Ca ZFA 75.5 h-Ca ZFA	L	N/A	Wu et al. (2008)
Coal FA*	Power plant	Cr(VI)	2.0	30	23.86	L	PFO, PSO	Bhattacharya et al. (2008)
Rice husk FA*	Rice mill boiler		3.0		25.64			
Rice hull ash	Rice mill boiler	Cr(III)	4.1	20–50	11.98 (20°C) 12.65 (30°C) 14.46 (40°C) 16.14 (50°C)	N/A	PSO	Wang and Lin (2008)
Poultry litter FA*	Fluidized-bed combustor	Cr(III)	5.3	20–40	52.60 (20°C) 60.20 (30°C) 106.38 (40°C)	L	PSO	Kelleher et al. (2002)
Bagasse FA*	Sugar factory	Cr(III)	5.0	30–50	4.35 (30°C) 4.30 (40°C) 4.25 (50°C)	L, F	FD, IPD	Gupta and Ali (2004)
FA porous pellets	Coal power plant	Cr(III)	7.0	25	22.88	L	PSO	Papandreou et al. (2011)

(continued)

Table 12.2. Continued.

Adsorbent	Adsorbent origin	Element	pH	T [°C]	Adsorption capacity [mg g^{-1}]	Isotherm model	Kinetic model	Reference
Agricultural by-products								
Rice straw	Local source	Cr(VI)	2.0	30	12.17	F	PSO	Singha and Das (2011)
Rice bran			2.0		12.34	F		
Rice husks			1.5		11.40	L		
Hyacinth roots			2.0		15.28	F		
Neem leaves			2.0		15.95	L		
Coconut shell			2.0		18.70	L		
Wool	Local source	Cr(VI)	2.0	30	41.15	L, F	PFO	Dakiky et al. (2002)
Olive cake					33.44			
Sawdust					15.82			
Pine needles					21.50			
Almond shells					10.62			
Cactus leaves					7.08			
Pinus roxburghii bark	Joinery mills	Cr(VI)	6.5 / 3.0	30	4.15 / 4.81	F	N/A	Ahmad et al. (2005)
Neem bark	Local source	Cr(VI)	2.0	30	19.60	L	PFO, PSO	Bhattacharya et al. (2008)
Tea factory waste	Tea factory	Cr(VI)	2.0	25–60	30.00 (25°C) 35.60 (45°C) 39.62 (60°C)	L	PFO	Malkoc and Nuhoglu (2007)
Sugarcane bagasse	Sugarcane industry	Cr(VI)/ Cr(III)	1.9	30	22.98	N/A	N/A	Krishnani et al. (2009)
Sugar beet pulp	Sugar industry	Cr(VI)/ Cr(III)	5.5	20	10.04	L, F	N/A	Reddad et al. (2003)
Pomace olives-cake (WPOOF*)	Olive oil factory	Cr(VI)	2.0	25–60	12.15(25°C) 13.02 (45°C) 16.49 (60°C)	L, F	N/A	Malkoc et al. (2006)
Sugarcane bagasse	Sugar cane processing	Cr(VI)	2.0	25	5.75	L, F	N/A	Garg et al. (2007)
Maize corncobs	Corn milling process				3.0			
Jatropha oil cake	Biodiesel recovery				11.75			
Rice straw	Rice mill	Cr(VI)	2.0	27	3.15	L	N/A	Gao et al. (2008)
Teak sawdust	Saw mill	Cr(VI)	3.0	N/A	1.48	F	N/A	Sumathi et al. (2005)
Rice husk	Rice mill				0.06			
Coir pith	Coconut mill				0.16			

Adsorbent	Source	Metal			Capacity	Isotherm	Kinetics	Reference
Banana peel	Fruit waste	Cr(VI)	2.0	20–40	131.56 (20°C) 96.20 (25°C) 84.24 (30°C) 79.04 (35°C) 60.32 (40°C)	L, D-R	PFO	Memon et al. (2009)
Maize bran	Flour mill	Cr(VI)	2.0	20–40	9.00 (20°C) 9.49 (30C) 10.00 (40°C)	L	PSO	Hasan et al. (2008)
Sorghum straw Oats straw Agave bagasse	Local field	Cr(III)	4.0	25	9.35 12.10 28.72	L, F	PSO, FD	Garcia-Reyes and Rangel-Mendez (2010)
Rubber wood sawdust	Wood factory	Cr(VI)	2.0	25–45	0.92 (25°C) 4.87 (45°C)	L	N/A	Zakaria et al. (2009)
Sawdust	Joinery workshop	Cr(VI)	1.0	30	41.52	L	PSO	Gupta and Babu (2009)
Teak-wood sawdust	Local sawmill	Cr(VI)	3.0	30	20.70	L	PFO, PSO	Bhattacharya et al. (2008)
Sawdust of poplar tree FM-modified peanut husk	Local source	Cr(III)	4.0	25	5.52 7.67	L, F	PSO	Li et al. (2007)
SP-treated pine sawdust	Timber mill	Cr(VI)	2.0	40	121.95	L	PFO	Uysal and Ar (2007)
HCl-treated meranti sawdust	Timber industry	Cr(III)	6.0	30	37.9	L, F, D-R	PSO	Rafatullah et al. (2009)
Rice husk FM-treated rice husk	Rice mill	Cr(VI)	2.0	25	8.5 10.4	L, F, D-R	N/A	Bansal et al. (2009)
Coir pith AA-grafted coir pith	Coconut mattress production	Cr(VI)	2.0	30 30–50	165.00 (30°C) 196.00 (30°C) 259.76 (40°C) 310.56 (50°C)	L	N/A	Suksabye and Thiravetyan (2012)
Coconut coir	Coconut production	Cr(VI)	3.0	25	70.4	N/A	PFO	Shen et al. (2012)
CA-treated walnut shells	Local factory	Cr(VI)	2.0	25	8.0 (untreated), 31.0 (CA-treated)	L, F, D-R	N/A	Altun and Pehlivan (2012)
FM-treated sunflower heads	Local field	Cr(VI)	2.0	25	7.85	L	N/A	Jain et al. (2013)
Wheat bran TA-treated wheat bran	Flour milling	Cr(VI)	2.0 2.2	25 5.28	4.53	F	N/A	Kaya et al. (2014)

*FA – fly ash; Class C-FA – high calcium oxide content; Class F-FA – high silica and alumina content; ZFA – zeolites from fly ash; WPOOF – the waste pomace of olive oil factory; SP – 1,5-disodium hygrogen phosphate; AA – acrylic acid; CA – citric acid; FM – formaldehyde; TA – tartaric acid.

Bagasse fly ash (BFA) is a by-product of the sugarcane industry and this material mostly contains Al_2O_3, SiO_2, CaO, Fe_2O_3 and MgO. Ali *et al.* (2014) showed that this material can be used to remove both arsenite and arsenate species from aqueous solution. The BET surface area and pH_{ZPC} was $450.0 \, m^2 \, g^{-1}$ and 8.23, respectively. It was observed that the optimal conditions for As(V) and As(III) removal were as follows: initial concentration of $0.05 \, mg \, L^{-1}$, pH 7.0, adsorbent dose of $3.0 \, g \, L^{-1}$ and temperature of 20°C. The maximum adsorption capacities are presented in Table 12.1. The presence of calcium, magnesium, sodium, potassium, phosphates, nitrate and sulfate ions caused a decrease in the adsorption capacity of As(III) and As(V) (competitive adsorption). Thermodynamic data showed that adsorption of As(III) and As(V) onto the BFA surface was exothermic. Kinetic data followed the pseudo-first-order model equation. The equilibrium condition was reached after 50 minutes for both As(III) and As(V) ions.

Another example of using by-product of the sugarcane industry to remove As species was presented by Roy *et al.* (2013). In their studies they applied sugarcane carbon modified with thioglycolic acid. This material was prepared by carbonization of bagasse (SCC) and impregnation of acids in a structure of SCC. The specific surface area and pH_{ZPC} were $5690 \, m^2 \, g^{-1}$ and 4.68, respectively. The authors performed batch and column tests. The optimal removal of As in batch tests was observed at pH 6.0 for initial concentration of $0.1 \, mg \, L^{-1}$ and dosage of $2.5 \, g \, L^{-1}$ at 25°C. The maximum adsorption density was 0.0177 and $0.0176 \, mg \, g^{-1}$ for As(III) and As(V), respectively. Equilibrium was reached within 30 minutes. It was observed that adsorption is not effective below and above pH 6.0. Thermodynamic studies showed that adsorption of As(III) and As(V) on this material increased with temperature, however, above 35°C the adsorption capacity decreased due to denaturation. In column tests the maximum retained amounts of As(III) and As(V) were 0.085 and $0.085 \, mg \, g^{-1}$ at initial concentration of $1500 \, \mu g \, L^{-1}$, adsorbent dosage of 6.0 g, flow rate of $3.0 \, mL \, min^{-1}$ and pH 6.0. To analyze the column experimental data the authors used the Thomas model.

The fly ash from bagasse, a waste material form the sugar industry, was studied as a potential adsorbent for Cr(III) removal from wastewater (Gupta and Ali, 2004). The composition of the adsorbent was SiO_2 (60.5%); Al_2O_3 (15.4%); CaO (3.0%); Fe_2O_3 (4.90%); MgO (0.81%). The loss on ignition was found to be 18.01% by weight. It was found that the equilibrium time was 40 min for Cr(III) removal (98%). The adsorption of Cr was studied within a pH range of 2.0–9.0 and the maximum uptake took place at pH 5.0. The adsorption capacity decreased slightly with increasing temperature, showing the exothermic nature of Cr(III) adsorption. The influence of other ions was studied and the results indicated a non-selective nature of the adsorbent. The equilibrium data were described by using the Langmuir and Freundlich isotherm models. The maximum adsorption monolayer was found to be 4.35, 4.30, and $4.25 \, mg \, g^{-1}$ at 30, 40 and 50°C, respectively.

The removal of Cr(VI) from aqueous solution by the batch adsorption technique using rice husk ash and coal fly ash was studied by Bhattacharya *et al.* (2008). Rice husk ash is a product of burning the husk in a rice mill boiler. The coal fly ash was a product of incineration of coal in a thermal power plant and no more details were given. Over a pH range of 2–11, the adsorption process was found to be strongly dependent on pH. The optimum pH range for removal of Cr(VI) was found to be between 2.0 and 3.0. For rice husk ash, at pH 3 the maximum 94.8% removal of Cr was observed. For coal fly ash, the optimal pH was found at 2.0 and the maximum removal 89.2%. For rice husk ash the adsorption equilibrium was achieved within 2 h, and for coal fly ash 3 h was needed. Rate kinetics for the adsorption of Cr(VI) was best fitted with the pseudo-second-order kinetic model. The Langmuir and Freundlich adsorption isotherms were applicable to the adsorption process and their constants were evaluated. The adsorption capacity calculated from the Langmuir isotherm was found to be 23.86 and $25.64 \, mg \, g^{-1}$ for coal fly ash and rice husk fly ash, respectively.

12.2.2   *Slag*

Steel making slag (SMS) is a by-product generated from steel manufacturing plants. The slag consists mainly of iron and aluminum oxides and calcium hydroxides. Ahn *et al.* (2003) tested slag materials of different composition. These materials, taken from the iron and steel making processes, were evaporation cooler dust (ECD), oxygen gas sludge (OGS), basic oxygen furnace slag (BOFS), blast furnace slag (BFS) and electrostatic precipitator dust (EPD). The studies of removal of As were performed in batch mode. The authors noticed that for initial concentration of 25 mg L^{-1} As(V) and As(III) at natural pH (9–12) ECD, OGS and BOFS reduced the concentration of As to values below 0.5 mg L^{-1} and the sorption capacity was below 1.25 mg g^{-1} at dosage of 20 g L^{-1} after 24 and 72 hours, respectively. For EPD the final concentration of As(V) and As(III) was 3.0 and 2.4 mg L^{-1}, and the adsorbed amounts of As were 1.10 and 1.13 mg g^{-1}, respectively. The worst removal effect was observed for BFS and the sorption capacity was 0.15 and 0.25 mg g^{-1} for As(V) and As(III), respectively. However, at lower pH (pH 6) the removal of As(V) and As(III) by BFS was enhanced, and the sorption capacity of As(V) and As(III) was about 0.90 and 0.50 mg g^{-1}, respectively. Differences in efficiency of adsorbents in removal of As at natural pH (pH 9–12) were due to varying content of CaO and calcium hydroxide. According to data presented by the authors, the content of calcium oxide decreased in the following order: BOFS > ECD > OGS > EPD > BFS. Moreover, in batch tests at high pH (pH 9-12) the samples of ECD, EDP, BOFS and OGS released calcium ions into solution due to dissolution of calcium hydroxides. The lowest concentration of Ca^{2+} in suspension after leaching was for ECD.

Zhang and Itoh (2005) in their studies used incinerator melted slag, which was dried at 105°C for 24 hours before use. The slag used in this study mainly contained Ca (14.9%), Si, Al and Fe, which mainly existed in the form of oxides. The BET surface area was found to be 2.59 m^2 g^{-1}. It was reported that for initial As concentration of 150 mg L^{-1} at pH 2.5, dosage of 4 g L^{-1} and room temperature, the adsorption density of As(V) was about 7 mg g^{-1}, while in those conditions the slag sample was not effective for removal of As(III) due to the neutral form of As(III) at pH 0–9.

Jovanović et al. (2011) in their paper observed that the waste iron slag maximum As(III) and As(V) adsorption capacity was 0.82 and 4.04 mg g^{-1}, respectively, at initial concentration of 0.50– 100 mg L^{-1}, 20°C, and pH 7. The BET surface area was 2.9 m^2 g^{-1}. The kinetics data showed that equilibrium was reached after 6 hours. The adsorption data follow the pseudo-second-order model equation and the Langmuir isotherm model. This was explained by the presence of high amounts of iron oxides. XRD analysis showed that iron slag is a mixture of silica, iron and calcium oxides. In those conditions, the iron was the main active component of slag.

Lekić et al. (2013) used raw and modified furnace slag (BFS) to remove As(V) and As(III) from aqueous solution. Chemical analysis showed that the BFS used in these studies was rich in silica, iron and calcium oxides. The specific surface area was 2.90 and 17.17 m^2 g^{-1} for raw and modified furnace slag, respectively. During the modification the raw BFS was treated with sulfuric acid and dried in an oven at 70°C for 24 hours, and it resulted in gypsum formation. Equilibrium was reached within 360 and 240 minutes for BFS and modified-BFS, respectively. The kinetic data followed the pseudo-second-order equation. The adsorption data fitted the Langmuir isotherm model, and the maximum adsorption capacity on BFS was found to be 0.82 and 4.04 mg g^{-1} for As(III) and As(V), respectively. In the case of modified BFS, the maximum amount adsorbed was 0.7 and 2.79 mg g^{-1} for As(III) and As(V). The conditions for the adsorption studies were as follows: initial concentration 0.5–100 mg L^{-1}, pH 7 and dosage of 10 g L^{-1}. Modification did not improve the adsorption capacity of slag adsorbent toward As ions. However, the modified slag had minimal impact on the water properties and ensured the safe use of it in removal of As from water.

Zhang and Itoh (2005) in their studies modified melted steel slag by loading amorphous FeOOH on the surface of the slag, FeOOH-loaded steel slag, to increase efficiency of As removal. This modification was conducted by mixing the slag with NaOH. Then FeCl$_3$ was added to the slurry and on the surface of the slag porous structures were created via chemical reaction. The new sorbent possessed a greater surface (196 m^2 g^{-1}) as compared to raw steel slag. The adsorption data revealed that FeOOH-loaded steel slag effectively removed As(V) at acidic pH. The best removal was observed at pH 1.5. In the case of As(III), there were no significant changes in adsorption capacities at pH 1–9. The best removal was observed at pH 10. Kinetic studies showed that equilibrium was reached within 8 and 15 hours for As(III) and As(V), respectively. The maximum adsorption capacity was found to be 30.1 and 78.5 mg g^{-1} for As(III) and As(V). The adsorption studies conditions were as follows: 20–300 mg L^{-1} of As, 4 g L^{-1} of sorbent, pH

2.5 and 20°C. From data presented by the authors, it can be said that 15 g of modified-steel slag is sufficient to remove 200 mg of As(V) from 1 L of solution. In the same conditions, 40 g of FeOOH-loaded steel slag removed 75 mg of As(III). Leaching experiments of ions from the adsorbent showed that this material is environmentally acceptable.

12.2.3 Red mud

Red mud (RM) is a waste product from the process of alumina extraction from bauxite by the Bayer process (Brunori *et al.*, 2005). Accumulation and deposition of this material cause a serious problem for the environment due to its high alkalinity and huge amount. Several researches all over the world have tried to use this waste material in wastewater treatment processing. Altundoğan *et al.* (2000) found that red mud obtained from aluminum plants can be used as an alternative adsorbent for As removal. The material was washed with water and dried in an oven at 105°C. The XRD analysis showed that the material mostly contains different forms of iron and aluminum oxides, silicates, calcite, and titanium dioxides, which can interact with As species. Adsorption experiments revealed that red mud effectively removed As(III) and As(V) species from water at initial concentrations of 2.5–30 mg L^{-1}, dosage of 20 g L^{-1}, 25°C, and pH 9.5 and 3.2, respectively (Altundogan *et al.*, 2000). The maximum adsorption densities in those conditions were 0.66 and 0.51 mg g^{-1} for As(III) and As(V), respectively. The authors observed that the adsorption of As(III) decreased sharply at lower and higher values of pH. In the case of As(V) the adsorption decreased at pH above 3.2. Kinetic studies showed that the equilibrium concentration of As(III) and As(V) was reached at 45 and 90 minutes, respectively, and the data obtained fitted the first-order-rate expression of Lagergren well. The adsorption isotherm data fitted the Langmuir equation well. Based on the thermodynamic value calculation it was concluded that adsorption of As(III) and As(V) on the red mud surface is exothermic and endothermic, respectively.

Due to high alkalinity the red mud application in water treatment is limited, and some modifications of RM are proposed in the literature. Altundogan *et al.* (2002) prepared an activated red mud by heat and acid treatment. They observed that adsorption of As(III) on heat-treated red mud was dependent on pH and the temperature of heating. The maximum adsorption of As(III) was achieved at pH 9.5 when the temperature of heating was 400°C. These pH conditions are the same as for raw red mud, but better results were obtained with red mud activated by acid treatment. Increasing the concentration of acid up to 1 M during the activation process led to increase in adsorption of As(III) at pH below 9.5. Addition of more acid in the activation of the material caused worse adsorption due to decreasing the surface area. However, activation of red mud at low acid concentration (below 0.5 M) resulted in production of material with low As adsorptivity due to blocking of active oxidic sites by silicic acid. The concentration of acid added to activation of red mud used in further studies was 1 M. The optimal range of pH for effective adsorption of As(III) and As(V) onto acid-activated red mud was 7.25–5.8 and 1.8–3.0, respectively. Based on thermodynamic data, the authors stated that adsorption of As(III) and As(V) was exothermic and endothermic, respectively. The kinetic data showed that equilibrium was attained within 45 minutes for As(III) and 60 minutes for As(V). The adsorption data fitted the pseudo-first-order rate expression and the Langmuir isotherm model. The calculated values of maximum adsorption capacities at 25°C were 0.88 and 0.94 mg g^{-1} in the following conditions: 2.5–30 mg L^{-1} of As, dosage of 20 g L^{-1}, pH 7.25 and 3.5 for As(III) and As(V), respectively.

Genç *et al.* (2003) presented data of As adsorption onto seawater-neutralized red mud (Bauxsol). Bauxsol was obtained by stirring red mud in seawater solution until equilibrium pH was reached. After neutralization this material mostly contains Fe_2O_3, Al_2O_3, hydroxide and hydrocarbonates. The BET surface area was 30 m^2 g^{-1}. They observed that As adsorption is not dependent on ionic strength, indicating chemisorption as the mechanism of adsorption. The highest adsorption capacity was obtained at pH 6.3 and the adsorption density decreased with increasing pH up to 12 (ligand-like adsorption). It was found that 5.0 g L^{-1} of sorbent is sufficient to achieve the WHO standard at pH 7.3 (WHO, 2008). Equilibrium was reached within 3 hours at pH 7.5. The adsorption data followed the Langmuir isotherm model. The maximum adsorption capacity was found to be 1.08 mg g^{-1} at initial concentration of 0.052–3.0 mg L^{-1}, dosage of 5 g L^{-1}, 0.01 M NaCl, 23°C, and 6.3. The presence of HCO$_3^-$ has a negative effect on arsenate adsorption, while addition of Ca^{2+} increased removal of arsenate from solution. NaCl had a minimal effect on adsorption.

Genç-Fuhrman *et al.* (2004a, 2004b) used activated neutralized seawater red mud. After neutralization the material was activated through acid treatment, combined acid and heat treatment, and the addition of ferric sulfate or aluminum sulfate. According to earlier studies it can be concluded that addition of seawater to red mud led to neutralization of high basicity. Treatment with heat increases the surface area of the sorbent. Activation by acid removes solid compounds. Moreover, to retain solubilized Al and Fe in the form of oxides combined heat and acid treatment was applied. The BET surface area of not activated, acid-treated, acid and heat-treated neutralized red mud was 30, 60 and 130 m^2 g^{-1}, respectively. Activation by addition of ferric sulfate or aluminum sulfate did not improve removal of As. This was because of competition of sulfate with As, saturation of the surface with Fe^{3+}/Al^{3+} and precipitation of Al(OH)$_3$ on the surface of activated red mud. It was observed that the ionic strength of the solution did not change the As(V) adsorption capacity of red mud activated by combined heat and acid treatment. The experimental data showed that equilibrium was reached within 3 and 6 hours for As(V) and As(III), respectively. The maximum As(V) adsorption capacity was 7.6 mg g^{-1} at initial concentration of 0.5–16.5 mg L^{-1}, pH 4.5, 0.01 M NaCl, dosage of 1 g L^{-1} and 23°C. In the case of As(III), the monolayer adsorption capacity was 0.54 mg g^{-1} at initial concentration 0.15–11.7 mg L^{-1}, pH 6.8, dosage of 5 g L^{-1} and 23°C. The kinetic data for both As(V) and As(III) fitted the pseudo-first-order equation well (Genç-Fuhrman *et al.*, 2004b). The authors observed that the addition of sulfate, phosphate and bicarbonate had an insignificant effect on As(V) adsorption when the sorbent dosage was 5 g L^{-1}, whereas when 0.5 g L^{-1} of sorbent were used, a decrease in removal of As(V) was observed due to anion competition. The strongest influence was shown by phosphate anions (similar chemistry with As(V)) (Genç-Fuhrman *et al.*, 2004a). Investigating the toxicity of activated red mud, the authors stated that this sorbent is not hazardous. Thermodynamic studies showed that the adsorption of As(V) was endothermic and spontaneous in nature (Genç-Fuhrman *et al.*, 2004b).

Zhang *et al.* (2008) studied FeCl$_3$-modified red mud (MRM). After modification the new sorbent possessed higher specific surface area (192 m^2 g^{-1}) than raw red mud (115 m^2 g^{-1}). X-ray analysis showed that FeCl$_3$-modified red mud has its Fe content increased and is a mixture of Fe and Al oxides. The authors observed that As(V) showed better removal from solution at lower pH (below 6). The adsorption density of As(V) was 68.5, 50.6 and 23.2 mg g^{-1} at pH 6, 7 and 9, respectively, when the initial concentration of As(V) was 1 mg L^{-1}, and dosage was 100 mg L^{-1}. The adsorption data fitted the Langmuir model well and also the pseudo-second-order kinetic model. Equilibrium was attained within 24 hours. The removal of arsenate using MRM from aqueous solution was enhanced by addition of Ca^{2+}. The presence of HCO$_3^-$ had a negative effect on adsorption of As(V), while the presence of NO$_3^-$ had no obvious effect on removal of arsenate. Desorption studies showed that modified red mud can be regenerated using NaOH solution, and can be used once again in the adsorption process with the same capacity.

Activated red mud was studied as a potential adsorbent for removal of hexavalent Cr from aqueous solution (Pradhan *et al.*, 1999). Activated red mud was prepared by refluxing red mud in 20% HCl for 2 h. The specific surface area of the sample was found to be 249 m^2 g^{-1}. The equilibrium of adsorption was attained in about 2 h. In the pH range 3.5–7.0, as the pH was increasing, there was a little increase in the adsorption and the maximum adsorption was achieved at pH 5.2 and then a sudden decrease was observed above this value. The percentage of adsorption of Cr(VI) on activated red mud decreased with increasing temperature in the range 25–50°C. The equilibrium data fitted both the Langmuir and Freundlich isotherm models well. The Langmuir maximum capacity was calculated and found to be 30.74 mg g^{-1}. Also industrial effluent containing a very high concentration of Cr(VI) was treated by activated red mud. More than 90% removal of Cr was achieved.

Use of activated red mud for removing hexavalent Cr from aqueous solution was also studied by Gupta *et al.* (2001). Red mud was first treated with hydrogen peroxide to oxidize organic matter and washed with water and dried at 100°C. It was then activated in air in a muffle furnace at 500°C for 3 h. The surface area of prepared adsorbent was found to be 108 m² g⁻¹. The maximum Cr uptake took place at pH 2.0. The presence of Na^+, Zn^{2+}, Cd^{2+} and Pb^{2+} ions as well as anionic surfactants (sodium diisobutyl sulfosuccinate) and a cationic one (cetyltrimethylammonium bromide) did not caused significant disturbing effects. The equilibrium data fitted both the Freundlich and Langmuir isotherm models well. The maximum monolayer capacity calculated from the Langmuir model was found to be 22.67, 21.58, 21.06 mg g⁻¹, at 30, 40 and 50°C, respectively. Negative values of the Gibbs free energy change (ΔG^0) indicated feasibility of the adsorption process and its spontaneous nature without any induction period. The negative enthalpy change (ΔH^0) confirmed its exothermic nature and the possibility of strong bonding between metal ions and red mud. A negative entropy change (ΔS^0) reflected the affinity of adsorbent towards Cr. Adsorption of Cr(VI) followed the first-order kinetics model.

12.2.4   *Sludges*

Wu *et al.* (2013a) used Fe-based backwashing sludge in removal of As(III). The BET surface was 151.9 m² g⁻¹ and the pH_{ZPC} was 7.7. The SEM and XRD analysis showed that the adsorbent contains sulfate inter-layered with Fe hydroxide, lepidocrocite, quartz and calcium carbonate. Equilibrium was reached within 18 hours. The kinetics data fitted well both Elovich and power equations. Experimental data showed that the best removal efficiency of As(III) was at pH 8.0. Below and above pH 8, the adsorption capacity decreased with pH. The dosage of 6 g L⁻¹ was sufficient to remove almost 100% of 5 mg of As from 1L of solution. The adsorption data followed the Langmuir isotherm model, and the maximum adsorption capacity was calculated to be 59.7 mg g⁻¹ at dosage of 0.6 g L⁻¹, pH 7.0, 0.01 M NaCl, initial concentration ranging from 1 to 120 mg L⁻¹ and 25°C. The authors observed that sulfate and chloride ions did not influence adsorption of As(III) onto adsorbent. Phosphate ions, due to similar chemistry with As(III) ions and formation of complexes with Fe(III), significantly influenced the As removal in the investigated system. FTIR and XPS studies showed that in adsorption of As(III) hydroxyl and sulfate groups participated. Also, the precipitation of Fe(III) was involved. Desorption studies showed that the retrieval of As(III) from As-loaded adsorbent occurred by alkaline or phosphate eluent.

In other studies Fe-based backwashing sludge was used for removal of As(III) (Wu *et al.*, 2013b). The BET surface area was found to be 148.4 m² g⁻¹. The adsorption data followed the Elovich model and the maximum adsorption capacity was found to be 43.20 mg g⁻¹ under the same conditions as in their earlier paper (Wu *et al.*, 2013a). Equilibrium was reached within 17 hours. The authors observed that the process was endothermic in nature.

Corillo and Pedroza (2014) used blast furnace residues for removal of As from wastewater. The specific surface area of adsorbent was 21.7 m² g⁻¹. XRD analysis showed that the adsorbent was a mixture of magnetite, franklinite, wustite, iron and calcite. It was said that the adsorption of As(V) on the adsorbent was efficient at pH below 8. Experimental data showed that 20 g L⁻¹ of adsorbent removed completely 1.174 mg of As(V) from 1L of wastewater. The adsorption data fitted the Langmuir isotherm well. The maximum adsorption capacity was 0.21 mg g⁻¹ at As concentration 1.174 mg L⁻¹, pH 4 and 25°C. The mechanism of adsorption can be explained by the formation of Fe(II) and Fe(III) corrosion products on the adsorbent surface and their interaction with As species.

The removal of hexavalent Cr from industrial effluent using distillery sludge was studied by Selvaraj *et al.* (2003). The composition of the sludge was not provided by the authors. The raw sludge was obtained from a distillery unit. The experimental results showed that the removal of Cr increased as the pH decreased and reached the maximum 93% at pH 3.0. The equilibrium of Cr adsorption was attained in 105 minutes. The adsorption dynamic was described by the PFO Lagergren kinetic model. Both Langmuir and Freundlich isotherm models fitted the equilibrium data well and the maximum adsorption capacity was found to be 5.7 mg g⁻¹ at optimal pH. Above pH about 6.0, the adsorption of Cr(VI) was very low and constant up to pH 10. In a comparative study, the removal of Cr-plating wastewater by distillery sludge has also been studied and 91% removal of Cr from the effluent was achieved.

The removal of trivalent Cr from synthetic aqueous solution was studied using a wine pro-cessing waste sludge (Li *et al.*, 2004). The sludge is generated from the final clarifier and coagulation/settling reservoir of the wastewater treatment process of the wine processing fac-tory. EDX and chemical analysis indicated the major components of the sludge were organic materials and nitrogen, 40.5 and 23.4%, respectively. The equilibrium adsorption data fitted both the Langmuir and Freundlich isotherm models well. The adsorption capacity increased from 10.46 mg g^{-1} at 20°C to 16.36 mg g^{-1} at 50°C, showing the process to be endothermic in nature. The calculated Gibbs free energy changed from -6.0 to -6.8 kJ mol^{-1} and was in the range of physical adsorption. Reducing the sludge particle size increased the uptake capacity at equilibrium. The Cr(III) adsorption followed the Lagergren pseudo-first kinetic model and the intraparticle diffusion model.

Blast furnace sludge (BFS) is a by-product of the steel industry. It is produced when gases generated during the manufacture of pig iron are cleaned before their atmospheric emission in a wet process. The effluent consists of a sludge concentrated in a Dorr thickener. A huge amount of sludge is dumped in landfill without recycling. Due to small particle size (<0.1 mm) and to the high content of iron oxide and coke, the sludge could be an efficient adsorbent of metals, such as lead, copper, zinc, cadmium, and Cr. A blast furnace sludge was applied for Cr(III) removal from multi-cationic aqueous solution (López-Delgado *et al.*, 1998). XRD analysis indicated that the sludge is composed mainly of hematite and coke with minor quantities of other iron oxides (wustite, maghemite, magnetite), calcium ferrite, quartz, and calcium and aluminum silicates. The specific surface area was found to be 27.43 m^2 g^{-1} and the adsorbent was a mesoporous material. The adsorption equilibrium was achieved within 5 hours at pH 7.2. The equilibrium experimental data fitted the Langmuir isotherm model better than that of Freundlich. For various temperatures (20–80°C), the maximum Cr(III) adsorption capacity changed from 9.55 to 16.05 mg g^{-1}. The positive value of ΔH^0 indicated that adsorption of Cr(III) on BFS may be considered an endothermic process.

The removal of hexavalent Cr from aqueous solution by a batch adsorption technique using a clarified sludge was investigated by Bhattacharya *et al.* (2008). Clarified sludge is a steel industry waste material generated in huge quantity and disposed of in landfill. The optimum pH range for removal of Cr(VI) was found to be between 2.0 and 3.0. The maximum 97.4% removal of Cr was observed at pH 3. Optimal adsorbent dosage was found to be 10 g L^{-1} and initial Cr(VI) concentration 50 g L^{-1}. Adsorption of Cr(VI) by clarified sludge may be attributed to the combined effect of its main constituents, silica, metal oxides and carbon. For clarified sludge the adsorption equilibrium was achieved within 2 h. The rate kinetics for the adsorption of Cr(VI) best fitted the pseudo-second-order kinetic model. The Langmuir and Freundlich adsorption isotherms were applicable to the adsorption process. The adsorption capacity value (26.31 mg g^{-1}) showed that clarified sludge was a very effective adsorbent for the removal of Cr(VI) from aqueous solution.

12.2.5   *Carbon slurry*

Carbon slurry, a waste material from fertilizer plants, has been used without any pretreatment as an adsorbent for hexavalent Cr from aqueous solution (Singh and Tiwari, 1997). Carbon slurry is generated from the scrubbers of the gasification section of ammonia plants based on partial oxi-dation of naphtha. The equilibrium of Cr(VI) adsorption was attained within 60 min. On changing the initial concentration from 52 to 130 mg L^{-1}, the adsorption increased from 11.9 mg g^{-1} (95% removal) to 24.9 mg g^{-1} (76% removal). It was found that Cr removal was higher at low pH, and the maximum adsorption was achieved at pH 2.5, indicating that it is the HCrO$_4^-$ species of Cr(VI) which is adsorbed preferentially on carbon. The adsorption kinetics followed the pseudo-first-order Lagergren model. The equilibrium data fitted the Langmuir isotherm model well. The maximum adsorption capacity was found to increase slightly with increasing temperature and was 24.05, 25.15 and 25.64 mg g^{-1} at 30, 45 and 60°C, respectively. The negative values of Gibbs free energy change (ΔG^0) over the investigated temperature range indicated the spontaneous nature of the adsorption process. The positive value of enthalpy change (ΔH^0) confirmed the endothermic nature of Cr adsorption on carbon slurry.

Activated carbon slurry was used to remove hexavalent Cr from aqueous solution in batch experiments as well as in column mode (Gupta *et al.*, 2010). The raw material produced in generators of fuel oil-based industrial generators was chemically treated with hydrogen peroxide to oxidize the organic matter and then subjected to heating at 200°C. The activation was realized in a muffle furnace for 2 h at 450°C in air. The activated product was then treated with 1.0 N HCl to remove ash content and washed with distilled water. The specific surface area was determined as 388 m^2 g^{-1}. The Cr(VI) removal by activated carbon slurry was found to be pH-dependent. Acidic pH in the range 1–3 was favored for adsorption and the maximum adsorption was observed at pH 2.0. At low pH, the reduction of Cr(VI) to Cr(III) is also possible and chemical analysis confirmed the presence of trivalent Cr at the surface of activated carbon slurry during the adsorption process at very low pH. It was found that the equilibrium of adsorption was attained in about 70 minutes. The equilibrium data fitted both Langmuir and Freundlich isotherm models well. Maximum monolayer capacity was found to be 15.24 mg g^{-1} at 30°C, optimum pH 2.0, 4.0 g L^{-1} adsorbent dose and 100 mg L^{-1} Cr(VI) and decreased to 5.08 mg g^{-1} at 50°C. The calculated thermodynamic parameters revealed the exothermic, feasible and spontaneous nature of the process as well as increasing randomness at the adsorbent-adsorbate interface during the fixation of Cr to the active sites. The kinetics of Cr adsorption onto activated carbon slurry followed the pseudo-second-order model. 5% HNO$_3$ efficiently desorbed metal from adsorbent. Studies in column mode showed that the adsorbent could also be used on an industrial scale.

12.2.6   *Spent activated clay*

Adsorption of hexavalent Cr onto spent activated clay (SAC) was investigated for its use of water treatment (Weng *et al.*, 2008). The raw spent activated clay was a waste product from an edible oil refinery company. The commercial virgin activated clay originates from montmorillonite activated by sulfuric acid. It is used for color removal from oil. To recover more activated sites, the RSAC was many times boiled with distilled water in a high pressure cooker and dried. The surface area of treated and untreated clay was 7.42 and 6.84 m^2 g^{-1}, respectively. From the zeta potential, the isoelectric point was found at pH around 3.8. After 1.5 h of contact time the equilibrium was reached. The adsorption kinetics followed the pseudo-first-order model and the rate of removal was found to increase with decreasing pH and increasing temperature. The activation energy (E_a) calculated from the Arrhenius equation suggested that the adsorption was diffusion controlled and was governed by interactions of a physical nature. The Langmuir isotherm model fitted the equilibrium data well and the maximum adsorption capacity ranged from 0.743 to 1.422 mg g^{-1} at temperatures between 4 to 40°C at pH 2.0. The negative values of ΔG^0 indicated the spontaneity of the process and more energetically favorable conditions for Cr(VI) removal. As the values of Gibbs free energy ranged from -6.9 to -8.56 kcal mol^{-1} (≈ -28.9 to -35.8 kJ mol^{-1}), the adsorption mechanism is not attributed to ion exchange. It was suggested that the weak surface complex reaction is the major mechanism for chromate ions adsorption at pH below pH$_{iep}$. The ΔH^0 and ΔS^0 values were positive, suggesting that the adsorption reaction was spontaneous at high temperature and endothermic in nature.

12.2.7   *Pisolite*

Pisolite is a waste material which mostly contains iron and manganese hydroxides. This waste product has an ability to remove cationic metal species from aqueous media (Helfferich, 1995). Pereira *et al.* (2007) presented the experimental data of As(V) adsorption on pisolite in batch and column tests. The BET surface area was 61.4 and 90.45 m^2 g^{-1} for raw and activated pisolite, respectively. The pisolite was activated by heating the material at 400°C for 4 hours. The maximum As loading was observed at pH 6.5 and it was 1.5 and 3.5 mg g^{-1} for pisolite and activated pisolite in column tests, respectively. In the case of batch mode experiments, the maximum density was

1.29 and 3.17 mg g^{-1} for pisolite and activated pisolite, respectively. The initial concentration of As(V) was 50 mg L^{-1}. The dosage of sorbent was 1 g L^{-1}.

12.2.8   *Fe(III)/Cr(III) hydroxide*

Fe(III)/Cr(III) hydroxide waste is a by-product obtained in industry from treatment of Cr(VI)-containing wastewaters using Fe(II) (Namasivayam and Senthikumar, 1998). In the reaction of Cr(VI) with Fe(II) the reduction of Cr(VI) to Cr(III) and oxidation of Fe(II) to Fe(III) occurs. In that system ferric hydroxide is formed rapidly and sorption of Cr(III) and/or coprecipitation with ferric hydroxide takes place, resulting in formation of Fe(III)/Cr(III) hydroxide waste material (Guan *et al.*, 2011; Lee and Haring, 2003). Namasivayam and Senthikumar (1998) used this material to remove As(V) from water. The dry waste was washed with distilled water, and dried at 60°C for 10 h before use. The specific surface area was 424 m^2 g^{-1}. The ZPC point was found to be at pH 8.30. The best removal, 97.8%, was observed at sorbent dosage of 8.0 g L^{-1}. Adsorption experiments showed that the removal of As(V) onto Fe(III)/Cr(III) hydroxide waste was not sensitive to changes in pH below 10. Equilibrium was reached within 180, 220 and 240 minutes, depending on initial concentration of As(V), which was 20, 30 and 40 mg L^{-1}, respectively. The kinetics data followed the pseudo-first-order equation. The adsorption data fitted both the Langmuir and Freundlich isotherm models well, and maximum adsorption capacity was found to be 11.02 mg g^{-1} at initial concentration ranging from 20 to 100 mg L^{-1}, dosage of 10 g L^{-1}, pH 4.0 and 32°C. Thermodynamic data showed that adsorption of As(V) onto Fe(III)/Cr(III) hydroxide waste was spontaneous and endothermic in nature.

12.2.9   *Laterite*

Laterite by-product is a low cost material which is a waste product generated from companies producing ferric aluminum sulfate, a valuable coagulant. This low cost material contains aluminum, silica, iron and titanium oxides. Glocheux *et al.* (2013) employed laterite by-product as a sorbent for removal of As from aqueous solution. The authors observed that laterite by-product completely removed As(V) from solution at pH below 6. The best adsorption of As(III) was observed at pH 8.5. The removal efficiency decreased below and above pH 8.5. Comparing data Glocheux *et al.* (2013) could state that laterite by-product has a higher affinity to arsenate than to arsenite due to the presence of sulfate groups. The maximum adsorption capacity of laterite by-product at 1 mg L^{-1} of As and pH 7 was 0.30 and 0.92 mg g^{-1} for As(III) and As(V) respectively. The adsorption data better fitted the Freundlich equation than the Langmuir model. Kinetic studies showed that equilibrium was reached after 48 hours. The data fitted the pseudo-second-order model. The authors tried to regenerate the sorbent with 0.1 M NaOH. They observed that after desorption studies, the efficiency in removal of As(V) was decreased due to loss of sulfate surface groups.

12.2.10   *Foundry sand*

Foundry sands are waste from both ferrous and non-ferrous metal casting. In the casting process, the sands are recycled and used many times. However, when degraded sands can no longer be reused, they are landfilled. They consist of 93–99% silica and 1–3% binder and a catalyst which initializes a reaction that cures and hardens the mass and residual iron particles (Siddique and Noumowe, 2008). The foundry sands were evaluated as a potential adsorbent for hexavalent Cr removal from aqueous solution (Campos *et al.*, 2013). In the pH range 1.6–6.0 Cr removal of 100–40% was achieved. It was found that the adsorption kinetics followed a pseudo-second-order kinetic model and the equilibrium data fitted the Langmuir isotherm model well. The maximum monolayer capacity was calculated as 1.99, 2.40, 2.50 and 3.14 mg g^{-1} at 25, 35, 45 and 55°C, respectively. The obtained data revealed an endothermic physisorption mechanism spontaneous in nature and increasing randomness of Cr(VI) removal by foundry sands.

12.2.11   *Biomass*

The waste material *Mucor meihi* biomass was studied as a potential sorbent for the removal of Cr from industrial tanning effluent (Tobin and Roux, 1998). The biomass was washed with water to constant pH, dried at 60°C, ground and sieved. The tannery effluent was characterized by the deep blue color solution of Cr(III) with pH 4.0 and the concentrations of Cr 1.77 g L^{-1} and sodium 30.1 g L^{-1}, respectively. The adsorption capacity was found to be 36.4 mg g^{-1} at pH 2.0 and 59.8 mg g^{-1} at pH 4.0 while the precipitation effect increased this value at higher pH (108 mg g^{-1} and 250 mg g^{-1} at pH 5.5 and 7.0, respectively). Furthermore, the Cr desorption increased with decreasing eluent pH to about 30% at approximately zero pH. It was found that combined elution treatment was able to increase recovery up to 100% at low metal loadings, when 1 N NaOH followed by 1 N H$_2$SO$_4$ are used. At higher biomass metal loading, desorption decreased up to 80–60%. The binding of Cr to *Mucor* biomass was found to be reversible, but the total recovery is possible only with harsh acid/base treatment.

12.3   AGRICULTURAL BY-PRODUCTS

Solid agricultural wastes have been applied for removal of tri- and hexavalent Cr from water and wastewater. There are also numerous examples in the literature of using agricultural by-products as adsorbent for As(III) and As(V). Bioremediation is a cheap and effective method for the removal of heavy metals. The most popular agricultural by-products commonly applied as adsorbents of metal ions comprise sawdust and bark of different origins, sugar beet pulp, rice bran/hulls/husk/straw, coconut shell/coirpith, sugarcane bagasse, olive cake, leaves, tea factory waste, etc. (Miretzky and Fernandez Cirelli, 2010). The characteristics of these adsorbent and the mechanisms of both As and Cr removal from aqueous solutions are discussed below.

12.3.1   *Husk, bran and straw*

Rice husk is a by-product of the rice milling industry and dry rice husk consists of 70–85% of organic matter (cellulose, hemicelluloses, lignin, sugar), ash and silica (Lata and Samadder, 2014). In the literature this sorbent was used for removal of various heavy metals and metalloids from aqueous media (Lata and Samadder, 2014). Amin *et al.* (2006) applied rice husk in column tests to remove As(III) and As(V) from water. Rice husk was washed with pure water and dried in a hot-air oven at 60°C for 24 h. The adsorption density of As(III) was 0.016 mg g^{-1} for initial concentration of 0.5 mg L^{-1}, flow rate 6.7 mL min^{-1}, dosage of 1 g. For As(V) the adsorption capacity was 0.013 mg g^{-1} at the same initial concentration, dosage of 2 g, flow rate 1.7 mL min^{-1}. The authors noticed that the most effective removal of As(V) was in the range of pH 8–12, while As(III) was effectively removed at pH below 10. Desorption studies showed that potassium hydroxide solution can be useful for desorption of As from rice husk surface and after desorption this material can be used as a fuel source.

Also Ranjan *et al.* (2009) used untreated rice polish for the removal of As from water in batch mode. The material was washed with pure water and dried at 60°C for 24 h before use. The authors observed that the capacity of As(III) and As(V) adsorption increased with an increase of pH up to the maximum value, and then decreased with further increase in the pH. The maximum As(V) and As(III) adsorption was reached at pH 4.0 and 7.0, respectively. In these conditions, the maximum sorption capacity was found to be 0.139 and 0.147 mg g^{-1} for As(III) and As(V), respectively at initial concentration of 1 mg L^{-1}, biosorbent dosage of 20 g L^{-1} and 20°C. Equilibrium was attained within 60 and 40 minutes for As(III) and As(V), respectively. The adsorption data fitted three isotherms well: Langmuir, Freundlich and the Dubinin-Radushkevich. The kinetics data followed the pseudo-second-order kinetics model. The authors observed that sorption of As(III) and As(V) was feasible, spontaneous and exothermic in nature.

Kamsonlian *et al.* (2012a) investigated the influence of various parameters on adsorption of As(III) onto rice husk in batch tests. Analysis of the materials showed that the rice husk investigated contains 61.3% volatile, 7.03% ash and 1.3% fixed carbon. The authors observed that the best removal was achieved at pH 6 and sorbent dosage of 6–10 g L^{-1}. Equilibrium was reached within 150 minutes. The maximum adsorption capacity was 224.4 mg g^{-1} at dosage of 8 g L^{-1}, pH 6, initial concentration of 20–1000 mg L^{-1} and 30°C. The adsorption data fitted the Freundlich isotherm model well. Based on adsorption diffusion studies the authors concluded that adsorption of As(III) onto rice husk is very complex and consists of both surface adsorption and intraparticle transport within the pores of the material. The authors observed a favorable effect of temperature on As(III) sorption (endothermic process). The kinetics data were well described by the pseudo-second-order kinetic model. Desorption studies showed that nitric acid and EDTA were the best as an eluent.

Lee *et al.* (1999) used modified rice husk to remove As(V) from water. The modification was performed by attaching quaternary ammonium groups to the surface of rice husk. The authors conducted batch and column tests. Equilibrium was reached after 20 minutes. The sorption of As(V) on modified rice husk (QWR) increased in the pH range 2–10, and above pH 10 decrease in removal was observed (ion-exchange process). The maximum sorption density was found to be for QWR dosage of 1 g L^{-1} in the given conditions; initial concentration of 100 mg L^{-1}, pH 7.5, agitation speed of 150 rpm, contact time of 2 hours, temperature of 25°C, and it was 18 mg g^{-1}. The adsorption data fitted the Langmuir model. Thermodynamic data revealed that sorption was an endothermic process. The presence of NO$_3^-$ had no effect on removal of As(V), however addition of SO$_4^{2-}$ and CrO$_4^{2-}$ reduced removal of As(V) (competitive effect). In the column studies, breakthrough depends on bed depth but not on the flow rate, as the sorption was very rapid.

Chaudhuri and Mohammed (2012) used QWR to remove As(III) from water. The authors conducted batch tests. The maximum adsorption was reached in the range pH of 7–8. The equilibrium of adsorption was attained in 2 hours. The calculated maximum adsorption capacity was 0.77 mg g^{-1} at total concentration of 1.0 mg L^{-1}, sorbent dosage of 10 g L^{-1}, pH 7.0 and room temperature (22°C). The adsorption data were better described by the Langmuir isotherm model than by that of Freundlich.

The use of formaldehyde-pretreated rice husk to remove Cr(VI) from synthetic wastewater was investigated under different experimental conditions (Bansal *et al.*, 2009). Raw rice husk was subjected to boiling (BRH) and formaldehyde (FRH) treatment to immobilize the color and water-soluble substances. Maximum metal removal was observed at pH 2.0. The efficiencies of boiled and formaldehyde-treated rice husk for Cr(VI) removal were 71.0% (8.5 mg g^{-1}) and 76.5% (10.4 mg g^{-1}), respectively, for dilute solutions at 20 g L^{-1} adsorbent dose. The equilibrium experimental data were found to fit well both the Freundlich and Dubinin-Radushkevich isotherm models.

Modified peanut husk by-product was used as a potential adsorbent to remove Cr(III) from aqueous solution (Li *et al.*, 2007). The peanut husk was modified with formalin to reduce organic pigments. It was observed that adsorption equilibrium was achieved within 1 h. Results of kinetic experiments showed that the adsorption of Cr(III) was effective and rapid. It was found that the adsorption of the heavy metals investigated on modified peanut husk was not controlled by intraparticular diffusion. The kinetic adsorption data were described by the pseudo-second-order equation and the adsorption might be a rate-limiting phenomenon. The adsorption data were described well by both the Langmuir and the Freundlich isotherm models. The maximum Cr(III) adsorption at fixed temperature 25°C and pH 4.0 was found to be 7.67 mg g^{-1} for modified peanut husk.

The easily available and biodegradable biosorbent, maize bran, was utilized for the removal of Cr(VI) from aqueous solution (Hasan *et al.*, 2008). Raw maize bran was obtained from a flour milling plant, washed with distilled water, dried at 60°C, crushed and sieved. The maximum Cr uptake of 312.52 mg g^{-1} was found at pH 2.0, initial Cr(VI) concentration of 200 mg L^{-1}, and temperature of 40°C. It was found that maize bran not only removed Cr(VI) from solution but also reduced it to less toxic Cr(III). The adsorption process followed the pseudo-second order kinetic rate model. The adsorption equilibrium data were described by the Langmuir isotherm model and the uptake changed from 9.00 to 10.0 with the rise in temperature from 20 to 40°C. Thermodynamic parameters revealed that the Cr sorption process on maize bran adsorbent was feasible, spontaneous and endothermic in nature. Desorption experiments were also performed and it was found that complete desorption of Cr is possible at pH 9.5.

Wheat bran (WB) and wheat bran modified with tartaric acid (M-WB) were experimented with and Cr(VI) adsorption was investigated by changing various parameters (Kaya *et al.*, 2014). The adsorption increased with contact time and became optimum at 180 min for WB and 200 min for M-WB. When the pH of the solution increased, some of the toxic Cr(VI) reduced into less toxic Cr(III) on the WB surface. The maximum removal of Cr(VI) from the solution having an initial Cr(VI) concentration of 200 mg L^{-1} was obtained at pH 2.0 as 51.0% and 90.0% for WB and M-WB, respectively. Isotherm data of Cr(VI) adsorption on WB and M-WB were described by the Freundlich adsorption model. The desorption study showed the highest Cr recovery in strongly basic conditions with 1 M NaOH as stripping agent.

Rice straw was used as a Cr(VI) adsorbent from aqueous solution (Gao *et al.*, 2008). The air-dried rice straw was obtained from a local rice mill. The optimal pH was 2.0 and the Cr(VI) removal rate increased with decreasing initial Cr(VI) concentration and with increasing temperature. The adsorption equilibrium was achieved within 48h. Isotherm tests showed that the equilibrium sorption data fitted the Langmuir model better than that of Freundlich and the sorption capacity of rice straw was found to be 3.15 mg g^{-1}. It was concluded that rice straw may be a new type of low-cost adsorbent which could convert Cr(VI) to the less toxic and more stable Cr(III) and could be used economically in Cr(VI)-contaminated wastewater treatment.

The adsorption kinetics of trivalent Cr ions using sorghum straw and oats straw, water-washed agro-waste adsorbents, was studied (Garcia-Reyes and Rangel-Mendez, 2010). The equilibrium adsorption data were fitted well by both Langmuir and Freundlich isotherm models, however, that of Langmuir gave higher correlation coefficients. The maximum monolayer adsorptions were calculated and the values were found to be 9.35 and 12.1 mg g^{-1} for sorghum straw and oats straw, respectively at pH 4.0 and 25°C. It was found that a film diffusion model predicted adequately the Cr(III) adsorption kinetics on agro-waste adsorbents. It was concluded that diffusion models (as the film diffusion) could be more helpful to understand the adsorbate mass transfer to the adsorbent surface than empirical models.

12.3.2   *Nut shells and coconut residues*

Walnut shell (WNS) (*Juglans regia*) has been utilized as an adsorbent for the removal of Cr(VI) ions from aqueous solutions after treatment with citric acid (CA-WNS) (Altun and Pehlivan, 2012). Adsorption of Cr(VI) was found to be in all cases pH-dependent, showing a maximum at pH values between 2.0 and 3.0 for citric acid modified walnut shell (CA-WNS). The Langmuir, Freundlich and D-R adsorption isotherm models have been tested with the equilibrium data. Maximum adsorption capacities of CA-WNS and untreated WNS under experimental conditions were 31.0 and 8.0 mg g^{-1} for Cr(VI) ions, respectively. The positive values of ΔH^0 and ΔS^0 showed that adsorption process was endothermic in nature. The negative ΔG^0 value expressed the feasibility and spontaneous nature of the adsorption process. The ΔG^0 value became more negative with increasing temperature, suggesting that Cr(VI) adsorption on CA-WNS was favored by the increase in temperature.

Cr(VI) adsorption from electroplating plating wastewater chemically modified by grafting with acrylic acid coir pith was investigated (Suksabye and Thiravetyan, 2012). At pH of 2, Cr(VI) in the HCrO$_4^-$ form could be adsorbed on acrylic acid-grafted coir pith via electrostatic attraction. The adsorption equilibrium data showed a good fit with the Langmuir isotherm model. The maximum Cr(VI) removal of 196 mg g^{-1} was obtained for 2M acrylic acid-grafted coir pith at pH 2, contact time of 22 h, and 30°C, compared with 165 mg g^{-1} for non-modified sorbent. Thermodynamic study indicated that the overall adsorption process was endothermic, spontaneous and random. Additionally, the adsorption process was favored by high temperatures.

Coconut shell was used for the removal of hexavalent Cr from aqueous solution (Singha and Das, 2011). In batch studies the optimum pH for Cr(VI) removal was found to be 2.0 for $10 \, g \, L^{-1}$ adsorbent dosage. The equilibrium time was found to be 3 h for the adsorbent investigated. The maximum adsorption capacity was found to be $18.7 \, mg \, g^{-1}$. The adsorption kinetics data best fitted the pseudo-second order model. It was found that the adsorption process was controlled by both film and intraparticle diffusion. The equilibrium adsorption data were well described by the Langmuir equation isotherm model. The sorption energy calculated using the Dubinin-Radushkevich isotherm model confirmed chemisorption as a possible sorption mechanism. Thermodynamic parameters revealed that the Cr(VI) adsorption on biosorbent was a spontaneous process, endothermic in nature. Additionally, an application study using wastewater from an electroplating unit containing among other Cr(VI) showed that coconut shell gave high Cr removal (95%).

12.3.3 *Sawdust and bark*

Sawdust is a by-product obtained from the wood industry. This material modified by urea and ferric oxides was used in removal of As(V) (Urík *et al.*, 2009). In this paper biosorption of As from natural and model waters by non-modified and modified spruce sawdust was presented. Four types of biosorbent, unmodified, modified by urea, modified by Fe(III) and modified by mixture of urea and Fe(III) were investigated. It was shown that the most effective adsorbent was Fe(III) modified sawdust, which removed 100% of As from natural contaminated water and the adsorption density was $6 \, mg \, g^{-1}$ at initial concentration of $0.230 \, mg \, L^{-1}$, dosage of $20 \, mg \, L^{-1}$. In those same conditions, the adsorption capacity for unmodified and sawdust modified with urea was about $0.2 \, mg \, g^{-1}$ and $0.3 \, mg \, g^{-1}$, respectively. After modification of materials by urea adsorption, removal decreased due to occupation of adsorption sites on the material surface (complexation). In further studies the authors used Fe(III) modified sawdust, and presented adsorption and kinetic analysis of the data obtained. The kinetic data showed that equilibrium was attained at 120 minutes. The adsorption data well fitted the pseudo-second-order kinetic model and the Langmuir adsorption isotherm model. To check the probability of adsorption by a diffusion mechanism Urík *et al.* (2009) applied the integrated intraparticle diffusion model (Webber and Morris, 1963). As a result they showed that intraparticle diffusion is not the rate limiting step during As(V) adsorption onto modified sawdust. The maximum capacity was found to be $9.26 \, mg \, g^{-1}$ at initial concentration changing from 20 up to $500 \, mg \, L^{-1}$, sorbent dosage of $10 \, g \, L^{-1}$ and pH 8. There is no information about the temperature at which the process was conducted and no thermodynamic data.

Lopez-Leal *et al.* (2012) in their paper presented data about application of un-modified (UN-SW) and Fe(III)-modified pine sawdust for As(V) removal. Modification was done using two different compounds, $FeCl_3$ (FeCl-SW) and $Fe(NO)_3$ (FeNit-SW). All experiments were conducted as batch tests. The results obtained showed that equilibrium was attained within 40 minutes and the adsorption capacities were 0.032, 0.031 and $0.025 \, mg \, g^{-1}$ at pH 7.0, initial concentration of $2 \, g \, L^{-1}$ and dosage of $50 \, g \, L^{-1}$ for FeCl-SW, FeNit-SW and UN-SW, respectively. The kinetic data followed the pseudo-second-order equation. The data presented revealed that adsorption of As(V) onto all the investigated biosorbents was higher at pH 4 than at pH 7.0. The adsorption data fitted the Langmuir-Freundlich model well, and the calculated maximum adsorption capacities were 4.4, 12.8, and $6 \, mg \, g^{-1}$ at pH 4.0 and 2.6, 5.9 and $4.6 \, mg \, g^{-1}$ at pH 7 for UN-SW, FeCl-SW and FeNit-SW, respectively. The conditions during adsorption studies were as follows: initial concentration ranging from $6.4 \cdot 10^{-4}$ to $8 \cdot 10^{-3} \, mol \, L^{-1}$, dosage of $16 \, g \, L^{-1}$. The highest removal of As(V) was observed when FeCl-SW was used as biosorbent. The authors explained this as because the FeCl-SW biosorbent possesses a higher amount of Fe than the UN-SW and FeNit-SW.

Untreated rubber wood sawdust was investigated as a potential Cr(VI) adsorbent (Zakaria *et al.*, 2009). Rubber wood sawdust (RWS) was collected from a wood-finishing factory and was used without any pretreatment. Complete Cr removal was achieved at pH less than 2.0, initial Cr(VI) concentration of $100 \, mg \, L^{-1}$ and adsorbent dosage greater than 1.5% (w/v). The analysis of both hexavalent and total Cr concentration provided evidence that both adsorption and reduction reactions are involved in Cr(VI) removal by rubber wood sawdust. The point of zero charge (pH$_{PZC}$) of 4.90 explained a decrease of Cr(VI) removal at pH above this value. It was observed that the removal capacity increased from 0.92 mg g^{-1} to 4.87 mg g^{-1}, when the temperature increased from 25 to 45°C. The calculated thermodynamic parameters suggested that Cr(VI) removal by RWS was a non-spontaneously occurring endothermic process with positive entropy.

By using sawdust of different origins higher Cr(VI) capacity was achieved (Gupta and Babu, 2009). Raw sawdust was collected from a workshop and washed with distilled water to remove dust and soluble impurities. The equilibrium adsorption data were described well by the Langmuir isotherm model. The maximum adsorption monolayer capacity was 41.5 mg g^{-1}. The Cr(VI) adsorption process followed the pseudo-second order kinetic model. Since the adsorption of Cr(VI) onto sawdust was pH-dependent, the regeneration of adsorbent is possible by increasing the solution pH and 1 N NaOH was recommended as eluent.

The maximum adsorption of Cr(III) ions from synthetic aqueous solutions was found to be 37.9 mg g^{-1} of Cr(III) at pH 6 for meranti sawdust (Rafatullah *et al.*, 2009). The percentage of Cr(III) adsorption increased with an increase of temperature. The adsorption equilibrium was established within 120 minutes. The adsorption isotherm fitted the Langmuir model well, while the Dubinin-Radushkevich (D-R) isotherms predicted the physical nature of the Cr(III) adsorption process onto sawdust. It was found that the Cr(III) adsorption kinetics followed the pseudo-second-order model with the high R^2 value of 0.999.

Chemical treatment of pine sawdust improved the removal of hexavalent Cr from aqueous solution (Uysal and Ar, 2007). Raw pine sawdust was immersed in 1,5-disodium hydrogen phosphate solution and the mixture was left at 21°C, for 24 h. Optimum conditions for adsorption were determined as $T = 40°C$, sawdust dose 4 g, and pH 2. The Cr removal was found to be pH-dependent. The experimental data fitted both the Langmuir and Freundlich isotherm models well. However, the Langmuir isotherm suited better. The maximum adsorption in uncontrolled pH conditions, at 20°C and adsorbent dosage 2 g, was found to be 30.49 mg g^{-1}, while in optimum conditions it was 121.95 mg g^{-1}, respectively. It was found that the adsorption process followed first-order kinetics. The positive value of the enthalpy change showed that the process is endothermic chemical adsorption, while the positive value of the adsorption entropy and the negative value of free energy change supported the conclusion that the process is spontaneous in nature.

The potential of *Pinus roxburghii* (Himalayan long-leaf pine) bark as an adsorbent for the removal of Cr(VI) was investigated (Ahmad *et al.*, 2005). In a preliminary study, the adsorption capacity was found to be 4.15 mg g^{-1} for Cr(VI) at an initial metal ion concentration of 50 mg L^{-1} at pH 6.5 and 30°C. The adsorption equilibrium was established within 1 h. The adsorption equilibrium data fitted the Freundlich isotherm well in the concentration range studied. The maximum Cr adsorption (96.2% – 4.81 mg g^{-1}) was achieved at pH 3 and decreased steadily to 78.8% as pH increased to 8.0. Additionally, the removal and recovery of Cr(VI) from industrial wastewater was carried out by both batch and column experiments. The column operation was found to be more effective compared to the batch process and was found to be 85.8 and 65%, respectively. The adsorbed Cr could be recovered with 0.1 M HCl solution.

Both Langmuir and Freundlich fitted the equilibrium data well and the maximum adsorption capacity for neem bark at pH 2.0 was 19.60 mg g^{-1} within 3 h (Battacharya *et al.*, 2008). A better fit of PSO with the experimental data than that of the Lagergren PFO was found.

12.3.4   *Sugar beet pulp*

The Cr(III) and Cr(VI) removal form aqueous solution by a sugar beet pulp was investigated by Reddad *et al.* (2003). This agricultural by-product is mainly used as animal feed. It was found that 72.5% of the dry mass was accounted for by polysaccharides. It was confirmed that the majority of Cr(III) ions were removed within 1 h and the equilibrium was established after 120 min of contact time. The equilibrium time was fixed at 24 h and initial pH at 5.5 to provide the maximum Cr(III) adsorption capacity of polysaccharide. Both the Langmuir and Freundlich isotherm models fitted the experimental data well. The maximum monolayer adsorption capacity was found to be 10.04 mg g^{-1} of Cr(III). Kinetic experiments of Cr(VI) removal performed at pH 2.0 revealed that the concentration of Cr(VI) decreased slowly and Cr(III) appeared in the acidic solution, simultaneously.

12.3.5   *Olive industry wastes*

Adsorption of hexavalent Cr on pomace, an olive oil industry waste, was studied in batch and column mode (Malkoc *et al.*, 2006). The waste pomace of olive oil factories (WPOOF) is an oil-cake produced by the olive oil mills in Mediterranean countries. It is a solid residue after pressing the olives. This residue includes olive (6–8%), water (20–33%), seeds and pulps (59–74%) depending on the olive oil extraction processes. The removal of Cr(VI) from aqueous solutions strongly depends on the pH of the solution, adsorbent mass, initial Cr concentration and temperature. The adsorption capacity increased from 12.15 to 16.49 mg g^{-1} with increasing temperatures from 25 to 60°C. The Langmuir and Freundlich adsorption models were applied to the experimental data and that of Langmuir fitted better. The calculated monolayer adsorption capacity was found to be 13.95, 14.07, and 18.69 mg g^{-1} at 25, 45 and 60°C, respectively (optimum pH 2.0, 5 g L^{-1} adsorbent dosage and 120 min contact time). The calculated thermodynamic parameters showed that the Cr(VI) sorption process by WPOOF was endothermic and spontaneous in nature. The investigated adsorbent was found to be a low-cost adsorbent when compared with the commercially available adsorbents.

12.3.6   *Tea factory wastes*

The potential of tea factory waste (TFW) for Cr(VI) removal from aqueous solution was investigated (Malkoc and Nuhoglu, 2007). The maximum adsorption was observed at pH 2.0 and it decreased from 9.9 mg g^{-1} to 3.7 mg g^{-1} when the pH increased to 5.0. The adsorption data followed the Langmuir model better than that of Freundlich and the maximum adsorption capacity of 54.65 mg g^{-1} of Cr(VI) ions on TFW at 60°C was obtained. The adsorption capacity increased from 30.00 to 39.62 mg g^{-1} with an increase in temperature from 25 to 60°C at 400 mg L^{-1} of initial Cr(VI) ion concentration, indicating the endothermic nature of the adsorption process and it was confirmed by a positive value of ΔH^0. Using the first-order Lagergren kinetic constants, the activation energy of adsorption (E_a) was calculated as 18.57 kJ mol^{-1} using the Arrhenius equation, indicating the physical adsorption and the process is controlled by intraparticle diffusion.

12.3.7   *Bagasse*

The use of a lignocellulosic biosorbent for the adsorption and ion exchange of nine different heavy metal ions was studied by Krishnani *et al.* (2009). However, adsorption of Cr(VI) among them was considered only as a process of detoxification and both Cr(VI) and total Cr concentration was analyzed. Lignocellulosic bagasse was treated under steam by immersing in water in an autoclave at 121°C for 20 minutes. The detoxification of Cr(VI) was performed in acidic conditions (pH 1.9) at 30°C. It was found the Cr(VI) concentration decreased slowly and simultaneously. The Cr(III) appeared in solution and about 50% of the maximum adsorption capacity of Cr(III) was reached within 24 h. The Cr maximum capacity was found to be 22.98 mg g^{-1}. It was concluded that Cr adsorption occurred with the reduction of Cr(VI). The biosorbent has been found to be a good electron donor for the reduction of Cr(VI) in an acidic medium.

Hexavalent Cr removal by sugarcane bagasse was also investigated by Garg *et al.* (2007). It was observed that there was a sharp decrease in Cr removal with increasing pH. The maximum adsorption was obtained at pH 2.0 and was found to be 5.75 mg g^{-1} for SCB. It has been postulated that under acidic conditions, Cr(VI) could be reduced to Cr(III) in the presence of the adsorbent, and the Cr^{3+} ion being small in size, it was easily replaced by the positively charged species. The equilibrium of Cr adsorption was established within 60 minutes. The optimal stirring speed

250 rpm and adsorbent dose 20 g L^{-1} were found. The equilibrium adsorption data fitted both the Langmuir and Freundlich isotherm models well.

The adsorption kinetics of Cr(III) ions was studied using agave bagasse (Garcia-Reyes and Rangel-Mendez, 2010). The equilibrium adsorption data fitted the Langmuir isotherm model well. The maximum monolayer adsorption was calculated and the value was found to be 28.7 mg g^{-1} at pH 4.0 and 25°C. The high capacity of agave bagasse was attributed to the presence of carboxyl groups. It was found that a film diffusion model predicted adequately the Cr(III) adsorption kinetics on this agricultural by-product. It was concluded that diffusion models (as the film diffusion) could be more helpful to understand the adsorbate mass transfer to the adsorbent surface than empirical models.

12.3.8   *Fruit remains*

Orange waste is a by-product obtained after juicing. In the literature it can be found that this material can be applied as a sorbent for removal of heavy metals in a form of gel (Dhakal *et al.*, 2005). This biomass consists of pectin, cellulose, hemicelluloses, limonene and other low molecular weight polymers (Nagy *et al.*, 1977). The active binding sites for metals are the functional groups of pectin (Biswas *et al.*, 2008a; Dhakal *et al.*, 2005; Ghimire *et al.*, 2003). Pectin is an intercellular cementing material, which possesses free carboxyl groups and methylated ester groups in its polymer chains (Dhakal *et al.*, 2005). The gel is produced by saponification with calcium hydroxide or by crosslinking with epichlorohydrin (Dhakal *et al.*, 2005; Ghimire *et al.*, 2003). To enhance the capacity of orange waste gel for heavy metal ions the gel is loaded with ions such as Fe(III), Ce(III), La(III) and Zr(IV) (Biswas *et al.*, 2008a; 2008b; Dhakal *et al.*, 2005; Ghimire *et al.*, 2003).

Ghimire *et al.* (2003) in their study used Fe(III)-load phosphorylated crosslinked orange waste. Phoshorylation of orange waste resulted in conversion of alcoholic hydroxyl groups into phosphoric groups in the cellulose content. This modification enhanced the Fe(III) loading capacity and created more sites to bind As. The authors tested the sorbent efficiency in batch and column tests. It was observed that equilibrium was attained within 18 hours. The best removal of arsenite from water was found to be in the range of pH 6–9 and it was about 80%. In the case of arsenate adsorption was effective under acidic conditions (pH 2–5) and it was almost 99%. The optimal pH was 3 and 9 for arsenate and arsenite, respectively. Maximum adsorption capacity for both As(III) and As(V) was 68.1 mg g^{-1} (0.91 mmol g^{-1}) at initial concentration of 15 mg L^{-1}, dosage of 1667 g L^{-1}, 30°C and optimal pH. In column tests the bed was completely saturated at 377 of bed volume and As(V) amount of As adsorbed on packed gel bed was 1.1 mg for dosage of 0.1 g of gel at pH 4, initial concentration of 15 mg L^{-1}. In desorption studies, 1 M HCl was applied. It was observed that elution of As(V) was completed within 52 bed volume of HCl, and the amount of As(V) eluted was 0.62 mg (60% of total adsorbed As(V)).

Biswas *et al.* (2008a) investigated the effect of various parameters on adsorption of As(III) and As(V) using Zr(IV)-loaded orange waste gel. Equilibrium was attained within 15 hours. The kinetic data followed the pseudo-second-order equation. The optimal pH condition for removal of As(V) was in the range of 2-6. In the case of As(III) the maximum adsorption occurred at pH 9-10. The adsorption data followed the Langmuir adsorption isotherms. The maximum adsorption capacities of As(V) and As(III) were 88 mg g^{-1} and 130 mg g^{-1} at pH 3 and 10, respectively at initial concentration of 15 mg L^{-1}, dosage of 1.56 g L^{-1} and 30°C. In column tests the bed gel was completely saturated at 1400 and 840 bed volume for As(V) and As(III), respectively. The adsorption capacities obtained from the breakthrough profiles were 36.3 mg g^{-1} for As(V) and 35.1 mg g^{-1} for As(III) at initial concentration of 15 mg L^{-1}, weight of gel 0.05 g, flow rate 3.2 mL h^{-1} and at pH the same as in a batch test. Biswas *et al.* (2008a) observed that adsorption of As(V) and As(III) onto Zr(IV)-loaded orange waste gel was not affected by anions such as chloride, sulfate or carbonate. Desorption studies showed that 1 M NaOH is a good eluent. The gel after elution can be used once again as a sorbent with no losses in adsorption capacity of As. Biswas *et al.* (2008b) presented application of La(III)- and Ce(III)-loaded orange waste gels as a biosorbent for removal of As(III) and As(V). Experiments were conducted as batch and column tests. Using La(III)-loaded orange waste gel equilibrium was attained within 8 hours for As(III) and 15 hours for As(V). The adsorption capacities in the equilibrium condition were 8, 22 and 28 mg g^{-1} at pH 6.5, dosage of 1.67 g L^{-1} and initial As(V) concentration of 15, 30, 60 mg L^{-1}, respectively. In the case of As(III) the adsorption densities were 10, 16, 19 mg g^{-1} at pH 10, dosage of 1.67 g L^{-1} and initial concentration of 15, 30, 60 mg L^{-1}, respectively. The kinetic data fitted the pseudo-second-order equation. As(V) can be effectively removed by La(III)- and Ce(III)-loaded orange waste gels in the pH range 6-9.5. The most effective adsorption of As(III) was found to be at pHs from 9 to 11. The adsorption data followed the Langmuir equation. The maximum adsorption of La(III) and Ce(III)-loaded orange waste gel was the same, 43 mg g^{-1} for As(III) and 42 mg g^{-1} for As(V) at dosage of 1.67 g L^{-1}, 30°C, and pH 10 and 6.5, respectively. The concentration of As changed from 0.97 to 899 mg L^{-1}. In column tests saturation of the gel bed was achieved at 690 bed volume for As(V) and 280 bed volumes for As(III). The adsorption capacities in column tests was found to be 14.2 mg g^{-1} and 10.5 mg g^{-1} at weight of La(III)-loaded gel 0,150 g, feed rate 6.44 mL h^{-1}, and pH 6.5 and 9 for As(V) and As(III), respectively. For effective eluation of As(V) and As(III) from La(III)-loaded orange waste gel 1 M HCl and 1 M NaOH were used, respectively.

Banana peel is a low-cost biosorbent, which consists of proteins and polysaccharides inside the biomass cell cover (Kamsonlian *et al.*, 2012b). Kamsonlian *et al.* (2012) used banana peels as a potential sorbent of As(III). XRD analysis showed the presence of functional groups such as –NH and –OH. Sorption studies were performed in a batch system. The authors observed that efficiency of As(III) adsorption slightly increased with pH up to the maximum value (86%) at pH 7. The optimal biosorbent dosage was observed at 8 g. Equilibrium was reached within 90 minutes. The maximum adsorption density was found to be 0.843 mg g^{-1} at pH 7, initial concentration of 10 mg L^{-1} and 35°C. The adsorption data followed the Freundlich isotherm model (biosorption on a heterogeneous surface). The kinetic data fitted the pseudo-second-order model well (chemisorption as a rate limiting step).

Banana peel was also used as potential adsorbent for Cr(VI) removal from both artificial Cr solution and several industrial effluents (Memon *et al.*, 2009). The surface area of the adsorbent was measured using the BET method and found to be 13 m^2 g^{-1}. The dry matter, moisture, fat, crude fiber, crude protein, and ash in the dried banana peel were found to be 90.4, 9.6, 5.0, 11.0, 10.1 and 19.0% respectively. The adsorption experiments were carried out within the pH range of 1–9, the maximum Cr(VI) adsorption was recorded at pH 2. A sharp decline was observed with increasing pH. The sorption kinetics followed the pseudo-first-order Lagergren rate equation. The equilibrium data fitted both the Langmuir and Dubinin-Radushkevich isotherm models well. The negative values of ΔH^0 and ΔG^0 revealed the exothermic and spontaneous nature of the sorption phenomenon. The adsorption capacity decreased from 131.56 to 60.32 mg g^{-1} when the temperature changed from 20 to 40°C. It was found that only Fe^{2+} ions suppressed the Cr uptake significantly.

12.3.9 *Miscellaneous*

White chicken feathers are a by-product provided by the poultry industry, which contain a high amount of fibrous protein. This material is rich in keratin, which consists of cysteine amino acid residues. Teixeria and Ciminelli (2005) used activated biomass in removal of As(III) and As(V). Activation was made by treatment of biomass with ammonium thioglycolate, that reacted with the disulfide bridge, restoring –SH groups. In that paper the removal efficiency of As(III) and As(V) was compared with untreated and treated biomass. The experimental data showed that the raw biomaterial was not effective in removal of As from water. The optimal removal of As(III) was observed at activated-biomass dosage of 2.0 g L^{-1}. Equilibrium was reached within 10 minutes. Arsenic uptake on modified biomass decreased with increasing pH. The best removal was observed at pH 2 (dehydration of H$_3$AsO$_3$). The adsorption data fitted the Langmuir adsorption isotherm model well. The maximum capacity was 13.0 mg g^{-1} at pH 2.0, dosage of 2 g L^{-1}, initial concentration of 1.34 mmol L^{-1} and 25°C. The authors observed that phosphate ions did not compete with arsenite for active sites on the biomaterial.

The development of formaldehyde-treated, deseeded sunflower head agricultural waste biosorbent (FSH) for the biosorption of Cr(VI) from aqueous solution and industrial wastewater was investigated (Jain *et al.*, 2013). The maximum sorption occurred at pH 2.0, biosorbent dose 4.0 g L^{-1}, concentration of 100 mg L^{-1} at 25°C at 180 rpm after 2 h contact time. The adsorption capacity of FSH was found to be 7.85 mg g^{-1} for Cr(VI) removal at pH 2.0. The equilibrium sorption data fitted the Langmuir isotherm model well. The removal of heavy metals from industrial wastewater was significantly affected by the pH of the effluent and due to competition between different metal ions for adsorption sites was less than with artificial monometal solutions.

Hyacinth roots and neem leaves were also used for the removal of Cr(VI) (Singha and Das, 2011). Hyacinth roots were boiled in water for 6 h to remove the color. Neem leaves were treated with 0.1 N sodium hydroxide and subsequently with 0.1 N sulfuric acid to remove lignin-based colored matter. Batch studies were conducted with synthetic solutions. The optimum pH for Cr(VI) removal was found to be 2.0 for the adsorbents under investigation. Maximum uptake was obtained for 10 g L^{-1} adsorbent dosage and was found to be 15.2 mg g^{-1} for hyacinth roots and 16 mg g^{-1} for neem leaves, respectively. The adsorption kinetic data best fitted the pseudo-second order model. It was found that the adsorption process was controlled by both film and intraparticle diffusion. The equilibrium adsorption data were well described by the Langmuir equation for neem leaves and the Freundlich isotherm model for hyacinth roots, respectively. The sorption energy calculated using the Dubinin-Radushkevich isotherm model confirmed that the process is chemisorption. The thermodynamic parameters revealed that Cr(VI) adsorption on biosorbents is a spontaneous process, endothermic in nature.

Maize corn cobs and jatropha oil cake have been studied for the removal of hexavalent Cr from aqueous solution (Garg *et al.*, 2007). The maximum adsorption was obtained at pH 2.0 and was found to be 3.0 mg g^{-1} for JOC and 11.75 mg g^{-1} for MCC, respectively. The equilibrium adsorption was established within 1h. The equilibrium adsorption data fitted both the Langmuir and Freundlich isotherm models well.

Other natural adsorbents were studied for Cr(VI) removal from laboratory-simulated synthetic samples as well as from true wastewater samples from an aluminum powder coating factory (Dakiky *et al.*, 2002). Wool, pine needles and cactus leaves were cleaned, dried to constant weight and ground. Wool was washed with water and detergent, dried and cut into 1 cm fibers. No further characterization or analysis was performed on the adsorbents. For all adsorbents, the percentage removal of Cr(VI) was dramatically affected by initial pH of solution. The maximum removal from 100 mg L^{-1} Cr(III) solution and adsorbent dosage 8 g L^{-1} was achieved at pH approximately 2.0 and sharply decreased with increasing pH. The adsorption equilibrium was established within 1.5 h for wool and 2 h for other adsorbents, respectively. It was found that adsorption of Cr(VI) followed the first-order Lagergren kinetic model. Both Langmuir and Freundlich isotherm models were applied to describe the equilibrium data. The maxima calculated from the Langmuir equation are depicted in Table 12.2. It can be seen that wool is the best adsorbent in this group, followed by pine needles. Cactus had the lowest adsorption. The Gibbs free energy change was found to be negative for wool, and confirmed the feasibility of these adsorbents and the spontaneity of adsorption. Comparing Cr removal from synthetic solution and real effluent, both containing 100 mg L^{-1} Cr(VI), by wool, which was found to be the most effective adsorbent, it was observed that the percent removal of Cr(VI) was higher for the synthetic solution. The difference was attributed to the presence of other ions and impurities in the industrial effluent (Al, Mg, Ca and B).

12.4   ARSENIC AND CHROMIUM REMOVAL MECHANISMS

In the following paragraphs, the mechanisms of As and Cr removal by the adsorbents described above are discussed in detail. Only the mechanisms of the most popular and used adsorbents were presented.

12.4.1   *The sorption mechanisms of arsenic ions*

Salim *et al.* (1994) reported that the removal of metal ions from aqueous solutions may be due to adsorption on the surface and in pores, and to complexation by biosorbents. The influences on adsorption mechanisms of As(III) and As(V) have various parameters such as pH, temperature, initial As concentration in solution. But very important is also the type of sorbent and the modification to which the sorbent was subjected.

The process of adsorption onto porous materials (as fly ash or red mud) can be governed by the four following steps (Ho, 2000):

- Transport in the bulk solution;
- Film diffusion;
- Pore diffusion;
- Sorption and desorption of the adsorbate at the adsorption site.

As shown in the literature, ions of As can be chemically fixed into the cementation environment of the solidified/stabilized matrices by three important immobilization mechanisms (Phenrat *et al.*, 2005):

- Sorption onto the surface of C-S-H;
- Immobilization of oxyanions (especially As(V) ions) by replacing of ettringite;
- Reaction with cement components to from Ca-As compounds.

Vandecasteele *et al.* (2002) reported that precipitation of $Ca_3(AsO_4)_2$ and $CaHAsO_3$ follows in the presence of $Ca(OH)_2$ and it governs immobilization of As ions on fly ash. Solid calcium arsenate hexahydrate can be formed, according to the following reaction (Balsamo *et al.*, 2010):

$$3Ca^{2+} + 2(AsO_4^{3-}) \xrightarrow{H_2O} Ca_3(AsO_4)_2 \cdot 6H_2O \tag{12.1}$$

The dissociated $H_2AsO_3^-$ ion may interact with the other surface species C-S-H to undergo complex formation reactions in the following manner (Kundu and Gupta, 2007):

$$(Si-O^-)_2Ca^{2+} + H_3O^+ \rightarrow 2(Si-OH) + Ca^{2+} + OH^- \tag{12.2}$$

$$2Si-OH + 2H_2AsO_3^- \rightarrow 2Si-H_2AsO_3 + OH^- \tag{12.3}$$

The surface of fly ash has a very varied chemical composition. The surface acidity of fly ash results from several mineral oxides (including silicon dioxide, aluminum oxides and iron oxides), which influence the sorption mechanism. The adsorption reactions of different As species are expressed as (Wang *et al.*, 2008):

$$SOH_2^+ + H_2AsO_4^- \rightleftarrows S-H_2AsO_4 + H_2O \tag{12.4}$$

$$SOH_2^+ + HAsO_4^{2-} \rightleftarrows S-HAsO_4^- + H_2O \tag{12.5}$$

$$SOH_2^+ + AsO_4^{3-} \rightleftarrows S-AsO_4^{2-} + H_2O \tag{12.6}$$

The predominant mechanism usually involves ion-exchange reactions between positively charged groups present in the biomaterial structure, like: carboxyl, hydroxyl, amide, amine, imidazole, phosphate, sulfate, phenol and carbonyl (Al-Asheh and Duvnjak, 1997). All these are potential reactive sites to form adsorptive complexes with negatively charged ions, such as arsenate or arsenite (Teixeira and Ciminelli, 2005; Veglio and Beolchini, 1997).

The major components of phosphogypsum and sugar foam are gypsum and calcium carbonate, respectively. The sorption mechanism onto these sorbents can be attributed to chemical bonding and/or hydrogen-bonding between $H_2AsO_4^-$ (or $HAsO_4^{2-}$) ions and surface hydroxyls. These reactions can be written as (Aguilar-Carrillo *et al.*, 2006):

$$\equiv R-OH + H^+ + H_2AsO_4^- \rightleftarrows \equiv R-H_2AsO_4 + HOH \tag{12.7}$$

$$\equiv R-OH + H_2AsO_4^- \rightleftarrows \equiv R-H_2AsO_4 + OH^- \tag{12.8}$$

where: $\equiv R$ are the reactive surface phases (Al or Fe oxy-hydroxides).

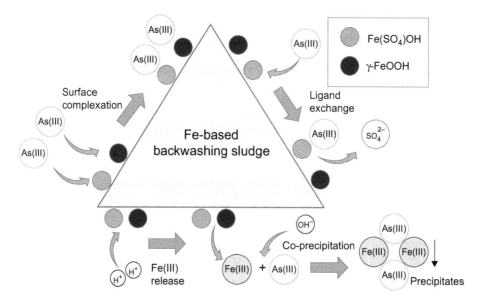

Figure 12.1.   Mechanism of As(III) adsorption on Fe-based backwashing sludge (reproduced with permission from Wu *et al.*, 2013a, copyright of Elsevier).

Considering the adsorption of As on blast furnace residues, which mainly consist of zero-valent Fe, FeO and Fe_3O_4, the following reactions were proposed by Carrillo-Pedroza *et al.* (2014):

$$Fe + H_2AsO_4^- + H^+ + \tfrac{3}{4}O_2 \rightleftarrows FeAsO_4 + \tfrac{3}{2}H_2O \text{ (g)} \tag{12.9}$$

$$FeO + H_2AsO_4^- + H^+ + \tfrac{1}{4}O_2 \rightleftarrows FeAsO_4 + \tfrac{3}{2}H_2O \tag{12.10}$$

$$Fe_3O_4 + 3H_2AsO_4^- + 3H^+ + \tfrac{1}{4}O_2 \rightleftarrows 3FeAsO_4 + \tfrac{9}{2}H_2O \tag{12.11}$$

Fe-based blackwashing sludge contains an inner layer of $Fe(SO_4)OH$, γ-FeOOH, SiO_2 and $CaCO_3$. FTIR and XPS analyses of spectra before and after As(III) adsorption onto this showed that the mechanism of adsorption involved:

- Ligand exchange reaction between sulfate and arsenite,
- Fe(III) precipitation,
- Adhesion to the hydroxyl groups of the sorbent.

These reactions can be written as follows (Wu *et al.*, 2013a):

$$\text{a)} \quad \equiv Fe(SO_4)OH + H_3AsO_3 \rightleftarrows \equiv FeH_2AsO_3^{2+} + SO_4^{2-} + H_2O \tag{12.12}$$

$$\text{b)} \quad \equiv FeOOH + H_3AsO_3 + 2H^+ \rightleftarrows \equiv FeH_2AsO_3^{2+} + 2H_2O \tag{12.13}$$

$$\text{c)} \quad \equiv FeOOH + H_3AsO_3 \rightleftarrows FeOH_2AsO_3 + H_2O \tag{12.14}$$

The proposed mechanism can be presented as the scheme in Figure 12.1.

Chicken feathers are a cysteine-rich biomass. The lateral group of each cysteine molecule is the sulfhydryl group (–SH). The removal of As(III) from aqueous solutions can proceed as irreversible enzymatic inhibition by the reaction of As(III) with cysteine sulfhydryl groups. The following

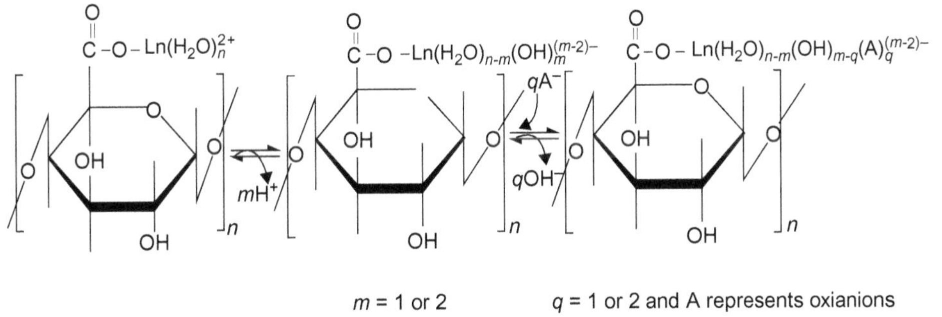

$$\overset{O}{\overset{\|}{C}} - O - Ln(H_2O)_n^{2+} \qquad \overset{O}{\overset{\|}{C}} - O - Ln(H_2O)_{n-m}(OH)_m^{(m-2)-} \qquad \overset{O}{\overset{\|}{C}} - O - Ln(H_2O)_{n-m}(OH)_{m-q}(A)_q^{(m-2)-}$$

$$m = 1 \text{ or } 2 \qquad\qquad q = 1 \text{ or } 2 \text{ and A represents oxianions}$$

Figure 12.2. Mechanism of ligand exchange for arsenic adsorption onto the metal-loaded SOW gel (reproduced with permission from Biswas 2007, copyright of Japan Society of Ion Exchange).

equation is proposed to describe arsenite adsorption by the fibrous protein biomass (Teixera and Ciminelli, 2005):

$$H_3AsO_3 + 3B-CysSH \rightarrow As(B-CysS-)_3 + 3H_2O \tag{12.15}$$

where B is the biomaterial matrix. The sulfhydryl groups are the active groups and strong enough to involved in As biosorption onto this material (Teixera and Ciminelli, 2005).

Electrostatic attractions between the negative surface charge of biochar and metal cations as well as metal precipitation for the case of metal anions may occur. Yet, high surface area and pore volume facilitate faster mass transfer of pollutants into the biochar pores and provide more opportunities for metal-active site binding. The mechanism for adsorption of As ions onto bone char may proceed as follows (Chen et al., 2008):

$$Ca_{10}(PO_4)_6(OH)_2 + Ca^{2+} + HAsO_4^{2-} \rightarrow Ca_{10}(PO_4)_6(OH)_2 \cdot Ca(HAsO_4) \tag{12.16}$$

or ion exchange:

$$Ca_{10}(PO_4)_6(OH)_2 + HAsO_4^{2-} \rightarrow Ca_{10}(PO_4)_6(HAsO_4) + 2OH^- \tag{12.17}$$

Ghimire et al. (2003) used orange juice residue to remove As ions. This material was transformed into the form of gel by phosphorylation followed by loading with Fe(III), treated with La(III), Ce(III) (Biswas et al., 2008b) or modification with Ca(OH)$_2$, which led to conversion of methy-lated ester groups in pectin into carboxyl groups (Dhakal et al., 2005). Orange waste consists of cellulose, hemicellulose, pectins and other lower molecular weight compounds (Dhakal et al., 2005; Nagy et al., 1977). This material can bind metal ions by the carboxylic groups of pectins (Biswas et al., 2008a; Ghimire et al., 2003). Arsenate or arsenite have been considered to be adsorbed by the mechanism of ligand exchange, for example see Figure 12.2.

12.4.2 The sorption mechanisms of chromium ions

The major parameters which govern the adsorption mechanism of Cr(VI) and Cr(III) are: pH, temperature, initial Cr concentration in solution, adsorbent dose, time of contact with sorbent, particle size and type of adsorbent (Mohan and Pittman, 2006). The type of adsorbent has important effects for the sorption of Cr, which can proceed as: chemisorption, complexation, precipitation, ion exchange, chelation, diffusion on the surface adsorbent or in the pores of the adsorbent and also electrostatic attraction (Sud et al., 2008).

As for other heavy metals, the adsorption of Cr is governed by the four steps mentioned above, and if the adsorption of Cr(VI) or Cr(III) proceeds on porous sorbents such as fly ash, the last step can be chemical reaction. This process is considered very rapid and not process controlling. The overall sorption process rate is governed by the slowest stage and it would be the film diffusion or the pore diffusion (Wang and Wu, 2006). Regarding the composition of fly ash and the oxidation state of Cr, the removal of Cr(VI) was found to be higher at low pH for both, class C and class F fly ash (Bayat, 2002). This was explained by the fact that at pH range 2.0–4.0 the Cr(VI) is present mainly as $Cr_2O_7^{2-}$ and the fly ash surface was positively charged, thus the adsorption capacity of both types of fly ash could be attributed to the electrostatic interaction of the adsorbate with surface iron and alumina sites. The adsorption of Cr(VI) on fly ash surfaces can be explained as a surface complex formation and ligand exchange reactions and the adsorption mechanism may be expressed by the following reactions (Banerjee *et al.*, 2004):

$$-SEOH + H^+ \rightleftarrows -SEOH_2^+ \qquad (12.18)$$

$$-SEOH_2^+ + HCrO_4^- \rightleftarrows -SEOH_2(HCrO_4) \qquad (12.19)$$

or

$$-SEOH_2^+ + HCrO_4^- \rightleftarrows [SE(HCrO_4) + H_2O] \qquad (12.20)$$

where: $-SEOH$ is surface hydroxyl sites and SE=Si, Al or Fe.

Additionally, it was confirmed that the surface of class F coal fly ash contains three types of acidic sites with the acidity constants (pK_H) of 2.7, 7.8, and 11.0 (Wang *et al.*, 2004). However, the Cr(III) adsorption results showed that only the acid site with 7.8 pK_H is responsible for metal adsorption. Zeolites from fly ash had greater ability than fly ash to remove trivalent Cr, due to their high cation-exchange capacity (*CEC*) and the high acid neutralizing capacity (*ANC*) of ZFAs (Wu *et al.*, 2008). The authors suggested that the mechanism of Cr(III) removal by ZFAs involved both ion exchange and precipitation. A high calcium content in both fly ash and ZFAs resulted in a high ANC value and a high immobilization ability toward Cr. Cr(III) removal by ZFAs was found to be strongly pH-influenced. Within the solubility range, the removal of Cr increased with increasing pH. On the other hand, at pH values above the solubility zone, the efficacy of Cr removal by the ZFAs approached 100% due to the precipitation of Cr(OH)$_3$ on the sorbent surface.

Among the various components of red mud, the pH at the point of zero charge of SiO$_2$ is around 2.3 and that of Fe$_2$O$_3$ around 8.6, thus the composite pH$_{pzc}$ for red mud is 3.2 (Gupta *et al.*, 2001). As the maximum Cr uptake took place at pH 2.0, it was attributed to the interaction of the positively charged surface of red mud below pH 3.2 with Cr anions. The higher adsorption of Cr at lower pH may be also due to the neutralization of surface charge by an excess of hydrogen ions, facilitating the diffusion of dichromate ions and their adsorption. A significant reduction of Cr removal at higher pHs can be the result of the abundance of OH$^-$ ions hindering the diffusing species.

Due to ionization of functional groups the surface charge of clay is pH-dependent (Weng *et al.*, 2008). Below pH$_{iep}$ the adsorbent is positively charged and favors adsorption of Cr anions. The active surface functional groups at the clay edges are silanol (Si-O-) and aluminol (Al-O-) sites, as the main components of this clay are SiO$_2$ and Al$_2$O$_3$. As the solution pH increases, the number of protonated functional groups decreases and more OH$^-$ ions compete with Cr ions for the active surface. The formation of complexes becomes difficult and the level of Cr adsorption decreases. The formation of the bonding surface complexes Si,Al-HCrO$_4$ and Si,Al-CrO$_4^-$ below and above pH$_{iep}$, respectively, was proposed as a mechanism of Cr(VI) adsorption onto spent activated clay.

If biomaterial is used as adsorbent, the adsorption mechanism can be divided into four categories (Saha and Orvig, 2010):

- Anionic adsorption, in which Cr(VI) is adsorbed by biomaterial through electrostatic attraction and the adsorbed Cr(III) is not reduced by the biosorbent.
- Adsorption-coupled reduction, in which Cr(VI) is adsorbed and completely reduced to Cr(III) by the biomaterial.
- Anionic and cationic adsorption, in which a fraction of the Cr(VI) is reduced to Cr(III) by the biomaterial and the resultant Cr(III) is adsorbed by the biomaterial.
- Reduction and anionic adsorption, in which a fraction of the Cr(VI) is adsorbed and the residue is reduced to Cr(III).

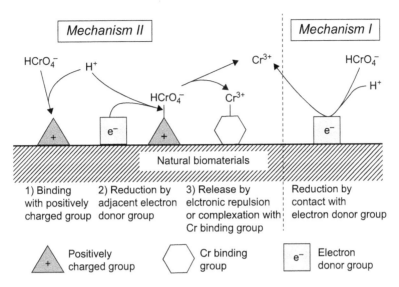

Figure 12.3.   Mechanism of Cr(VI) sorption by biomaterials (reproduced with permission from Park *et al.*, 2007, copyright of Elsevier).

The functional groups on the surface of the biosorbent also influence the sorption mechanism. They are among others: carbonyl groups (R-COOH), amine groups (R-NH$_2$), hydroxyl groups (R-OH), sulfhydryl groups (R-SH) (Bansal *et al.*, 2009; Li *et al.*, 2007; Zakaria *et al.*, 2009). It was confirmed that the functional groups of the wine processing sludge were R-COOH and R-NH$_2$, which were similar to chitosan-based sorbent (Li *et al.*, 2004). The hydroxyl (R-OH) group and nonprotonized amino group can serve as coordination and electrostatic interaction binding sites for heavy metals. At pHs below 4.0 the Cr(III) removal increased abruptly and also slightly above 4.0. Increasing the pH favored the attractive electrostatic force from the R-OH functional group and cationic Cr species.

Park *et al.* (2005; 2007) proposed that Cr ions can be removed from aqueous solution by biomaterials through both direct and indirect reduction mechanisms (Fig. 12.3).

- Mechanism I (direct reduction mechanism) – Cr(VI) is directly reduced to Cr(III) in the aqueous phase by contact with the electron-donor groups of the biosorbent. The reduced Cr(III) forms complexes with the biosorbent or remains in the aqueous phase.
- Mechanism II (indirect reduction mechanism) consists of three steps:
  a) the binding of anionic Cr(VI) to the positively-charged groups present on the biosorbent surface;
  b) the reduction of Cr(VI) to Cr(III) by adjacent electron-donor groups;
  c) the release of the reduced Cr(III) into the aqueous phase due to electronic repulsion between the positively-charged groups and the Cr(III) or the complexation of the reduced Cr(III) with adjacent groups.

The main mechanism of Cr(VI) removal by different biomaterials is the reduction reaction of Cr(VI) to Cr(III).

Cr(VI) is reduced to Cr(III) by reductants present at the banana peel surface, such as polysaccharides, glycoproteins, glucolipids and nucleic acids. Thus, a proportion of bioreduced Cr(III) is released from the banana peel surface and the reaction mechanism may be represented by (Memon *et al.*, 2009):

$$HCrO_4^- + \text{banana peel} \rightarrow HCrO_4^- - \text{banana peel}$$

$$\downarrow$$

$$Cr^{3+} + H_2O + \begin{matrix} \text{banana peel} \\ \text{(oxidized)} \end{matrix} \qquad (12.21)$$

The mechanism of Cr adsorption can involve carbonyl groups (C=O) and methoxy groups (OCH$_3$). Cr(VI) is reduced to Cr(III) by lignin in coir pith and the reaction between Cr(VI) and biomaterial can be followed as (Suksabye and Thiravetyan, 2012):

$$\begin{matrix} \text{coir pith} \\ \text{grafted with} + HCrO_4^- \xrightarrow{\text{reduction}} Cr^{3+} \\ \text{acrylic acit} \end{matrix} \begin{matrix} \nearrow Cr^{3+} + C=O & \begin{matrix} \text{in coir pith grafted} \\ \text{with acrylic acid} \end{matrix} \\ \\ \searrow Cr^{3+} + O-CH_3 & \begin{matrix} \text{in coir pith grafted} \\ \text{with acrylic acid} \end{matrix} \end{matrix} \qquad (12.22)$$

The reduction of Cr(VI) to Cr(III) occurred in solution after contact with biosorbent (Reddad *et al.*, 2003). The predominant ion-exchange mechanism involved Ca^{2+} neutralizing the carboxyl groups of the material in Cr(III) adsorption, while for Cr(VI) the optimum removal resulting from the reduction mechanism was achieved at acidic pH values. The carboxylic groups of the sugar beet pulp biosorbent were found to be the main reduction sites of Cr(VI) species. Lignocellulosic agricultural waste may have a high Cr(VI) reduction capacity depending on the kind and amount of functional organic groups. The reduction reaction can be illustrated as below (Reddad *et al.*, 2003):

$$\begin{matrix} HCrO_4^- \\ \text{or} \qquad \qquad + \\ CrO_4^{2-} \end{matrix} \begin{matrix} \text{organic} \\ \text{biomaterial} \end{matrix} + H^+ \rightarrow Cr^{3+} + H_2O + CO_2 \begin{matrix} \text{(and/or oxidizes} \\ \text{products)} \end{matrix} \qquad (12.23)$$

For rice straw the adsorption equilibrium was achieved within 48 h and the presence of Cr(III) in the solution indicated that both reduction and adsorption were involved in the Cr(VI) removal (Gao *et al.*, 2008). The increase of the solution pH suggested that protons were needed for the Cr(VI) removal. The promotion of rice straw modified by tartaric acid and the slight inhibition of the esterified rice straw on Cr(VI) removal indicated that carboxyl groups present on the biomass played an important role in the removal of Cr but were not fully responsible for it.

Sawdust is a cellulose-based material containing tannin- and lignin-based organic compounds (Zakaria *et al.*, 2009). The surface of sawdust has functional groups (chiefly –OH), which influence the sorption of Cr ions. Different mechanisms, such as electrostatic attraction, ion exchange, chemical complexation, can be taken into account while discussing the sorption mechanism (Gupta and Babu, 2009). It was observed that the pH of the solution increased with increasing Cr(VI) initial concentration and contact time with sawdust. This was attributed to the fact that sawdust is a carbonaceous material and on its surface oxo groups (C$_x$O and C$_x$O$_2$) are present and hydrolyze water molecules and the pH of the solution increases (Gupta and Bubu, 2009). The Cr(VI) ions are adsorbed onto the positively charged surface, as given by the following (Dakiky *et al.*, 2002):

$$\equiv OH_2^+ + HCrO_4^- \rightleftarrows \equiv OH_2^+(HCr_4^-) \qquad (12.24)$$

Adsorption on these biomaterials may be due to physical adsorption, complexation with functional groups, ion exchange, surface precipitation and chemical reaction with surface sites. Probably between the functional groups of rice husk and Cr(VI) ions there is complexation (Bansal *et al.*, 2009). The functional groups like surface O–H, aliphatic, C–H, unsaturated groups like alkene, aromatic C–NO$_2$, carboxylate anion, Si–O groups, –SO$_3$ and sulfonic acid groups are responsible for sorption of Cr(VI) onto rice husk, rice straw, rice bran, hyacinth roots, neem bark, saw dust of teakwood origin, neem leaves and coconut shell (Singha and Das, 2011). It was observed that with increasing pyrolysis temperature the surface area of the coconut coir chars increased, but the amount of oxygen-containing functional groups of the chars decreased (Shen *et al.*, 2012). The Cr(VI) removal was primarily attributed to the reduction of Cr(VI) to Cr(III). The ability of CC and CC-derived chars to reduce the Cr(VI) content of polluted waters was found to be determined by the type and amount of the oxygen-containing functional groups (particularly phenolic groups) of the materials rather than by their specific surface area. However, the non-modified coconut coir had the highest content of functional groups that can reduce toxic Cr(VI) to the less toxic Cr(III).

12.5   CONCLUSIONS

Solid waste materials have been used as adsorbents without any pretreatment or after little processing, such as washing with distilled water or hydrogen peroxide, boiling in water, drying, grinding, and finally sieving to obtain the desired fractions. To improve adsorbent capacity, some modifications such as impregnation (e.g. Al and Fe-impregnated fly ash), treatment with acid (e.g. sulfuric, hydrochloric, tartaric or citric acid) or formaldehyde, and activation in air in furnace at high temperature (usually 400–500°C) are used. The organic matter of biosorbents can be subjected to carbonization and then air-activation. Activated carbons prepared from solid waste have been widely used for removal of As(III)/As(V) as well as Cr(III)/Cr(VI). However, this type of adsorbent was not of concern in this work. Specific surface area and porosity usually are the significant factors in the case of removal of organic compounds by industrial and agricultural by-products. The bigger the surface area and porosity, the higher pollutant uptake. However, for adsorption of heavy metals ions the number and type of functional groups of the sorbent are most important, since ion exchange, reduction and precipitation may contribute or dominate (Mohan and Pittman, 2006). The values were achieved in various experimental conditions, such as initial concentrations, adsorbent dosage, pH of solution and the temperature. The metal removal investigations were also performed for both synthetic and real wastewaters. The latter usually contain not only considered heavy metal ions but also are rich in various ions, e.g. sulfates, nitrates, phosphates, chlorides, sodium, and water hardness ions. Some of the investigated adsorbents were selective towards As or Cr ions. However, most of them possess high affinity to all ions present in wastewater. Among the industrial and agricultural by-products reported, a high adsorption capacity for hexavalent Cr was shown by activated red mud, carbon slurry, banana peel and coconut coir, but modification of sawdust by sodium phosphate and acrylic acid grafting of coir pith gave the highest capacities. Trivalent Cr was efficiently removed by using *Mucor meihi* biomass, fly ash from poultry litter, zeolites from class C coal fly ashes and HCl-treated sawdust. The highest adsorption uptake toward As(V) was shown by $FeCl_3$-modified red mud, Fe_2O_3 nanoparticles obtained from red mud, FeOOH-loaded steel slag, class C fly ash and Zr(IV)-load orange waste. For trivalent As, Zr(IV)-load orange waste, Fe(III)-load orange waste, rice husk, and class C fly ash were found to be the most effective adsorbents.

ACKNOWLEDGEMENTS

The work was financially supported by a statutory activity subsidy from the Polish Ministry of Science and Higher Education for the Faculty of Chemistry of Wrocław University of Technology.

REFERENCES

Agrafioti, E., Kalderis, D. & Diamadopoulos, E. (2014) Arsenic and chromium removal from water using biochars derived from rice husk, organic solid wastes and sewage sludge. *Journal of Environmental Management*, 133, 309–314.

Aguilar-Carrillo, J., Garrido, F., Barrios, L. & Garcia-Gonzales, M.T. (2006) Sorption of As, Cd and Ti as influenced by industrial by-products applied to an acidic soil: equilibrium and kinetics experiments. *Chemosphere*, 65, 2377–2387.

Ahmad, R., Rao, R.A.K. & Masood, M.M. (2004) Removal and recovery of Cr(VI) from synthetic and industrial wastewater using bark of *Pinus Roxburghii* as an adsorbent. *Water Quality Research Journal of Canada*, 40 (4), 462–468.

Ahn, J.S., Chon, C.-M., Moon, H.-S. & Kim K.-W. (2003) Arsenic removal using steel manufacturing byproducts as permeable reactive materials in mine tailing containment systems. *Water Research*, 37, 2478–2488.

Akin, I., Arslan, G., Tor, A., Ersoz, M. & Cengeloglu, Y. (2012) Arsenic(V) removal from underground water by magnetic nanoparticles synthesized from waste red mud. *Journal of Hazardous Materials*, 235–236, 62–68.

Al-Asheh, S. & Duvnjak, Z. (1997) Sorption of cadmium and other heavy metals by pine bark. *Journal of Hazardous Materials*, 56, 35–51.

Ali, I., Al-Othman, Z.A., Alwarthan, A., Asim, M. & Khan, T.A. (2014) Removal of arsenic species from water by batch and column operations on bagasse fly ash. *Environmental Science and Pollution Research*, 21, 3218–3229.

Altun, T. & Pehlivan, E. (2012) Removal of Cr(VI) from aqueous solutions by modified walnut shells. *Food Chemistry*, 132, 693–700.

Altundoğan, H.S., Altundoğan, S., Tümen, F. & Bildik, M. (2000) Arsenic removal from aqueous solutions by adsorption on red mud. *Waste Management*, 20, 761–767.

Altundoğan, H.S., Altundoğan, S., Tümen, F. & Bildik, M. (2002) Arsenic adsorption from aqueous solutions by activated red mud. *Waste Management*, 22, 357–363.

Amin, M.N., Kaneco, S., Kitagawa, T., Begum, A., Katsumata, H., Suzuki, T. & Ohta, K. (2006) Removal of arsenic in aqueous solutions by adsorption onto waste rice husk. *Industrial and Engineering Chemistry Research*, 45, 8105–8110.

Babel, S. & Kurniawan, T.A. (2003) Low-cost adsorbents for heavy metals uptake from contaminated water: a review. *Journal of Hazardous Materials*, B97, 219–243.

Bailey, S.E., Olin, T.J., Bricka, R.M. & Adrian, D.D. (1999) A review of potentially low-cost sorbents for heavy metals. *Water Research*, 33 (11), 2469–2479.

Balsamo, M., Natale, F., Erto, A., Lancia, A., Montagnaro, F. & Santoro, L. (2010) Arsenate removal from synthetic wastewater by adsorption onto fly ash. *Desalination*, 263, 58–63.

Banerjee, S.S., Joshi, M.V. & Jayaram, R.V. (2004) Removal of Cr(VI) and Hg(II) from aqueous solutions using FA and impregnated FA. *Separation Science and Technology*, 39 (7), 1611–1629.

Bansal, M., Garg, U., Singh, D. & Garg, V.K. (2009) Removal of Cr(VI) from aqueous solutions using pre-consumer processing agricultural waste: a case study of rice husk. *Journal of Hazardous Materials*, 162, 312–320.

Bayat, B. (2002) Comparative study of adsorption properties of Turkish fly ashes, II. The case of chromium (VI) and cadmium (II). *Journal of Hazardous Materials*, B95, 275–290.

Bertocchi, A.F., Ghiani, M., Peretti, R. & Zucca, A. (2006) Red mud and fly ash for remediation of mine sites contaminated with As, Cd, Cu, Pb and Zn. *Journal of Hazardous Materials*, B134, 112–119.

Bhattacharya, A.K., Naiya, T.K., Mandal, S.N. & Das, S.K. (2008) Adsorption, kinetics and equilibrium studies on removal of Cr(VI) from aqueous solutions using different low-cost adsorbents. *Chemical Engineering Journal*, 137, 529–541.

Biswas, B.K., Ghimire, K.N., Inoue, K., Kawakita, H., Ohto, K. & Kawakita, H. (2007) Adsorptive separation of arsenic and phosphorus from an aquatic environment using metal-loaded orange waste. *Journal of Ion Exchange*, 18 (4), 428–433.

Biswas, B.K., Inoue, J., Inoue, K., Ghimire, K.N., Harada, H., Ohto, K. & Kawakita, H. (2008a) Adsorptive removal of As(V) and As(III) from water by a Zr(IV)-loaded orange waste gel. *Journal of Hazardous Materials*, 154, 1066–1074.

Biswas, B.K., Inoue, K., Ghimire, K.N., Kawakita, H., Ohto, K. & Harada, H. (2008b) Effective removal of arsenic with lanthanum(III)- and cerium(III)-loaded orange waste gels. *Separation Science and Technology*, 43, 2144–2165.

Brunori, C., Cremisini, C., Massanisso, P., Pinto, V. & Torricelli, L. (2005) Reuse of a terated red mud bauxite waste: studies on environmental compatibility. *Journal of Hazardous Materials*, B117, 55–63.

Campos, I., Álvarez, J.A., Villar, P., Pascual, A. & Herrero, L. (2013) Foundry sands as low-cost adsorbent material for Cr(VI) removal. *Environmental Technology*, 34 (10), 1267–1281.

Carrillo-Pedroza, F.R., Soria-Aguilar, M.J., Martínez-Luevanos, A. & Narvaez-Garcia, V. (2014) Blast furnace residues for arsenic removal from mining-contaminated ground water. *Environmental Technology*, 35 (21–24), 2895–2902.

Chaudhuri, M. & Mohammed, M.A. (2012) Arsenic(III) immobilization on rice husk. *Journal of Science and Technology*, 4, 47–54.

Chen, Y., Chai, L. & Shu, Y. (2008) Study of arsenic(V) adsorption on bone char from aqueous solution. *Journal of Hazardous Materials*, 160, 168–172.

Dakiky, M., Khamis, M., Manassra, A. & Mereb, M. (2002) Selective adsorption of chromium(VI) in industrial wastewater using low-cost abundantly available adsorbents. *Advances in Environmental Research*, 6, 533–540.

Demirbas, A. (2008) Heavy metal adsorption onto agro-based waste materials: a review. *Journal of Hazardous Materials*, 157, 220–229.

Dhakal, R.P., Ghimire, K.N. & Inoue, K. (2005) Adsorptive separation of heavy metals from an aquatic environment using orange waste. *Hydrometallurgy*, 79, 182–190.

Diamadopoulos, E., Ioannidis, S. & Sakellaropoulos, G.P. (1993) As(V) Removal from aqueous solutions by fly ash. *Water Research*, 27 (12), 1773–1777.

Fu, F. & Wang, Q. (2011) Removal of heavy metal ions from wastewaters: a review. *Journal of Environmental Management*, 92, 407–418.

Gao, H., Liu, Y., Zeng, G., Xu, W., Li, T. & Xia, W. (2008) Characterization of Cr(VI) removal from aqueous solutions by a surplus agricultural waste – rice straw. *Journal of Hazardous Materials*, 150, 446–452.

Garcia-Reyes, R.B. & Rangel-Mendez, J.R. (2010) Adsorption of chromium(III) ions on agro waste-materials. *Bioresource Technology*, 101, 8099–8108.

Garg, U.K., Kaur, M.P., Garg, V.K. & Sud, D. (2007) Removal of hexavalent chromium from aqueous solution by agricultural waste biomass. *Journal of Hazardous Materials*, 140, 60–68.

Genç, H., Tjell, J.C., McConchie, D. & Schuiling, O. (2003) Adsorption of arsenate from water using neutralized red mud. *Journal of Colloid and Interface Science*, 264, 327–334.

Genç-Fuhrman, H., Tjell, J.C. & McConchie, D. (2004a) Increasing the arsenate capacity of neutralized red mud (Bauxsol). *Journal of Colloid and Interface Science*, 271, 313–320.

Genç-Fuhrman, H., Tjell, J.C. & McConchie, D. (2004b) Adsorption of arsenic from water using activated neutralized red mud. *Environmental Science & Technology*, 38, 2428–2434.

Ghimire, K.N., Inoue, K., Yamaguci, H., Makino, K. & Miyajima, T. (2003) Adsorptive separation of arsenate and arsenite anions from aqueous medium by using orange waste materials. *Water Research*, 37, 4945–4953.

Glocheux, Y., Pasarín, M.M., Albadarin, A.B., Allen, S.J. & Walker, G.M. (2013) Removal of arsenic from groundwater by adsorption onto an acidified laterite by-product. *Chemical Engineering Journal*, 228, 565–574.

Guan, X., Dong, H., Ma, J. & Lo, I.M.C. (2011) Simultaneous removal of chromium and arsenate from contaminated groundwater by ferrous sulfate: batch uptake behavior. *Journal of Environmental Science*, 23 (3), 372–380.

Gupta, S. & Babu, B.V. (2009) Removal of toxic metal Cr(VI) from aqueous solutions using sawdust as adsorbent: equilibrium, kinetics and regeneration studies. *Chemical Engineering Journal*, 150, 352–365.

Gupta, V.K. & Ali, I. (2004) Removal of lead and chromium from wastewater using bagasse FA – a sugar industry waste. *Journal of Colloid and Interface Science*, 271, 321–328.

Gupta, V.K., Gupta, M. & Sharma, S. (2001) Process development for the removal of lead and chromium from aqueous solutions using red mud – an aluminum industry waste. *Water Research*, 35 (5), 1125–1134.

Gupta, V.K., Rastogi, A. & Nayak, A. (2010) Adsorption studies on the removal of hexavalent chromium from aqueous solution using a low cost fertilizer industry waste material. *Journal of Colloid and Interface Science*, 342, 135–141.

Hasan, S.H., Singh, K.K., Prakash, O., Talat, M. & Ho, Y.S. (2008) Removal of Cr(VI) from aqueous solutions using agricultural waste 'maize bran'. *Journal of Hazardous Materials*, 152, 356–365.

Hegazi, H.A. (2013) Removal of heavy metals from wastewater using agricultural and industrial wastes as adsorbents. *HBRC Journal*, 9, 276–282.

Helfferich, F. (1995) *Ion exchange*. Dover Publications Inc., New York, NY. p. 624.

Hermandez-Ramirez, O. & Holmes, S.M. (2008) Novel and modified materials for wastewater treatment applications. *Journal of Materials Chemistry*, 18, 2751–2761.

Ho, Y.S., Ng, J.C.Y. & McKay, G. (2000) Kinetics of pollutant sorption by biosorbents: review. *Separation and Purification Technology*, 29, 189–232.

Jainm, M., Garg, V.K. & Kadirvelu, K. (2013) Chromium removal from aqueous system and industrial wastewater by agricultural wastes. *Bioremediation Journal*, 17 (1), 30–39.

Jovanović, B.M., Vukašinović-Pešić, V.L., Veljović, D.N. & Rajaković, L.V. (2011) Arsenic removal from water using low-cost adsorbents – a comparative study. *Journal of the Serbian Chemical Society*, 76 (10), 1437–1452.

Kamsonlian, S., Suresh, S., Ramanaiah, V., Majumde, C.B., Chand, S. & Kumar, A. (2012a) Biosorptive behaviour of mano leaf powder and rice husk for arsenic (III) from aqueous solutions. *International Journal of Environmental Science & Technology*, 9, 565–578.

Kamsonlian, S., Balomajumeder, C. & Chand, S. (2012b) A potential of biosorbent derived from banana peel for removal of As(III) from contaminated water. *International Journal of Chemical Sciences and Applications*, 3, 269–275.

Kaya, K., Pehlivan, E., Schmidt, C. & Bahadir, M. (2014) Use of modified wheat bran for the removal of chromium(VI) from aqueous solutions. *Food Chemistry*, 158, 112–117.

Kelleher, B.P., O'Callaghan, M.N., Leahy, M.J., O'Dwyer, T.F. & Leahy, J.J. (2002) The use of FA from the combustion of poultry litter for the adsorption of chromium(III) from aqueous solution. *Journal of Chemical Technology and Biotechnology*, 77, 1212–1218.

Krishnani, K.K., Meng, X. & Dupont, L. (2009) Metal ions binding onto lignocellulosic biosorbent. *Journal of Environmental Science and Health*, Part A, 44, 688–699.

Kundu, S. & Gupta, A.K. (2007) Adsorption characteristics of As(III) from aqueous solution on iron oxide-coated cement (IOCC). *Journal of Hazardous Materials*, 142, 97–104.

Lata, S. & Samadder, S.R. (2014) Removal of heavy metals using rice husk: a review. *International Journal of Environmental Research and Development*, 2, 165–170.

Lee, C.K., Low, K.S., Liew, S.C. & Choo, C.S. (1999) Removal of arsenic(V) from aqueous solution by quaternized rice husk. *Environmental Technology*, 20, 971–978.

Lee, G. & Hering, J.G. (2003) Removal of chromium(VI) from drinking water by redox-assisted coagulation with iron(II). *Journal of Water Supply: Research & Technology – Aqua*, 52 (5), 319–332.

Lekić, B.M., Marković, D.D., Rajaković-Ognjanović, V.N., Đukić, A.R. & Rajaković, L.V. (2013) Arsenic removal from water using industrial by-products. *Journal of Chemistry*, 2013, 1–9.

Li, Q., Zhai, J., Zhang, W., Wang, M. & Zhou, J. (2007) Kinetic studies of adsorption of Pb(II), Cr(III) and Cu(II) from aqueous solution by sawdust and modified peanut husk. *Journal of Hazardous Materials*, 141, 163–167.

Li, Y., Liu, C. & Chiou, C. (2004) Adsorption of Cr(III) from wastewater by wine processing waste sludge. *Journal of Colloid and Interface Science*, 273, 95–101.

Li, Y., Zhang, F.-S. & Xiu, F.-R. (2009) Arsenic (V) removal from aqueous system using adsorbent developed from a high iron-containing fly ash. *Science of the Total Environment*, 407, 5780–5786.

Lopes, G., Guilherme, L.R.G., Costa, E.T.S., Curi, N. & Penha, H.G.V. (2013) Increasing arsenic sorption on red mud by phosphogypsum addition. *Journal of Hazardous Materials*, 262, 1196–1203.

López-Delgado, A., Pérez, C. & López, F.A. (1998) Sorption of heavy metals on blast furnace sludge. *Water Research*, 32 (4), 989–996.

López Leal, M.A., Cortés Martínez, R. & Cuevas Villanueva, R.A. (2012) Arsenate biosorption by iron-modified pine sawdust in batch systems: kinetics and equilibrium studies. *BioResources*, 7, 1389–1404.

Malkoc, E. & Nuhoglu, Y. (2007) Potential of tea factory waste for chromium(VI) removal from aqueous solutions: thermodynamic and kinetic studies. *Separation and Purification Technology*, 54, 291–298.

Malkoc, E., Nuhoglu, Y. & Dundar, M. (2006) Adsorption of chromium(VI) on pomace – an olive oil industry waste: batch and column studies. *Journal of Hazardous Materials*, B138, 142–151.

Memon, J.R., Memon, S.Q., Bhanger, M.I., El-Turki, A., Hallam, K.L. & Allen, G.C. (2009) Banana peel: a green and economical sorbent for the selective removal of Cr(VI) from industrial wastewater. *Colloids and Surfaces*, B70, 232–237.

Miretzky, P. & Fernandez Cirelli, A. (2010) Cr(VI) and Cr(III) removal from aqueous solution by raw and modified lignocellulosic materials: a review. *Journal of Hazardous Materials*, 180, 1–19.

Mohan, D. & Pittman, C.U. (2006) Activated and low cost adsorbents for remediation of tri- and hexavalent chromium from water. *Journal of Hazardous Materials*, B137, 762–811.

Nagy, S., Shaw, P. & Veldhuis, M.K. (1977) In nutrition, anatomy, chemical composition and bioregulation. *Circus science and technology*, Volume 1. The AVI Publishing Company, Inc., Westport, CT. pp. 74–479.

Namasivayam, C. & Senthilkumar, S. (1998) Removal of arsenic(V) from aqueous solution using industrial solid waste: adsorption rates and equilibrium studies. *Industrial and Engineering Chemistry Research*, 37, 4816–4822.

Namasivayam, C. & Yamuna, R.T. (1995) Adsorption of chromium(VI) by a low-cost adsorbent: biogas residual slurry. *Chemosphere*, 30 (3), 561–578.

Owlad, M., Aroua, M.K., Daud, W.A.W. & Baroutian, S. (2009) Removal of hexavalent chromium-contaminated water and wastewater: a review. *Water, Air, & Soil Pollution*, 200, 59–77.

Papandreou, A.D., Stournaras, C.J., Oanias, D. & Paspaliaris, I. (2011) Adsorption of Pb(II), Zn(II) and Cr(III) on coal FA porous pellets. *Minerals Engineering*, 24, 1495–1501.

Park, D., Yun, Y.S., Jo, J.H. & Park, J.M. (2005) Mechanism of hexavalent chromium removal by dead fungal biomass of *Aspergillus niger*. *Water Research*, 39, 533–540.

Park, D., Lim, S.-R., Yun Y.-S. & Park, J.M. (2007) Reliable evidences that the removal mechanism of hexavalent chromium by natural biomaterials is adsorption-coupled reduction. *Chemosphere* 70, 298–305.

Pereira, P.A.L., Dutra, A.J.B. & Martins, A.H. (2007) Adsorptive removal of arsenic from river waters using pisolite. *Minerals Engineering*, 20, 52–59.

Phenrat, T., Marhaba, T.F. & Rachakornkij, M. (2005) A SEM and X-ray study for investigation of solidified/stabilized arsenic-iron hydroxide sludge. *Journal of Hazardous Materials*, 118, 185–195.

Polowczyk, I., Bastrzyk, A., Koźlecki, T., Rudnicki, P., Sawiński, W. & Sadowski, Z. (2007) Application of fly ash agglomerates in the sorption of arsenic. *Polish Journal of Chemical Technology*, 9, 37–41.

Polowczyk, I., Bastrzyk, A., Sawiński, W., Koźlecki, T., Rudnicki, P., Sadowski, Z. & Sokołowski, Z. (2010a) Sorptive properties of coal-burning fly ash/Właściwości sorpcyjne popiołów ze spalania węgla, *Inżynieria i Aparatura Chemiczna*, 49 (1), 93–94.

Polowczyk, I., Bastrzyk, A., Koźlecki, T., Sawiński, W., Rudnicki, P., Sokołowski, A. & Sadowski, Z. (2010b) Use of fly ash agglomerates for removal of arsenic. *Environmental Geochemistry and Health*, 32, 361–366.

Polowczyk, I., Bastrzyk, A., Sawiński, W., Koźlecki, T., Rudnicki, P. & Sadowski, Z. (2011) Sorption properties of fly ash from brown coal burning towards arsenic removal. *Technical Transaction*, 8, 135–142.

Pradhan, J., Das, S.N. & Thakur, R.S. (1999) Adsorption of hexavalent chromium from aqueous solution by using activated red mud. *Journal of Colloid and Interface Science*, 217, 137–141.

Rafatullah, M., Sulaiman, O., Hashim, R. & Ahmad, A. (2009) Adsorption of copper (II), chromium (III), nickel (II) and lead (II) ions from aqueous solutions by meranti sawdust. *Journal of Hazardous Materials*, 170, 969–977.

Ranjan, D., Talat, M. & Hasan, S.H. (2009) Biosorption of arsenic from aqueous solution using agricultural residue "rice polish". *Journal of Hazardous Materials*, 166, 1050–1059.

Reddad, Z., Gerente, C., Andres, Y. & Le Cloirec, P. (2003) Mechanisms of Cr(III) and Cr(VI) removal from aqueous solutions by sugar beet pulp. *Environmental Technology*, 24 (2), 257–264.

Roy, P., Mondal, N.K., Bhattacharya, S., Das, B. & Das, K. (2013) Removal of arsenic(III) and arsenic(V) on chemically modified low-cost adsorbent: batch and column operations. *Applied Water Science*, 3, 293–309.

Saha, B. & Orvig, C. (2010) Biosorbents for hexavalent chromium elimination from industrial and municipal effluents. *Coordination Chemistry Reviews*, 254, 2959–2972.

Salim, R., Al-Subu, M. & Dawod, E. (2008) Efficiency of removal of cadmium from aqueous solutions by plant leaves and the effects of interaction of combinations of leaves on their removal efficiency. *Journal of Environmental Management*, 87, 521–532.

Selvaraj, K., Manonmani, S. & Pattabhi, S. (2003) Removal of hexavalent chromium using distillery sludge. *Bioresource Technology*, 89, 207–211.

Shen, Y.-S., Wang, S.-L., Tzou, Y.-M., Yan, Y.-Y. & Kuan W.-H. (2012) Removal of hexavalent Cr by coconut coir and derived chars – the effect of surface functionality. *Bioresource Technology*, 104, 165–172.

Siddique, R. & Noumowe, A. (2008) Utilization of spent foundry sands in controlled low-strength materials and concrete. *Resources, Conservation, and Recycling*, 53 (1–2), 27–35.

Singh, A.P., Srivastava, K.K. & Shekhar, H. (2009) Arsenic (III) removal from aqueous solutions by mixed sorbents. *Indian Journal of Chemical Technology*, 16, 136–141.

Singh, V.K. & Tiwari, P.N. (1997) Removal and recovery of chromium(VI) from industrial waste water. *Journal of Chemical Technology and Biotechnology*, 69, 376–382.

Singha, B. & Das, S.K. (2011) Biosorption of Cr(VI) ions from aqueous solutions: kinetics, equilibrium, thermodynamics and desorption studies. *Colloids and Surfaces*, B84, 221–232.

Sud, D., Mahajan, G. & Kaur, M.P. (2008) Agricultural waste materials as potential adsorbent for sequestering heavy metal ions from aqueous solution – a review. *Bioresource Technology*, 99, 6017–6027.

Suksabye, S. & Thiravetyan, P. (2012) Cr(VI) adsorption from electroplating plating wastewater by chemically modified coir pith. *Journal of Environmental Management*, 102, 1–8.

Sumathi, K.M.S., Mahimairaja, S. & Naidu, R. (2005) Use of low-cost biological wastes and vermiculite for removal of chromium from tannery effluent. *Bioresource Technology*, 96, 309–316.

Teixeira, M.C. & Ciminelli, V.S.T. (2005) Development of a biosorbent for arsenite: structural modeling based on X-ray spectroscopy. *Environmental Science & Technology*, 39, 895–900.

Tobin, J.M. & Roux, J.C. (1998) Mucor biosorbent for chromium removal from tanning effluent. *Water Research*, 32 (5), 1407–1416.

Ulatowska, J., Polowczyk, I., Sawiński, W., Bastrzyk, A., Koźlecki, T. & Sadowski, Z. (2014) Use of fly ash and fly ash agglomerates for As(III) adsorption from aqueous solution. *Polish Journal of Chemical Technology*, 16 (1), 21–27.

Urík, M., Littera, P., Ševc, J., Kolenčík, M. & Čerňaský, S. (2009) Removal of As(V) from aqueous solution using chemically modified sawdust of spruce (*Picea abies*): kinetics and isotherm studies. *International Journal of Environmental Science & Technology*, 6, 451–456.

Uysal, M. & Ar, I. (2007) Removal of Cr(VI) from industrial wastewaters by adsorption. Part I: Determination of optimum conditions. *Journal of Hazardous Materials*, 149, 482–491.

Vandecasteele, C., Dutre, V., Geysen, D. & Wauters, G. (2002) Solidification/stabilization of arsenic bearing fly ash from the metallurgical industry. Immobilisation mechanism of arsenic. *Waste Management*, 22, 143–146.

Vassilev, V.V. & Vassileva, C.G. (2005) Methods for characterization of composition of fly ashes from coal-fired power stations: a critical overview. *Energy Fuels*, 19, 1084–1098.

Veglio, F. & Beolchini, F. (1997) Removal of metals by biosorption: a review. *Hydrometallurgy*, 44, 301–316.

Wan Ngah, W.S. & Hanafiah, M.A.K.M. (2008) Removal of heavy metal ions from wastewater by chemically modified plant wastes as adsorbents: a review. *Bioresource Technology*, 99, 3935–3948.

Wang, J., Teng, X., Wang, H. & Ban, H. (2004) Characterizing the metal adsorption capability of a class F coal FA. *Environmental Science & Technology*, 38, 6710–6715.

Wang, J., Wang, T., Burken, J.G., Chusuei, Ch.C., Ban, H., Ladwig, K. & Huang, C.P. (2008) Adsorption of arsenic(V) onto fly ash: a speciation-based approach. *Chemosphere*, 72, 381–388.

Wang, S. & Wu, H. (2006) Environmental-benign utilisation of fly-ash as low-cost adsorbents. *Journal of Hazardous Materials*, B136, 482–501.

Weber, W.J. & Morris, J.C. (1963) Kinetics of adsorption on carbon from solution. *Journal of the Sanitary Engineering Division American Society of Civil Engineers*, 86, 31–60.

Weng, C.-H., Sharma, Y.C. & Chu, S.-H. (2008) Adsorption of Cr(VI) from aqueous solutions by spent activated clay. *Journal of Hazardous Materials*, 155, 65–75.

WHO (ed.) (2008) *Guidelines for Drinking-Water Quality*. World Health Organization, Geneva, Switzerland.

Wu, D., Sui, Y., He, S., Wang, X., Li, C. & Kong, H. (2008) Removal of trivalent chromium from aqueous solution by zeolite synthesized from coal FA. *Journal of Hazardous Materials*, 155, 415–423.

Wu, K., Liu, R., Li, T., Liu, H., Peng, J. & Qu, J. (2013a) Removal of arsenic(III) from aqueous solution using a low-cost by-product in Fe-removal plants-Fe-based backwashing sludge. *Chemical Engineering Journal*, 226, 393–401.

Wu, K., Liu, T. & Peng, J.M. (2013b) Adsorption behaviors of arsenic(V) onto Fe-based backwashing sludge produced from Fe(II)-removal plants. *Applied Mechanics and Materials*, 295–298, 1321–1326.

Zakaria, Z.A., Suratman, M., Mohammed, N. & Ahmad, W.A. (2009) Chromium(VI) removal from aqueous solution by untreated rubber wood sawdust. *Desalination*, 244, 109–121.

Zhang, F.-S. & Itoh, H. (2005) Iron oxide-loaded slag for arsenic removal from aqueous system. *Chemosphere*, 60, 319–325.

Zhang, S., Liu, C., Luan, Z., Peng, X., Ren, H. & Wang, J. (2008) Arsenate removal from aqueous solutions using modified red mud. *Journal of Hazardous Materials*, 152, 486–492.

Part III
Innovative methods for removal
of arsenic and chromium

CHAPTER 13

Emerging technologies in the removal of arsenic from polluted waters

Juan Saiz, Eugenio Bringas & Inmaculada Ortiz

13.1 INTRODUCTION

Several methods of As removal are already available including precipitation, electrochemical reduction, adsorption, ion exchange, solvent extraction, nanofiltration and reverse osmosis (Mayo et al., 2007). Also, the presence of competing ions such as phosphate, silicate, nitrate, chloride, carbonate, and sulfate affects the removal efficiency, and thus, the selection of a suitable treatment strategy is influenced by the groundwater composition. Highly efficient, cheap and sustainable technology which could be used by rural populations and colonies without any common treatment facilities is therefore of great urgency and high priority. Besides the importance of the removal step to assure the quality of drinking water, the way in which secondary wastes are finally managed is equally important in order to avoid future As recycling in aquifers.

Arsenic contamination of groundwater is often due to natural phenomena causing high concentrations of As in deeper levels of groundwater. The level of As in natural waters generally ranges from 1 and $2 \mu g L^{-1}$ (Hindmarsh and McCurdy, 1986; USNRC, 1999). However, concentrations may be elevated in areas with volcanic rocks and sulfide mineral deposits (Hindmarsh and McCurdy, 1986) and in geothermal waters ($500 \mu g L^{-1}$) (USNRC, 1999). The estimated exposed population, based on the $10 \mu g L$ limit, in the most affected areas and the principal water characteristic are reported in Table 13.1.

The chapter takes stock of the global scenario of As-polluted groundwaters and discusses several removal alternatives with special attention to emerging technologies.

Table 13.1. Summary of the physicochemical parameters of typical groundwaters (Kar et al., 2010).

Country/region	Population exposed [millions]	Concentration range [$\mu g L^{-1}$]	Properties
Argentina	2	1–5300	Oxidizing conditions; neutral pH
Australia	0.2	1–80	Neutral pH
Bangladesh	57	0.5–2500	Strongly reducing conditions; neutral pH
Canada	–	117	High particulate matter
Chile	0.5	100–1000	Oxidizing conditions; high pH
China	8	1–5300	Strongly reducing conditions
Finland	–	1–166	Estuary recharged water
Ghana	0.3	2–6000	High TDS
Germany	0.02	10–150	Low pH; mine water
Hungary	1.5	2–1760	Reducing conditions; presence of humic acids
India	78	10–3200	Strongly reducing conditions; neutral pH
Mexico	2	8–620	Oxidizing conditions, neutral pH
USA	9	2.1–4400	Wide range of conditions
Vietnam	10	1–3050	Reducing conditions; high alkalinity

Table 13.2.   Alternatives for arsenic mitigation (USEPA, 2003).

	Ion exchange	Activated Alumina	Iron-based sorbents	Reverse osmosis	Precipitation/ coagulation
Removal efficiency [%]	95	95	up to 98	>95	90
Total water loss [%]	1–2	1–2	1–2	15–75	0–5
Pre-oxidation required	YES	YES	YES	YES	YES
Recommendable conditions	pH = 6–9 <5 mg NO_3^- L^{-1} <50 mg SO_4^{2-} L^{-1} <500 mg TDS L^{-1} <0.3 NTU	pH = 5.5–8.3 <250 mg Cl^- L^{-1} <2 mg F^- L^{-1} <1 g TDS L^{-1} <0.3 NTU	pH = 6–8.5 <1 mg PO_4^{3-} L^{-1} <0.3 NTU	No particulates	pH = 5.5–8.5
Other considerations	Possible pH adjustment; Prefiltration required	Possible pH adjustment; Prefiltration required	Prefiltration required	High water loss	Possible pH adjustment
Cost	Medium	Medium	Medium	High	Low–medium

13.2   ARSENIC REMOVAL FROM GROUNDWATER

13.2.1   *Conventional technologies for arsenic removal from polluted waters*

From all the previous discussion, it is obvious that there is a crucial need to develop efficient processes to bring down the As concentration to permissible limits in waters. Table 13.2 summarizes
the main characteristics of the most commonly used As removal technologies namely chemical
precipitation/coagulation, adsorption with activated alumina or iron-based sorbents, ion exchange
or membrane filtration by reverse osmosis.

13.2.1.1   *Chemical precipitation/coagulation*
Arsenic can be removed by precipitation as arsenic sulfide, calcium arsenate or ferric arsenate.
However, it is difficult to achieve final concentrations below 1 mg L^{-1} (Robins, 2001) and a
relatively large volume of As containing sludge is formed representing a potential source of
contamination (Litter *et al.*, 2010). Additionally, As(III) removal by coagulation is considerably
less efficient than As(V) under similar conditions and pre-oxidation is required to convert As(III)
to As(V). It is also known that As precipitation with ferric salts is more efficient than others;
working at pH 3–4 sludge generation is reduced but the amount of iron needed for equivalent
As removal increases (Camacho *et al.*, 2011). This technology is simple, with low installation
costs and only common chemicals are used. Moreover, it can be easily applied to large water
volumes to serve for large communities (Choong *et al.*, 2007). Figure 13.1 shows a typical
coagulation/precipitation process for As removal.

13.2.1.2   *Ion exchange*
The application of synthetic ion-exchange resins allows the removal of As species very effectively
from drinking water as described by USEPA (An *et al.*, 2011). An extensive review published
by Dambies (2005) reported different ion exchangers useful in the removal of As from aqueous
media. In particular, commercially available strongly basic resins in the chloride form (Purolite,
Bayer or Dow Chem) have been proposed by EPA as efficient ion exchangers to attain values

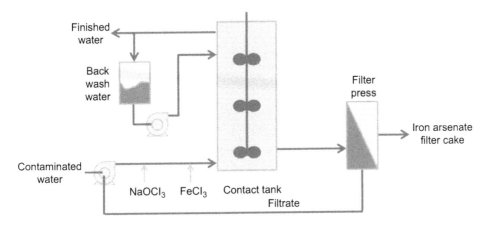

Figure 13.1.   Schematic diagram of coagulation/precipitation process (DOW Chemical, 2013).

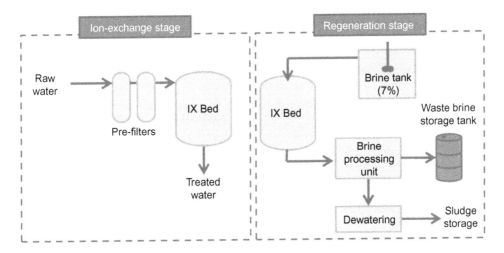

Figure 13.2.   Schematic diagram of ion-exchange treatment process.

of As concentration below $10\,\mu g\,L^{-1}$ (Höll, 2010). As(V) removal is efficient by ion exchange, producing effluents with less than $1\,\mu g\,L^{-1}$ of As, while As(III), being uncharged, is not effectively removed, and a previous oxidation step is necessary (Litter *et al.*, 2010). Figure 13.2 shows a typical ion-exchange technology for As removal.

13.2.1.3 *Membrane technology*

Membrane technologies, especially reverse osmosis (RO) have been reported as reliable As removal alternatives. RO has been identified as a likely best available technology which can completely purify water and meet the water legislation. Both bench and pilot-scale experiments demonstrated that, As(V) removal efficiency rates higher than 95% and As(III) removal efficiency rates of about 74% were achieved (Höll, 2010; Katsoyiannis and Zouboulis, 2006). Electrodialysis (ED) exhibits similar efficiency than RO process, mainly in treating water with high total dissolved solids (TDS), such as groundwater. Electrodialysis with reversion of polarity of the electrodes (EDR) can be also employed for removal of As with removal percentages varying from 28 to 86% (USEPA, 2000).

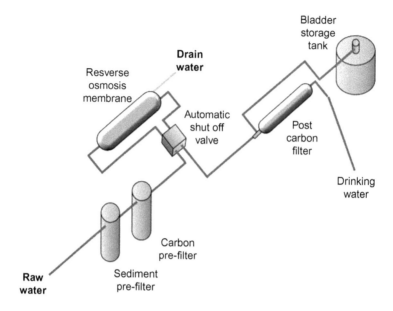

Figure 13.3.   Schematic diagram of membrane treatment process (H2O Distributors Inc.).

An overview of As removal on pressure driven membrane processes has been published by Shih (2005). It illustrates the parameters that may influence the As removal efficiency by membrane technologies such as the source water parameters, the membrane material, and the membrane characteristics. Figure 13.3 shows a typical domestic reverse osmosis-based device for As removal.

13.2.1.4 *Adsorption*

As per USEPA classification, adsorption is one of the best available technologies for As removal in drinking water (Dambies, 2005). The main selected adsorbents for As removal are iron oxides, modified iron oxides and hydroxides, aluminum oxide and hydroxide, and also silica, carbon, and organic polymers either as adsorbents in themselves or as supports for other adsorbents. Activated carbon, usually prepared from coconut shells, wood char, lignin, petroleum coke, etc., has also been extensively studied for the removal of As in water (Chuang *et al.*, 2005; Gu *et al.*, 2005). Adsorption of As on different sorbents (Chowdhury and Yanful, 2010) such as iron, iron coated sand, iron coated activated carbon and granular ferric hydroxides have also been investigated. Naturally occurring ores and minerals, namely kaolinite, feldspar, magnetite, hematite and maghemite have also been used for the adsorption of As though not as extensively as other materials (Mohan and Pittman, 2007).

13.2.2   *Nanomaterials-based emerging technologies for arsenic removal from polluted waters*

In recent years, nanoscale solid materials have become very important because of their different chemical and physical properties as compared to what they exhibit on the microscale. Nanomaterials have an elevated number of surface atoms compared to inner atoms, a fact that increases the surface area and therefore the rate of interfacial phenomena when nanomaterials are employed. If the parent material exhibits magnetic properties and the particle size is small enough, superparamagnetic properties are also expected.

Thus, the design, synthesis, characterization and applications of nanostructures are critical aspects for the emerging field of nanomaterials. In particular, nanomaterials as adsorbents exhibit an array of properties such as the ease and variety of functionalization possibilities and a vastly increased ratio of surface area to volume which make them appropriate for the design of novel selective separation processes (Hristovski *et al.*, 2007). If magnetic properties are added to the structure of the adsorbent, the adsorption process might be performed in fluidized bed reactors or stirred tanks by incorporating magnetic separation steps to recover the solid from the liquid media (Hubbuch *et al.*, 2001).

Several researchers have studied the application of superparamagnetic nanomaterials like magnetite (Fe_3O_4) nanoparticles as adsorbent for As removal, due to their strong adsorption activities and their easily separation by an external magnetic field (Feng *et al.*, 2012). However, these adsorbents are difficult to recycle because of both their tendency to oxidation when exposed to an oxidant atmosphere and the ease of aggregation caused by the anisotropic attraction (Fan *et al.*, 2012). Therefore, the development of composite materials incorporating magnetite nanoparticles is a possible solution to avoid the limitations of iron oxide. Mesoporous silica has been known to be one of the most ideal coating layers due to its reliable chemical stability, high surface area and facile surface modification (Fan *et al.*, 2012). Different pollutants have been removed from water employing mesoporous silica core-shell nanoparticles containing magnetite ($Fe_3O_4@SiO_2$): lead, mercury, cadmium, copper, uranium and hexavalent chromium.

Tables 13.3 and 13.4 summarize respectively, different non-magnetic and magnetic nanomaterials employed in the removal of As from polluted waters. The main available properties of the nanomaterials, the initial characteristics of the treated water and the adsorption capacities of As(III) and/or As(V) are reported.

13.2.2.1   Synthesis, characterization and As adsorption/desorption performance of functionalized magnetic nanoparticles

This section reports the methodology for the synthesis and characterization of magnetic $Fe_3O_4@SiO_2$ composite nanoparticles and their functionalization with aminopropyl groups incorporating Fe^{3+} as a model nanomaterial to be effectively employed in the removal or As from polluted groundwater. In addition the viability of the adsorption/desorption process was analyzed in terms of equilibrium and kinetic behavior.

Figure 13.4 depicts the different steps followed in the synthesis of superparamagnetic mesoporous silica nanoparticles functionalized with amino groups coordinated with Fe^{3+}:

- Fe_3O_4 (M1): superparamagnetic iron oxide nanoparticles.
- $Fe_3O_4@SiO_2$ (M2): mesoporous silica nanoparticles incorporating Fe_3O_4 cores.
- $Fe_3O_4@SiO_2$-Fe^{3+} (M3): Fe^{3+}-coordinated to amine acid-functionalized $Fe_3O_4@SiO_2$ nanoparticles.

Different techniques are traditionally employed in for the characterization of nanomaterials: X-ray diffraction (XRD), scanning electron microscopy (SEM), transmission electron microscopy (TEM), atomic force microscopy (AFM), Fourier transform infrared spectroscopy (FTIR), surface area (Brunauer–Emmett-Teller (BET-method)), thermogravimetric analysis (TGA) and vibrating sample magnetometry (VSM). These methods aim at determining the morphology, chemical composition, phase identification, surface area, magnetic behavior, etc.

13.2.2.1.1   Characterization techniques

X-ray diffraction (XRD)

X-ray diffraction is one of the most important non-destructive tools to analyze all kinds of matter. From research, XRD is an indispensable method for structural materials characterization and quality. This technique uses X-ray diffraction on powder or microcrystalline samples (Bhardwaj and Gupta, 2013). Figure 13.5 shows low angle X-ray diffraction patterns of material M3. The low angle powder XRD pattern shows three low-angle reflections typical of a hexagonal array that can be indexed as (100), (110) and (200) Bragg peaks.

Scanning electron microscopy (SEM)

SEM uses a focused electron probe to extract structural and chemical information point-by-point from a region of interest in the sample. The high spatial resolution of a SEM analysis makes

Table 13.3.   Non-magnetic nanoparticles (NPs) for arsenic water treatment.

Material	Size [nm]	Area [$m^2 g^{-1}$]	Operation conditions	As(III) capacity [$mg g^{-1}$]	As(V) capacity [$mg g^{-1}$]	Reference
Mesoporous silica functionalized with amines	–	–	$T = 25°C$; $[NPs] = 1 g L^{-1}$	–	140	Fryxell et al. (1999)
Activated carbon NPs with impregnated Fe	–	840	$pH = 7$; $T = 22°C$; $[NPs] = 2 g L^{-1}$; $[As] = 1 mg L^{-1}$	4.7	4.5	Reed et al. (2000)
Mesoporous silica functionalized with amines	–	1283–310	$T = 25°C$; $pH = 7–9$; $[NPs] = 0.5 g L^{-1}$	–	62.8–226	Yoshitake et al. (2002); Yoshitake et al. (2003); Yokoi et al. (2004)
Impregnated silica nanoparticles with metals	–	155–412	$pH = 7$; $T = 25°C$;	–	21	Jang et al. (2003)
Granulated active carbon	0.6–2	1000–650	$pH = 4.7$; $[NPs] = 3 g L^{-1}$; $[As] = 0–200 mg L^{-1}$	–	40	Gu et al. (2005)
Nanocrystalline titanium dioxide	30–5.0	55–330	$T = 25°C$; $pH = 7$	57	57	Pena et al. (2005)
Alumina nanoparticles with Fe(II)	–	23	$T = 20°C$; $pH = 5.5$; $[NPs] = 0.02 g L^{-1}$ $[As] = 0.15 mg L^{-1}$	–	22.5	Dousova et al. (2006)
Thiol functionalized activated alumina	5	200	$T = 25°C$; $pH = 7$; $[NPs] = 1 g L^{-1}$; $[As] = 2–20 mg L^{-1}$	11.5	–	Hao et al. (2009)
MCM-41 from rice	–	650–250	$T = 25°C$; $pH = 6$; $[NPs] = 3 g L^{-1}$; $[As] = 3 g L^{-1}$	–	1	Wantala et al. (2010)
Nanoporous titanium adsorbents (NTAs)	–	588	$T = 25°C$; $pH = 6$; $[NPs] = 1 g L^{-1}$; $[As] = 0.046 mg L^{-1}$	–	–	Han et al. (2010)
Activated carbon with iron nanoparticles on its surface	67	1058–806	$T = 25°C$; $pH = 7$; $[NPs] = 1 g L^{-1}$	–	4.4	Nieto-Delgado and Rangel-Mendez (2012)
MCM-41-Fe from rice	–	–	$pH = 4$; $[NPs] = 2 g L^{-1}$	–	0.015	Wantala et al. (2012)

Table 13.4. Magnetic nanoparticles for arsenic water treatment.

Material	Size [nm]	Area [m^2 g^{-1}]	Operation conditions	As(III) capacity [mg g^{-1}]	As(V) capacity [mg g^{-1}]	Reference
Nanoparticles of zero-valent iron	20	25–37.3	$T = 25°C$; pH $= 7$; [NPs] $= 0.5$ g L^{-1}	1	~100% removal	Kanel et al. (2005); Kanel et al. (2006)
Akaganeite nanocrystals	–	231	$T = 25°C$; [NPs] $= 0.5$ g L^{-1}; [As] $= 200$ mg L^{-1}	82.3	–	Deliyanni et al. (2006)
Magnetite	20–300	–	$T = 25°C$; [NPs] $= 0.5$ g L^{-1}; [As] $= 0$–30 mg L^{-1}	>98% removal	>98% removal	Mayo et al. (2007)
Coated magnetic functionalized MCM-41	1150	800	$T = 25°C$; [NPs] $= 0.25$ g L^{-1}	–	60	Chen et al. (2009)
Maghemite NPs	18	50	$T = 23°C$; [NPs] $= 1$ g L^{-1}	–	55	Tuutijärvi et al. (2009)
Fe$_3$O$_4$ nanocrystals	12–300	–	$T = 25°C$; pH $= 7$; [NPs] $= 0.5$ g L^{-1}; [As] $= 0.5$ mg L^{-1}	–	–	Yavuz et al. (2010)
Commercial magnetite NPs	20	60	$T = 25°C$; pH $= 8$	0.05	0.05	Shipley et al. (2010)
Functionalized magnetite	10.0–40.0	–	$T = 30°C$; pH $= 2$–12; [NPs] $= 1.25$ g L^{-1}	–	–	Singh et al. (2011)
α-Fe$_2$O$_3$ nanoparticles	–	162	$T = 25°C$; [NPs] $= 0.1$–0.6 g L^{-1}	9	9	Tang et al. (2011)
Fe$_3$O$_4$ – γFe$_2$O$_3$ nanoparticles	20	60	$T = 25°C$; pH $= 5$–7; [NPs] $= 1$ g L^{-1}; [As] $= 3$ mg L^{-1}	5	5	Chowdhury et al. (2011)
Fe$_3$O$_4$/Fe$_2$O$_3$ nanoparticles	12.0–17.0	–	$T = 30°C$; pH $= 6$; [NPs] $= 0.1$ g L^{-1}; [As] $= 0.3$–100 mg L^{-1}	1.25	8.2	Luther et al. (2012)
Fe$_3$O$_4$ nanoparticles covered with ascorbic acid	10	179	$T = 20°C$; [As] $= 0.1$ mg L^{-1}	46	16.5	Feng et al. (2012)

it a powerful tool to characterize a wide range of specimens at the nanometer to micrometer length scales (Goldstein, 2003). The images have a greater depth of field and resolution than optical micrographs making it ideal for rough specimens such as fracture surfaces and particulate materials. Figure 13.6 shows a typical SEM analysis of Fe$_3$O$_4$@SiO$_2$ nanoparticles. Some

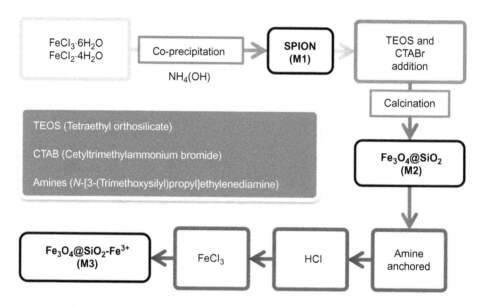

Figure 13.4.   Synthesis of solids M1, M2 and M3.

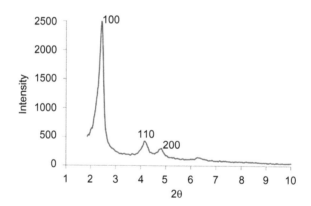

Figure 13.5.   Typical XDR analysis of $Fe_3O_4@SiO_2$ nanoparticles.

agglomerates are detected due to the magnetic behavior of the solids, the image evidence the practically spherical morphology of the particles and their average diameter around 100 nm.

Transmission electron microscope (TEM)
TEM is a powerful characterization tool that gives information about the morphology, crystallography and elemental composition for innovative materials (Bhardwaj and Gupta, 2013). TEM forms a major analysis method in a range of scientific fields, in both physical and biological sciences. Figure 13.7 shows a typical TEM analysis of $Fe_3O_4@SiO_2$ nanoparticles; it is observed that a combination of single core and multiple-core NPs are obtained under the selected synthesis conditions.

Atomic force microscopy (AFM)
In AFM analysis a cantilever with a sharp tip at its end is used to scan the specimen surface. The cantilever is typically silicon or silicon nitride with a tip radius of curvature on the order of nanometers (Sugimoto *et al.*, 2007). Change in the tip specimen interaction are often monitored

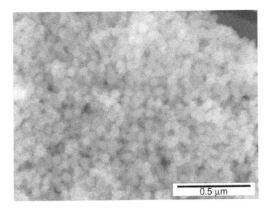

Figure 13.6. Typical SEM analysis of $Fe_3O_4@SiO_2$ nanoparticles.

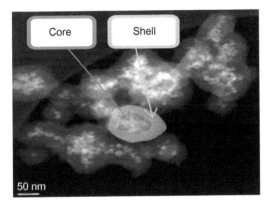

Figure 13.7. Typical TEM analysis of $Fe_3O_4@SiO_2$ nanoparticles.

using an optical lever detection system, in which a laser is reflected off of the cantilever and onto a position sensitive photodiode. It can achieve a resolution of 1 nm. AFM can be used to explore the nanostructures, properties and surface of solid materials. Figure 13.8 shows a typical AFM analysis of $Fe_3O_4@SiO_2$ nanoparticles in which the sphericity of the material is confirmed.

Fourier transform infrared spectroscopy (FTIR)
Fourier transform infrared spectroscopy is a technique which is used to obtain an infrared spectrum of absorption, emission, photoconductivity or Raman scattering of a solid, liquid or gas (Griffiths and de Hasseth, 2007). Some of the infrared radiation is absorbed by the sample being part of the radiation transmitted. The resulting spectrum represents the molecular absorption and transmission, creating a molecular fingerprint of the sample. This makes infrared spectroscopy useful for several types of analysis. Figure 13.9 shows a typical FTIR analysis of $Fe_3O_4@SiO_2$ nanoparticles (M2) where several representative transmittance peaks are identified at different wavelengths: 590 cm^{-1} is for Fe-O stretching, the broad peak between 1050 and 1250 cm^{-1} is for Si-O-Si and Fe-O-Si. The moderate peak at 1375 cm^{-1} is due to the C–H symmetric deformation of –CH$_3$ group. The typical infrared signals of mesoporous silica materials (i.e. MCM-41) at 940 and 3740 cm^{-1} corresponding to Si-OH moieties are very low due to the grafting of aminopropyls on the material surface.

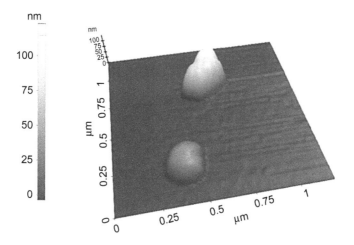

Figure 13.8. Typical AFM analysis of $Fe_3O_4@SiO_2$ nanoparticles.

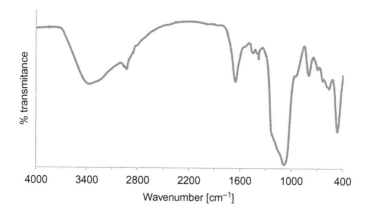

Figure 13.9. Typical FTIR analysis of $Fe_3O_4@SiO_2$ nanoparticles.

BET method

Brunauer–Emmett-Teller (BET) theory aims to explain the physical adsorption of gas molecules on a solid surface and serves as the basis for an important analysis technique for the measurement of the specific surface area of a material. The gas adsorption method measures the amount of gas (usually N_2) adsorbed on the surface of a powder sample as a function of the partial pressure of the adsorbate; from the correlation of adsorption isotherm data the specific mesoporous and microporous surface area and porosity of the material is determined (Brunauer *et al.*, 1938). The BET surface area of solid M2 is $912\,m^2\,g^{-1}$ being reduced up to $273\,m^2\,g^{-1}$ after the material functionalization (M3).

Thermogravimetric analysis (TGA)

Thermogravimetric analysis (TGA) is a method of thermal analysis in which the changes in properties of materials are measured as a function of increasing temperature (with constant heating rate), or as a function of time (with constant temperature) (Coats and Redfern, 1963). TGA provides information about physical phenomena, such as phase transitions and about chemical phenomena including chemisorptions, desolvation, or decomposition. Figure 13.10 shows a typical TGA analysis of solids with different levels of functionalization (M1, M2 and M3). Non-functionalized materials (M1 and M2) exhibit weight losses lower than 5% related to the

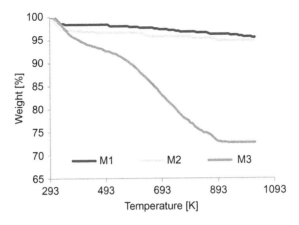

Figure 13.10. TGA of the solids with different degree of functionalization (M1, M2 and M3).

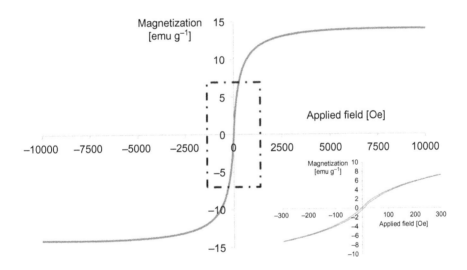

Figure 13.11. Magnetization curves of material M3.

decomposition of volatile matter adsorbed on the material surface during the synthesis procedure. In the selected range of temperature, the functionalized material M3 exhibits weight losses around 25%, resulting from the decomposition of the aminopropyl groups grafted to the silica surface.

Vibrating sample magnetometry (VSM)
Vibrating sample magnetometry (VSM) is a method that measures magnetic properties of solid materials. A sample is placed inside a uniform magnetic field where it is magnetized, and then is physically vibrated sinusoidally using a piezoelectric material (Foner, 1959). A typical VSM analysis of $Fe_3O_4@SiO_2$ nanoparticles is shown in Figure 13.11; M3 material exhibits saturation magnetization values of 14.0 emu g^{-1}. It should be noticed that the saturation value of iron oxide nanoparticles is around four times higher ($\sim$60 emu g^{-1}) than the value obtained for the composite material; this behavior is in good agreement with the molar ratio SiO_2:$Fe_3O_4 = 4$ obtained for materials M2 and M3.

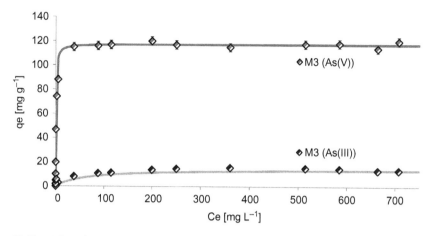

Figure 13.12.   Adsorption isotherms of As(V) and As(III) with solid M3. Dots represent equilibrium data and solid lines are predicted isotherms using the Langmuir model.

Table 13.5.   Parameters of the Langmuir equilibrium model.

Adsorbate/adsorbent	Langmuir model		
	q_m [mg g^{-1}]	K_L [L mg^{-1}]	R^2
As(V)/M3	121	0.383	0.97
As(III)/M3	14.7	0.016	0.96

13.2.2.1.2   *Analysis of the adsorption equilibrium*

Once the material was successfully synthesized and characterized, the adsorption equilibrium, which depends on both the adsorbate/adsorbent characteristics and temperature, was evaluated in order to quantify the maximum amount of the goal species potentially collected by the solid. The equilibrium state is thus described by the respective adsorption isotherms at 298 K between solid M3 and As(V) and As(III) which are depicted in Figure 13.12.

The good arsenate adsorption performance of material M3 can be explained as follows:

- The positively charged sorption sites caused by incorporation of iron enhances the columbic interaction between arsenate and Fe^{3+} and,
- The Lewis interactions between Fe^{3+}, a relatively soft Lewis acid, and the basic arsenate oxyanions.

The shape of the equilibrium isotherms – vertical and close to the ordinate axis – indicates, from a qualitative point of view, the favorable adsorption equilibrium especially at low As concentration levels (Saiz *et al.*, 2013). On the other hand, the removal of arsenite is less efficient than the oxidized species uptake due to the neutral state of As(III) species. Although the solid M3 exhibits a certain affinity towards arsenite species, the treatment of polluted groundwater contained in confined aquifers under anaerobic conditions needs pre-oxidation stages.

Equilibrium data were satisfactorily described by the Langmuir adsorption model being the values of the specific parameters obtained from the fitting of experimental data to the Langmuir model (Equation (13.1)) summarized in Table 13.5.

$$q_e = \frac{q_m \, K_L C_e}{1 + K_L C_e} \tag{13.1}$$

Table 13.6. Maximum Langmuir adsorption capacities (q_m) of As(III) and/or As(V) with different nanomaterials reported in literature.

Adsorbent	q_m [mg g^{-1}]		K_L [L mg^{-1}]	
	As(III)	As(V)	As(III)	As(V)
Amine functionalized mesoporous silica (MCM-41) (Fryxell *et al.*, 1999)	–	75	–	–
Amine functionalized mesoporous silica (MCM-48/SBA-41) (Yoshitake *et al.*, 2003)	–	121–189	–	–
Amine functionalized mesoporous silica (MCM-41) Yokoi *et al.* (2004)	–	90–117	–	–
Iron-containing mesoporous carbon Gu *et al.* (2007)	5.5	5.5	–	–
Maghemite nanoparticles Tuutijärvi *et al.* (2009)	–	55	–	–
Carboxil coated Fe$_2$O$_3$ nanoparticles Goswami *et al* (2011)	10.5	–	0.004	–
Agglomerated Fe^{3+}/Al^{3+} mixed oxide nanoparticles Basu and Ghosh (2011)	58.3	–	0.0126	–
Nano zero-valent iron Tanboonchuy *et al.* (2011)	102	118	0.0024	0.42
Magnetite nanoparticles Chowdhury *et al.* (2011)	4.8	5	–	–
Ascorbic acid coated Fe$_3$O$_4$ nanoparticles Feng *et al.* (2012)	46.1	16.6	0.168	1.42
Amino-functionalized cellulose Anirudhan *et al.* (2012)	–	123.9	–	0.21

where q_m is the maximum adsorption capacity of the solid [mg g^{-1}] and K_L is the Langmuir adsorption constant.

The information listed in Table 13.5 confirms that the maximum As(V) and As(III) capacities of solid M3, 121 and 14.7 mg g^{-1} respectively, are similar to the values reported in literature for magnetic nanoparticles and functionalized mesoporous nanomaterials (see Table 13.6). In a similar way, the values of the affinity parameter K_L that quantifies the affinity between the target species and the adsorbent follow a similar trend than the observed for the maximum capacity.

13.2.2.1.3 *Arsenic desorption from nanomaterials*

Adsorbents have a finite capacity to remove species from the fluid phase; when it is achieved, it is necessary to regenerate the adsorbent or to dispose it. In certain applications it may be more economical to discard the adsorbent after use. In the majority of applications, the disposal of adsorbents as waste is not an economic option and therefore, regeneration is carried out allowing the reuse of the solid.

Arsenate desorption is affected by certain variables like desorption time, pH, desorption agent concentration and adsorbent/adsorbate characteristics (Ghosh *et al.*, 2004; Jackson and Miller, 2000). For this reason, the desorption of As(V) from previously loaded material M3 was evaluated employing either water, 0.05 M HCl or 0.05 M NaOH (Tuutijärvi *et al.*, 2012).

Figure 13.13 shows the desorption percentages obtained after contacting As(V)-loaded materials with the different regeneration solutions during different operation times. Under the selected conditions, the best desorption efficiency was achieved with NaOH, which exhibits desorption efficiencies up to 5 times higher than the obtained with HCl.

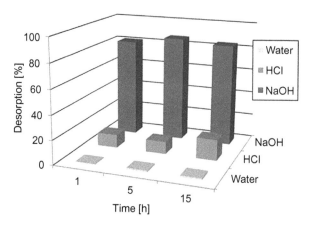

Figure 13.13.   Performance of M3 regeneration step.

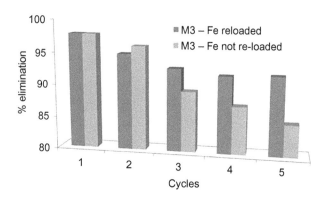

Figure 13.14.   Desorption cycles from As(V) loaded solid M3. [NaOH] = 0.001 M.

As it is reported in bibliography most As desorption processes are based on the use of strong basic solutions. However, silica-based materials may quite easily degrade in strong acid or alkali solutions, which might be a limiting step for the process applicability (Balaji *et al.*, 2005; Di Natale *et al.*, 2013; Tuutijärvi *et al.*, 2012). Therefore, the desorption of As from M3 with low NaOH concentrations (10^{-3} mol L^{-1}) is evaluated. In addition the reusability of the regenerated material M3 is evaluated after its protonation with HCl and the contact with a FeCl$_3$ solution to re-introduce the Fe^{3+} adsorption sites.

Successive adsorption/desorption tests were carried out working with a solution containing 5 mg As(V) L^{-1}. The results shown in Figure 13.14 demonstrate that after five cycles, the As removal efficiency decreases around 6% (as compared with that of synthesized material) working with the Fe^{3+} re-loaded material. The material without Fe^{3+} leads to removal percentages of As around 85% after 5 cycles which are lower than the obtained with the material containing Fe^{3+}.

13.3   CONCLUDING REMARKS AND FUTURE WORK

Nanomaterials and nanotechnology have known an increasing number of applications related to the recovery of compounds from groundwater and industrial effluents due to their potential to replace conventional processes by achieving selective elimination. In particular, adsorption based on functionalized magnetic nanomaterials offers additional advantages due to their selective adsorption of the target species. However, simply removing the contaminant is not enough because it is necessary to get an integrated process for the possible reuse of the material, contributing to the process economy. Therefore, the development of integrated processes consisting of adsorption/desorption an employing nanomaterials is a promising alternative.

From the analysis previously undertaken, it is concluded that the process design needs the assessment of the adsorption performance working at low concentration of As as well as the influence of the complex water matrix as a preliminary step in the process design and application. The experimentally obtained results so far, place this technology among the promising options to remediate As polluted groundwaters in an effective and efficient way.

ACKNOWLEDGEMENTS

Financial support from projects CTQ2008-00690 (MICINN, Spain), CTQ2012-31639 (MINECO, Spain-FEDER) and INDIGO-DST1-017 (PRI-PIMNIN-2011-1462, MINECO, Spain) is gratefully acknowledged.

REFERENCES

An, B., Liang, Q. & Zhao, D. (2011) Removal of arsenic(V) from spent ion exchange brine using a new class of starch-bridged magnetite nanoparticles. *Water Research*, 45 (5), 1961–1972.

Anirudhan, T.S., Divya, L. & Parvathy, J. (2012) Arsenic adsorption from contaminated water on Fe(III)-coordinated amino-functionalized poly(glycidylmethacrylate)-grafted TiO_2-densified cellulose. *Journal of Chemical Technology and Biotechnology*, 88 (5), 878–886.

Ayoob, S., Gupta, A.K. & Bhat, V.T. (2008) A conceptual overview on sustainable technologies for the defluoridation of drinking water. *Critical Reviews in Environmental Science & Technology*, 38, 401–470.

Balaji, T., Yokoyama, T. & Matsunaga, H. (2005) Adsorption and removal of As(V) and As(III) using Zr-loaded lysine diacetic acid chelating resin. *Chemosphere*, 59 (8), 1169–1174.

Basu, T. & Ghosh, U.C. (2011) Arsenic(III) removal performances in the absence/presence of groundwater occurring ions of agglomerated Fe(III)-Al(III) mixed oxide nanoparticles. *Journal of Industrial and Engineering Chemistry*, 17 (5), 834–844.

Bhardwaj, R. & Gupta, V. (2013) Characterization techniques for nano-materials in nanoelectronics: a review. *2nd International Conference on Role of Technology in Nation Building*. Meerut, India. pp. 43–47.

Brunauer, S., Emmett, P.H. & Teller E. (1938) Adsorption of gases in multimolecular layers. *Journal of the American Chemical Society*, 60 (2), 309–319.

Camacho, L.M., Gutierrez, M., Alarcon-Herrera, M.T., Villalba, M.L. & Deng, S. (2011) Occurrence and treatment of arsenic in groundwater and soil in northern Mexico and southwestern USA. *Chemosphere*, 83 (3), 211–225.

Chen, X., Lam, K.F., Zhang, Q., Pan, B., Arruebo, M. & Yeung, K.L. (2009) Synthesis of highly selective magnetic mesoporous adsorbent. *Journal of Physical Chemistry*, 113 (22), 9804–9813.

Choong, T.S.Y., Chuah, T.G., Robiah, Y., Koay, F.L.G. & Azni, I. (2007) Arsenic toxicity, health hazards and removal techniques from water: an overview. *Desalination*, 217 (1), 139–166.

Chowdhury, S.R. & Yanful, E.K. (2010) Arsenic and chromium removal by mixed magnetite-maghemite nanoparticles and the effect of phosphate on removal. *Journal of Environmental Management*, 91 (11), 2238–2247.

Chowdhury, S.R., Yanful, E.K. & Pratt, A.R. (2011) Arsenic removal from aqueous solutions by mixed magnetite-maghemite nanoparticles. *Environmental Earth Sciences*, 64 (11), 411–423.

Chuang, C.L., Fan, M., Xu, M., Brown, R.C., Sung, S., Saha, B. & Huang, C.P. (2005) Adsorption of arsenic(V) by activated carbon prepared from oat hulls. *Chemosphere*, 61 (4), 478–483.

Coats, A.W. & Redfern, J.P. (1963) Thermogravimetric analysis. A review. *Analyst*, 88 (1053), 906–924.

Dambies, L. (2005) Existing and prospective sorption technologies for the removal of arsenic in water. *Separation Science and Technology*, 39 (3), 603–627.

Deliyanni, E.A., Nalbandian, L.K. & Matis, A. (2006) Adsorptive removal of arsenites by a nanocrystalline hybrid surfactant-akaganeite sorbent. *Journal of Colloid and Interface Science*, 202, 458–466.

Di Natale, F., Erto, A. & Lancia, A.J. (2013) Desorption of arsenic from exhaust activated carbons used for water purification. *Journal of Hazardeous Materials*, 260, 451–458.

Dousova, B., Grygar, T., Martaus, A., Fuitova, L., Kolousek, D. & Machovi, V. (2006) Sorption of As(V) on aluminosilicates treated with Fe(II) nanoparticles. *Journal of Colloid and Interface Science*, 302 (2), 424–431.

DOW Chemical. Dow Water & Process Solutions. Available from: http://www.dowwaterandprocess.com/ [accessed 19th December 2013].

Fan, F.L., Qin, Z., Bai, J., Rong, W.D., Fan, F.Y., Tian, W., Wu, X.L., Wang, Y. & Zhao, L. (2012) Rapid removal of uranium from aqueous solutions using magnetic $Fe_3O_4@SiO_2$ composite particles. *Journal of Environmental Radioactivity*, 106, 40–46.

Feng, L., Cao, H., Ma, X., Zhu, Y. & Hu, Y. (2012) Superparamagnetic high-surface-area Fe_3O_4 nanoparticles as adsorbents for arsenic removal. *Journal of Hazardous Materials*, 217–218, 439–446.

Foner, S. (1959) Versatile and sensitive vibrating-sample magnetometer. *Review of Scientific Instruments*, 30 (7), 548–557.

Fryxell, G.E., Liu, J., Hauser, T.A., Nie, Z., Ferris, K.F., Mattigod, S., Gong, M. & Hallen, R.T. (1999) Design and synthesis of selective mesoporous anion traps. *Chemistry of Materials*, 11 (8), 2148–2154.

Ghosh, A., Mukiibi, M. & Ela, W. (2004) Leaching of arsenic from solid residuals under landfill conditions. *Environmental Science & Technology*, 38 (17), 4677–4682.

Goldstein, J. (2003) *Scanning electron microscopy and X-ray microanalysis*. Kluwer Adacemic/Plenum Pulbishers, Dordrecht, The Netherlands.

Goswami, R., Deb, P., Thakur, R., Sarma, K.P. & Basumallick, A. (2011) Removal of As(III) from aqueous solution using functionalized ultrafine iron oxide nanoparticles. *Separation Science and Technology*, 46 (6), 1017–1022.

Griffiths, P. & de Hasseth, J.A. (2007) *Fourier transform infrared spectrometry*. 2nd edition. Wiley-Blackwell, Hoboken, NJ.

Gu, Z. & Deng, B. (2007) Use of iron-containing mesoporous carbon (IMC) for arsenic removal from drinking water. *Environmental Engineering Sciences*, 24 (1), 113–121.

Gu, Z., Fang, J. & Deng, B. (2005) Preparation and evaluation of GAC-based iron-containing adsorbents for arsenic removal. *Environmental Science & Technology*, 39 (10), 3833–3843.

H2O Distributors. Available from: http://www.h2odistributors.com/ [accessed 19th December 2013].

Han, D.S., Abdel-Wahab, A. & Batchelor, B. (2010) Surface complexation modeling of arsenic(III) and arsenic(V) adsorption onto nanoporous titania adsorbents (NTAs). *Journal of Colloid and Interface Science*, 348 (2), 591–599.

Hao, J., Han, M.J. & Meng, X. (2009) Preparation and evaluation of thiol-functionalized activated alumina for arsenite removal from water. *Journal of Hazardous Materials*, 167 (1), 1215–1221.

Hindmarsh, J.T. & McCurdy, R.F. (1986) Clinical and environmental aspects of arsenic toxicity. *CRC Critical Reviews in Clinical Laboratory Sciences*, 23 (4), 315–347.

Höll, W.H. (2010) Mechanisms of arsenic removal from water. *Environmental Geochemistry and Health*, 32 (4), 287–290.

Howard, G., Bartram, J., Pedley, S., Schmoll, O., Chorus, I. & Berger, P. (2006) Groundwater and public health. In: *Protecting groundwater for health*. World Health Organization, Cornwall, UK.

Hristovski, K., Baumgardner, A. & Westerhoff, P. (2007) Selecting metal oxide nanomaterials for arsenic removal in fixed bed columns: from nanopowders to aggregated nanoparticle media. *Journal of Hazardous Materials*, 147 (1), 265–274.

Hubbuch, J.J., Matthiessen, D.B., Hobley, T.J. & Thomas O.R.T. (2001) High gradient magnetic separation versus expanded bed adsorption: a first principle comparison. *Bioseparation*, 10 (1), 99–112.

Jackson, B.P. & Miller, W.P. (2000) Effectiveness of phosphate and hydroxide for desorption of arsenic and selenium species from iron oxides. *Soil Science Society of American Journal*, 64 (5), 1616–1622.

Jang, M., Shin, E.W., Park, J.K. & Choi, S.I. (2003) Mechanisms of arsenate adsorption by highly-ordered nano-structured silicate media impregnated with metal oxides. *Environmental Science & Technology*, 37 (21), 5062–5070.

Jha, A.K., Bose, A. & Downey, J.P. (2006) Removal of As(V) and Cr(VI) from aqueous solution using a continuous hybrid field-gradient magnetic separation device. *Environmental Science & Technology*, 41 (15), 3297–3312.

Kanel, S.R., Charlet, B. & Choi, L. (2005) Removal of As(III) from groundwater by nanoscale zerovalent iron. *Environmental Science & Technology*, 39 (5), 1291–1298.

Kanel, S.R., Greneche, J.M. & Choi, H. (2006) As(V) removal from ground water using nanoscale zerovalent iron as a colloidal reactive barrier material. *Environmental Science & Technology*, 15 (40), 2045–2050.

Kar, S., Maity, J.P., Jean, J.-S., Liu, C.C., Nath, B., Yang, H.J. & Bundschuh, J. (2010) Arsenic-enriched aquifers: occurrences and mobilization of arsenic in groundwater of Ganges Delta Plain, Barasat, West Bengal, India. *Applied Geochemistry*, 25 (12), 1805–1814.

Katsoyiannis, I.A. & Zouboulis, A.I. (2006) Comparative evaluation of conventional and alternative methods for the removal of arsenic from contaminated groundwaters. *Reviews on Environmental Health*, 21 (1), 25–41.

Kim, S.H., Kim, K., Ko, K.S., Kim, Y. & Lee, K.S. (2012) Co-contamination of arsenic and fluoride in the groundwater of unconsolidated aquifers under reducing environments. *Chemosphere*, 87, 851–856.

Li, J., Wang, Y., Xie, X. & Su, C. (2012) Hierarchical cluster analysis of arsenic and fluoride enrichments in groundwater from the Datong basin, N. China. *Journal of Geochemical Exploration*, 118, 77–89.

Litter, M.I., Morgada, M.E. & Bundschuh, J. (2010) Possible treatments for arsenic removal in Latin American waters for human consumption. *Environmental Pollution*, 158 (5), 1105–1118.

Liu, T.Z., Rao, P.H., Mak, M.S.H., Wang, P. & Lo, I.M.C. (2009) Removal of co-present chromate and arsenate by zero-valent iron in groundwater with humic acid and bicarbonate. *Water Research*, 43 (9), 2540–2548.

Luther, S., Borgfeld, N., Kim, J. & Parsons, J.G. (2012) Removal of arsenic from aqueous solution: a study of the effects of pH and interfering ions using iron oxide nanomaterials. *Microchemical Journal*, 101, 30–36.

Mandal, B.K. & Suzuki, K.T. (2002) Arsenic round the world: a review. *Talanta*, 58, 201–235.

Mara, D.D. (2003) Water, sanitation and hygiene for the health of developing nations. *Public Health*, 117 (6), 452–456.

Mayo, J.T., Yavuz, C., Yean, S., Cong, L., Yu, W. & Colvin, V.L. (2007) The effect of nanocrystalline magnetite size on arsenic removal. *Science and Technology of Advanced Materials*, 8 (1), 71–75.

Mohan, D. & Pittman, C.U., Jr. (2007) Arsenic removal from water/wastewater using adsorbents – a critical review. *Journal of Hazardous Materials*, 142 (1), 1–53.

Morris, B.L., Lawrence, A.R.L., Chilton, P.J.C., Adams, B., Calow R.C. & Klinck, B.A. (2003) Groundwater and its susceptibility to degradation: a global assessment of the problem and options for management. *Early Warning and Assessment Report Series*. United Nations Environment Programme, Nairobi, Kenya.

Murcott, S. (2012) *Arsenic contamination in the world*. IWA Publishing, London, UK.

Nieto-Delgado, C. & Rangel-Mendez, J.R. (2012) Anchorage of iron hydro(oxide) nanoparticles onto activated carbon to remove As(V) from water. *Water Research*, 46 (9), 2973–2982.

Pena, M.E., Koratis, G.P., Patel, M., Lippincott, L. & Meng, X. (2005) Adsorption of As(V) and As(III) by nanocrystalline titanium dioxide. *Water Research*, 39 (11), 2327–2337.

Pierce, M.L. & Moore, C.B. (1980) Adsorption of arsenite on amorphous iron hydroxide from dilute aqueous solution. *Environmental Science & Technology*, 14 (2), 214–216.

Qin, G., Mcguire, M.J., Blute, N.K., Seidel, C. & Fong, L. (2005) Hexavalent chromium removal by reduction with ferrous sulfate, coagulation, and filtration: a pilot-scale study. *Environmental Science & Technology*, 39 (16), 6321–6327.

Reed, B.E., Vaughan, R. & Jiang, L. (2000) As(III), As(V), Hg, and Pb removal by Fe-oxide impregnated activated carbon. *Journal of Environmental Engineering* 126 (9), 869–873.

Robins, R.G. (2001) Some chemical aspects relating to arsenic remedial technologies. *Proceedings of the USEPA Workshop on Managing Arsenic Risks to the Environment, 1–3 May, Denver, CO, USA*.

Saiz, J., Bringas, E. & Ortiz, I. (2014) Functionalized magnetic nanoparticles as new adsorption materials for arsenic removal from polluted waters. *Journal of Chemical Technology and Biotechnology*, 89 (6), 909–918.

Shih, M.C. (2005) An overview of arsenic removal by pressure-driven membrane processes. *Desalination*, 172, 85–97.

Shipley, H.J., Yean, S., Kan, A.T. & Tomson, M.B. (2010) A sorption kinetics model for arsenic adsorption to magnetite nanoparticles. *Environmental Science and Pollution Research*, 17 (5), 1053–1062.

Singh, S., Barick, K.C. & Bahadur, D. (2011) Surface engineered magnetic nanoparticles for removal of toxic metal ions and bacterial pathogens. *Journal of Hazardous Materials*, 192 (3), 1539–1547.

Smedley, P.L. & Kinniburgh, D.G. (2002) A review of the source, behaviour and distribution of arsenic in natural waters. *Applied Geochemistry*, 17 (5), 517–568.

Sugimoto, Y., Pou, P., Abe, M., Jelinek, P., Pérez, R., Morita, S. & Custance, O. (2007) Chemical identification of individual surface atoms by atomic force microscopy. *Nature*, 446 (7131), 64–67.

Tanboonchuy, V., Hsu, J.C., Grisdanurak, N. & Liao, C.H. (2011) Impact of selected solution factors on arsenate and arsenite removal by nanoiron particles. *Environmental Science and Pollution Research*, 18 (6), 857–864.

Tang, W., Li, Q., Gao, S. & Shang, J.K. (2011) Arsenic (III,V) removal from aqueous solution by ultrafine α-Fe$_2$O$_3$ nanoparticles synthesized from solvent thermal method. *Journal of Hazardous Materials*, 192, 131–138.

Tchounwou, P.B., Patlolla, A.K. & Centeno, J.A. (2003) Carcinogenic and systemic health effects associated with arsenic exposure – a critical review. *Toxicologic Pathology*, 31 (6), 575–588.

Tuutijärvi, T., Lu, J., Sillanpää, M. & Chen, G. (2009) As(V) adsorption on maghemite nanoparticles. *Journal of Hazardous Materials*, 166 (2), 1415–1420.

Tuutijärvi, T., Vahalaa, R., Sillanpitää, M. & Chen, G. (2012) Maghemite nanoparticles for As(V) removal: desorption characteristics and adsorbent recovery. *Environmental Technology*, 33 (16–18), 1927–1936.

USEPA (2000) Technologies and costs for removal of arsenic from drinking water. US Environmental Protection Agency, Washington, DC.

USEPA (2003) Arsenic treatment technology evaluation handbook for small systems. US Environmental Protection Agency, Washington, DC.

USNRC (1999) Arsenic in drinking water. United States National Research Council, National Academy Press, Washington, DC.

Wantala, K., Sthiannopkao, S., Srinameb, B.O., Grisdanurak, N. & Kim, K.W. (2010) Synthesis and characterization of Fe-MCM-41 from rice husk silica by hydrothermal technique for arsenate adsorption. *Environmental Geochemistry and Health*, 32 (4), 261–266.

Wantala, K., Khongkasem, E., Khlongkarnpanich, N., Sthiannopkao, N. & Kim, K.W. (2012) Optimization of As(V) adsorption on Fe-RH-MCM-41-immobilized GAC using Box-Behnken design: effects of pH, loadings, and initial concentrations. *Applied Geochemistry*, 27 (5), 1027–1034.

Yavuz, C.T., Mayo, J.T., Suchecki, C., Wang, J., Ellsworth, A.Z., D'Couto, H., Quevedo, E., Prakash, A., Gonzalez, L., Nguyen, C., Kelty, C. & Colvin, V.L. (2010) Pollution magnet: nano-magnetite for arsenic removal from drinking water. *Environmental Geochemistry and Health*, 32 (4), 327–334.

Yokoi, T., Tatsumi, T. & Yoshitake, H. (2004) Fe^{3+} coordinated to amino-functionalized MCM-41: an adsorbent for the toxic oxyanions with high capacity, resistibility to inhibiting anions, and reusability after a simple treatment. *Journal of Colloid and Interface Science*, 274 (2), 451–457.

Yoshitake, H., Yokoi, T. & Tatsumi, T. (2002) Adsorption of chromate and arsenate by amino functionalizated MCM-41 and SBA-1. *Chemistry of Materials*, 14 (11), 4603–4010.

Yoshitake, H., Yokoi, T. & Tatsumi, T. (2003) Adsorption behavior of arsenate at transition metal cations captured by amino-functionalized mesoporous silicas. *Chemistry of Materials*, 15 (8), 1713–1721.

CHAPTER 14

Arsenic removal from geothermal water with inorganic adsorbents

Kazuharu Yoshizuka & Syouhei Nishihama

14.1   INTRODUCTION

Geothermal energy is renewable heat energy from deep in the earth which is generated in many places, such as hot spring areas. Geothermal power generation based on geothermal energy is quite attractive, even compared with other renewable energies, because the energy is hardly affected by weather, thus allowing a stable energy supply. For example, there are currently 18 operational geothermal power plants in Japan, providing 0.2% of total electricity supply. Since geothermal energy is a renewable resource with very low CO_2 emissions, electricity providers are planning to construct several more geothermal power plants in the next decade, to increase the contribution up to 1% of electricity supply (Yoshizuka *et al.*, 2010).

Some geothermal water contains large concentrations of dissolved minerals, such as sodium, calcium, sulfate, chloride, or iron containing minerals. Geothermal water also contains a variety of valuable minerals, such as lithium ($\sim$10 mg L^{-1}) and boron ($\sim$30 mg L^{-1}), as a result of hot leaching from the rocks in the aquifer by geothermal water (Kabay *et al.*, 2004; Tyrovola *et al.*, 2006; Xu *et al.*, 2001; Yanagase *et al.*, 1983).

Geothermal water has, however, the disadvantage that concentrations of some constituents such as boron, arsenic, and fluoride, usually exceed those in the standards recommended for drinking water by the water authorities. Thus, geothermal water that is withdrawn and used can become a disposal problem (Bunschuh and Maity, 2015). Therefore, in many areas, reinjection of geothermal water is required after being used for geothermal power generation.

14.2   ARSENIC IN GEOTHERMAL WATER

There are a lot of separation techniques of As removal from aqueous solution, e.g. co-precipitation (Ladeira *et al.*, 2002), flotation, ion exchange (Henke, 2009), and adsorption (Carlo and Thomas, 1985). Selecting an arsenic (As) treatment technology for remediation of geothermal waters depends on several key factors. Among these, speciation of As, initial As concentration, regulatory requirements and target treatment levels must be considered. Due to variations in As speciation and large differences in the chemistry and physical properties of geothermal waters, no single technology will adequately meet the needs of every project. Furthermore, successful remediation often requires a combination of two or more treatment technologies, There are several inorganic arsenite species (inorganic As(III); e.g., H_3AsO_3, $H_2AsO_3^-$, $HAsO_3^{2-}$ and AsO_3^{3-}) and inorganic arsenate species (inorganic As(V); e.g., H_3AsO_4, $H_2AsO_4^-$, $HAsO_4^{2-}$ and AsO_4^{3-}) in geothermal water.

Table 14.1 lists the As content of several geothermal waters in the world. Though large variation of As content is indicated in geothermal sites, extremely high As content is detected in some hot spring areas. This is linked to the mineralogical, chemical and physical characteristics of the soils, sediments and rocks in contact with these waters. In addition, the As is significantly leached from aquifers under the extremely high temperature conditions. Depending on oxidation-reduction (redox) conditions and biological activity, groundwaters and geothermal waters may

Table 14.1. Worldwide arsenic contents in geothermal waters.

Site	Total dissolved As [$\mu g \, L^{-1}$]	Other information	Reference
Kalloni drainage basin, Lesvos, Greece	41.1–90.7	pH = 6.56 (6.33–6.79)	Aloupi et al. (2009)
Kalikratia, Chalkidiki, Greece	4.2–35.8	pH = 6.73 (6.65–6.8)	Aloupi et al. (2009)
Yangbajing geothermal power plant, Tibet, China	5700	[B] = 119 mg L^{-1} [F] = 19.6 mg L^{-1}	Guo et al. (2008)
Rehai geothermal field, Yunnan Province, China	43.6–687 (91% of As(III))	[Sb] = 0.38–23.8 $\mu g \, L^{-1}$	Zhang et al. (2008)
Aksios area, northern Greece	3.0–68.8 (As(III) predominant)	pH = 7.5–8.2 [HCO_3] = 4–6 mM [B] = 0.14–0.28 mg L^{-1}	Katsoyiannis et al. (2007)
Kalikratia area, northern Greece	3.6–74.6 (As(V) predominant)	pH = 6.7–7.5 [HCO_3] = 6–12 mM [B] = 0.11–1.58 mg L^{-1}	Katsoyiannis et al. (2007)
Chalkidiki, area, northern Greece	1–1843 (As(V) predominant)	pH = 5.8–8.1 [HCO_3] = 6–17 mM [B] = 0.04–6.50 mg L^{-1}	Kouras et al. (2007)
Bath Spring, Yellowstone National Park, USA	1560 (32% of trithioarsenate)	pH = 9.0	Planer-Friedrich et al. (2007)
Hot spring Ojo Caliente, Yellowstone National Park, USA	1500 (51% of trithioarsenate)	pH = 7.5	Planer-Friedrich et al. (2007)
Kawerau geothermal field, New Zealand	38	[Cl] = 50 mg L^{-1}	Mroczek (2005)
Norris Geyser Basin, Yellowstone National Park, USA	2500 (As(III) predominant)	pH = 3.1 [B] = 8.7 mg L^{-1}	Langner et al. (2001)
Hot Creek, eastern Sierra Nevada, USA	750–1400 (max 73% of As(III))	[B] = 10 mg L^{-1}	Wilkie and Hering, (1998)
Beppu hot spring, Oita, Japan	210–1360	pH = 2.6–7.9 [F] = 0.53–1.64 mg L^{-1}	Yoshizuka et al. (2010)
Hachoubaru geothermal power plant, Oita, Japan	3230	pH = 8.0 [B] = 33.9 mg L^{-1} [F] = 3.76 mg L^{-1}	Yoshizuka et al. (2010)
Obama hot spring, Nagasaki, Japan	550	pH = 7.9 [B] = 14.9 mg L^{-1} [F] = 1.27 mg L^{-1}	Yoshizuka et al. (2010)
Kizildere geothermal power plant, Denizli, Turkey	853	pH: 9.0–9.2 [B] = 18–20 mg L^{-1} [F] = 15.36 mg L^{-1}	

contain As(V) and the more toxic As(III) forms (US Environmental Protection Agency (USEPA, 2002)). Considering that As contamination can originate from geological materials, the remediation of these materials is usually necessary to reduce As concentrations in associated geothermal waters. In some cases, however, geothermal water contamination is so severe that affordable and effective remediation is not possible. The physical and chemical characteristics of geothermal waters will affect the selection of reliable treatment technologies to work effectively at high temperatures. Alaerts and Khouri identified several factors that affect the costs and feasibility of treating As in geothermal water (Alaerts and Khouri, 2004). The lowering of As drinking water standards (maximum contaminant level, MCL) from 50 to $10\,\mu g\,L^{-1}$ in many countries has resulted in increasing demands for additional removal technologies when geothermal water is used for drinking and cooking.

14.3   TECHNOLOGIES FOR ARSENIC REMOVAL

Adsorption involves the removal of contaminants by causing them to attach to the surfaces of solid materials (adsorbents) and is widely used in water treatment technologies. Sometimes the adsorbed solute is called the adsorbate (Krauskopf and Bird, 1995). Adsorption often involves ion exchange (Eby, 2004). For example, adsorbed As oxyanions will replace other anions on the surface of the adsorbent.

Adsorbents are sometimes suspended in the separator to adsorb As contaminants from water. More commonly, however, 0.3–1.0 mm diameter adsorbents or ion-exchange granules, fibers, or other materials are packed into columns or filters (Clifford and Ghurye, 2002; Greenleaf *et al.*, 2006). The materials have large enough granules to facilitate permeability and water flow while still providing sufficient surface area for numerous adsorption and ion-exchange sites. Other desirable properties of adsorbents and ion-exchange media include (i) ability to rapidly and effectively remove large amounts of both As(III) and As(V) before regeneration or disposal, (ii) capability of being regenerated, (iii) high durability in water, and (iv) reasonable costs. Nevertheless, few, if any, adsorption and ion-exchange systems adequately achieve all of these goals.

Selecting an appropriate adsorption or ion-exchange technology strongly depends on the characteristics of the contaminated water, including its total dissolved solids (TDS) content, pH, redox conditions, microbial activity, the presence of any organic and inorganic species that might interfere with treatment, and the concentrations of the As species. For example, very acidic waters may corrode alumina adsorbents, thus decreasing the number of active adsorption sites and reducing the effective life span of the adsorbents (Prasad, 1994; USEPA, 2002).

For As, adsorption onto inorganic solids is generally more convenient than chemical precipitation/co-precipitation methods and less expensive than ion-exchange resins or membrane filtration (Kim *et al.*, 2004). In contrast to precipitation/co-precipitation, methods that produce large volumes of sludge that are difficult to dewater (Deliyanni *et al.*, 2003), adsorption and ion-exchange columns usually produce relatively little waste and can often be regenerated. Adsorption and ion-exchange technologies are most often used with low TDS waters where As is the only significant contaminant or as the final step after precipitation/co-precipitation (USEPA, 2002). Adsorbents and ion-exchange materials are more vulnerable to chemical interferences when compared to precipitation/co-precipitation methods for the removal of As from water (USEPA, 2002). Some interfering chemicals directly compete with As species for adsorption and ion-exchange sites. Two prime examples are phosphate (PO_4^{3-}) and silicate (SiO_4^{4-}), which have the same tetrahedral structure as arsenate (AsO_4^{3-}). Due to these similarities, phosphate and silicate desorb As(V) from sorbents such as iron (hydr)oxide, aluminum (hydr)oxide and silica over a wide pH range, or at least hinder the adsorption of As(V) onto these materials (Clifford and Ghurye, 2002; Smith and Edward, 2005; Violante *et al.*, 2006; Zhang *et al.*, 2004). While carbonates (H_2CO_3, HCO_3^- and CO_3^{2-}) mainly present in geothermal waters often have little or no effect on As(V) adsorption, evidence suggests that they may interfere with As(III) adsorption due to their similar trigonal molecular structures. The cost of adsorption and ion-exchange systems depend on the concentration and speciation of the As contaminants, regeneration and disposal requirements for the spent media, water flow rates, and the overall chemistry of the water, which controls fouling mainly caused by organic materials and silica (SiO_2).

Table 14.2.   Methods for arsenic removal from geothermal water.

Method	Separation media	Species	Comment	Reference
Adsorption/ coprecipitation	Zero-valent iron	As(III), As(V)	Laboratory study, effect of NO_3^- and PO_4^{3-}	Tyrovola *et al.* (2007)
Adsorption	Hydroxyapatite-based ceramics	As(III), As(V)	Laboratory study, XANES measurement	Nakahira *et al.* (2006)
Adsorption/ coprecipitation	Zero-valent iron	As(III), As(V)	Laboratory and field pilot studies, effect of temperature, NO_3^- and PO_4^{3-}	Tyrovola *et al.* (2006)
Adsorption/ magnetic separation	Magnetite	As(III), As(V)	Laboratory study	Okada *et al.* (2004)
Adsorptive bubble flotation	Colloidal ferric hydroxide	As(III), As(V)	Quantitative removal above pH 6.5	De Carlo and Thomas (1985)
Adsorption	Chelating resin with mercapto groups	As(III)	Laboratory study	Egawa *et al.* (1985)
Adsorptive bubble flotation	Iron flocculant	As(V)	Laboratory study	Buisson *et al.* (1979)
Adsorption	Granulated magnetite	As(III), As(V)	Laboratory study of chromatographic removal	Yoshizuka *et al.* (2010)

Table 14.2 lists the studies of As removal from geothermal waters. Adsorption by inorganic materials is mostly used for removal of As. Tyrovola *et al.* (2006; 2007) carried out the adsorptive removal of As(III) and As(V) with zero-valent iron. In the presence of water and oxygen, zero-valent iron rapidly oxidizes. Depending on pH, the amount of oxygen (O_2), the presence of bacteria and other conditions of the water, the oxidation of zero-valent iron may produce a wide variety of compounds, including lepidocrocite (γ-FeO(OH)), sulfate ($4Fe(OH)_2 \cdot 2FeOOH \cdot FeSO_4 \cdot 4H_2O$) and carbonate (possibly $Fe(II)_4Fe(III)_2(OH)_{12}(CO_3 \cdot 2H_2O)$), mackinawite ($Fe_9S_8$), magnetite ($Fe_3O_4$), maghemite ($\gamma$-$Fe_2O_3$), and amorphous ferrous sulfide (FeS) (Gu *et al.*, 1999). These oxidation products are largely responsible for sorbing and/or coprecipitating As from water. De Carlo *et al.* (1985) and Buisson *et al.* (1979) carried out As removal by adsorptive bubble flotation using iron flocculant. They successfully removed both As(III) and As(V) from geothermal waters by the adsorptive bubble flotation technology.

Okada *et al.* used a superconducting magnet as part of a treatment system that reduced As concentrations from 3400 (mostly as As(III)) to 15 μg L^{-1} in geothermal waters from Kakkonda, Japan (Okada *et al.*, 2004). Before magnetic treatment, the As was entirely oxidized to As(V) with hydrogen peroxide. Fe(III) sulfate was then added to the water at pH 4 to coprecipitate the As(V) with paramagnetic iron (oxy)(hydr)oxides and magnetic wire meshes were used to capture the precipitates.

14.4   ARSENIC REMOVAL FROM GEOTHERMAL WATER

There are several typical inorganic adsorbents for As, i.e., magnetite (Fe_3O_4), titania (TiO_2), and alumina (Al_2O_3). We have evaluated the behavior of adsorption equilibrium properties of As on these adsorbents, together with column separation of As from geothermal water (Park *et al.*, 2012). Evaluation of adsorption of arsenic (As(III) and As(V)) was conducted by batchwise adsorption using the adsorbents in powder form.

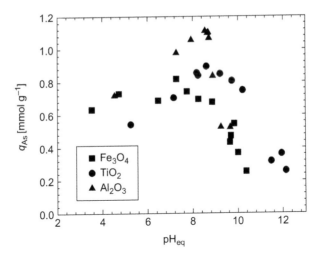

Figure 14.1. Effect of pH on adsorption of As(III) with Fe_3O_4, TiO_2 and Al_2O_3 adsorbents (reproduced with permission from Park *et al.*, 2012a, copyright of Japan Society of Ion Exchange).

The adsorption amount of As on adsorbent, q_{As} [mmol g^{-1}], was determined by:

$$q_{As} = \frac{(C_{As,0} - C_{As})\, L}{W} \tag{14.1}$$

where $C_{As,0}$ and C_{As} are initial and equilibrium concentrations of As in the aqueous phase [mmol dm^{-3}], respectively, L is volume of aqueous solution [dm^3], and W is weight of adsorbent [g].

The adsorbents were applied for the batchwise adsorption of As. Figure 14.1 shows the effect of equilibrium pH on the adsorption of As(III) on Fe_3O_4, TiO_2 and Al_2O_3. The adsorption of As(III) with all adsorbents is slightly increased with pH in the acidic region, and then is decreased at pH > 9. The adsorption behavior obtained in the present work is almost consistent with the previous works for Fe_3O_4 (Ohe *et al.*, 2005), TiO_2 (Pena *et al.*, 2005), and Al_2O_3 (Singh and Pant, 2004).

Figure 14.2a shows the species distribution of As(III) in aqueous solution (Egawa *et al.*, 1985). Comparing the adsorption behavior and the species distribution, the nonionic species of H_3AsO_3 is a possible species adsorbed on all three adsorbents. Since the surface of Fe_3O_4 is positively charged at pH < 6.6, while negatively charged at pH > 6.6, H_3AsO_3 may be adsorbed by hydroxyl groups of Fe_3O_4 as in Equations (14.2) and (14.3):

$$|\text{-OH} + H_3AsO_3 \quad \rightarrow \quad |\text{-}H_2AsO_3 + H_2O \tag{14.2}$$

$$|\text{-O} + H_3AsO_3 \quad \rightarrow \quad |\text{-}H_2AsO_3 + OH \tag{14.3}$$

The adsorption abilities of OH and O$^-$ on Fe_3O_4 for H_2AsO_3 are estimated to be comparable, because the adsorption curve shown in Figure 14.1 corresponds to the species distribution of H_3AsO_3 as shown in Figure 14.2. In the case of TiO_2, the same adsorption schemes as that of Fe_3O_4, Equations (14.2) and (14.3) progress with boundary pH = 5.7. Maximum adsorption at pH = 8–9 is due to stronger adsorption ability of O$^-$ than OH on the TiO_2 surface. In the case of Al_2O_3, adsorption progresses only by Equation (14.2), since the IP value of Al_2O_3 is 9.5. Dramatic decrease in the adsorption amount in pH > 9 is due to the decrease in OH on Al_2O_3 surface.

The adsorption of As(V) at various pHs was also carried out with the adsorbents. Figure 14.3 shows the effect of equilibrium pH on the adsorption of As(V). Adsorption of As(V) decreases with pH value in all cases. Considering the species distribution of As(V) in aqueous solution shown in

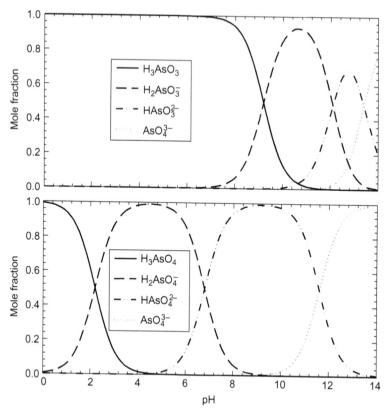

Figure 14.2.   Mole fraction of (a) As(III) and (b) As(V) species in aqueous solution (Egawa *et al.*, 1985).

Figure 14.2b, H_3AsO_4 and H_2AsO_4 are the possible species adsorbed as in Equations (14.4)–(14.6) (Eby, 2004).

$$|\text{-OH} + H_3AsO_4 \quad \rightarrow \quad |\text{-}H_2AsO_4 + H_2O \tag{14.4}$$

$$|\text{-OH} + H_2AsO_4 \quad \rightarrow \quad |\text{-}HAsO_4^- + H_2O \tag{14.5}$$

$$|\text{-O} + H_2AsO_4^- \quad \rightarrow \quad |\text{-}HAsO_4^- + OH \tag{14.6}$$

Figure 14.4 shows adsorption isotherms of As(III) and As(V) with the adsorbents. The adsorption of both As(III) and As(V) progresses via the Langmuir monolayer adsorption mechanism. The maximum adsorption amount ($q_{As,0}$) and the adsorption equilibrium constant (K) calculated with the linear relationship of Langmuir adsorption isotherms are listed in Table 14.3. The calculated lines of the adsorption amounts of As(III) and As(V) based on the maximum adsorption amounts and adsorption equilibrium constants are also shown in Figure 14.4. In the case of As(III), the maximum adsorption amounts are in the order of $Al_2O_3 \approx TiO_2 > Fe_3O_4$. The lower adsorption amount of Fe_3O_4 is caused by the smaller BET surface area. In the cases of TiO_2 and Al_2O_3, similar adsorption amounts of As(III) are obtained, though the BET surface area of TiO_2 is much smaller than that of Al_2O_3. This may be because the acidity of Ti-OH on the TiO_2 surface is higher than that of OH on the Al_2O_3 surface (Pena *et al.*, 2005). In the case of As(V), the maximum adsorption amounts are in the order of $Al_2O_3 > TiO_2 > Fe_3O_4$, which corresponds to the order of their respective BET surface areas. The adsorption amount of As(III) is much higher than that of As(V) for all adsorbents. These results are entirely consistent with our previous research (Ohe *et al.*, 2005). Considering the pH value of geothermal water (about 6–8) (Yoshizuka *et al.*, 2010), all the adsorbents are applicable for As removal, especially for As(III) having higher toxicity than As(V), from geothermal water.

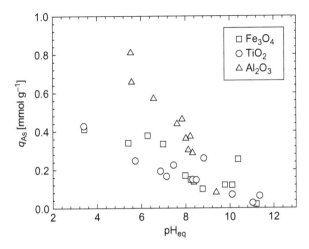

Figure 14.3. Effect of pH on adsorption of As(V) with Fe_3O_4, TiO_2 and Al_2O_3 adsorbents (reproduced with permission from Park *et al.*, 2012a, copyright of Japan Society of Ion Exchange).

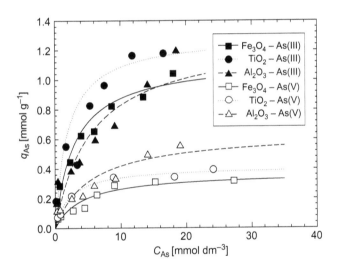

Figure 14.4. Adsorption isotherms of As(III) and As(V) with Fe_3O_4, TiO_2 and Al_2O_3 adsorbents (reproduced with permission from Park *et al.*, 2012a, copyright of Japan Society of Ion Exchange).

Table 14.3. The $q_{As,0}$ and K of As(III) and As(III) with powder-form Fe_3O_4, TiO_2 and Al_2O_3 adsorbents (Park *et al.*, 2012a).

		Fe_3O_4	TiO_2	Al_2O_3
BET surface area [$m^2 g^{-1}$]		94.3	167	271
Isoelectric point		6.6	5.7	9.5
As(III)	$q_{As,0}$ [$mmol\ g^{-1}$]	1.135	1.296	1.292
	K [$dm^3\ mmol^{-1}$]	0.336	0.516	0.179
As(V)	$q_{As,0}$ [$mmol\ g^{-1}$]	0.368	0.414	0.640
	K [$dm^3\ mmol^{-1}$]	0.245	0.444	0.189

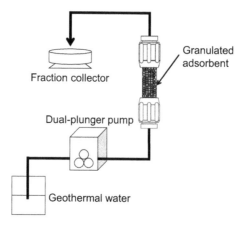

Figure 14.5.   Schematic diagram of the column adsorption apparatus.

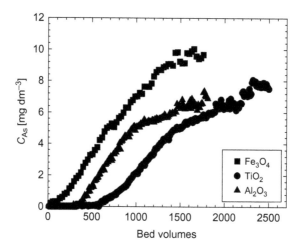

Figure 14.6.   Breakthrough profiles of As(III) with granulated Fe_3O_4, TiO_2 and Al_2O_3 adsorbents (reproduced with permission from Park *et al.*, 2012a, copyright of Japan Society of Ion Exchange).

Based on the batchwise adsorption, the column operation for aqueous solution containing As(III) was conducted.

The geothermal water was fed into the adsorption column packed with granulated adsorbent as shown in Figure 14.5.

Figure 14.6 shows the breakthrough profiles of As in aqueous solution with granulated adsorbents Fe_3O_4, TiO_2 and Al_2O_3. Here, the bed volume (BV) was calculated by:

$$BV = vt/V \qquad (14.7)$$

where v is the flow rate of the feed solution [cm^3 min^{-1}], t is the supply time of the feed solution [min], and V is wet volume of the adsorbent [cm^3].

In all systems, chromatographic removal of As can be achieved to be completely adsorbed until $BV = 100$ (Fe_3O_4), 500 (TiO_2) and 280 (Al_2O_3), respectively. Especially, the value obtained by granulated Al_2O_3 is lower than that by granulated TiO_2, although the maximum adsorption amounts of arsenic (As(III) and As(V)) with powder-form Al_2O_3 and powder-form TiO_2 are almost the same as shown in Table 14.3. This may be due to the lower adsorption kinetics of granulated Al_2O_3. Granulated TiO_2 is therefore the superior adsorbent of the three kinds of adsorbents investigated for chromatographic removal of As from geothermal water.

14.5 CONCLUSIONS

Geothermal waters may contain several toxic elements such as As, which are mostly at higher concentrations than recommended for drinking water by the water authorities; thus, geothermal water that is withdrawn and used can become a disposal problem. In this paper, various methods for removal of As from water have been reviewed. Selecting an appropriate technology for As removal strongly depends on the characteristics of the contaminated water, including its total dissolved solids (TDS) content, pH, redox conditions, microbial activity, the presence of any organic and inorganic species that might interfere with treatment, and the concentrations of the As species. Adsorption by inorganic materials has been mostly used for As removal.

Adsorption of arsenic (As(III) and As(V)) has been investigated, employing magnetite (Fe_3O_4), titania (TiO_2), and alumina (Al_2O_3) as adsorbents. Three kinds of adsorbents possess high adsorption ability for both As(III) and As(V) in the acidic and neutral pH regions, which indicates that all adsorbents are applicable for the As removal from common geothermal waters. The maximum adsorption amounts determined by Langmuir isotherms are in the order of $Al_2O_3 \approx TiO_2 > Fe_3O_4$ in the case of As(III), and $Al_2O_3 > TiO_2 > Fe_3O_4$ in the case of As(V), respectively. These adsorption abilities are affected by both hydroxyl groups on the surface and the specific surface area of the adsorbents in powder form. Chromatographic removal of As from geothermal water was performed with the granulated adsorbents of Fe_3O_4, TiO_2 and Al_2O_3. Considering these three, the most effective removal of As(III) from aqueous solution can be achieved using granulated TiO_2, due to its high adsorption kinetics.

ACKNOWLEDGEMENTS

The authors would like to thank Marek Bryjak for his kind invitation to write this chapter and their co-workers cited in the various references for their efforts in making this all possible. The authors acknowledge financial support by Grant-in-Aid for Scientific Research (B) (No. 23360344) and Grant-in-Aid for Challenging Exploratory Research (24656476) from the Japan Society for the Promotion of Science.

REFERENCES

Alaerts, G.L. & Khouri N. (2004) Arsenic contamination of groundwater: mitigation strategies and policies. *Hydrogeology Journal*, 12, 103–114.
Aloupi, M., Angelidis, M.O., Gavriil, A.M., Koulousaris, M. & Varnavas, S.P. (2009) Influence of geology on arsenic concentrations in ground and surface water in central Lesvos, Greece. *Environmental Monitoring and Assessment*, 151, 383–396.
Awuah, E., Morris, R.T., Owusu, P.A., Sundell, R. & Lindstrom, J. (2009) Evaluation of simple methods of arsenic removal from domestic water supplies in rural communities. *Desalination*, 248, 42–47.
Buisson, D.H., Rothbaum, H.P. & Shannon, W.T. (1979) Removal of arsenic from geothermal discharge waters after absorption on iron floc and subsequent recovery of the floc using dissolved air flotation. *Geothermics*, 8, 97–110.
Bundschuh, J. & Maity, J.P. (2015) Geothermal arsenic: occurrence, mobility and environmental implications. *Renewable & Sustainable Energy Reviews*, 42, 1214–1222.
Carlo, E.H.D. & Thomas, D.M. (1985) Removal of arsenic from geothermal fluids by adsorptive bubble flotation with colloidal ferric hydroxide. *Environmental Science & Technology*, 19, 538–544.

Clifford, D.A. & Ghurye, G.L. (2002) Metal-oxide adsorption, ion exchange, and coagulation – microfiltration for arsenic removal from water. In: Frankenberger, W.T., Jr. (ed.) *Environmental chemistry of arsenic*. Marcel Dekker, New York, NY. pp. 217–245.

Deliyanni, E.A., Bakoyannakis, D.N., Zouboulis, A.I. & Peleka, E. (2003a) Removal of arsenic and cadmium by akaganeite fixed-beds. *Separation Science and Technology*, 38, 3967–3981.

Deliyanni, E.A., Bakoyannakis, D.N., Zouboulis, A.I. & Matis, K.A. (2003b) Sorption of As(V) ions by akaganéite-type nanocrystals. *Chemosphere*, 50, 155–163.

Eby, G.N. (2004) *Principles of environmental geochemistry*. Thomson Brooks/Cole, Pacific Grove, CA. p. 514.

Egawa, H., Nonaka, T. & Maeda, H. (1985) Studies of selective adsorption resins. XXII. Removal and recovery of arsenic ion in geothermal power waste solution with chelating resin containing mercapto groups. *Separation Science and Technology*, 20, 653–664.

Ferguson, J.F. & Gavis, J. (1972) A review of the arsenic cycle in natural waters. *Water Research*, 6, 1259–1265.

Greenleaf, J.E., Lin, J.-C. & SenGupta, A.K. (2006) Two novel applications of ion exchange fibers: arsenic removal and chemical-free softening of hard water. *Environmental Progress*, 25, 300–311.

Gu, B., Phelps, T.J., Liang, L., Dickey, M.J., Roh, Y., Kinsall, B.L., Palumbo, A.V. & Jacobs, G.K. (1999) Biogeochemical dynamics in zero-valent iron columns: implications for permeable reactive barriers. *Environmental Science & Technology*, 33, 2170–2177.

Guo, Q., Wang, Y. & Liu, W. (2008) B, As, and F contamination of river water due to wastewater discharge of the Yangbajing geothermal power plant, Tibet, China. *Environmental Geology*, 56, 197–205.

Henke, K. (2009) Waste treatment and remediation technologies for arsenic. In: Henke, K. (ed.) *Arsenic – environmental chemistry, health threats and waste treatment*. Wiley, Chichester, UK. pp. 351–430.

Kabay, N., Yilmaz, I., Yamac, S., Yuksel, M., Yuksel, U., Yildirim, N., Aydogdu, O., Iwanaga, T. & Hirowatari, K. (2004) Removal and recovery of boron from geothermal wastewater by selective ion-exchange resins – II. Field tests. *Desalination*, 167, 427–438.

Katsoyiannis, I.A., Hug, S.J., Ammann, A., Zikoudi, A. & Hatziliontos, C. (2007) Arsenic speciation and uranium concentrations in drinking water supply wells in northern Greece: correlations with redox indicative parameters and implications for groundwater treatment. *Science of the Total Environment*, 383, 128–140.

Kim, S.O., Kim, W.S. & Kim, K.W. (2005) Evaluation of electrokinetic remediation of arsenic-contaminated soils. *Environmental Geochemistry and Health*, 27, 443–453.

Kouras, A., Katsoyiannis, I. & Voutsa, D. (2007) Distribution of arsenic in groundwater in the area of Chalkidiki, northern Greece. *Journal of Hazardous Materials*, 147, 890–899.

Krauskopf, K.B. & Bird, D.K. (1995) *Introduction to geochemistry*. 3rd edition. McGraw-Hill, Boston, MA. p. 647.

Ladeira, A.C.Q., Ciminelli, V.S.T. & Nepomuceno, A.L. (2002) Seleção de solos para a imobilização de arsênio. *Revista Escola de Minas*, 55, 215–220.

Langner, H.W., Jackson, C.R., Mcdermott, T.R. & Inskeep, W.P. (2001) Rapid oxidation of arsenite in a hot spring ecosystem, Yellowstone National Park. *Environmental Science & Technology*, 35, 3302–3309.

Mroczek, E.K. (2005) Contributions of arsenic and chloride from the Kawerau geothermal field to the Tarawera River, New Zealand. *Geothermics*, 34, 218–233.

Nakahira, A., Okajima, T., Honma, T., Yoshioka, S. & Tanaka, I. (2006) Arsenic removal by hydroxyapatite-based ceramics. *Chemistry Letters*, 35, 856–857.

Ohe, K., Tagai, Y., Nakamura, S., Oshima, T. & Baba, Y. (2005) Adsorption behavior of arsenic(III) and arsenic(V) using magnetite. *Journal of Chemical Engineering Japan*, 38, 671–676.

Okada, H., Kudo, Y., Nakazawa, H., Chiba, A., Mitsuhashi, K., Ohara, T. & Wada, H. (2004) Removal system of arsenic from geothermal water by high gradient magnetic separation-HGMS reciprocal filter. *IEEE Transactions on Applied Superconductivity*, 14, 1576–1579.

Park, J.E., Sato, H., Shibata, K., Shao, S., Nishihama, S. & Yoshizuka, K. (2012a) Adsorption behavior of arsenic by Fe_3O_4, TiO_2, and Al_2O_3 adsorbents. *Journal of Ion Exchange*, 23, 82–87.

Park, J.E., Sato, H., Nishihama, S. & Yoshizuka, K. (2012b) Lithium recovery from geothermal water by combined adsorption methods. *Solvent Extraction Ion Exchange*, 30, 398–404.

Pena, M.E., Korfiatis, G.P., Patel, M., Lee, L. & Meng, X. (2005) Adsorption of As(V) and As(III) by nanocrystalline titanium dioxide. *Water Research*, 39, 2327–2337.

Planer-Friedrich, B., London, J., Mccleskey, R.B., Nordstrom, D.K. & Wallschläger, D. (2007) Thioarsenates in geothermal waters of Yellowstone National Park: determination, preservation, and geochemical importance. *Environmental Science & Technology*, 41, 5245–5251.

Prasad, G. (1994) Removal of arsenic(V) from aqueous systems by adsorption onto some geologic materials. In: Nriagu, J.O. (ed.) *Arsenic in the environment*. Part I: *Cycling and characterization*. John Wiley & Sons Inc., New York, BT. pp. 133–154.

Smith, S.D. & Edward, M. (2005) The influence of silica and calcium on arsenate sorption to oxide surfaces. *Journal of Water Supply: Research Technology Aqua*, 54, 201–211.

Tokuyama, H., Maeda, S. & Takahashi, K. (2004) Development of a novel moving bed with liquid-pulse and experimental analysis of nickel removal from acidic solution. *Separation and Purification Technology*, 38, 139–147.

Tyrovola, K., Nikolaidis, N.P., Veranis, N., Kallithrakas-Kontos, N. & Koulouridakis, P.E. (2006) Arsenic removal from geothermal waters with zero-valent iron-effect of temperature, phosphate and nitrate. *Water Research*, 40, 2375–2386.

Tyrovola, K., Peroulaki, E. & Nikolaidis, N.P. (2007) Modeling of arsenic immobilization by zero valent iron. *European Journal of Soil Biology*, 43, 356–367.

USEPA (2002) *Arsenic treatment technologies for soil, waste and water*. EPA-540-F-98-054, Environmental Protection Agency, Office of Solid Wastes and Emergency (5102G), Washington, DC.

Violante, A., Ricciardella, M., Del Gaudio, S. & Pigna, M. (2006) Coprecipitation of arsenate with metal oxides: nature, mineralogy, and reactivity of aluminum precipitates. *Environmental Science & Technology*, 40, 4961–4967.

Wilkie, J.A. & Hering, J.G. (1998) Rapid oxidation of geothermal arsenic(III) in streamwaters of the eastern Sierra Nevada. *Environmental Science & Technology*, 32, 657–662.

Xu, T., Pruess, K., Pham, M., Klein, C. & Sanyal, S. (2001) Reactive chemical transport simulation to study geothermal production with mineral recovery and silica scaling. *Geothermal Resources Council Transactions*, 25, 513–517.

Yanagase, K., Yoshinaga, T., Kawano, K. & Matsuoka, T. (1983) The recovery of lithium from geothermal water in the Hatchobaru area of Kyushu, Japan. *Bulletin of the Chemical Society Japan*, 56, 2490–2498.

Yoshizuka, K., Nishihama, S. & Sato, H. (2010) Analytical survey of arsenic in geothermal waters from sites in Kyushu, Japan, and a method for removing arsenic using magnetite. *Environmental Geochemistry and Health*, 32, 297–302.

Zhang, G., Liu, C.-Q., Liu, H., Jin, Z., Han, G. & Li, L. (2008) Geochemistry of the Rehai and Ruidian geothermal waters, Yunnan Province, China. *Geothermics*, 37, 73–83.

Zhang, W., Singh, P., Paling, E. & Delides, S. (2004) Arsenic removal from contaminated water by natural iron ores. *Minerals Engineering*, 17, 517–524.

CHAPTER 15

Arsenic removal by advanced oxidation assisted by solar energy

Héctor D. Mansilla, Jorge Yáñez, David R. Contreras & Lorena Cornejo

15.1 INTRODUCTION

Several procedures have been proposed as suitable technologies for arsenic (As) removal in developing countries based on simple procedures ranging from conventional technologies (redox precipitation, coagulation, membrane, among others), emerging technologies (zero-valent iron and biological methods) and photochemical technologies. The latter include those activated by visible or solar light such as solar removal of As procedure (SORAS), photo-Fenton reaction and TiO_2 photocatalysis. This chapter summarizes photochemical methods for oxidation of As(III) and removal of As(V) in aqueous solutions driven by visible or solar light.

15.2 SOLAR REMOVAL OF ARSENIC

The processes commonly used for As removal from water are based on the formation of insoluble compounds followed by separation techniques. Iron is the most used agent for As precipitation as a result of the formation of insoluble Fe(III)(hydr)oxides compounds which absorb As. As(V) ($H_2AsO_4^-$ and $HAsO_4^{2-}$) adsorbs almost quantitatively on the precipitated Fe(III)(hydr)oxides allowing its easy removal. On the other hand, As(III) (H_3AsO_3) adsorbs only weakly on Fe(III) (hydr)oxides (Haque *et al.*, 2008; Hug *et al.*, 2001; Manning *et al.*, 2002). Since pentavalent As is easier to remove, As(III) must be oxidized before separation procedures. Various conventional agents have been used for the oxidation of As(III), such as hydrogen peroxide, chlorine or permanganate. Furthermore, solar light can be used to promote As(III) oxidation through the formation of reactive oxygen species (ROS), offering a great potential for low cost water purification (Bundschuh *et al.*, 2010; Cornejo *et al.*, 2008; García *et al.*, 2004a; Hug *et al.*, 2001; Lara *et al.*, 2006).

The SORAS procedure (solar removal of As) was first published by Hug *et al.* (2001) for As removal of As from water for human consumption. SORAS is an effective way of removing As(V) from waters through formation of insoluble iron compounds using citrate and solar light for the oxidation of As(III). The ROS promoted oxidation is characterized by fast reaction rates compared to conventional oxidation procedures (Bundschuh *et al.*, 2010; Cornejo *et al.*, 2008; García *et al.*, 2004a; 2004b; Hug *et al.*, 2001; 2003; Lara *et al.*, 2006; Wegelin *et al.*, 1999). To understand the mechanism of the SORAS procedure it is essential to know the behavior of inorganic As species, iron and citrate in aqueous solution at neutral pH.

The equilibrium of inorganic As species in water depends fundamentally on the redox potential and pH value. As(V) is the predominant species in oxygenated surface waters and occurs primarily as $H_2AsO_4^-$ and $HAsO_4^{2-}$ (pK_{a2} 6.8) in neutral waters. On the other hand, As(III) abounds in anoxic waters as H_3AsO_3 (pK_{a3} 9.2) (Bednar *et al.*, 2005; Mandal and Suzuki, 2002).

Regarding iron, the most important species in aqueous solution are Fe(II) and Fe(III). In aqueous solution, Fe(II) to Fe(III) can be readily oxidized by dissolved oxygen, forming superoxide anions and, subsequently, hydrogen peroxide (Equations (15.1) and (15.2)) (Bang *et al.*, 2005; Bednar *et al.*, 2005; Hug *et al.*, 2001, 2003; Lara *et al.*, 2006):

$$Fe(II) + O_{2(ac)} \rightarrow Fe(III)(OH)_2^+ + O_2^{\bullet-} \tag{15.1}$$

$$Fe(II) + O_2^{\bullet-} \rightarrow Fe(III)(OH)_2^+ + H_2O_2 \tag{15.2}$$

Moreover, iron (hydr)oxide complexes appear when the pH increases. The main Fe(III) species in aqueous media are $Fe(OH)^{++}$, $Fe(OH)_2^+$, $Fe(OH)_{3(ac)}$, $Fe(OH)_{3(s)}$ and $Fe(OH)_4^-$. These complexes have the property of absorbing UV and visible light being highly photosensitive, especially $Fe(OH)_2^+$ (Bang *et al.*, 2005; Hug *et al.*, 2001)

While SORAS in its original design uses Fe(III) naturally dissolved in water, the addition of soluble Fe (II) has been reported (Hug *et al.*, 2001; Lara *et al.*, 2006). In addition, the incorporation of iron in the form of Fe(0) either in the form of steel wool or as nanoparticles has been published (Bang *et al.*, 2005; Bhowmick *et al.*, 2014; Cornejo *et al.*, 2008; Morgada *et al.*, 2009).

The presence of citrate, or other organic complexing agents, plays an important role increasing the rate of As oxidation in the SORAS procedure. The mechanism of citrate and its optimal concentrations for As removal has been studied in detail by various authors (Cornejo *et al.*, 2008; Hug *et al.*, 2001; Lara *et al.*, 2006; Majumder *et al.*, 2013). The complexes formed between iron and citrate are dependent on pH. Because of the high stability of the citrate-iron complexes, the hydroxyl groups are replaced, forming the $FeOHCit^-$ complex in a wide pH range. This compound presents a high extinction coefficient in the UV-Vis range (Hug *et al.*, 2001).

The chemical equilibrium between As species, iron species, citrate and dissolved oxygen in water generate conditions for a large number of chemical reactions, such as complexes formation, precipitation, acid-base and redox. Thus, a fairly complex SORAS mechanism could be foreseen. In addition, photochemical processes are involved when the system is illuminated with UVA or solar light increasing the complexity of reaction mechanism.

The oxidation of As(III) is thermodynamically favorable in the presence of dissolved oxygen. However, it is very slow, because of low dissolved oxygen concentrations (less than $5 \, mg \, L^{-1}$ at 15°C) and by the presence of other species that can compete with oxygen consumption. The incorporation of ROS such as $HO^\bullet$, $O_2^{\bullet-}/HO_2^\bullet$ promoted by UVA or solar radiation favors the oxidation rate. As established by Hug *et al.* (2001), the photo-decomposition of $FeOHCit^-$ and $FeOH_2^+$ complexes, generates those radicals. The photo-decomposition of $FeOHCit^-$ generates the radical 3-hydroxiglutarate ($3HGA^{\bullet 2-}$), which in turn breaks down rapidly in 3-oxogluratare ($3\text{-}OGA^{2-}$), generating superoxide anion ($O_2^{\bullet-}$) (Equations (15.3) and 15.4)). Superoxide anion can, in consecutive reactions produce hydrogen peroxide (Equations (15.5) and 15.6)). Once hydrogen peroxide is generated, Fenton reaction can be initiated forming hydroxyl radicals (Equation (15.7)) (Hug and Leupin, 2003):

$$FeCitOH^- + h\nu \rightarrow Fe(II) + 3\text{-}HGA^{\bullet 2-} \tag{15.3}$$

$$3\text{-}HGA^{2-} + O_2 \rightarrow 3\text{-}OGA^{2-} + O_2^{\bullet-} \tag{15.4}$$

$$O_2^{\bullet-} + H^+ \rightarrow HO_2^\bullet \tag{15.5}$$

$$HO_2^\bullet + O_2^{\bullet-} \rightarrow H_2O_2 \tag{15.6}$$

$$Fe(II) + H_2O_2 \rightarrow Fe(III) + HO^\bullet + OH^- \tag{15.7}$$

Simultaneously, under irradiation with visible light, the complex $FeOH^{2+}$ absorbs light restoring Fe(II) and generating additional hydroxyl radicals ($HO^\bullet$), in a reaction known as photo-Fenton process (Equation (15.8)) (Hug and Leupin, 2003):

$$FeOH^{2+} + h\nu \rightarrow Fe(II) + HO^\bullet \tag{15.8}$$

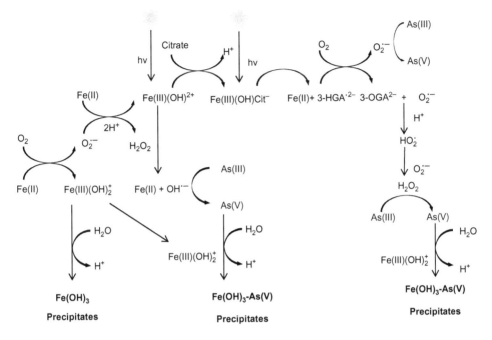

Figure 15.1. Schematic outline of the chemical reactions involved in SORAS.

ROS are highly reactive species being able to rapidly oxidize As(III) according to Equations (15.9)–(15.11):

$$As(III) + O_2^{\bullet -} \rightarrow As(V) + 1/2O_2 \tag{15.9}$$

$$As(III) + H_2O_2 \rightarrow As(V) \tag{15.10}$$

$$As(III) + HO^{\bullet} \rightarrow As(V) \tag{15.11}$$

The reaction between As(III) and hydroxyl radical occurs quickly; however, experimental evidence indicates that the main oxidation of As(III) occur with $O_2^{\bullet -}/HO_2^{\bullet}$ and H_2O_2. Indeed, the use of $HO^{\bullet}$ scavengers, such as 2-propanol, does not inhibit the As(III) oxidation (Hug et al., 2001). In summary, the reactions involved in SORAS process may be outline as shown in Figure 15.1.

The As removal efficiency by SORAS is remarkably dependent on the initial conditions, such as total As concentration, As(III):As(V) ratio, iron and citrate concentrations (Litter et al., 2010). Thus, the As removal efficiency is different in synthetic solutions compared with natural contaminated waters. A comparative data of SORAS effectiveness is shown in Table 15.1. In conclusion, the SORAS system can be considered a simple and suitable procedure to remove As using low cost and available materials. Because of these features, SORAS can be implemented in rural and remote areas highly irradiated by the sun. Therefore, SORAS have been applied successfully in Asia and Latin America as presented in Table 15.1.

In field, the best conditions for As removal appears to be low concentration of iron, low concentration of citrate (to avoid the ROS quenching) and high solar irradiance. Additionally, the SORAS system allows the simultaneous disinfection of water from *E. coli,* in the process known as Solar Disinfection (SODIS) (McGuigan et al., 2012).

Usually, in SORAS and in SORAS modified technology (also called also as *zero-valent iron technology* (ZVIT) runs are done in PET bottles (Fig. 15.2a). In the ZVIT, Fe(II) is replaced by zero-valent iron. Concerning the scaling up of SORAS and ZVIT, different approaches can be

Table 15.1.   Comparative arsenic removal yields by SORAS under different experimental conditions.

Initial conditions	Added reagents	Removal [%] (time)	Radiation	Reference
Synthetic waters As(III) 500 $\mu g\,L^{-1}$ As(tot) 500 $\mu g\,L^{-1}$	Citrate	80–90% (2–3 h)	UV-A light	Hug *et al.* (2001)
Well water Bangladesh As(III) 6–720 $\mu g\,L^{-1}$ As(tot) 61–1240 $\mu g\,L^{-1}$	Lemon juice	45–78% (12–15 h)	Solar light	Hug *et al.* (2001)
River water Chile As(tot) 1000 $\mu g\,L^{-1}$	Fe(0) Lemon juice	>99% (2 h)	Solar light	Cornejo *et al.* (2008)
Synthetic waters As(III) 500 $\mu g\,L^{-1}$ As(tot) 500 $\mu g\,L^{-1}$	Fe(II) Citrate	95% (1 h) 80% (4 h)	UV-A light	Lara *et al.* (2006)
River water Chile As(tot) 1000 $\mu g\,L^{-1}$	Fe(II) Lemon juice	>99% (4 h)	Solar light	Lara *et al.* (2006)
Synthetic waters As(tot) 189–297 $\mu g\,L^{-1}$	Fe(III) Lemon juice	>86% (5 h)	Solar light	García *et al.* (2004a)
Groundwaters Argentina As(tot) 86–1023 $\mu g\,L^{-1}$	Fe(III) Lemon juice	26–58% (5 h)	Solar light	García *et al.* (2004b)
Synthetic waters As(III) 100–500 $\mu g\,L^{-1}$	Tomate juice	78–98% (4 h)	Solar light	Majumder *et al.* (2013)
Groundwaters West Bengal As(tot) 111–430 $\mu g\,L^{-1}$	Tomate juice	90–97% (4 h)	Solar light	Majumder *et al.* (2013)

(a)   (b)

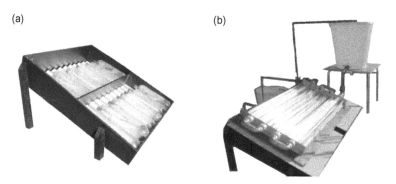

Figure 15.2.   In field SORAS devices (a) PET bottles containing contaminated water and (b) semi-continuous pilot plant (Litter *et al.*, 2012).

attempted. For instance, a semi-continuous and simple device designed to treat small volumes of As contaminated waters in the Atacama Desert (Arica, Chile) is shown in the Figure 15.2b (Litter *et al.*, 2012).

15.3   HOMOGENEOUS PHOTOCATALYSIS

To the best of our knowledge, there is a lack of information regarding As removal in homogeneous photocatalysis assisted by solar or visible light systems. Most of the reported studies use ultraviolet irradiation with the purpose of produce free radicals in order to improve As oxidation. The most reported systems are UV/H$_2$O$_2$ (Yang *et al.*, 1999; Yoon *et al.*, 2008), UV/ultrasound (Lee *et al.*, 2012), UV/Fe(III)-oxalate (Kocar and Inskeep, 2003), UV/Fe(II)/H$_2$O$_2$ (Buchmann *et al.*, 2005) and UV/S$_2$O$_8^{2-}$ (Kanel *et al.*, 2005; Nepolian *et al.*, 2008; Yoon *et al.*, 2011). Also, visible light can be used for *in situ* radical production. Woods *et al.* (1963) studied the oxidation of As(III) using Fe/S$_2$O$_8^{2-}$ system illuminated by a tungsten lamp where it was demonstrated that the effectiveness of peroxydisulfate in As removal is less efficient in absence of iron. In addition, Nishida and Kimura (1990), using a tungsten lamp, studied the oxidation of As(III) using S$_2$O$_8^{2-}$ assisted by tri(2,2′-bipyridine) ruthenium(II), used as a sensitizer to enhance the decomposition of S$_2$O$_8^{2-}$.

Regarding solar assisted homogeneous photocatalysis, Nepolian *et al.* (2008) compare different lamps and natural solar light as an irradiation source to oxidize As(III) by S$_2$O$_8^{2-}$. In this report about 50% oxidation of As(III) (0.135 mM) was attained within 30 min at pH 3.

15.4   HETEROGENEOUS PHOTOCATALYSIS

Heterogeneous photocatalysis is an advanced oxidation process widely studied for removal of pollutants in water or air (Lizama *et al.*, 2002; Muruganandham *et al.*, 2014; Thu *et al.*, 2005). It is based in the simultaneous formation of valence band holes (h$_{vb}^{+}$) and conduction band electrons (e$_{cb}^{-}$) on the surface of the catalyst promoted by energy absorption (h$\nu \geq$ band gap energy) (Equation (15.12)). Usually the catalysts are solid semiconductors with a narrow band gap. For instance, Evonik TiO$_2$ P-25 (former Degussa) with a band gap around 3.2 eV (387 nm) is one of the most studied catalysts. This implies that catalyst activation could be promoted by solar light that accounts for around 5% of energy in that range. Subsequently, several reduced oxygen species (ROS), such as superoxide anion (O$_2^{-}$) and hydroxyl radical (HO$^{\bullet}$), are readily produced. Hydroxyl radicals are promoted by holes, as a result of the electron abstraction from catalyst-adsorbed water or hydroxide anions (Equations (15.13) and 15.14)). In addition, the photo-generated holes can directly oxidize organic compounds (RH) (Equation (15.15)). Conversely, the slightly reductive electrons (e$_{cb}^{-}$) contribute to the production of superoxide anions. The reductive sites have also been reported as responsible for the reduction of heavy metals such as chromium and mercury (Litter, 1999; 2005; Miranda *et al.*, 2013). On the other hand, the oxygen-based radicals are responsible for the oxidation of several organic and inorganic compounds (Equation (15.16)). A very simplified description of the possible events that occurs when the photocatalyst absorbs a photon can be summarized as follows:

$$TiO_2 + h\nu \rightarrow e_{cb}^{-} + h_{vb}^{+} \tag{15.12}$$

$$h_{vb}^{+} + H_2O \rightarrow HO^{\bullet} + H^{+} \tag{15.13}$$

$$h_{vb}^{+} + HO^{-} \rightarrow HO^{\bullet} \tag{15.14}$$

$$h_{vb}^{+} + RH \rightarrow RH^{\bullet+} \tag{15.15}$$

$$e_{cb}^{-} + O_2 \rightarrow O_2^{\bullet-} \tag{15.16}$$

It is well known that As(III) to As(V) oxidation could be rapidly carried out by oxidants such as chlorine, ozone and MnO$_4^{-}$ while chloramines, hydrogen peroxide and chlorine dioxide are less effective (Sharma *et al.*, 2007). However, the main advantage of using semiconductor photocatalysis is based on the low cost of the catalyst, the possibility of activation by solar light and no need for addition of supplementary reagents. For these reasons many reports have been recently focused on the oxidation of As(III) to As(V) followed by a filtration step to remove the oxidized As (Litter *et al.*, 2010). The oxidation of arsenite to arsenate is highly desirable because the latter is less toxic and easily removable by adsorption on solid surfaces (iron or aluminum) while As(III) is mobile, toxic and has low affinity to mineral surface.

Regarding the photocatalytic As(III) oxidation mechanism, superoxide anion (O$_2^{\bullet-}$) has been originally proposed as the primary radicals responsible for the TiO$_2$-assisted oxidation of arsenite in aqueous solution (Lee and Choi, 2002; Ryu and Choi, 2006). However, considering that around 90% of As species are adsorbed onto the TiO_2 (Dutta *et al.*, 2004), the role of O_2, H_2O_2, $HO^\bullet$ and $O_2^{\bullet-}$ in the oxidation of arsenite on TiO_2 surface was evaluated using UV-A irradiation (350 nm) (Xu *et al.*, 2005). Both arsenite, As(III), and arsenate, As(V), adsorb extensively on TiO_2 surface where the reaction take place. It was concluded that the reaction of superoxide anion radical does not contribute to the conversion of As(III) when compared to the reaction of As(III) with $HO^\bullet$ during TiO_2-assisted photocatalysis.

The photo-oxidation of arsenite can be completely attained in few minutes using the hybrid system TiO_2/UV/zero-valent iron. As(III) removal was accelerated when zero-valent iron was added to the photo-reactor following a second-order reaction mechanism (Nguyen *et al.*, 2008).

In a recent report Li and Leng (2013) showed that As(III) can be efficiently oxidized to As(V) using a ruthenium dye-synthesized titania under irradiation with visible light ($\lambda \geq 420$ nm). This finding opens the options to use solar light to oxidize As(III) at low cost. The oxidation was conducted mainly by reactive oxygen species (67%), such as superoxide anion, and by the photo-generated dye cation (S^+) (33%). The sensitized photocatalytic oxidation could be significantly enhanced by I^- in aqueous solution, because I^- might accelerate the regeneration rate of S^+ which could be the only oxidant of As(III) since iodine ion deactivate the ROS ($O_2^{\bullet-} + 2\,I^- \rightarrow I_2^- + O_2^{2-}$) (Li and Leng, 2012; Li *et al.*, 2013).

The photocatalyst immobilization on inert surfaces has been suggested as an appropriate solution to avoid the separation step of the catalyst at the end of reaction. Fostier *et al.* (2008) showed that As(III) can be efficiently oxidized in the presence of immobilized TiO_2 on PET bottles irradiated by solar light. The oxidation is attributed to ROS formed during irradiation. The residual As(V) produced was completely removed by coprecipitation with Fe(III) followed by filtration resulting in a simple low cost technology, which could be easily implemented in developing countries (Fostier *et al.*, 2008).

The combination of TiO_2-photocatalysis and adsorption on granulated activated carbon was recently proposed (Yao *et al.*, 2012). The reaction was carried out in suspension under illumination with UV light. The oxidation of As(III) is favored at pH 3 reaching 100% oxidation yield after 4 h illumination. In a similar approach, the oxidation of As(III) was followed by membrane filtration with satisfactory performance (Yuksel *et al.*, 2014). Complete oxidation of 10 mg L^{-1} was achieved in 30 min using 0.5 g L^{-1} of TiO_2 at pH 2. Afterwards, the As(V) was quantitatively removed by a water-soluble polymer containing quaternary ammonium groups.

Arsenic bonded to organic groups is commonly found in agricultural applications. Although these compounds have been defined as harmless to the environment, their biotransformation produces inorganic As compounds which contributes to water pollution, and must be eliminated by appropriate water treatment (Castlehouse *et al.*, 2003). For example, the photocatalytic degradation of phenylarsonic acid (PA, Ph-AsO_3H) was carried out using TiO_2 and visible light (UVA) (Zheng *et al.*, 2010). The oxidation was promoted by ROS ($HO^\bullet$, $O_2^{\bullet-}$, $^1O_2$) were hydroxyl radicals play a major role. Complete PA removal was achieved in around 10 min showing that heterogeneous photocatalysis could be effectively employed to removing organoarsenical compounds from contaminated water. Similarly, cacodylic acid (CH_3-AsO_3H) has been efficiently removed from aqueous solution producing As(V) and methanol as main products. TiO_2 has shown better performance than ZnO as a catalyst under UVA irradiation (Miranda *et al.*, 2012).

In order to improve the efficiency of utilizing solar energy in the photocatalytic process of As oxidation, new materials have been synthesized. For instance, $Bi_2Sn_2O_7$ semiconductor was prepared by hydrothermal methodology obtaining a catalyst active in the solar region ($\lambda \geq 420$ m) which was capable to completely oxidize As(III) to As(V) in one hour by means of superoxide ion generated in the process (Tien *et al.*, 2012).

In a different approach, As(III) can be removed by complete reduction to As(0) using heterogeneous photocatalysis. The reduction of As(III) and As(V) to As(0) can be conducted by TiO_2 heterogeneous photocatalysis in N_2-saturated solutions with the addition of a sacrificial reagent such as methanol (Levy *et al.*, 2012). The drawback of this procedure is the simultaneous formation of arsine (AsH_3) which is one of the most toxic forms of As.

Table 15.2.   Comparison of As(III) removal extent under irradiation at different experimental conditions.

Experimental conditions	Removal [%] (time)	Radiation source	Reference
As(III) 10 mg L^{-1} TiO$_2$ 0.5 g L^{-1}; pH 2	100% (0.5 h)	UVA (365 nm)	Yuksel *et al.* (2014)
As(III) 2 mg L^{-1} Bi$_2$Sn$_2$O$_7$ 0.08 g L^{-1}; neutral pH	100% (1 h)	Visible light (420–850 nm)	Tian *et al.* (2012)
As(III) 1 mg L^{-1} TiO$_2$ immobilized on glass rods; Fe(II) 7.0 mg L^{-1}; pH 7	100% (1 h)	Solar light	Fostier *et al.* (2008)
As(III) 0.5 mg L^{-1} TiO$_2$ 0.05 g L^{-1}; pH 6.7	95% (10 min)	UV (254 nm)	Nguyen *et al.* (2008)
As(III) 17.5 mg L^{-1} K$_2$S$_2$O$_8$ 27 mg L^{-1} pH 3.0	50% (30 min)	Solar light	Neppolian *et al.* (2008)

Finally, Table 15.2 summarizes removal yields under different experimental conditions of different photon-driven procedures. It can be concluded that SORAS procedure or heterogeneous photocatalytic process, activated by visible light or directly by solar light, provide suitable tools for oxidation of As(III) in the treatment of As contaminated groundwater in developing countries featuring high levels of As in water sources.

ACKNOWLEDGEMENTS

The authors thank the financial support of the National Commission for Scientific and Technological Research (CONICYT), through the programs Fondecyt (Grants 1130502, 1121128), Associative Research Program (Grant PIA ACT-130) and Conicyt/Fondap (Grant 15110019). Furthermore, the support of Red Doctoral REDOC.CTA, Mineduc (Grant UCO 1202) and of FIC-R project 2011, BIP code N° 30110571-0 is also appreciated.

REFERENCES

Bang, S., Johnson, M.D., Korfiatis, G.P. & Meng, X. (2005) Chemical reactions between arsenic and zero-valent iron in water. *Water Research*, 39, 763–770.
Bednar, A.J., Garbarino, J.R., Ranvillea, J.F. & Wildemana, T.R. (2005) Effects of iron on arsenic speciation and redox chemistry in acid mine water. *Journal of Geochemical Exploration*, 85, 55–62.
Bhowmick, S., Chakraborty, S., Mondal, P., Van Renterghem, W., Van den Berghe, S., Roman-Ross, G., Chatterjee, D. & Iglesias, M. (2014) Montmorillonite-supported nanoscale zero-valent iron for removal of arsenic from aqueous solution: kinetics and mechanism. *Chemical Engineering Journal*, 243, 14–23.
Bundschuh, J., Litter, M., Ciminelli, V.S.T., Morgada, M.E., Cornejo, L., Garrido-Hoyos, S., Hoinkis, J. & Alarcón-Herrera, M.T. (2010) Emerging mitigation needs and sustainable options for solving the arsenic problems of rural and isolated urban areas in Latin America: a critical analysis. *Water Research*, 44, 5828–5845.
Bushmann, J., Canonica, S., Lindauer, U., Hug, S. & Sigg, L. (2005) Photoirradiation of dissolved humic acid induced arsenic(III) oxidation. *Environmental Science & Technology*, 39, 9541–9546.
Castlehouse, H., Smith, C., Raab, A., Deacon, C., Meharg, A.A. & Feldman, J. (2003) Biotransformation and accumulation of arsenic in soil amended with seaweed. *Environmental Science & Technology*, 37, 951–957.
Cornejo, L., Lienqueo, H., Arenas, M., Acarapi, J., Yañez, J. & Mansilla, H.D. (2008) In field arsenic removal from natural water by zero valent iron. *Environmental Pollution*, 156, 827–831.

Dutta, P.K., Ray, A.K., Sharma, V.K. & Millero, F.J. (2004) Adsorption of arsenate and arsenite on titanium dioxide suspensions. *Journal of Colloid and Interface Science*, 278, 270–275.

Fostier, A.H., Pereira, M.S.S., Rath, S. & Guimarães, J.R. (2008) Arsenic removal from water employing heterogeneous photocatalysis with TiO_2 immobilized in PET bottles. *Chemosphere*, 72, 319–324.

García, M.G., d'Hiriart, J., Giullitti, J., Lin, H., Custo, G., Hidalgo, M., Litter, M.I. & Blesa, M.A. (2004a) Solar light induced removal of arsenic from contaminated groundwater: the interplay of solar energy and chemical variables. *Solar Energy*, 77, 601–613.

García, M.G., Lin, H.J., Custo, G., d'Hiriart, J., Hidalgo, M., Litter, M.I. & Blesa, M.A. (2004b) Advances in arsenic removal by solar oxidation from water of Tucumán, Argentina. OAS AE 141 Project. In: Litter, M.I. & Jiménez-González, A. (eds.) *Advances in economic solar technologies for disinfection, decontamination, and removal of arsenic from water of rural communities of Latin America*. Digital Graphic Publisher, La Plata, Argentina. pp. 43–56.

Haque, N., Morrison, G., Cano-Aguilera, I. & Gardea-Torresdey, J.L. (2008) Iron-modified light expanded clay aggregates for the removal of arsenic(V) from groundwater. *Microchemical Journal*, 88, 7–13.

Hug, S. & Leupin, O. (2003) Iron-catalyzed oxidation of arsenic(III) by oxygen and by hydrogen peroxide: pH-dependent formation of oxidants in the fenton reaction. *Environmental Science & Technology*, 37, 2734–2742.

Hug, S., Canonica, L., Wegelin, M., Gechter, D. & Von Guten, U. (2001) Solar oxidation and removal of arsenic at circumneutral in iron containing waters. *Environmental Science & Technology*, 35, 2114–2121.

Kanel, S., Manning, B., Charlet, L. & Choi, H. (2005) Removal of arsenic(III) from groundwater by nanoscale zero-valent iron. *Environmental Science & Technology*, 39, 1291–1298.

Kocar, B. & Inskeep, W. (2003) Photochemical oxidation of As(III) in ferrioxalate solutions. *Environmental Science & Technology*, 37, 1581–1588.

Lara, F., Cornejo, L., Yañez, J., Freer, J. & Mansilla, H.D. (2006) Solar light assisted removal of arsenic from natural water: effect of iron and citrate concentrations. *Journal of Chemical Technology and Biotechnology*, 81, 1282–1287.

Lee, H. & Choi, W. (2002) Photocatalytic oxidation of arsenite in TiO_2 suspension: kinetics and mechanisms. *Environmental Science & Technology*, 36, 3872–3878.

Lee, S., Cui, M., Na, S. & Khim, J. (2012) Arsenite oxidation by ultrasound combined with ultraviolet in aqueous solution. *Japanese Journal of Applied Physics*, 51, 7–15.

Levy, I.K., Misrahi, M., Ruano, G., Zampieri, G., Requejo, F.G. & Litter, M.I. (2012) TiO_2-photocatalytic reduction of pentavalent and trivalent arsenic: production of elemental arsenic and arsine. *Environmental Science & Technology*, 46, 2299–2308.

Li, X. & Leng, W. (2012) Highly enhanced dye sensitized photo catalytic oxidation of arsenite over TiO_2 under visible light by I^- as an electron relay. *Electrochemistry Communications*, 22, 185–188.

Li, X. & Leng, W. (2013) Regenerated dye-sensitized photocatalytic oxidation of arsenite over nanostructured TiO_2 films under visible light in normal aqueous solutions: an insight into the mechanism by simultaneous (photo)electrochemical measurements. *Journal of Physical Chemistry C*, 117, 750–762.

Li, X., Leng, W. & Cao, C. (2013) Quantitatively understanding the mechanism of highly enhanced regenerated dye sensitized photooxidation of arsenite over nanostructured TiO_2 electrodes under visible light by I^-. *Journal of Electroanalytical Chemistry*, 703, 70–79.

Litter, M.I. (1999) Heterogeneous photocatalysis. Transition metal ions in photocatalytic systems. *Applied Catalysis B: Environmental*, 23, 89–114.

Litter, M.I. (2005) Introduction to photochemical advanced oxidation processes for water treatment. In: Boule, P., Bahnemann, D.W. & Robertson, P.K.J. (eds.) *The handbook of environmental chemistry*. Part M *Environmental photochemistry*. Part II, Volume 2. Springer-Verlag, Berlin, Heidelberg, Germany. pp. 325–366.

Litter, M.I., Morgada, M.E. & Bundschuh, J. (2010) Possible treatments for arsenic removal in Latin America waters for human consumption. *Environmental Pollution*, 158, 1105–1118.

Litter, M.I., Alarcón-Herrera, M., Arenas, M., Armienta, M., Avilés, M., Cáceres R., Nery, H., Cornejo, L., Dias, L., Fernández, A., Farfán, E., Garrido, S., Lorenzo, L., Morgada, M. Olmos-Márquez, M. & Pérez-Carrera, A. (2012) Small-scale and household methods to remove arsenic from water for drinking purposes in Latin America. *Science of the Total Environment*, 429, 107–122.

Lizama, C., Freer, J., Baeza, J. & Mansilla, H.D. (2002) Optimized degradation of Reactive Blue on TiO_2 and ZnO suspensions. *Catalysis Today*, 76, 235–246.

Majumder, S., Nath, B., Sarkar, S., Islam, S.M., Bundschuh, J., Chatterjee, D. & Hidalgo, M. (2013) Application of natural citric acid sources and their role on arsenic removal from drinking water: a green chemistry approach. *Journal of Hazardous Materials*, 262, 1167–1175.

Mandal, B.K. & Suzuki, K.T. (2002) Arsenic around the world: a review. *Talanta*, 58, 201–235.

Manning, B.A., Hunt, M.L., Amrhein, C. & Yarmoff, J.A. (2002) Arsenic(III) and arsenic(V) reactions with zerovalent iron corrosion products. *Environmental Science & Technology*, 36, 5455–5461.

McGuigan, K.G., Conroy, R.M., Mosler, H.J., du Preez, M., Ubomba-Jaswa, E. & Fernandez-Ibanez, P. (2012) Solar water disinfection (SODIS): a review from bench-top to roof-top. *Journal of Hazardous Materials*, 235–236, 29–46.

Miranda, C., Yáñez, J., Matschullat, J., Daus, B., Fierro, J.L.G. & Mansilla, H.D. (2012) Degradation of cacodylic acid by heterogeneous photocatalysis assisted by UV-A radiation. *Proceedings of 7th European Meeting on Solar Chemistry and Photocatalysis: Environmental Applications, Porto, Portugal, 17–20 June 2012.* p. 77.

Miranda, C., Yáñez, J., Contreras, D. & Mansilla, H.D. (2013) Removal of phenylmercury assisted by heterogeneous photocatalysis. *Journal of Environmental and Science Health*, Part A, 48, 1642–1648.

Morgada, M.E., Levy, I.K., Salomone, V., Farías, S.S., López, G. & Litter, M.I. (2009) Arsenic (V) removal with nanoparticulate zerovalent iron: effect of UV light and humic acids. *Catalysis Today*, 143, 261–268.

Muruganandham, M., Suri, R.P.S., Sillanpaa, M., Wu, J.J., Ahmmad, B., Balachandran, S. & Swaminathan, M. (2014) Recent developments in heterogeneous catalyzed environmental processes. *Journal of Nanoscience and Nanotechnology*, 14, 1898–1910.

Neppolian, B., Celik, E. & Choi, H. (2008) Photochemical oxidation of arsenic(III) to arsenic(V) using peroxydisulfate ions as an oxidizing agent. *Environmental Science & Technology*, 42, 6179–6184.

Nguyen, T.V., Vigneswaran, Ngo, H.H., Kandasamy, J. & Choi, H.C. (2008) Arsenic removal by photocatalysis hybrid system. *Separation and Purification Technology*, 61, 44–50.

Nishida, S. & Kimura, M. (1990) Kinetics of the oxidation reaction of arsenious acid by peroxydisulfate ion, induced by irradiation with visible light of aqueous solutions containing tris(2, 2′-bipyridine) ruthenium(II) ion. *Inorganica Chimica Acta*, 174, 231–235.

Ryu, J. & Choi, W. (2006) Photocatalytic oxidation of arsenite on TiO$_2$: understanding the controversial oxidation mechanism involving superoxides and the effect of alternative electron acceptors. *Environmental Science & Technology*, 38, 2928–2933.

Sharma, V.K., Dutta, P.K. & Ray, A.K. (2007) Review of kinetics of chemical and photochemical oxidation of arsenic(III) as influenced by pH. *Journal of Environmental Science and Health*, Part A, 42, 997–1004.

Thu, H.B., Karkmaz, M., Puzenat, E., Guillard, C. & Herrmann, J.M. (2005) From the fundamentals of photocatalysis to its applications in environment protection and in solar purification of water in arid countries. *Research on Chemical Intermediates*, 31, 449–461.

Tien, Q., Zhuang, J., Wang, J., Xie, L. & Liu, P. (2012) Novel photocatalyst, Bi$_2$Sn$_2$O$_7$, for photooxidation of As(III) under visible-light. *Applied Catalysis* A: *General*, 425–426, 74–78.

Wegelin, M., Gechter, D., Hug, S., Mahmud, A. & Motaleb, A. (1999) SORAS – a simple arsenic removal process. Eawag/SANDEC, Duebendorf, Switzerland. Available from: http://www.eawag.ch/forschung/sandec/index_EN [accessed July 2015].

Woods, R., Cutoff, I. & Meehan, E. (1963) Arsenic(IV) as an intermediate in the induced oxidation of arsenic(III) by the iron(II)-persulfate reaction and the photoreduction of iron(III). I. Absence of oxygen. *Journal of the American Chemical Society*, 85, 2385–2390.

Xu, T., Kamat, P.V. & O'Shea, K.E. (2005) Mechanistic evaluation of arsenite oxidation assisted photocatalysis. *Journal of Physical Chemistry A*, 109, 9070–9075.

Yang, H., Lin, W. & Rajeshwar, K. (1999) Homogeneous and heterogeneous photocatalytic reactions involving As(III) and As(V) species in aqueous media. *Journal of Photochemistry Photobiology* A: *Chemistry*, 123, 137–143.

Yao, S.H., Jia, Y.F. & Zhao, S.L. (2012) Photocatalytic oxidation and removal of arsenite by titanium dioxide supported on granular activated carbon. *Environmental Technology*, 33, 983–988.

Yoon, S.H., Lee, J., Oh, S. & Yang, J. (2008) Photochemical oxidation of As(III) by vacuum-UV lamp irradiation. *Water Research*, 42, 3455–3463.

Yoon, S.H., Lee, S., Kim, T.K., Lee, M. & Yu, S. (2011) Oxidation of methylated arsenic species by UV/S$_2$O$_8^{2-}$. *Chemical Engineering Journal*, 173, 290–295.

Yuksel, S., Rivas, B.L., Sánchez, J., Mansilla, H.D., Yáñez, J., Kochifas, P., Kabay, N. & Bryjak, M. (2014) Water-soluble polymer and photocatalysis for arsenic removal. *Journal of Applied Polymer Science*, 131 (19), DOI: 10.1002/app.40871.

Zheng, S., Cai, Y. & O'Shea, K.E. (2010) TiO$_2$ photocatalytic degradation of phenylarsonic acid. *Journal of Photochemistry and Photobiology* A: *Chemistry*, 210, 61–68.

CHAPTER 16

Arsenic removal by means of membrane techniques

Krystyna Konieczny & Michał Bodzek

16.1 INTRODUCTION

The main methods in drinking water production used for removal of arsenic (As) are (Ann., 2000; Hering *et al.*, 1997; Kartinen and Martin, 1995; Kołtuniewicz and Drioli, 2008; Vrijenhoek and Waypa, 2000):

- Precipitation processes, including coagulation/filtration, coagulation with microfiltration, advanced coagulation, lime softening and advanced lime softening;
- Adsorption processes, especially on an activated alumina;
- Ion-exchange processes, mainly on anionites;
- Membrane filtration, including reverse osmosis (RO) and nanofiltration (NF), ultra- and microfiltration, most common combined in integrated/hybrid systems;
- Electrodialysis (ED), electrodialysis reversal (EDR) and Donnan dialysis.

Two directions in the search for effective methods of As removal from water are clearly indicated. The first concerns the technology implemented on a large scale, which could be used in industrial solutions cleaning systems and water treatment stations. There, the research is focused on improving and perfecting known processes used during water treatment (coagulation, water softening, deferrization, adsorption on activated alumina, ion exchange and membrane technologies) for removing As. Most of the methods of this group are included in the *best available technology* (BAT), developed by the USEPA (Kociołek-Balawejder and Ociński, 2006). The second direction is the search for easy-to-use, low-cost and effective technologies, which could be installed in individual intakes of drinking water (adsorption methods based on iron compounds).

Arsenic compounds can be removed by membrane processes according to the screening mechanism, electrostatic repulsion and adsorption. If the molecules/particles containing As are larger than the pore sizes of the membrane, retention will be according to the screening mechanism. Several studies have shown that some of the membranes could remove As compounds of diameter one to two orders less than the size of the membrane pores, which indicated a different mechanism than the screening separation (Ann., 2000). The shape and chemical characteristics of As compounds play an important role in its retention. Membranes can also remove As compounds by electrostatic repulsion or adsorption on the membrane surface. It depends on the chemical properties of both membrane material, mainly its charge and hydrophobicity, but also on the impact of other compounds present in the water. A lot of research has been carried out to assess membrane processes for As removal. Table 16.1 shows the mean values of retention rates obtained during As removal using different technologies (Ann., 2000).

16.2 REMOVAL OF ARSENIC WITH REVERSE OSMOSIS

High pressure membrane techniques can be used for the removal of As from waters which are a source of drinking water to a level established in the relevant standards (Braundhuber and Amy, 1998; Kartinen and Martin, 1995; Oh *et al.*, 2000a; 2000b; Shih, 2005).

Table 16.1.   The maximum removal rates of As obtained for the different available technologies.

Technology	Maximum removal [%]
Coagulation/filtration	95
Advanced coagulation/filtration[1]	95
Coagulation/microfiltration	90
Lime softening (pH > 10.5)	90
Advanced lime softening (pH > 10.5)[1]	90
Ion exchange (sulfates $50\,mg\,L^{-1}$)	95
Activated alumina	95
Reverse osmosis	>95
Filtration on the "greensand" (20:1 iron:arsenic)	80
Household equipment-activated alumina	90
Household equipment-ion exchange	90

[1] Advanced processes assume that existing technology allows achieving 50% removal without modification.

The application of reverse osmosis (RO) for this purpose has been proven in a number of laboratory and pilot scale tests. In groundwater, As is present in 80 to 90% in the dissolved form, and thus RO is a suitable technology for this form of contamination removal. However, the RO effectiveness in removal of most of the solutes should be taken into consideration.

In the 1990s in the USA a number of studies on the removal of As in laboratory and pilot scale installations were carried out. The results obtained during those researches are summarized in Table 16.2 (Ann., 2000; Fox, 1989; Fox and Sorg, 1987). They indicate that the type of membrane and operating conditions of the process have a significant impact on the process efficiency, and their appropriate selection determines the success of the project. The reverse osmosis process reveals greater ability for As(V) retention than for As(III), and thus the assurance of oxidizing conditions is an important indication of the proper conduct of the process.

Short-term studies with the use of single spiral and flat modules Desal DK2540F RO membrane were also carried out. Two types of feed water i.e. natural lake water and simulated water, prepared on the basis of deionized water were studied (Ann., 2000). The results of these tests are shown in Table 16.3. The study revealed high retention of As(V), but low of As(III) at neutral pH. This confirmed the fact that the oxidizing conditions of the process improve process efficiency and the charge of the membrane surface as well as electrostatic repulsion play a significant role in the mechanism of As removal. Furthermore, it was found that the retention of As for flat membranes was comparable with the results obtained for the spiral modules.

Long-term (over 34 days) studies on As removal from the groundwater collected near the silver mine in Park City, Utah, USA performed with the use of Hydranautics RO membrane module ESPA2-4040 and Koch TFC-ULP 4 module have also been carried out (Ann., 2001; Kołtuniewicz and Drioli, 2008; Pawlak et al., 2006). The Koch module of a spiral configuration with a polyamide membrane characterized with cut-off of 100 Da. The system worked at a pressure of 1×10^3 kPa, the water recovery was 15%, and the capacity of $10.5\,L\,m^{-2}\,day^{-1}$ (25°C) during the first days of the operation was obtained. The Hydranautics module was equipped with hollow-fiber polyamide composite membrane with cut-off 300–500 Da. The installations worked under the same conditions as the Koch module with permeate flux of $6.5\,L\,m^{-2}\,day^{-1}$. Due to the high turbidity of raw water, filter bags with mesh size of $5\,\mu m$ were installed before the modules, and they required frequent replacement. The characteristics of both modules used for As removal are given in Table 16.4 (Pawlak et al., 2006). In the tests with the Koch installation, the mean concentration of total As in the raw water was $60\,\mu g\,L^{-1}$, and the system enabled the reduction of the contaminant concentration to the level of $0.9\,\mu g\,L^{-1}$ in permeate, which corresponded to the retention of 98.5%, while membranes in the Hydranautics unit reduced the concentration of

Table 16.2.   Summary of the results of studies on arsenic removal by RO performed in the United States in the 1990s.

Location	Type of installation	Operational parameters	As removal
Eugene, OR	Household filters	• 11.3–18.7 L day^{-1} • Water recovery 90% • Pressure 0.014–0.7 MPa	50%
Eugene, OR	Household filters	• 11.4–18.8 L day^{-1} • Water recovery 67% • Pressure 1.4 MPa	Below recommendation
Fairbanks, AL	Household filters	• Pressure < 0.689 MPa	50%
San Ysidro, NM	Pilot (hollow-fiber, cellulose acetate)	• Pater recovery 50% • pH correction to 6.3 • Antiscalant addition • Qater recovery 50%	93–99% 99%
San Ysidro, NM	Pilot (hollow-fiber, polyamide)	• pH correction to 6.3 • Antiscalant addition	
San Ysidro, NM	Household filters Pilot (FilmTec BW30)	• Water recovery 10–15% • 57 L day^{-1}	91%
Tarrytown, NY	Hydranautics (NCM1, Fluid Systems TCFL)	• Water recovery 10%	As in the permeate < recommendation
Tarrytown, NY	Household filters	No data	86%
Charlotte Harbor, FL	Household filters (several types of membranes)	3785 L day^{-1} Water recovery 10–60%	As(V) 96–99% As(III) 46–84%
Cincinnati, OH	Household filters	No data	As(III) 73%
Hudson, NH	Household filters	No data	40%

Table 16.3.   The removal of arsenic by RO on laboratory scale using single spiral module with DK2540F membranes by Desal Company (cut-off 180 Da).

Raw water	As speciation	pH	As retention rate [%]
Deionized water	V	6.8	96
Lake water	V	6.9	96
Deionized water	III	6.8	5
Lake water	III	6.8	5
Deionized water (flat membrane)	V		88

total As on average to 0.5 µg L^{-1} in the permeate (the retention rate equal to 99.2%) (Ann., 2001; Kociołek-Balawejder and Ociński, 2006; Kołtuniewicz and Drioli, 2008).

In the case of the Koch module, analysis of the raw water showed that the predominant form of arsenic was As(V), for which the concentration of 32 mg L^{-1} in raw water was reduced to ca. 0.8 mg L^{-1} in permeate. Similarly, As(III) was also removed by the membrane from 8 µg L^{-1} in raw water to 0.6 µg L^{-1} in permeate. The content of dissolved As compounds, corresponding to 70% of the total As, was reduced by 97% (Ann., 2001). It was found that Koch module membranes were not affected by the chemical fouling at the end of the testing period, and the application of the hydraulic cleaning was enough to recover initial operating parameters of the system.

In the case of the Hydranautics module the concentration of total As was reduced on average to 0.5 µg L^{-1} (retention rate 99.2%), and dissolved As content was decreased from ca. 42 µg L^{-1}

Table 16.4.   Characteristics of the RO systems.

The membrane type	Concentration of As in raw water [μg L^{-1}]	Concentration of As in permeate [μg L^{-1}]	As retention rate [%]
Module no. 1, (Koch TFC-ULP 4)	60.0	0.9	98.5
Module no. 2, (Hydronautics ESPA 2-4040)	60.0	0.5	99.2

Table 16.5.   The average value of the concentration and the removal of arsenic in pilot studies (Hydranautics).

Arsenic speciation	Raw water [μg L^{-1}]	Retentate [μg L^{-1}]	Permeate [μg L^{-1}]	Retention [%]
As total	65	62	0.5	99
As dissolved	42	47	0.8	98
As(V)	35	40	0.5	99
As(III)	7	6.2	0.5	84

to less than 0.8 μg L^{-1} in the permeate (treated water) (retention rate 98%) (Table 16.5) (Ann., 2001). Dissolved As accounted for 65% of the total As in the raw water. The average concentration of As(V) in raw water was 35 μg L^{-1} and it was reduced to ca. 0.5 μg L^{-1}, while the concentration of As(III) was decreased from 7 μg L^{-1} in the feed to 0.5 μg L^{-1} in permeate.

During the study with ES-10 polyamide membrane and polyvinyl alcohol NTR-729HF membrane, both by Nitto-Japan, the lower removal of As(III) than As(V) at the pH range 3–10, was observed. The removal of As(V) obtained for ES-10 membrane was equal to 95% in the whole pH range, while for NTR-729HF the retention rate was 80% for pH=3 and 95% for pH range 5–10. The removal rate of As(III) was equal to 75% for ES-10 membrane at acidic conditions and it increased to 90% at pH level ca.10, whereas in the case of NTR-729HF membrane the observed removal rate did not exceed 20% (Gorenfl et al., 2002). Nevertheless, a number of other laboratory and pilot researches on As removal using reverse osmosis membranes have also been performed (Braundhuber and Amy, 1998; Kang et al., 2000; Kartinen and Martin, 1995; Ning, 2002; Pawlak et al., 2006).

In the natural environment, As is usually accompanied by antimony. Ning (2002) and Kang et al. (2000) studied the effect of pH on the separation of arsenic and antimony using reverse osmosis. They stated that the removal of As(V) and Sb(V) is much greater than the As(III) and Sb(III) at overall range of applied pH (i.e. pH=3–10). Removal of As compounds strongly depended on the pH of the solution, more in the case of As(III), than of antimony. The removal of Sb compounds does not depend on pH, since its oxidation from Sb(III) to Sb(V) takes place in a very narrow range of pH. It can therefore be concluded that raw water pH control is an essential condition for the effective removal of As compounds and, to a lesser extent, antimony compounds. In addition, it has been shown that antimony removal during the production of drinking water using RO membrane is more efficient than of As compounds, regardless of the pH. In contrast, the removal of As(III) compounds by means of RO method at pH characteristic for natural waters is lower than As(V) (Ning, 2002). Higher retention of As(V) arises from the fact that As(V) exists in the solution in the ionic form, while As(III) appears mainly in molecular form. The results of the work (Ann., 2001) showed that the antimony concentration in permeate was lower than 3.0 μg L^{-1} in all samples. The highest retention rate was 67% calculated basing on the maximum element concentration in raw water equal to 9.2 μg L^{-1}.

The content of dissolved organic carbon (DOC) has a great influence on As removal. Series of pilot reverse osmosis tests using groundwater with a low and a high content of DOC and

Table 16.6. The removal of arsenic by a pilot scale RO system.

Membrane type	Water type	As speciation	As retention [%]
TFCL-HR	Groundwater (high DOC)	As total	>90%
	Groundwater (low DOC)	As total	>80%
RO 1	Surface water	V	>95%
		III	60%
RO 2	Surface water	V	>95%
		III	75%
RO 3	Surface water	V	>95%
		III	68%
RO 4	Surface water	V	>95%
		III	85%

surface waters containing As as a feed, have been carried out (Table 16.6) (Hering *et al.*, 1997). The slightly higher retention of As was found at higher NOM content in water (90%) than in one of lower NOM concentration (80%). The results determined for four membranes operated on surface water containing As also showed significant contaminant retention. For all membranes the removal of As(V) exceeded 95%, whereas the retention of As(III) was only 74%.

The removal of arsenate (As(V)) and arsenite (As(III)) from water using reverse osmosis (RO) SWHR (seawater high rejection) and BW-30 (brackish water) membranes by Filmtec was investigated by Akin *et al.* (2011). The effect of pH and concentration of the feed water and operating pressure on the rejection of both As species was examined. The experimental results indicated that rejection of both As(V) and As(III) depended on the investigated parameters in the following order: the membrane type, pH of the feed water and operating pressure. For both membranes, As(V) rejection was maximal at pH above 4.0, while As(III) could be effectively removed from water at pH above 9.1. The rejection of both As species increased with the operating pressure increase. The lowest permeate concentration of As(III) and As(V) was obtained for SWHR membrane. Finally, a natural (ground)water sample containing $50 \, \mu g \, L^{-1}$ of As(V) and $12 \, \mu g \, L^{-1}$ of As(III) was treated using RO SWHR membrane. The obtained concentration of total As in permeate fulfilled WHO and EU standards and general permeate quality was evaluated considering the chemical composition of the groundwater sample.

With the rapid growth of membrane desalination processes, the removal of ionic contaminants, such as arsenic or boron, is of main interest. Those two contaminants often occur at the same time in groundwater resources. Compared to other contaminants boron and arsenic are the most difficult components to remove in desalination membrane processes. Teychene *et al.* (2013) in their study compared the rejection properties of several reverse osmosis membranes toward arsenic and boron. "Seawater" membranes were found to be more efficient than "brackish water" membranes. Rejection of As(III) and boron increased with transmembrane pressure (TMP) and pH. As(III) and boron permeabilities for each tested membrane were calculated at the operating conditions investigated. All membranes showed good rejection efficiency for both ions. Indeed, whatever the membrane or the operating conditions tested, the permeate quality always followed the WHO guidelines. Even in the worst case (lowest rejection, low TMP (2.4 MPa) and low pH (7.6)), the arsenic and boron concentrations in permeate were equal to $8.4 \, \mu g \, L^{-1}$ and $0.9 \, mg \, L^{-1}$, respectively. These results brought new membrane parameters regarding boron and As rejection.

Summarizing, it can be concluded that the reverse osmosis process enables achievement of a concentration of As(V) less than $0.002 \, mg \, L^{-1}$ in purified water. Retention of As(III) is not so high; however, its oxidation to As(V) can be easily performed in proper environment conditions.

16.3 REMOVAL OF ARSENIC WITH NANOFILTRATION

One of the most important properties of nanofiltration membranes (NF) is a combination of high retention of polyvalent ions with moderate retention of single-valence ones. This applies in particular to the composite NF membranes with a negative surface charge, which give higher selectivity and the possibility of obtaining higher recovery of water at a much lower operating pressure than RO membranes. NF is a promising method for a wide range of applications in water treatment technologies, including the removal of As to the standard level of $10\,\mu g\,L^{-1}$ (Ann., 2000; Baumann, 2006; Childress and Elimelech, 2000; Nguyen *et al.*, 2009; Oh *et al.*, 2000a,b; 2004; Peeters *et al.*, 1998; Schaep *et al.*, 1998; Seidel *et al.*, 2001; Urase *et al.*, 1998; Vrijenhoek and Waypa, 2000; Waypa *et al.*, 1997). The huge advantage is that As remaining after initial treatment of water prior to the introduction to RO/NF modules occurs commonly as a divalent anion ($HAsO_4^{2-}$), which should be effectively rejected by negatively charged NF membranes. The retention of compounds in nanofiltration depends on the nature of the contamination, i.e. the electric charge and the molecular weight, the cut-off and the nature of the NF membrane surface (electric charge and hydrophilic-hydrophobic properties). All of these properties require careful selection and optimization.

Nanofiltration membranes remove a substantial amount of soluble As compounds, including both As(V) and As(III), from natural waters. Therefore, NF is an appropriate process to remove As from groundwater, which contains up to 90% of the dissolved As (Ann., 2000). Considering As retention in the case of groundwater treatment with the NF method, one must notice that anions such as chlorides, nitrates (V) and sulfates (VI) occur at the level of the $mg\,L^{-1}$, whereas the concentration of As compounds is practically up to dozens of $\mu g\,L^{-1}$. Hence, in such a mixed solution, the retention of As compounds in nanofiltration should be interpreted in reference to other anions (Kołtuniewicz and Drioli, 2008). The application of nanofiltration to surface water treatment is possible after a preliminary removal of particles and dissolved components that cause fouling. The small size of membrane pores, however, makes NF membrane more susceptible to fouling than low pressure membrane filtration.

The extensive research conducted on the nanofiltration process shows that it is effective in As removal, but the effectiveness depends to a considerable extent on the operating parameters, membrane properties and speciation of As (Shih, 2005). Oh *et al.* (2004) presented results of the investigations of As(V), dimethylarsenic (V) acid (DMA), and chlorides, nitrates and sulfates removal, using low-pressure NF membranes. The results suggest that the interaction between the anions and the membrane are determined primarily by the surface charge density associated with ionic substance. Thus, the most important mechanism of the As ions removal is the interaction between the membrane and compounds containing As ions.

Studies on the removal of As(V) by the negatively charged composite membrane NE-90 (cut-off 220 Da) (Woongjin Chemical, South Korea) from synthetic water (pH = 7; $1\,mmol\,L^{-1}\,NaHCO_3$; $10\,mmol\,L^{-1}\,NaCl$) showed, that the retention rate of As increased in the range of 90% to 96% with an increase in the concentration of As(V) from 20 to $100\,\mu g\,L^{-1}$, but for As(III) it decreased from 44 to 41% when the initial concentration was increased from 20 to $100\,\mu g\,L^{-1}$ (Nguyen *et al.*, 2009). A slight influence of pressure in the range of 138 to 552 kPa on the As removal rate was also observed. At the initial concentration of As(V) and As(III) of $50\,\mu g\,L^{-1}$, the removal of As(V) increased for the all range of pH, but for As(III) the increase was observed significantly only in the parameter range from 8 to 10.

Laboratory scale research on As removal from simulated water and natural lake water using single spiral modules and flat NF membranes (by Osmonics) with a negative surface charge (Table 16.7) (Ann., 2000) was carried out. The removal of As(III) was low (12%), while for As(V) the efficiency of 89 and 85% respectively for the lake water and simulated water was obtained. The application of flat membranes resulted in the removal of As(V) at the level of 90%.

Seidel *et al.* (2001) used open (porous) NF membrane to investigate the differences between the retention of As(V) and As(III). The As(III) retention in this case was much less than As(V), while the latter depends on the pH and varied in the range of 60–90%, while the retention of As(III)

Table 16.7. The removal of As(V) and As(III) using Osmonics NF membranes.

Membrane type	Cut-off [Da]	Membrane charge	Water type	Valence of arsenic	pH	Retention rate [%]
A single module						
NF 45-2540	300	(−)	Simulated	V	6.7	85
NF 45-2540	300	(−)	Lake	V	6.9	89
NF 45-2540	300	(−)	Simulated	III	6.9	12
Flat membranes						
NF 45-2540	300	(−)	Simulated	V	No data	90

was always below 30%. Pilot studies carried out by Brandhuber and Amy (1998) confirmed the differences between retention of As(V) and As(III), with the exception of the compact NF membranes.

Urase *et al.* (1998) carried out systematic researches on the possibility of application of nanofiltration to the removal of As in groundwater treatment, using aromatic polyamide negatively charged NF membrane ES-10 by Nitto Denko Co. Ltd. Despite the dependence of the retention of As(V), As(III) and dimethylarsenic (V) acid (DMA) on the pH, it was found to be high regardless of the pH. As(V) retention increased with pH increase from 87% at pH $= 3$ to 93% at pH $= 10$. As(V), at such pHs, are either single-valence or bivalent anions. Retention rate of As(III) increased with pH change from 50% at pH $= 3$ to 89% at pH $= 10$. One must know that at pH 10, most of the As(III) occurs in the form of a monovalent anion, while at pH 3, 5 and 7 it occurs in the form of non-dissociated H_3AsO_3 acid according to its pK_a. Dimethylarsenic (V) acid, whose pK_a is 6.2, changes its form from an inert molecule to monovalent anion with pH increase. Unlike As(III), retention of DMA exceeded 98% of the overall range of pH.

Vrijenhoek and Waypa (2000) studied the removal of As from drinking water using the open composite NF membrane NF-45 (FilmTec), made from fully aromatic polyamide. The effect of pH, concentration of As in the raw water and the presence of electrolytes in the matrix on As retention were investigated. In general, As retention was favored by the increase of pH and As concentration in the raw water. It is accepted that the separation of As compounds is a combined mechanism involving steric (size) exclusion, preferable transport of more mobile ions and electrostatic repulsion. During the study it was observed that nonionic As(III) compounds, namely H_3AsO_3, were not removed in a high rate, regardless of the pH and the concentration of As(III) in raw water, which suggested that the separation mechanism of NF-45 membrane was solely based on the steric exclusion phenomenon. This was probably due to the neutral character of the membrane surface and the relatively low molecular weight of H_3AsO_3 (140 Da) compared to the NF-45 membrane cut-off (180–340 Da). Nevertheless, the dominant form of As that appears in natural waters is As(V). At pH 8.1 to 8.2, and in the presence of 10^{-2} mole NaCl L^{-1}, As(V) was removed by NF-45 membrane from simulated water containing 10, 30, 100, and 316 μg L^{-1} of the contaminant in 60–90%. Concentrations measured in the permeate were 4, 6, 10 and 25 μg L^{-1}, which mostly corresponded to the requirements on the quality of drinking water.

A series of pilot studies supervised by AWWA (USA) on As removal with the use of groundwater and simulated surface waters spiked with As at varying contents of DOC (1 and 11 mg L^{-1} and more) (Table 16.8) was also conducted (Ann., 2000). The research with groundwater was carried out using single spiral module and showed that in the case of low DOC level the removal of As was only 60%, whereas at high level DOC it was 80%. Such a behavior was probably caused by the adsorption of NOM on the surface of the membrane, resulting in the change of the electrostatic repulsion between the membrane and As ions. The retention of As(V) obtained during tests conducted with simulated surface water was above 95%, while for As(III) it was much lower (about 40%) for all three tested membranes. The authors found that these results confirmed the influence of the steric exclusion mechanism on the removal of As(III). The dimensions of the

Table 16.8.   The removal of As(V) and As(III) using NF membrane carried out on pilot-scale installation.

Membrane type	Cut-off [Da]	Membrane charge	Water type	Valence of arsenic	Retention ratio [%]
Single spiral module					
Set	400	(−)	Groundwater with high DOC	As total	80
		(−)	Groundwater with low DOC	As total	60
NF 1	No data	No data	Simulated surface water	V	>95
				III	52
NF 2	No data	No data	Simulated surface water	V	>95
				III	20
NF 3	No data	No data	Simulated surface water	V	>95
				III	30
A set of modules					
Accumen	400	(−)	Groundwater with high DOC	V	77 (initial) 3–16 (end)

As(III) molecules are very small, and they can easily diffuse even through the very small pores of the NF membrane and are not repulsed by the negatively charged membrane surface, as occurs in the case of As(V). Thus, a better solution for the removal As(III) is a combination of NF preceded with the oxidation of As(III) to As(V).

The results presented in the last line of Table 16.8 show the As retention rate decrease over time. At the beginning of the test retention rate was about 77% and within 60 days it decreased to 11%, and over the next 80 days remained constant at the level from 3 to 16%. This is very surprising because the membrane exhibits high As retention in a single module configuration. Analysis of samples taken from a set of modules indicated the change of the speciation of As from As(V) to As(III) took place inside the module. Since As(III) has a lower retention than As(V), the overall retention rate of As decreased. The decrease of retention in time suggests that negatively charged membrane is not able to maintain a high retention of As(V) over a long period of time without maintenance of As in the form of As(V) ion. Knowing that As retention rate depends on its oxidation state, the maintenance of oxidizing conditions should therefore be a necessity for improving the efficiency of the process. On the other hand, the addition of oxidants may, however, adversely affect the life-time of the membranes and shorten the duration of their use.

Xie et al. (2013) also investigated the influence of dissolved organic matter (DOM) on As removal by means of the nanofiltration process, performing a series of laboratory bench-level experiments. The introduction of humic acid (HA), used as substitute for DOM, to the feed solution increased the As removal due to the formation of humic/arsenic complexes. In the presence of high concentration of HA (10 mg total organic carbon (TOC) L^{-1}), the removal efficiency of As was almost 100% and was higher compared to the results (80%) obtained at low concentration of HA (2 mg TOC L^{-1}). The membrane flux during the separation process at the various concentrations of TOC decreased to 80% of the initial value after 400 min of the process carried out with the same cross-flow velocity (3.5 cm s^{-1}).

Chang and Liu (2012) performed a combined treatment of Ca-precipitation and nanofiltration to remove arsenate from water. The selected nanofiltration membrane was charged amphoterically. The arsenate and calcium removal efficiencies were the lowest at the isoelectric point of the nanofiltration membrane, attributed to the loosest steric hindrance and the weakest electrostatic repulsion. Above the isoelectric point, arsenate precipitated with calcium ions, forming the low solubility compound calcium arsenate, while steric hindrance was the main mechanism of arsenate removal. In contrast, below the isoelectric point, i.e. with the positively charged nanofiltration membranes, calcium ions were rejected by electrostatic repulsion. The high electrostatic shielding of calcium ion kept arsenate ions away from reaching the NF membrane surface. Either high feed arsenate concentration or high calcium oxide dose improved the removal rate of arsenate and its highest value was obtained at feed arsenate concentration 200 mg L^{-1}. The optimal transmembrane pressure was in a range of 0.5–0.7 MPa to restrict the formation of fouling cake on the nanofiltration membrane surface.

It was already shown that the mechanism of As removal in the nanofiltration process was very important. However, there is no research on the As removal mechanism using "compact" NF membrane. Even though Vrijenhoek and Waypa (2000) studied the mechanisms of As removal i.e. steric exclusion, exclusion associated with an electric charge (Donnan phenomenon) as well as preferential permeating of more mobile ions through the "compact" composite membrane NF-45, these studies did not explain in a systematic way the As removal mechanism. Particularly, they did not regard its speciation, as well as the presence of accompanying anions, such as SO$_4^{2-}$ and Cl$^-$.

It is known that As forms oxyanions, the structure of which depends on the pH and which are characterized by different mobility in oxidation and reduction conditions (Nguyen *et al.*, 2009). In oxidation conditions, at low pH (below 6.9), As(V) arsenic forms a monovalent ion (H$_2$AsO$_4^-$), while at higher pH, bivalent ion (HAsO$_4^{2-}$), which reveal different hydration and thereby differ in the diameter. In reduction conditions at pH less than 9.2, arsenic acid (III) (H$_3$AsO$_3$) forms electrically neutral particles (Nguyen *et al.*, 2009). All these relations have a great effect on As retention rate and separation mechanism.

For negatively charged NF membranes the Donnan equilibrium is significantly important in controlling speciation of As(V) anions in the membrane-solution interphase (Bowen *et al.*, 1997; Yeom *et al.*, 2002). When the charged membrane is in contact with the electrolyte solution, the concentration partition of co-ions and counter-ions occurs at the membrane solution interfaces on both the feed and permeate sides. At these boundaries potential differences, called Donnan potentials (ψ_D), are generated to maintain electrochemical equilibrium. The Donnan potential at the membrane-feed interface causes the rejection of co-ions whereas counter-ions are attracted. The Donnan potential is calculated as the difference between the electrical potential of the membrane and the bulk at the interface (Nguyen *et al.*, 2009; Peeters *et al.*, 1998):

$$\psi_D = \psi_m - \psi = \frac{RT}{z_i F} \ln \frac{C_i}{C_i^m} \qquad (16.1)$$

where ψ_D and ψ_m are the electric potentials of the membrane and bulk, respectively; C_i is the concentration of charged species "i" in the bulk, while C_i^m is the concentration in the membrane, and z_i is the valence of the species. As a simple case, for a single salt solution, consisting of B (co-ion) and A (counter-ion), Equation (16.1) is rewritten as follows:

$$\psi_D = \psi_m - \psi = \frac{RT}{z_A F} \ln \frac{C_A}{C_A^m} = \frac{RT}{z_B F} \ln \frac{C_B}{C_B^m} \qquad (16.2)$$

The partitioning of co-ion B at the membrane interface, with respect to the effective membrane charge density (X), combined with the electrically neutral conditions in the membrane and bulk, can be derived from Equation (16.2) (Peeters *et al.*, 1998).

$$\frac{C_B^m}{C_B} = \left(\frac{|z_B|}{X|z_B|C_B^m + X} \right)^{|z_B|/|z_A|} \qquad (16.3)$$

From Equation (16.3), it is inferred that the distribution of the co-ion B at the interfaces will depend on the salt concentration, effective membrane charge density and valence of the constituent ions. The concentration of the co-ion B inside the membrane increases with increasing bulk concentration of B or decreasing effective membrane charge. Consequently, lower salt-rejection of the co-ion is achieved. In addition, an increase or decrease in the counter-ion valence or co-ion valence, respectively, can result in an increase in the co-ion concentration in the membrane (Peeters *et al.*, 1998).

The removal of non-ionic species, such as H_3AsO_3, cannot be affected by Donnan exclusion and electrostatic repulsion. Therefore, nonelectric models must be considered to describe the transport of arsenites. In principle, the extended Nernst-Planck equation is universally applicable for the transport of all ionic and nonionic solutes. However, for simplification in the modeling of non-ionic transport, Spiegler and Kedem (Bodzek and Konieczny, 2005) proposed a membrane transport model describing the differential form of the solute flux, J_s:

$$J_s = P\Delta y \left(\frac{dc}{dy}\right) + (1 - \sigma)J_v c_s \qquad (16.4)$$

where P is the solute permeability, Δy the membrane thickness, c the solute concentration, σ the reflection coefficient and J_v the permeate flux.

Vrijenhoek and Waypa (2000) proposed mechanisms governing the removal of As on the basis of:

- Size exclusion;
- Charge exclusion;
- Preferential passage of more mobile ions.

Since As is removed from aqueous solutions containing various coexisting solutes, especially ionic species such as Cl^-, SO_4^{2-} and Na^+, mutual interactions between As and other ions must be of great concern. It is necessary to present the preferential passage phenomenon, in that the rejection of an ion is improved in the presence of a more mobile co-ion, but diminished in the presence of a less mobile co-ion (Childress and Elimelech, 2000; Vrijenhoek and Waypa, 2000). Nguyen *et al.* (2009) show the results of the efficiency of As removal by compact negatively charged NF composite membrane (NE-90) and presented a systematic attempt to explain the mechanisms governing the retention of As. The high individual rejections of Na_2SO_4 and Na_2HAsO_4 and the moderate individual rejections of $NaCl$ and NaH_2AsO_4 resulted from the steric and electric hindrance nature of ion transport through a compact and negatively charged membrane. However, the observed order of salt rejection ($CaCl_2 < NaCl < NaH_2AsO_4 < Na_2HAsO_4 < Na_2SO_4$) indicated the predominance of Donnan exclusion over the steric effect in governing solute removal. In addition, the removal of all individual salts decreased with the increase of the initial concentration of salts, which is in agreement with the Donnan equilibrium theory. The removal of As from an aqueous environment containing coexisting ions such as Cl^-, Na^+, SO_4^{2-} and HCO_3^-, was governed by steric exclusion and the preferential passage of more mobile ions, combined with Donnan exclusion. Separation of both monovalent and bivalent arsenate species was enhanced in the presence of more mobile ions, such as Cl^- and HCO_3^-, while the separation of monovalent arsenate diminished in the presence of a less mobile ion, SO_4^{2-}. Additionally, the retention of neutral arsenite, H_3AsO_3, is well described by the Spiegler and Kedem model, a non-electric model.

16.4   COMPARISON AND THE SUMMARY OF REVERSE OSMOSIS AND NANOFILTRATION APPLICATION IN ARSENIC REMOVAL

Numerous studies on As removal using NF/RO processes (Table 16.9) (Chang et al., 2014) have been presented. Generally, these studies showed that most of NF/RO membranes revealed a higher rejection of As(V) than As(III), especially in the circa-neutral pH range. As(III) is believed to be more mobile and toxic than As(V). According to the As chemistry (pK for H_3AsO_3 are 9.17 and 13.5, and for H_3AsO_3 are 2.26, 6.77 and 11.5, respectively) (Kang et al., 2000; Kartinen and Martin, 1995), the predominant species for As(III) and As(V) at pH 7 are uncharged and anionic forms, respectively. Anionic species are normally better rejected (Bandini and Vezzani, 2003; Bowen and Welfoot, 2002; Wang et al., 1997). Unfortunately, oxidation is not an easy way to enhance the efficiency of As removal, since oxidant could easily damage the applied NF/RO membranes.

A brief literature review of the previous studies showed that the rejection of As(III) by NF/RO differed largely, ranging from 5% only to more than 99% (Table 16.9). Normally, a RO membrane had a higher rejection than a NF membrane (Ann., 2000; Oh *et al.*, 2000a; 2000b; Sato *et al.*, 2002; Shih, 2005). This was apparently caused by the stronger steric effect of the former than that of the latter, as was generally also the case when comparing the various NF/RO membranes (Kang *et al.*, 2000; Oh *et al.*, 2004; Sato *et al.*, 2002; Teychene *et al.*, 2013). This clearly demonstrates the critical importance of the selected membrane in determining the As(III) rejection. As listed in Table 16.9, the NF/RO membranes that had been studied include that by Dow FilmTec, Nitto-Denko, Osmonics Desal, Fluid Systems, Hydranautics and other.

It is also generally accepted that the operational conditions play an important role in the As(III) rejection process using NF/RO membranes. The most notable operational condition is the feed pH, and almost always the removal efficiency of As(III) was higher at increased feed pH (Akin *et al.*, 2011; Kang *et al.*, 2000; Nguyen *et al.*, 2009; Teychene *et al.*, 2013; Urase *et al.*, 1998; Xia *et al.*, 2007). The feed pH controls both the dissociation extent of As(III) and the surface charge density of the NF/RO membranes.

The other operational conditions include the applied filtration pressure, the feed ionic strength, the feed As(III) concentration, the co-existing organic matter concentration and other factors (Ann., 2000; Akin *et al.*, 2011; Oh *et al.*, 2000a; 2000b; Shih, 2005; Vrijenhoek and Waypa, 2000; Xia *et al.*, 2007). However, the effect of these operational conditions on As(III) rejection significantly differs, depending on the applied NF/RO membrane type. This may be explained by the big difference in conditions of the As(III) rejection used in the various studies performed using the same NF/RO membranes (Table 16.9).

The removal of As from water and simulated surface water by reverse osmosis and nanofiltration was compared by Waypa *et al.* (1997). RO membrane ES-10 and compact composite NF membrane NF-70 (Dow-FilmTec) effectively removed both As(V) and As(III) from water in a wide range of operating parameters. The achieved retention was 99% for both types of membranes. Removal efficiency was approximately up to 99 and 55% for As(V) and As(III), respectively, and the performance slightly increased with the increase of applied pressure.

Chang *et al.* (2014) employed nanofiltration (NF) membrane and a low-pressure reverse osmosis (LPRO) membrane, both with an aromatic polyamide selective layer by the same manufacturer, for the comparison of their performance in terms of As(III) rejection and filtration flux at various operational conditions. In addition to the smaller membrane pore size, the LPRO membrane possesses much more dissociable functional groups than the NF membrane. When the feed pH was below the pK_1 value (9.22) of H_3AsO_3, for which steric hindrance is the only rejection mechanism, the removal efficiencies obtained for NF and LPRO were about 10% and 65%, respectively. When the feed pH was higher i.e. when electrostatic effect began, the removal efficiencies could reach 40 and 90% for NF and LPRO, respectively. The rejection performance of LPRO was marginally affected by the feed As(III) concentration or ionic strength, although ionic strength had a strong effect on the filtration flux. In contrast, feed As(III) concentration and ionic strength had little effect on the filtration flux, but great influence on the As(III) rejection performance of NF. The filtration flux was enhanced with the increase of transmembrane pressure for either NF or LPRO.

16.5 LOW-PRESSURE DRIVEN MEMBRANE PROCESSES

Microfiltration (MF) and ultrafiltration (UF) are membrane processes in which the separation mechanism is based on physical sieving. They are best used in the removal of suspended solids and the reduction of turbidity, as well as a pretreatment to desalination technologies such as nanofiltration or reverse osmosis. MF has the largest pore size (0.1–3 μm) of the wide variety among all membrane filtration systems, while UF membranes pore sizes range from 0.01 to 0.1 μm. In terms of pore size, MF fills in the gap between ultrafiltration and granular media filtration. Both processes require low transmembrane pressure to operate: up to 1.0 MPa for UF and 0.01–0.30 MPa for MF.

Table 16.9.   A brief summary of the discussed studies on As(III) rejection with the use of various NF/RO membranes.

Membrane	NaCl retention	As(III) retention
Fluid Systems TFCL	98.5% ($P = 1.52$ MPa)	N98% ($P = 1.72$ MPa, pH $= 8$)
Fluid Systems TFC-HR	99.5% ($P = 1.55$ MPa)	N98% ($P = 1.38$ MPa, pH $= 8$)
FilmTec NF-70	70% ($P = 0.5$ MPa)	N97% ($P = 0.56$ MPa, pH $= 8$)
Desalination Systems CE	95% ($P = 2.76$ MPa)	N97% ($P = 2.76$ MPa, pH $= 5.7$)
FilmTec NF-70 4040B	(MWCO $= 300$)	$\sim$50%
Desal HL-4040 F1550	(MWCO $= 300$)	$\sim$20%
Hydranautics 4040-UHA-ESNA	(MWCO $= 300$)	$\sim$30%
Fluid Systems TFC 4921	N/A	$\sim$60%
Fluid Systems TFC 4820-ULPT	N/A	$\sim$75%
Desal AG 4040	N/A	$\sim$70%
Hydranautics 4040 LSA-CPA2	N/A	$\sim$85%
Nitto-Denko ES-10	98.7% ($P = 0.24$ MPa)	50–60% ($P = 0.24$ MPa, pH $= 3.7$)
FilmTec NF-45-2540	(MWCO $= 300$)	12% (pH $= 6.9$)
Desal DK2540F	(MWCO $= 180$)	5% (pH $= 6.8$)
Nitto-Denko ES-10	>97% ($P = 0.75$ MPa)	>75% ($P = 0.75$ MPa, pH < 8)
Nitto-Denko NTR-729Hf	89% ($P = 0.75$ MPa)	20% ($P = 0.75$ MPa, pH < 8)
Nitto-Denko ES-10	84.3% ($P = 0.6$ MPa)	55% ($P = 0.6$ MPa, pH $= 8$)
Toyobo HR3155	99.9%	95% ($P = 4$ MPa, pH $= 8$)
FilmTec NF-45	N/A	<20% (pH $= 8.1$–8.2)
Osmonics BQ01	50% ($J = 8\,\mu\mathrm{m\,s^{-1}}$)	5–28% ($J = 8\,\mu\mathrm{m\,s^{-1}}$, pH $= 8.1$–8.2)
Nitto-Denko ES-10	99.6% ($P = 1.5$ MPa)	>75% ($P = 0.3$–1.1 MPa, pH $= 6.8$)
Nitto-Denko NTR-7250	93.0% ($P = 1.5$ MPa)	<22% ($P = 0.3$–1.1 MPa, pH $= 6.8$)
Nitto-Denko NTR-729Hf	70.0% ($P = 1.5$ MPa)	<22% ($P = 0.3$–1.1 MPa, pH $= 6.8$)
Nitto-Denko ES-10	N/A	$\sim$50% ($P = 0.25$ MPa, pH $= 7.8$)
Nitto-Denko NTR7250	N/A	$\sim$30% ($P = 0.25$ MPa, pH $= 7.8$)
Nitto-Denko NTR729HF	N/A	$\sim$50% (P $= 0.25$ MPa, pH $= 7.8$)
Toray NF	N/A	5%
FilmTec XLE-2521	98.2% ($P \approx 1$ MPa)	71% ($P \approx 1$ MPa, pH $= 7.2$)
FilmTec TW30-2521	99% ($P \approx 1$ MPa)	89% ($P \approx 1$ MPa, pH $= 7.2$)
FilmTec SW30-2521	99% ($P \approx 1$ MPa)	97% ($P \approx 1$ MPa, pH $= 7.2$)
Woongjin NE 90	87% ($P \approx 0.552$ MPa)	41–44% ($P = 0.14$–0.55 MPa, pH < 8)
FilmTec SWHR	99.6%	$\sim$75% (P $= 2.0$ MPa, pH $= 3.1$)
FilmTec BW-30	99.5%	$\sim$60% (P $= 2.0$ MPa, pH $= 3.1$)
FilmTec NF-270	(MWCO $= 600$)	<11% ($P = 3.5$ MPa, pH $= 5$)
FilmTec SW30HR	99.5% ($P = 4$ MPa)	>99% (pH $= 7.6$)
Hydranautics SCW5	99.3% ($P = 4$ MPa)	>99% (pH $= 7.6$)
FilmTec BW-30LE	99.9% ($P = 4$ MPa)	>99% (pH $= 7.6$)
Hydranautics ESPAB	99.3% ($P = 4$ MPa)	>99% (pH $= 7.6$)
Hydranautics ESPA2	99.1% ($P = 4$ MPa)	>99% (pH $= 7.6$)

Low pressure driven membrane processes (MF and UF) are very effective in retention of the As in the form of particles (even without using the pretreatment via coagulation), but they are not suitable for removal of soluble forms of this element (Baumann, 2006). Although the size of pollutants is an important factor affecting the retention, studies show sometimes significant retention of As compounds, despite the dimension of an ion (molecule) being one to two orders less than the pore diameter of MF or UF membrane. Such observations indicate a different mechanism of separation than physical sieving. Chemical characteristics, particularly electrostatic charge and hydrophobicity or hydrophilicity of both membrane material and components present in the indicates the electrostatic water, play an important role in the mechanism of separation (Shih, 2005). For example, membranes with a negative charge are more effective in the removal of As(V) anions than the membrane without charge, which indicates the electrostatic repulsion of ions with the same charge as the membrane (Braundhuber and Amy, 1998).

16.5.1 *Direct microfiltration and ultrafiltration*

The effectiveness of the use of microfiltration as a direct method of As removal is largely dependent on the size distribution of particles containing As in raw water, and, above all, the ratio of MF membrane pore size to the size of the particles. The size of the pores of MF membranes is too large to remove significantly dissolved or colloidal As. The rate of removal of As in such a case, will depend primarily on the initial concentration of this element and the percentage of its content in the form of particles, referring to the sieving mechanism of As removal. In groundwater, arsenic typically occurs in the form of fine suspensions in quantities of less than 10%, while in surface water its content can change from 0 up to 70% (Amy *et al.*, 1998). Arsenic particulate fraction amount is not connected with a specific type of water, which means there is no dependency of the size distribution of particles containing As size distribution on water turbidity or organic matter content (Amy *et al.*, 1998). Brandhuber and Amy (1998) studied the size distribution of particles containing As in the raw water designated for the production of drinking water, specifying the amount of dissolved, colloidal and particulate As. A lower proportion of colloidal and solids particles containing As (10–20%) in groundwater, compared to surface water, was found. McNeill and Edwards (1995) studied the speciation of As in many natural water sources and found that at least 15% of total As occurs in the form of particulate matter, which may be retained by filters with sieve mesh of the size 0.45 μm. Chen *et al.* (1999) found that more than half of the total As was removed through filters with sieve mesh of about 0.45 μm size used in installations for ground and surface water treatment. In order to increase the effectiveness of the MF process in removing As from waters with a low content of its particulate form, the MF can be combined with coagulation.

The ultrafiltration process is generally intended for the removal of colloids and impurities in the form of fine particles (Bodzek and Konieczny, 2005). As in the case of MF, UF as an independent process is not the appropriate technique for the removal of soluble As from groundwater. However, for surface waters with high concentration of colloids and particles, UF may be considered as a standalone unit process. Recent studies found (Ann., 2000) that electrostatic repulsion, increasing As retention regardless of screening mechanism, may play an important role in As retention with the UF method. The necessary condition is a negative surface charge on the membrane surface.

AWWA (Amy *et al.*, 1998) conducted pilot tests in the USA using two UF modules one with low cut-off by Osmonics (Desal GM and FV), and a second with flat sheet GM, FV and PM membranes by Desal to treat deionized water spiked with As salts. Such a simulated water did not contain colloidal As nor its particulate form. For negatively charged GM2540F membrane, retention of As(V) was high at neutral pH, but very low in acidic conditions, which is shown in Table 16.10 (Amy *et al.*, 1998). On the other hand, for the same membrane the retention of As(III) was high in the alkaline environment and negligible in the neutral one. The FV2540F charge-free membrane showed a low retention of both As(V) and As(III) at neutral pH. Nevertheless, high retention was observed in the case of some investigated membranes, even though their cut-off was about two orders of magnitude greater than the molecular weight of As compounds (McNeill and Edwards, 1995). This was due to electrostatic interaction between the negatively charged membrane surface and As ions. Interaction forces depended also on pH, because As(V) anions and the non-ionic As(III) changed the charge (protonation/deprotonation) depending on the pH. Thus, it was confirmed that both a charge on the membrane and pH played an important role in the retention of As. Research with the flat membranes showed comparable retention and, again, the negatively charged membrane was more effective than the membrane without charge.

UF pilot-scale investigations on the removal As were also carried out (Amy *et al.*, 1998). Tests using a single module were conducted for two groundwaters, which differ in dissolved organic carbon (DOC) concentration i.e. in one it was 11 mg DOC L^{-1} and in the second 1 mg DOC L^{-1} and for surface water to which As compounds were added (Table 16.11).

Table 16.10.   The removal of As(V) and As(III) using UF membranes.

Membrane type	Cut-off [Da]	Membrane charge	Valence of arsenic	pH	Retention rate [%]
Single module					
GM2540F	8000	(−)	V	6.9	63
GM2540F	8000	(−)	V	2.0	8
GM2540F	8000	(−)	III	7.2	<1
GM2540F	8000	(−)	III	10.8	53
FV2540F	10000	no data available	V	6.9	3
FV2540F	10000	no data available	III	6.8	5
Flat membranes					
GM	8000	(−)	V		52
FV	10000	no data available	V		–
PW	10000	no data available	V		5

Table 16.11.   Removal of As(V) and As(III) using UF in pilot scale studies.

Membrane type	Cut-off [Da]	Water type	Arsenic speciation	Retention rate [%]
Desal M2540F	8000	Groundwater with low DOC	As total	70
Desal M2540F	8000	Groundwater with high DOC	As total	30
		Surface water	V	47
			III	10

Results presented in Table 16.11 show that the removal of As from groundwater is much larger for water with higher DOC content (70%) than for water with low DOC concentration (30%). The authors postulate that this difference is due to the reduction of electrostatic repulsion induced by NOM adsorption onto the surface of the membrane. Any increase in the apparent size of the particles of As by "bridging" with humic substances did not affect the retention, because, in parallel, any increase in the removal of substances determined by UV_{254} was observed. As expected, during tests on the surface water treatment, greater removal of As(V) than As(III) was observed.

Recently, in decontamination of water from As, affinity membranes that combine adsorption and membrane filtration, to overcome the shortages of both, have been developed. In recent years, a great interest in preparation of PVDF and zirconium-based membrane, which act also as adsorbents, has been observed. Zheng *et al.* (2011) used in their study a PVDF/zirconium blend flat sheet self-prepared membrane so it could adsorb arsenate and remove other contaminants such as microorganisms. The crystal structure, hydrophobicity, porosity, water flux and As adsorption capacity of the membrane were investigated. The PVDF of α-form and the zirconia of amorphous nature were used for the membrane preparation. The addition of zirconia increased the hydrophilicity and porosity of the membrane, which led to a significant enhancement of the water flux through the membrane. The batch adsorption experiments demonstrated that the membrane effectively removed arsenate in a wide optimal pH ranging from 3 to 8. The adsorption equilibrium could be established within 25 h, and the maximum adsorption capacity was 21.5 mg g^{-1}, which was comparable to most of the available adsorbents. Co-anions such as fluoride did not obviously affect the adsorption. In addition, the membrane was efficient for removal of arsenate and bacteria (*Escherichia coli*), even during continuous filtration. The membrane saturated by arsenate could be easily regenerated using an alkaline solution and be subsequently reused.

16.5.2 *Coagulation – microfiltration/ultrafiltration integrated method*

Coagulation/co-precipitation with iron ions is the most commonly used method of As removal from ground and surface waters in drinking water production and in treatment of liquors, wastewaters, mine waters, as evidenced by the number of installations working on a pilot and a full industrial scale (Ann., 2005). Operational data indicate that this technology can reduce the concentration of As typically to less than $0.050\,mg\,L^{-1}$, and in some cases even to the concentration of $<0.010\,mg\,L^{-1}$ (Ann., 2000; Kociołek-Balawejder and Ociński, 2006; Kołtuniewicz and Drioli, 2008; McNeill and Edwards,1995). In order to improve the As removal operation, conventional gravity filters are increasingly being replaced with MF or UF (Kociołek-Balawejder and Ociński, 2006). The benefits of using membrane filtration in place of conventional filtration are (Kociołek-Balawejder and Ociński, 2006):

- More effective barrier to microorganisms;
- The ability to remove the flocks with smaller sizes (smaller quantities of coagulants are required);
- Increase of the total capacity (volume) of installation.

A hybrid process of coagulation and membrane filtration can be used to treat water of various qualities, which may significantly differ in turbidity, iron, manganese, sulfate and nitrate content (Baumann, 2006). The coagulation process integrated with the MF, as opposed to the traditional coagulation, does not require the flocculation phase (Kociołek-Balawejder and Ociński, 2006). For high performance of the process, flocks should have diameter of $2–10\,\mu m$, and this size can be obtained during the rapid mixing coagulation step (up to 20 s). In the method, iron-based coagulants, mainly $FeCl_3$ or $Fe_2(SO_4)_3$, are used. After the addition of coagulant and fast mixing, purified water is introduced to the membrane module, which is usually equipped with a membrane of pore size equal to $0.1\,\mu m$ (Chen *et al.*, 1999; Chwirka *et al.*, 2000). The use of MF/UF membranes eliminates overload of granular filters with coagulant, which is a great advantage of the integrated process. Doses of iron salts, pH and contact time are the main factors having an influence on the process efficiency. As for other membrane processes, the adequate pre-treatment of raw water, depending on its quality, the aim of which is to protect the membrane from fouling, is still important. Coagulant containing retentate may be sent to a sewer system. Membrane filters retain As(V), which is bonded with Fe(III) compounds at the stage of coagulation. The membrane must be subjected to a periodic backwashing in order to remove coagulation flocs, which contain As compounds. The addition of coagulant does not significantly affect the length of the filtration and membrane cleaning cycles, although it may increase the level of solids in the membrane system. In the case of the simultaneous removal of iron and manganese it is important to completely oxidize Fe and Mn, prior to the introduction to the membrane system, which also reduces fouling (Ann., 2000). Water, which is used for membrane washing is regarded as a wastewater with solid content usually below 1% depending on the type of coagulant, its dose, the length of the filtration cycle and the concentration of contaminants (Ann., 2005).

Tests carried out at water treatment stations showed that the coagulation/micro-filtration method enabled the reduction of As concentration to below $2.0\,\mu g\,L^{-1}$ (initial concentration at the level of $40\,\mu g\,L^{-1}$). The required coagulant dose was only $2.8\,mg\,L^{-1}$ (Ann., 2005; Chen *et al.*, 1999), although in other literature sources higher doses of coagulants (Ann., 2005; Chen *et al.*, 1999; Chwirka *et al.*, 2004) are mentioned. Chwirka *et al.* (2004) found that the particle size of iron(III) hydroxide, formed during rapid mixing (eliminating the flocculation stage) in the pilot test was mostly in the range of $2–10\,\mu m$, and thus As associated with these particles could be efficiently removed using the MF module. Coagulation with Fe(III) salts and direct UF for the removal of As from groundwater was also investigated by Ghurye *et al.* (2004). They stated that the size of the pores of the membrane below or equal to $0.2\,\mu m$ was sufficient for the effective removal of As

Table 16.12.   The comparison of arsenic removal by flocculation and microfiltration and by flocculation and sedimentation (dose of Fe as $Fe_2(SO_4)_3$: 7.5 mg L^{-1}, pH $= 6.8$, initial concentration of arsenic: 60 $\mu g L^{-1}$).

Time [min]		0	20	60
Flocculation and	As concentration	60	49	32
sedimentation	turbidity [NTU]	2.48	1.78	1.16
Flocculation and	As concentration	1	1	1
microfiltration	turbidity [NTU]	0.09	0.09	0.08

associated with iron(III) hydroxide sludge. In all these studies, Fe(III) chloride was used as the coagulant.

The ability to remove As by hybrid the coagulation/microfiltration method was confirmed by pilot studies, which showed that the level of As could be reduced below 2 $\mu g L^{-1}$ at pH between 6 and 7, and in the case when the initial concentration of Fe(III) was approximately 2.5 mg L^{-1} (Ann., 2000). These studies also found that the same rates of As removal could be achieved, even if the water contained high concentrations of sulfates (VI) and silica. In addition, the concentration of As could be reduced even more at slightly lower pH (about 5.5).

Floch et al. (2004) developed a new technology for As removal using immersed membrane module ZW-1000 (Zenon) and the feed pretreatment stage. Experiments were conducted on deep-well water which had a high concentration of As at the level of 200–300 $\mu g L^{-1}$. The pretreatment stage involved oxidation with $KMnO_4$, coagulation using Fe(III) sulfate ($Fe_2(SO_4)_3$), fast mixing of the chemicals, slow mixing flocculation and sedimentation. Next, the membrane separation took place. Such a treatment pathway enabled permeates with concentrations of As below the permissible level (10 $\mu g L^{-1}$). The results of the experiments proved that the new technology was suitable for the production of drinking water of the required quality from water containing high concentrations of As.

Han et al. (2002) conducted a study of the removal of As using coagulation, flocculation and microfiltration. Iron(III) sulfate and chloride were applied as coagulants and cationic polymer-flocculants assisting the coagulation were used. Table 16.12 shows the final concentration and turbidity of permeate after MF and in supernatants after flocculation and sedimentation (Han et al., 2002). It can be seen that the hybrid system with MF leads to the more rapid and complete removal of As and turbidity than the sedimentation-only based treatment.

Wickramasinghe et al. (2004) carried out pilot-scale experiments in order to evaluate the removal of As from groundwater by coagulation using iron (III) chloride and iron (III) sulfate and microfiltration. The authors established that the rate of retention of precipitated particles $(Fe(OH)_3 - As)$ by the membrane for a given dose of Fe(III) ions increased with pH decrease, because of greater adsorption at lower pH and formation of flocks of bigger size. They also observed a lower turbidity that a stream after coagulation with Fe(III) sulfates characterized with a lower turbidity, which suggested that sulfonic groups are involved in coagulation reactions. The presence of sulfonic groups led to the formation of larger flocks, which were more easily retained by the membrane in comparison with flocks formed with chloride ions.

Bray (2013) carried out pilot plant experiments on As removal from groundwater that contained 40 μg As L^{-1}, 0.7 mg Fe L^{-1}, and 0.1 mg Mn L^{-1}, using the integrated coagulation/microfiltration process. Iron (III) sulfate (PIX 112) in the form of aqueous solution was used as a coagulant in doses ranging from 0.5 to 19 mg Fe L^{-1}. For experimental purposes PVDF tubular microfiltration membrane with cut-off of 400 kDa was used. Dead-end mode of the process operation was applied and pressure ranging from 0.1 to 0.2 MPa was used. It was demonstrated that at coagulant dose of 2.0 mg Fe L^{-1} As content was brought down below the permissible level ($<10 \mu g$ As L^{-1}), established for potable water, while the retention coefficient exceeded 76%. Further increase in the coagulant dose led only to the further insignificant reduction of As content in the purified

Table 16.13. Capital and operational costs of arsenic removal with coagulation – classical pressure filtration and coagulation/microfiltration in Fallon Paiute Shoshone Tribe (USA).

Costs	Coagulation/filtration	Coagulation/MF
Capital costs [US$]	1,252,998	987,898
Operating costs [US$ year^{-1}]	71,436	82,392
Unit cost of water production		
[US$/ US gallon]	1.38	1.29
[US$/ L]	$\approx$0.365	$\approx$0.341

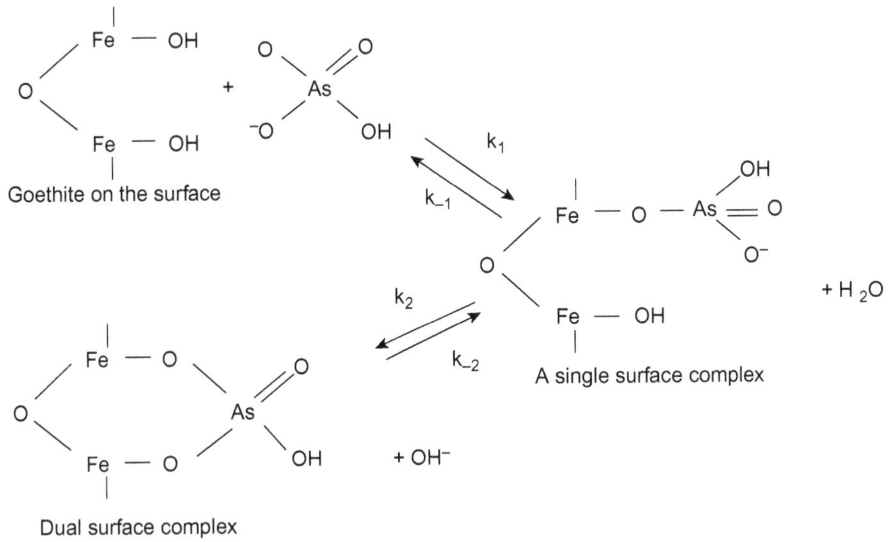

Figure 16.1. The structure of the arsenic-iron(III) hydroxide surface complexes.

water. For all applied coagulant doses, only trace amounts of iron could be found in the permeate stream. Moreover, As removal was accompanied by reduction of water color, UV-absorbance and chemical oxygen demand (COD). The process, however, was not effective in case of the Mn removal, which was necessary while treating groundwater.

The costs of the removal of As using a hybrid process of coagulation/microfiltration for all discussed cases (Ann., 2000; 2005) are presented in Table 16.13 (Ann., 2005).

In most cases, iron coagulants are used in As removal by the coagulation/membrane filtration process. This is because at pHs of 4–6, Fe(III) hydroxide forms positively charged colloids, which promote the As anion adsorption on the surface of flocks during coagulation. Both As(III) and As(V) are characterized by a high affinity to iron complexes. The adsorption of As on the surface of the sediment consists in exchange of OH-groups in goethite to As ions in surface coordination structure (Fig. 16.1) (Jain *et al.*, 1999).

A linear relationship between concentration of As and water turbidity was found, and hence the turbidity of permeate after flocculation and microfiltration could be a measure of the As removal. As concentration in permeate after MF may be therefore associated with the amount of flocks retained by the membrane. However, the amount of As, which is adsorbed on iron complexes, also depends on the pH of the suspension. Overall, in the pH range of 3–10, the adsorption of As(V) decreases with pH increase, while the adsorption of As(III) increases (Han *et al.*, 2002). Another advantage of MF application is the membrane compactness. The larger pores in the membrane have a smaller hydraulic resistance to the filtrate and the costs of pumping liquids for a given performance are lower. On the other hand, larger pores in the membrane cause more flocks with smaller radius to permeate. The results of some studies showed that for a given dose of Fe and pH, membrane with smaller pores (0.22 μm) was more effective in As removal than the membrane with the larger pores (1.2 μm) (Han et al., 2002). This shows that some flocks passed through the pores of the membrane with pore size of 1.2 μm, while they were already retained by the membrane with pore size of 0.22 μm. The size distribution of flocks can therefore be used as a measure of the choice of the membrane with the suitable pore size while designing the system of coagulation (flocculation)/microfiltration. Moreover, a small addition of cationic polymeric flocculants leads to a significant increase in the permeate flux and can also contribute to the extension of the intervals between successive water washing of the membrane. Flocculants, however, are much more expensive than ferrous compounds, so the use of such a solution depends on the quality of raw water.

Cecol et al. (2004) studied the removal of As(V) from simulated water using micellar surfactants and UF membrane (MEUF process) operated in the dead-end mode. The efficiency of the As removal and the permeate flux dependency on a membrane material (in this case regenerated cellulose and polyethersulfone were used), their cut-off (5 and 10 kDa), the initial concentration of As (0 to 221 mg L^{-1}) and pH of the raw water (5.5 and 8), and the presence of surfactant, were examined. It was found that the membrane made of regenerated cellulose (RC) more effectively removed As than polyethersulfone (PES) membrane, due to the negative charge on its surface. However, the concentration of As in permeate (produced water) was not reduced to the level required for drinking water (10 μg L^{-1}). After addition of 10 moles L^{-1} of cationic surfactant (cetylpyridinium chloride) to the raw water, both PES and RC membranes reduced As concentrations in permeate much below that established in standards. The complete removal of As was reached for raw water with initial contaminant concentration of 22 μg L^{-1} and 43 μg L^{-1} using the PES membrane (5 kDa) at pH 5.5 and at pH 8 for RC membrane (10 kDa). Regardless of the As concentration in the raw water, its total removal was obtained for PES membrane (5 kDa) at pH 8.

The necessity of effective As removal from water caused intensive research on the development of novel materials for that purpose and the water-soluble cationic polyelectrolytes that could be combined with membrane filtration in order to remove arsenates from aqueous solutions (Rivas and Aguirre, 2009; Rivas et al., 2006; 2007a; 2007b) were invented. This method is known as liquid-phase polymer-based retention (LPR) or polymer enhanced ultrafiltration (PEUF), and it involves the use of an ultrafiltration membrane that separates the ionic species interacting with the functional groups of water-soluble polymers with high molecular weights, and thus avoids their passage through the membrane. The LPR technique has a great capacity to separate arsenate oxyanions from solutions using the adequate anion-exchange polymer. Interaction occurs between the nitrogen of the ammonium group (positively charged) and the oxygen of arsenate anions, which form a dipole (Rivas et al., 2006). The interactions are mainly the result of the anion exchange between the counter ion of the quaternary ammonium salt and the arsenate anions at basic pH, as can be corroborated by the polymers' higher retention capacity at basic pH, at which divalent As(V) species are predominant (Rivas et al., 2007a; 2007b). On the other hand, the retention capacity is limited by the polymer concentration. Results obtained in the previous studies indicate an optimal polymer:arsenate molar ratio for complete separation equal to 20:1 (Rivas and Aguirre, 2009). The cationic polyelectrolytes remove arsenate ions more efficiently than arsenite ions in a wide pH range (Sánchez et al., 2010). In order to remove arsenite, it is necessary to combine the oxidation of As(III) to As(V) with LPR technique, which leads to the complete As removal (Rivas et al., 2012; Sanchez and Rivas, 2010; 2011; Wang et al., 1997). The LPR technique uses polymers with adequate molecular weights (i.e., higher than 50,000 Da) and membranes that have cut-off of 10,000 Da to interact with ions (Rivas and Aguirre, 2009). The membranes are typically composed of polycarbonate or cellulose esters, polyamides or polysulfones. Nevertheless, the fouling of the membrane is a disadvantage which must be minimized, because the permeate flux decreases and, therefore, the processing time increases (Rivas and Aguirre, 2009).

Arsenate retention obtained using poly[2-(acryloyloxy) ethyl] trimethylammonium chloride [P(ClAETA)] at a high arsenate concentration (47.6 mg L^{-1}) was 58% and this removal capacity increased gradually, reaching 100%, when the arsenate concentration in the cell was at minimum level (5.5 mg L^{-1}) at the previously determined optimal molar ratio (20:1) of polymer:As(V) (Sanchez and Rivas, 2011). Arsenic removal was also determined at low concentrations (in μg L^{-1}) and the results show P(ClAETA)-based retention rate of 65%. The regeneration process of polymers can be performed when the polymer-arsenate complex is in contact with the acidic solution. A study on the removal of As from the Camarones River (Chile) water using P(ClAETA) was also performed. The water-soluble polymer showed a high performance (100%) at the beginning of the process and then it decreased to 16% when the ratio of permeate to retentate volume was $Z = 10$. There is also a study i.e. Rivas *et al.* (2012) that shows the influence of Cl$^-$, SO$_4^{2-}$, NO$_3^-$, SiO$_3^{2-}$, Na$^+$, and Ca^{2+} ions on arsenate removal by anion-exchange polymers using the liquid-phase polymer-based retention (LPR) technique. The results show that in the presence of ionic mixtures, As(V) removal capacity decreases. Polymers with chloride exchange groups showed a higher ability to remove arsenate than the polymer that contains methyl sulfate as the anion-exchange group. At higher arsenate concentration (47.6 mg L^{-1}), arsenate retention by polymers was between 58 and 91%. This removal capacity increased gradually reaching 100% in the case of P(ClAETA) polymer at minimum arsenate concentration in the cell (5.5 mg L^{-1}).

Toledo *et al.* (2013) prepared and applied a new water-soluble polymer poly(glycidyl methacrylate-*N*-methyl-D-glutamine) [P(GMA-NMG)] to As retention using the LPR technique. The removal experiments were conducted using the washing method at varying pH, polymer to As(V) molar ratio and concentrations of interfering ions. P(GMA-NMG) showed a high affinity to bind arsenate species (82% of removal) at pH 3.0. The optimum molar ratio of polymer:As(V) was 70:1. Selectivity experiments showed that the presence of interfering substances gradually decreased the As removal capacity. The maximum retention capacity was determined by the method of enrichment, obtaining a value of 45.9 mg As g^{-1} of polymer. Finally, the results showed that the combination of P(GMA-NMG) with ultrafiltration membranes was a potential alternative for the removal of hazardous As(V) species from aqueous solutions at acidic pH.

In the study of Sanchez *et al.* (2013), regenerated cellulose membrane was used as a filtering liquid-phase in the polymer-based retention technique. Poly(4-vinyl-1-methylpyridinium bromide), P(BrVMP), was used as extracting reagent for As(V) removal. The role of pH, polymer:As(V) molar ratio, and influence of regenerated cellulose membrane were investigated during the process carried out with the use of the washing method. Efficient retention was obtained at pH $= 9$ and 20:1 polymer:As molar ratio and it was about 100% at $Z = 10$ for P(BrVMP). Experimental data showed that the regenerated cellulose membrane, in reference to poly(ethersulfone) membrane, had an affinity to As(V) ions. The maximum retention capacity of P(BrVMP) was determined by the enrichment method, and then, using alternately washing and enrichment methods, the charge-discharge process and recovery of P(BrVMP) were performed.

16.5.3   *Oxidation – microfiltration/ultrafiltration integrated method*

Removal of As using the precipitation/co-precipitation with iron method is usually connected with a pH correction and addition of chemical coagulants, but it can also include oxidants introduction (Ann., 2002). The oxidation of As(III) to the less soluble As(V) increases the efficiency of the precipitation/co-precipitation processes and can be performed in a separate stage or as part of the precipitation process. The methods of As(III) oxidation include aeration, ozonation, photooxidation, or addition of chemicals such as potassium permanganate, sodium perchlorate and hydrogen peroxide (Ann., 2002). In the treatment of groundwater, the oxidation of iron (II) and manganese (II) (Fe-Mn), followed by microfiltration, has been recognized as a cost-effective and acceptable technology, because at the same time, As removal takes place, which is very significant in the case of increased content of this element (Caniyilmaz, 2005). Hence, research on the effectiveness of As removal in the hybrid oxidation of Fe-Mn/microfiltration process in different process conditions (Fe:As ratio, pH, $KMnO_4$ dose, contact time, the status of the iron oxidation, process conditions) has been conducted.

It was found that the removal of As(III) did not depend on the water pH ranging from 6.5–8.0 using simple aeration as the initial oxidation step. At neutral pH and ratio Fe:As amounting to 60 it was possible to reduce As content from 25–250 $\mu g\,L^{-1}$ to below 10 $\mu g\,L^{-1}$ (standard for As) (even to 5 $\mu g\,L^{-1}$) in the aeration and microfiltration processes (Caniyilmaz, 2005). Iron in the second and third degree of oxidation is able to remove As to a comparable extent. However, if the concentration of iron is insufficient to reduce the As levels below the limited value, it is necessary to add iron ions to raw water.

Yoon et al. (2011) investigated the efficiency of As removal by UF and NF membrane preceded with ozonation. Furthermore, the role of iron in the removal of As was studied. They found that more than 99% of As(III) was oxidized into As(V) with a 1.0 $mg\,L^{-1}$ dose of ozone. The oxidized As(V) was partially removed (20% retention) by UF, and could be completely removed using NF. The effect of pH on As(V) removal by NF was also evaluated. It was found that As(V) total removal via NF was obtained at pH 9 and the retention rate decreased to 80% with pH decrease (pH 4). Variation of pH affected the predomination of As species, altering the overall removal rate observed for NF. Additionally, the effect of co-existing iron species on As removal was investigated. Thus, 5.0 $mg\,L^{-1}$ of $FeCl_3$ was added to the solution of initial As concentration of 100 $\mu g\,L^{-1}$. More than 40% of As(V) could be eliminated simply through co-precipitation with iron. After UF membrane filtration, up to 99% of As(V) could be removed. The enhanced removal efficiency of As(V) was due to the retention of precipitate of As(V) with Fe(III) hydroxide using the UF membrane. However, ultrafiltration carried out in the absence of iron ions did not separated As(V) ions effectively.

Arsenic removal from drinking water using low-pressure membranes together with chemical oxidation and adsorption on iron coagulants was deeply investigated by Elcik, (2013). Oxidation of solutions with 100 or 1000 $\mu g\,L^{-1}$ As(III) using ozone or NaClO followed with iron coagulation (0.5–20 $mg\,L^{-1}$) and next separation with five different ultrafiltration membranes (by Microdyn-Nadir) was performed. The rejection of As(III) or As(V) was not observed during direct filtration with the applied membranes. Although partial As(III) removal was possible in the presence of ferric ion, the effluent did not meet the drinking water standards. Oxidation of As(III) and Fe(II) to As(V) and Fe(III), respectively, significantly increased As retention during low-pressure membrane filtration. The removal rate of As depended on the type and concentration of oxidants, concentration of Fe(II), and the type of membrane. When the initial NaClO and Fe(II) concentrations were 2.5 $mg\,L^{-1}$, As(III) was decreased from 100 $\mu g\,L^{-1}$ to less than 5 $\mu g\,L^{-1}$ for the all studied membranes. When the initial As(III) concentration was 1000 $\mu g\,L^{-1}$ and 20 $mg\,L^{-1}$ of NaClO and Fe(II) was added, membrane filtration enabled the decrease of As concentration in the treated water to below 7 $\mu g\,L^{-1}$. Nevertheless, despite membrane filtration preceded with ozone oxidation and iron coagulation removed up to 85% (100 $\mu g\,L^{-1}$) and 96% (1000 $\mu g\,L^{-1}$) of As(III), the effluent target value (10 μg As L^{-1}) was not reached. Within the five studied different ultrafiltration membranes, UP150 turned out to be the most effective due to high flux (up to 565 $L\,m^{-2}\,h^{-1}$ at TMP of 0.3 MPa) and high contaminant rejection rates (up to 96.1% and 99.4% for 100 and 1000 $\mu g\,L^{-1}$ As(III), respectively). In general, the combination of hypochlorite + Fe(II) + membrane gave much better performance than the system comprised of ozone + Fe(II) + membrane.

Oxidation with potassium permanganate resulted in better removal of As(III) compared to aeration, which even in the presence of high concentrations of iron could only partially oxidize As(III) to As(V), while the total oxidation of As(III) using potassium permanganate was possible (Bodzek and Konieczny, 2011). The oxidation is crucial as highly dissociated As(V) anions easily connect with the iron (III) hydroxide in opposite to non-dissociated As(III). During oxidation with potassium permanganate the effect of pH on the efficiency of As removal was found to be significant. A higher As removal was observed for pH 7 than for pH 8.0 at the same dose of $KMnO_4$ (0.5 $mg\,L^{-1}$). It was found that pH affected both the efficiency of oxidation of iron (manganese) and adsorption of As. Higher removal of manganese (II), which one could reach at pH 5.0, improved the removal of As due to more effective adsorption of As(V), while the removal of As(III) at pH 7.0 was effective regardless of the removal of manganese. In studies of the simultaneous removal of Fe, Mn and As, the final concentration of Fe in permeate was always about 0.01 mg L^{-1} i.e. far below the standard established for iron (0.2 mg L^{-1}), while the permissible quantity for manganese (0.05 mg L^{-1}) was reached for initial Mn concentration 0.2 mg L^{-1}. Moreover, Fe(OH)$_3$ and MnO$_2$ particles were found as the main membrane foulants.

Since manganese oxides are able to oxidize As(III) to As(V) rapidly, they can act as detoxifying agents. Furthermore, the conversion of As(III) to As(V) decreases As mobility, which facilitates its adsorption onto mineral surfaces (Scott and Morgan, 1995). Manning *et al.* (2002) studied the surface reactions of As(III) and As(V) on synthetic MnO$_2$ and showed that MnO$_2$ possessed both oxidation and adsorption properties in the removal of As species. Current researches show that manganese oxides and their derivatives are widely used for the removal of As species (Camacho *et al.*, 2011; Chang *et al.*, 2009; Li *et al.*, 2010). However, the oxidation and adsorption capacities of bulk manganese oxides are relatively low due to their limited specific surface areas, which limit their performance in water treatment. Recently, nanostructured manganese oxides with high surface area have been developed in order to enhance As oxidation and adsorption capacities. However, separation or recovery of nanosized materials in heterogeneous systems continues to be a major challenge. Thus, studies on the application of membrane filtration, already found to be an efficient and economical method to separate nanomaterial, were performed (Bodzek and Rajca, 2013). Zhang and Sun (2013), developed multifunctional micro-/nano-structured MnO$_2$ spheres and applied them to remove As species from water. Morphology studies indicated that the synthesized material possesses microstructure and nanostructure, which endow the material with excellent oxidation, adsorption and separation properties. Experiments showed that As(III) species could be effectively oxidized by the synthesized MnO$_2$ followed by the adsorption of As(V) species. Investigations on process conditions revealed that removal of As(V) using the MnO$_2$ spheres was evidently depended on pH and ionic strength, while co-existing anions such as CO$_3^{2-}$, SO$_4^{2-}$ and PO$_4^{3-}$ induced suppressive effects. Furthermore, a dead-end microfiltration process was conducted to evaluate the separation property of the synthesized MnO$_2$ spheres from treated water. Membrane fouling studies showed that the synthesized MnO$_2$ spheres caused negligible blocking of the membrane pores. Besides, the material also formed a porous cake layer on the membrane, but its resistance to water did not affect the permeate flux level.

The technology with low-pressure membrane techniques is especially useful for groundwater treatment systems with high contents of As and iron in the raw water. It is much simpler in operation than the conventional coagulation process, and the need for chemicals is minimized through the use of iron contained in the raw water as adsorbent.

16.6   ION-EXCHANGE MEMBRANE PROCESSES

16.6.1   *Electrodialysis*

There are only a few publications concerning the application of electrodialysis (EDR) or electrodialysis reversal (EDR) in the process of As removal from water, which can be found in the literature. Nevertheless, the available data suggests that it is possible to remove As with efficiency exceeding 80% for As(V), and 50% for As(III) (Ann., 2000; Hering *et al.*, 1997; Kartinen and Martin, 1995; Kociołek-Balawejder and Ociński, 2006). The nature of the process, however, enables the removal of dissociated substances. Thus, the removal of As(III) is limited. On the one hand, the process can be improved by the addition of an oxidant to raw water. On the other hand, such an action may adversely affect the durability of the membranes (Kartinen and Martin, 1995).

Studies on the evaluation of the ED/EDR method in removal of As from water were performed by Ionics Inc., the leading EDR equipment manufacturer, in San Ysidro, New Mexico, USA (Ann., 2000). In EDR experiments (Hering *et al.*, 1997), the system comprised of water pretreatment via filtration (10-μm) and adsorption on granular activated carbon (GAC) followed by EDR and internal recirculation was applied and 85% of water recovery was reached. The installation was tested with the use of both surface water, contaminated with a mixture of As(III) and As(V), and groundwater, which contained mainly As(III). The obtained removal rate of As for groundwater was low, only 28%, and the concentration of the contaminant in the diluate was high and equal to $0.136\,mg\,L^{-1}$ (while its concentration in the feed water was $0.188\,mg\,L^{-1}$). In the case of surface water, the overall removal rate of As was estimated at 73%, including removal of As(III) at the level of 60%.

Another mobile installation consisted of RO, ED and EDR systems, and other necessary equipment was used to test eight types of waters in New Mexico, USA (Hering *et al.*, 1997). The study was carried out at the Bluewater New Mexico water treatment plant, USA. The final concentration of As in eluate equal to $0.003\,mg\,L^{-1}$ was obtained, at the initial level of $0.021\,mg\,L^{-1}$, which corresponded to the retention coefficient of 86%. Raw water introduced to EDR was taken off before the final chlorination. The system capacity was more than $18\,L\,min^{-1}$ at the water recovery rate of 80%.

The EDR unit of a capacity $113.5\,L\,day^{-1}$ was also used to treat mine water (Hering *et al.*, 1997). The water recovery of around 81% was obtained. The removal of As was about 59% at the initial concentration of the contaminant at the level of $0.022\,mg\,L^{-1}$.

The performed studies allow the conclusion that ED/EDR methods allow the removal of As to a low level. However, they have high energy consumption. Nevertheless, the required initial pretreatment of raw water is not demanding and the process can be run with a high capacity and satisfactory water recovery rate. Thus, ED/EDR systems are considered to be attractive and cost effective, but only for very small installations (Hering *et al.*, 1997).

16.6.2   *Donnan dialysis*

Donnan dialysis is an electrochemical potential driven membrane process based on the Donnan membrane equilibrium, which solely deals with completely ionized electrolytes. The specific Donnan membrane equilibrium arises by a typical Donnan dialysis process; an ion-exchange membrane is used to separate the treated solution into two streams. The ion-exchange membrane excludes co-ions (ions with the same electrical charge as the ion-exchange membrane) from the membrane permeation. Therefore, the flux of counter-ions (ions with the electrical charge opposite to the ion-exchange membrane) through the membrane caused by a concentration difference is always coupled with the transport of identical numbers of counter-ions in the opposite direction to maintain electroneutrality in either solution (Bodzek and Konieczny, 2011). The apparent flux of each counter-ion becomes zero when the Donnan membrane equilibrium is achieved. Donnan dialysis has already provided many environmental application opportunities for the recovery of valuable metal ions and target anions from water and wastewater and for the removal of ionic pollutants from drinking water (Bodzek and Konieczny, 2011). Electric fields or high pressures are not required, and the membrane is not susceptible to fouling. Thus, Donnan dialysis is straightforward to operate, easy to maintain, and affordable to implement. The study carried out on the process showed that the performance of Donnan dialyzers was reliable for the removal of arsenate (As(V)) from different water sources (Oehmen *et al.*, 2011; Velizarov, 2013; Zhao *et al.*, 2010; 2012). In Figure 16.2 the scheme of the Donnan dialysis process for the extraction of As ions is presented.

Zhao *et al.* (2010) investigated the effects of nitrate, sulfate, phosphate, bicarbonate, and silicate presence in the solution on the removal of arsenate (As(V)) by Donnan dialysis. The study was focused on the determination of both the simultaneous transfer of accompanying components and their influence on the ion-exchange reaction between the arsenate ions and the anion-exchange membrane. The competition of NO_3^-, SO_4^{2-} and PO_4^{3-} with As(V) for the functional groups in the membrane dominated the membrane average flux of As(V) within the first 2 h of the batch dialysis process. However, the overall removal of As(V) mainly depended on the dialysis kinetics of accompanying anions established during the 24 hours long dialysis run. The presence of CO_3^{2-} increased the solution pH and the fraction of $HAsO_4^{2-}$ ions; as a result, the transfer of As(V) was

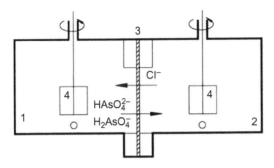

Figure 16.2.   The scheme of the Donnan dialysis of arsenic(V) ions: 1 – feed chamber; 2 – stripping chamber; 3 – anion-exchange membrane; 4 – stirrer.

accelerated. At neutral pH, non-dissociated silicate exhibited negligible inhibition on the As(V) removal. The As concentration remaining in feed solutions after 24 h was less than $50\,\mu g\,L^{-1}$ in all tests except for one with the presence of phosphate. The current results indicate that Donnan dialysis successfully removes the arsenate ions from groundwater containing various accompanying components. This feature makes it an attractive alternative for the As(V) removal point-of-use (POU) technologies in rural areas.

In another paper Zhao *et al.* (2012) developed a device based on the Donnan dialysis for the removal of As(V). The commercial anion-exchange membrane was used as a semipermeable barrier between the feed and stripping solution (As(V)-spiked groundwater and a $12\,g\,L^{-1}$ table salt solution, respectively). The proposed device could be operated 26 times before replacement of the stripping solution was required. In each batch, approximately 80% of the arsenate anions were transported across the membrane within 24 h, and the As concentration of the stripping solution was finally more than 180 times greater than in the treated water. Cations were well preserved in treated water, however, a slight increase in the sodium ion concentration was observed due to electrolyte leakage. The quality of the treated water was in compliance with drinking water standards. Membrane fouling was investigated, and a reduction in the As(V) removal rates was not observed when the membrane was used repeatedly. The results showed that the proposed Donnan dialysis device could effectively remove As from drinking water in rural areas in a sustainable manner.

Velizarov (2013) also studied the application of Donnan dialysis to the removal of As from water. The transport of arsenate across the main known types of polymeric anion-exchange membranes: heterogeneous, homogenous and monovalent-anion-permselective ones, was systematically studied in batch and continuous Donnan dialysis operating conditions (Velizarov, 2013). It was found that the Donnan dialysis process performance strongly depended on the properties of the anion-exchange membrane used in batch operating conditions, but the differences vanished for the continuous operation mode, and similar removal rates of As were obtained. The most appropriate anion-exchange membranes appeared to be those with a relatively open structure and low tortuosity of the polymeric matrix. The change in pH could be used as an indication of arsenate removal from the depleted feed solution (pH decreases) or accumulation in the stripping solution (pH increases). However, a proportional relation between the pH increase in the stripping solutions and the total amount of arsenate transported could not be established, which suggested that the intra-membrane conversions of monovalent arsenate into more charged divalent and trivalent forms proceeded to different degrees depending on the type of membrane used.

A novel combined process targeting arsenate removal from contaminated drinking water sources, consisting of arsenate transport through an anion-exchange membrane followed by its coagulation in a separate stripping compartment operated as a closed-circuit chamber, was developed (Oehmen *et al.*, 2011). The combined process concept was validated for tap water spiked with arsenate and fed in a single-pass mode to the feed water compartment of an anion-exchange membrane module during 3 months of continuous operation. The Ionics AR204-UZRA anion exchange membrane was chosen as the most suitable membrane for this process, due to the comparatively high flux of both monovalent ($H_2AsO_4^-$) and divalent ($HAsO_4^{2-}$) arsenate. The use of $AlCl_3$ as a coagulant presented an advantage over $FeCl_3$ in terms of reduced membrane scaling, and resulted in an increased As flux. The process was simple to operate and arsenate removal was effectively maintained over the whole duration of the experiment with minimal maintenance requirements, which makes it attractive, especially for rural areas located far from centralized drinking water treatment facilities. In conclusion, the process was successfully applied for As removal from contaminated drinking water sources and the level of the As in the product stream was below $10 \, \mu g \, L^{-1}$.

16.7   LIQUID MEMBRANES

A typical supported liquid membrane (SLM) consists of a microporous polymeric support impregnated with an organic solution (Gawroński, 2002) containing the carrier, thus contacting both the feed phase and the stripping phase. The extraction and stripping reactions, which take place at the interfaces of the aqueous solutions and the organic solution, generate a chemical pumping that enables the transport of the species through the liquid membrane.

Prapasawat *et al.* (2008) studied the simultaneous extraction and back-extraction of As(III) and As(V) from sulfate media using Cyanex-923 extractant in toluene as a carrier and water as a stripping agent. The commercial Liqui-Cel®Extra-Flow module was used. The rate of extraction was greater for As(V) than for As(III). The maximum metal extractions at 30% (v/v) of Cyanex-923 in the organic phase were 38% for As(III) and 46% for As(V). Since the extractant belongs to the group of solvating ones, the mechanism of As(V) extraction was considered to be the solvation of the uncharged As acid H_3AsO_4. The pH of the aqueous metal solutions was in the range of 0–2.

Cyanex-921 and Na_2SO_4 were found by Perez *et al.* (2007) to be highly efficient extractant and stripping agents, respectively, for As(V) uptake from sulfuric acid solutions. In the experiments a supported liquid membrane technology with flat commercial hydrophobic membranes made of polyvinylidene difluoride (PVDF) with trioctylphosphine oxide (Cyanex 921) in kerosene was used. Transport of As(V) from 2 M H_2SO_4 aqueous solutions was studied as a function of parameters such as the nature of the stripping solution, extractant concentration and stirring rate. Among all stripping solutions studied, Na_2SO_4 presented the highest As(V) transfer (close to 94%) within 120 min at concentrations at which the ionic strengths of the feed and stripping solutions were equal. Additionally, when the extractant concentration was increased from 0.05 to 0.12 M, higher As(V) transfer values were obtained. The highest As(V) transfer was found at a Cyanex concentration of 0.12 M. The study of stirring rate influence indicated that at 16.66 Hz (1000 rpm) the thickness of the limiting layer reached a minimum. Thus, it was considered that at this point the mass transfer coefficient was essentially constant.

Ballinas *et al.* (2004) studied the recovery of As(V) from copper electrolytic baths with a polymer inclusion membrane (PIM) in a sulfuric acid environment using di-butyl-butyl-phosphonate (DBBP). It was found that the extraction was controlled by the kinetic regime and 90% As recovery within 800 min was reported.

Guell *et al.* (2010) investigated the transport of As species through a supported liquid membrane (SLM) using Aliquat 336. To prepare the SLM, a Durapore polymeric support (polyvinyl difluoride) with a thickness of 125 μm, 75% porosity, and a 0.2 μm average pore size was impregnated and used. Preliminary liquid–liquid extraction experiments were performed and several conditions affecting the extraction i.e. the type of organic solvent and pH of the aqueous phase were evaluated. The best results were obtained for a mixture of dodecane modified with 4% dodecanol at pH 13. The effect of pH of the feed phase was also investigated, setting 0.1 M NaCl solution as a stripping phase (Guell *et al.*, 2011). It was observed that, in the case of SLM, As(V) was totally transported at pHs 10 and 7, whereas acidic pHs such as 3 or more basic, such as 13 resulted in poorer results. Moreover, the transport of As(III) was investigated under the best experimental conditions found for As(V) permeation, but any transport of As(III) was not achieved.

Considering the good results obtained in liquid–liquid experiments, a supported liquid membrane (SLM) system based on the best separation conditions was developed. As(V) was stripped using 0.1 M HCl. The SLM enabled the separation of As(V) and As(III) species based on their different kinetic behavior. The system also enabled the transport of As(V) at mg L^{-1} levels and the removal of As(V) from real matrices, such as spiked tap water and river water.

Bey *et al.* (2010) investigated the removal of As(V) from an aqueous stream by nondispersive solvent extraction in a hollow-fiber membrane contactor. The microporous hydrophobic poly(vinylidene fluoride) hollow-fibers were specially prepared for that purpose by the dry/wet spinning technique, obtaining pore diameters of about 0.2 μm and porosity of 80%. The produced fibers were used to perform As(V) extraction experiments in a membrane contactor device, using Aliquat-336 as extractant. At low velocity of both aqueous phase (0.48 mL s^{-1}) and organic phase (1.47 mL s^{-1}), the system showed a very high stability. In fact, no dispersion of any phase in the other one was observed. Different tests were carried out in order to study the effect of temperature, initial concentration, pH of the feed solution and membrane properties on the As separation performance. The results showed that the extraction of As(V) by non-dispersive solvent extraction was influenced by the pH of the feed, with an optimum reached for neutral values, increased with the As content in the feed and was not affected by temperature. The extraction was favored by working with thinner membranes. The highest removal of As achieved was around 70% after 6 h of operation for an As content in the feed of 100 mg L^{-1}. However, it has to be pointed out that the rate of extraction could be further increased if a back extraction, during which As is removed from the organic phase before its recirculation in the membrane contactor module, would be carried out together with the extraction step. The produced fibers kept their performance for all the experimental activity, giving reproducible results with time.

16.8   FINAL REMARKS

Two clear routes can be noted in the search for effective methods of As removal from water. The first concerns the technology implemented on a large scale, which can be used to treat industrial spent solutions and at drinking water treatment plants. Researchers are also focused on improving and perfecting the most popular processes used during water treatment (coagulation, softening, deferrization, adsorption, ion exchange, and pressure driven membrane processes) for As removal. Most of the mentioned methods are included in the list of the so-called *best available technology* (BAT). The second route is the search for easy-to-use, low-cost and effective technologies, which could be used in individual drinking water wells (adsorption with iron compounds, liquid membranes). Nevertheless, membrane techniques have a prominent place in both of those routes. High pressure reverse osmosis and nanofiltration and low pressure micro- and ultrafiltration combined with coagulation and oxidation seem to be an attractive solution for As removal from various water sources.

Kartinen *et al.* (1995) compared several methods of As elimination from water, namely, precipitation (for four coagulants), membrane separation (RO and EDR) and two adsorption processes. Seven criteria were taken into account and 3 evaluations were issued: (+) positive, (−) negative and zero as neutral (Table 16.14).

This statement can be considered subjective and disputed, in particular with regard to energy consumption. By comparing the energy consumption of membrane processes and adsorption, it seems that relationships are actually the reverse of those shown in Table 16.14.

Brandhuber and Amy (1998) presented comparative studies on the usefulness of various membrane processes (RO, NF, UF and MF) for the removal of As from groundwater dedicated to

Table 16.14.   The comparison of different methods used for arsenic removal from water.

Criterion	Al	Fe	CaOH	Fe/Mn	RO	EDR	AA	IX
Chemicals consumption					+	+	+/−	+
Energy consumption	+/−	+/−	+/−	+/−			+	+
Labor					+	+	+	+
Required area					+	+	+	+
Disposal of solid waste					+	+	+	+
Disposal of liquid waste	+	+	+	+				
Efficiency of removal					+	+	+	+

Table 16.15.   The comparison of arsenic removal with the use of various membranes.

Form of arsenic	RO	NF	UF	MF	Initial oxidation
As(III)	R	P	N	N	R
As(V)	R	R	P	N	N
Dissolved	R	P	N	N	N
Particulate	N	N	P	P	N
As + NOM	P	P	N	N	N
As- inorganic	R	P	N	N	N

the production of drinking water. These valuable tests were carried out with the pilot installation, using a number of membranes differing with As retention, but with similar permeate flux amounting to 17–25 L m^{-2} h^{-1}, namely:

- RO Fluid Systems TFC 4921 – composite membranes;
- RO Fluid Systems TFC 4820-ULPT – poly(ether urea) membranes;
- RO DESAL AG 4040 – composite polyamide membranes;
- RO Hydranautics 4040 LSA-CPA2 – composite membranes;
- NF FilmTec NF70 4040-B 0.3 kDa – aromatic polyamide membranes;
- NF DESAL HL-4040F1550 0.3 kDa – composite membranes;
- NF Hydranautics 4040-UHA-ESNA 0.3 kDa – composite polyamide membranes;
- UF DESAL GM-4040F-1020 8 kDa – sulfonated polysulfon membranes;
- MF Memcor 4M1W 0.2 μm.

The results of comparative tests are listed in the Table 16.15, in which general recommendations cover the following indications: R: recommended, P: possible, N: not recommended.

16.9   CONCLUSIONS

From the study several practical proposals can be concluded:

1. RO and NF can effectively eliminate As in the form of As(V).
2. NF membranes with a negative surface charge show As(V) retention similar to the reverse osmosis ones, and they have higher relative permeate flux than RO.

3. The impact of co-morbidity of soluble organic compounds on As retention is not unambiguous. However, high content of DOC is unfavorable due to membrane fouling. Co-occurrence of soluble inorganic compounds does not have a significant effect on As retention, although harmful effects should theoretically be related to the NF membranes. The presence of inorganic substances can affect the efficiency of coagulation of As by metal salts.

4. A combination of coagulation and microfiltration is a technically possible method to achieve As maximum concentrations of $5\ \mu g\,L^{-1}$. The use of Fe(III) salts as coagulants (coagulant dose $7\ mg\,L^{-1}$) enables an average reduction of As of 84% and 64% removal of turbidity at 90% of water recovery.

5. If the As occurs in the raw water primarily as As(III), the right solution to obtain a high level of As removal is the use of reverse osmosis membranes or compact nanofiltration membranes.

6. The removal of As(III) can be also performed with the use of the hybrid coagulation/microfiltration process with Fe(III) compounds, but larger doses of coagulant are required in comparison with ones required for As(V) or it is necessary to use the preliminary oxidation of As(III) to As(V) before the MF.

7. The use of electrodialysis, Donnan dialysis and liquid membranes for As removal from water is currently at the stage of laboratory tests.

REFERENCES

Akin, I., Arslan, G., Tor, A., Cengeloglu, Y. & Ersoz, M. (2011) Removal of arsenate [As(V)] and arsenite [As(III)] from water by SWHR and BW-30 reverse osmosis. *Desalination*, 281, 88–92.

Amy, G.L., Edwards, M., Benjamin, M., Carlson, K., Chwirka, J., Brandhuber, P., McNeill, L. & Vagliasindi, F. (1998) Arsenic treatability options and evaluation of residuals management issues. Draft report, April 1998, American Water Works Association Research Foundation and American Water Works Association, Denver, CO.

Anonymous (2000) Technologies and costs for removal of arsenic from drinking water. United States Environmental Protection Agency Report EPA 815-R-00-028.

Anonymous (2001) Removal of arsenic in drinking water at Park City. Utah, USA. Environmental Technology Verification Program (ETV). Reverse osmosis membrane filtration used in packaged drinking water treatment systems. Available from: http://www.membranes.com [accessed March 2001].

Anonymous (2002) Arsenic treatment technologies for soil, waste, and water. USEPA/National Service Center for Environmental Publications, Cincinnati, OH.

Anonymous (2005) Treatment technologies for arsenic removal. National Risk Management Research Laboratory EPA/600/S-05/006, USEPA, Cincinnati, OH.

Ballinas, M.L., De SanMiguel, E.R., De Jess Rodriguez, M.T., Silva, M., Muoz, O. & De Gyves, J. (2004) Arsenic(V) removal with polymer inclusion membranes from sulfuric acid media using BBP as carrier. *Environmental Science & Technology*, 38, 886–889.

Bandini, S. & Vezzani, D. (2003) Nanofiltration modeling: the role of dielectric exclusion in membrane characterization. *Chemical Engineering Science*, 58, 3303–3326.

Baumann, F. (2006) The removal of arsenic from potable water. *Water Conditioning & Purification*, May 2006.

Bey, S., Criscuoli, A., Figoli, A., Leopold, A., Simone, S., Benamor, M. & Drioli, E. (2010) Removal of As(V) by PVDF hollow fibers membrane contactors using Aliquat-336 as extractant. *Desalination*, 264, 193–200.

Bodzek, M. & Konieczny, K. (2005) Wykorzystanie procesów membranowych w uzdatnianiu wody [in Polish] (Application of membrane processes in water treatment). Bydgoszcz, Oficyna Wydawnicza Projprzem-Eko.

Bodzek, M. & Konieczny, K. (2011) Usuwanie zanieczyszczeń nieorganicznych ze środowiska wodnego metodami membranowymi [in Polish] (Removal of inorganic micropollutants from water environment by means of membrane methods). Warsaw, Wydawnictwo Seidel-Przywecki.

Bodzek, M. & Rajca, M. (2013) Fotokataliza w oczyszczaniu i dezynfekcji wody. Cz.II. Usuwanie metali i naturalnych substancji organicznych [in Polish] (Photocatalyse in water treatment and disinfection. Pt. II. Removal of metals and natural organic matter). *Technologia Wody*, 11, 18–30.

Bowen, W.R. & Welfoot, J.S. (2002) Modelling the performance of membrane nanofiltration – critical assessment and model development. *Chemical Engineering Science*, 57, 1121–1137.

Bowen, W.R., Mohammad, A.W. & Hilal, N. (1997) Characterization of nanofiltration membranes for predictive purposes – use of salts, uncharged solutes and atomic force microscopy. *Journal of Membrane Science* 126, 91–105.

Braundhuber, P. & Amy, G. (1998) Alternative methods for membrane filtration of arsenic from drinking water. *Desalination*, 117, 1–10.

Bray, R.T. (2013) Usuwanie arsenu z wody podziemnej w zintegrowanym procesie koagulacja/mikrofiltracja [in Polish] (Removal of arsenic from groundwater in an integrated process of coagulation/microfiltration). *Ochrona Środowiska*, 35 (4), 33–37.

Camacho, L.M., Parra, R.R. & Deng, S. (2011) Arsenic removal from groundwater by MnO$_2$-modified natural clinoptilolite zeolite: effects of pH and initial feed concentration. *Journal of Hazardous Materials*, 189, 286–293.

Caniyilmaz, S. (2005) *Arsenic removal from groundwater by Fe-Mn oxidation and microfiltration*. MSc Thesis, University of Pittsburgh, Pittsburgh, PA.

Cecol, H., Ergican, E. & Fuchs, A. (2004) Molecular level separation of arsenic(V) from water using cationic surfactant micelles and ultrafiltration membrane. *Journal of Membrane Science*, 241, 105–119.

Chang, F.F. & Liu, W.J. (2012) Arsenate removal using a combination treatment of precipitation and nanofiltration. *Water Science and Technology*, 65 (2), 296–302.

Chang, F.-F., Liu, W.-J. & Wang, X.-M. (2014) Comparison of polyamide nanofiltration and low-pressure reverse osmosis membranes on As(III) rejection under various operational conditions. *Desalination*, 334, 10–16.

Chang, Y.Y., Lee, S.M. & Yang, J.K. (2009) Removal of As(III) and As(V) by natural and synthetic metal oxides. *Colloid Surface A*, 346, 202–207.

Chen, H.W., Frey, M.M., Clifford, D., McNeill, L.S. & Edwards, M. (1999) Arsenic treatment considerations. *Journal of the American Water Works Association*, 91 (3), 74–85.

Childress, A.E. & Elimelech, M. (2000) Relating nanofiltration membrane performance to membrane charge (electrokinetic) characteristics. *Environmental Science & Technology*, 34, 3710–3716.

Chwirka, J., Thomson, B. & Stomp, J.M.I. (2000) Removing arsenic from groundwater. *Journal of the American Water Works Association*, 92 (3), 79–88.

Chwirka, J.D., Colvin, Ch., Gomez, J.D. & Mueller, P.A. (2004) Arsenic removal from drinking water using the coagulation/microfiltration process. *Journal of the American Water Works Association*, 96 (3), 106–114.

Elcik, H., Cakmakci, M., Sahinkaya, E. & Ozkaya, B. (2013) Arsenic removal from drinking water using low pressure membranes. *Industrial Engineering Chemistry Research*, 52 (29), 9958–9964.

Floch, J. & Hideg, M. (2004) Application of ZW-1000 membranes for arsenic removal from water sources. *Desalination*, 162, 75–83.

Fox, K.R. (1989) Field experience with point-of-use treatment systems for arsenic removal. *Journal of the American Water Works Association*, 81 (2), 94–101.

Fox, K.R. & Sorg, T.J. (1987) Controlling arsenic, fluoride, and uranium by point-of-use treatment. *Journal of the American Water Works Association*, 79 (10), 81–84.

Gawroński, R. (2002) Membrany ciekłe-pertrakcja (Liquid membranes-pertraction). *Proceedings of the Membrane School "Membrany i techniki membranowe w przemyśle – Stan obecny i postępy"* [in Polish] *(Membrane and membrane technology in the industry-current status and progress).* Politechnika Warszawska, Wydział Inżynierii Chemicznej i Procesowej, Warsaw, Poland. pp. 73–86.

Ghurye, G., Clifford, D. & Tripp, A. (2004) Iron coagulation and direct microfiltration to remove arsenic from groundwater. *Journal of the American Water Works Association*, 96 (4), 143–152.

Gorenfl, A., Valazquez-Padron, O.D. & Frimmel, F.H. (2002) Nanofiltration of a German groundwater of high hardness and NOM content: performance and costs. *Desalination*, 151, 253–265.

Güell, R., Fontàs, C., Salvadó, V. & Anticó, E. (2010) Modelling of liquid–liquid extraction and liquid membrane separation of arsenic species in environmental matrices. *Separation and Purification Technology*, 72, 319–325.

Güell, R., Fontàs, C., Anticó, E., Salvadó, V., Crespo, J.G. & Velizarov, S. (2011) Transport and separation of arsenate and arsenite from aqueous media by supported liquid and anion-exchange membranes. *Separation and Purification Technology*, 80 (3), 428–434.

Han, B., Runnells, T., Zimbron, J. & Wickramasinghe, R. (2002) Arsenic removal from drinking water by flocculation and microfiltration. *Desalination*, 145, 293–298.

Hering, J.G., Chen, B.Y., Wilkie, J.A. & Elimelech, M. (1997) Arsenic removal from drinking water during coagulation. *Journal of Environmental Engineering*, 9, 800–807.

Jain, A., Raven, K.P. & Loeppert, R.H. (1999) Arsenite and arsenate adsorption on ferrihydrate: surface charge reduction and and net OH^- release stoichiometry. *Environmental Science & Technology*, 33, 1179–1184.

Kang, M., Kawasaki, M., Tamada, S., Kamei, T. & Magara, Y. (2000) Effect of pH on the removal of arsenic and antimony using reverse osmosis membranes. *Desalination*, 131, 293–298.

Kartinen, E.O. & Martin, C.J. (1995) An overview of arsenic removal processes. *Desalination*, 103, 79–88.

Khandaker, N.R., Brady, P.V. & Krumhansl, J.L. (2009) Arsenic removal from drinking water, a handbook for communities. Sandia National Laboratories Albuquerque, Albuquerque, NM.

Kociołek-Balawejder, E. & Ociński, D. (2006) Przegląd metod usuwania arsenu z wód [in Polish] (Review of the methods of removing arsenic from water). *Przemysł Chemiczny*, 85 (1), 19–26.

Kołtuniewicz, A.B. & Drioli, E. (2008) *Membranes in clean technologies*. VCH Verlag GmbH, Weinheim, Germany.

Li, X.J., Liu, C.S., Li, F.B, Li, Y.T., Zhang, L.J., Liu, C.P. & Zhou, Y.Z. (2010) The oxidative transformation of sodium arsenite at the interface of α-MnO_2 and water. *Journal of Hazardous Materials*, 173, 675–681.

Mandal, B.K. & Suzuki, K.T. (2002) Arsenic around the world: a review. *Talanta*, 58, 201–235.

Manning, B.A., Fendorf, S.E., Bostick, B. & Suarez, D.L. (2002) Arsenic(III) oxidation and arsenic(V) adsorption reactions on synthetic birnessite. *Environmental Science & Technology*, 36, 976–981.

McNeill, L.S. & Edwards, M.A. (1995) Soluble arsenic removal at water treatment plants. *Journal of the American Water Works Association*, 87 (4), 105–113.

Ng, J.C., Wang, J. & Shraim, A. (2003) A global health problem caused by arsenic from natural sources. *Chemosphere*, 52, 1353–1359.

Nguyen, C.M., Bang, S., Cho, J. & Kim K.-W. (2009) Performance and mechanism of arsenic removal from water by a nanofiltration membrane. *Desalination*, 245, 82–94.

Ning, R.Y. (2002) Arsenic removal by reverse osmosis. *Desalination*, 143, 137–241.

Oehmen, A., Valerio, R., Llanos, J., Fradinho, J., Serra, S., Reis, M.A.M., Crespo, J.G. & Velizarov, S. (2011) Arsenic removal from drinking water through a hybrid ion exchange membrane – coagulation process. *Separation and Purification Technology*, 83 (1), 137–143.

Oh, J.I., Urase, T., Kitawaki, H., Rahman, M.M., Rahman, M.H. & Yamamoto, K. (2000a) Modeling of arsenic rejection considering affinity and steric hindrance effect in nanofiltration membranes. *Water Science and Technology*, 42 (3–4), 73–180.

Oh, J.I., Yamamoto, K., Kitawaki, H., Nakao, S., Sugawara, T., Rahman, M.M. & Rahman, M.H. (2000b) Application of low pressure nanofiltration coupled with a bicycle pump for treatment of arsenic-contaminated ground water. *Desalination* 31, 307–314.

Oh, J.I., Lee, S.-H. & Yamamoto, K. (2004) Relationship between molar volume and rejection of arsenic species in groundwater by low-pressure nanofiltration process. *Journal of Membrane Science*, 234, 167–175.

Pawlak, Z., Żak, S. & Zabłocki, L. (2006) Removal of hazardous metals from groundwater by reverse osmosis. *Polish Journal of Environmental Studies* 15, 579–583.

Peeters, J.M.M., Boom, J.P., Mulder, M.H.V. & Strathmann, H. (1998) Retention measurements of nanofiltration membranes with electrolyte solutions. *Journal of Membrane Science*, 145, 199–209.

Perez, M.E.M., Aguilera, J.A.R., Saucedo, T.I., Gonzalez, M.P., Navarro, R. & Rodriguez, M.A. (2007) Study of As(V) transfer through a supported liquid membrane impregnated with trioctylphosphine oxide (Cyanex 921). *Journal of Membrane Science*, 302, 119–126.

Prapasawat, T., Ramakul, P., Satayaprasert, C., Pancharoen, U. & Lothongkum, A.W. (2008) Separation of As(III) and As(V) by hollow fiber supported liquid membrane based on the mass transfer theory. *Korean Journal of Chemical Engineering*, 25 (1), 158–163.

Rivas, B.L. & Aguirre, M.C. (2009) Water-soluble polymers: optimization of arsenate species retention by ultrafiltration. *Journal of Applied Polymer Science*, 112, 2327–2333.

Rivas, B.L., Aguirre, M.C. & Pereira, E. (2006) Retention properties of arsenate anions of water-soluble polymers by a liquid-phase polymer-based retention technique. *Journal of Applied Polymer Science*, 102, 2677–2684.

Rivas, B.L., Aguirre, M.C., Pereira, E., Moutet, J.-C. & Saint-Aman, E. (2007a) Capability of cationic water-soluble polymers in conjunction with ultrafiltration membranes to remove arsenate ions. *Polymer Engineering Science*, 47, 1256–1261.

Rivas, B.L., Aguirre, M.C. & Pereira, E. (2007b) Cationic water-soluble polymers with the ability to remove arsenate through ultrafiltration technique. *Journal of Applied Polymer Science*, 106, 89–94.

Rivas, B.L., Pereira, E., Paredes, J. & Sánchez, J. (2012) Removal of arsenate from ionic mixture by anion exchanger water-soluble polymers combined with ultrafiltration membranes. *Polymer Bulletin*, 69 (9), 1007–1022.

Sanchez, J. & Rivas, B.L. (2010) Arsenic extraction from aqueous solution: electrochemical oxidation combined with ultrafiltration membranes and water-soluble polymers. *Chemical Engineering Journal*, 165, 625–632.

Sánchez, J. & Rivas, B.L. (2011) Arsenate retention from aqueous solution by hydrophilic polymers through ultrafiltration membranes. *Desalination*, 270 (1–3), 57–63.

Sánchez, J., Rivas, B.L., Pooley, A., Basaez, L., Pereira, E., Pignot-Paintrand, I., Bucher, C., Royal, G., Saint-Aman, E. & Moutet, J.-C. (2010) Electrocatalytic oxidation of As(III) to As(V) using noble metal-polymer nanocomposites. *Electrochimica Acta*, 55, 4876–4882.

Sánchez, J., Bastrzyk, A., Rivas, B.L., Bryjak, M. & Kabay, N. (2013) Removal of As(V) using liquid-phase polymer-based retention (LPR) technique with regenerated cellulose membrane as a filter. *Polymer Bulletin*, 70 (9), 2633–2644.

Sato, Y., Kang, M., Kamei, T. & Magara, Y. (2002) Performance of nanofiltration for arsenic removal. *Water Research*, 36, 3371–3377.

Schaep, J., Bruggen, B.V., Vandecasteele, C. & Wilms, D. (1998) Influence of ion size and charge in nanofiltration. *Separation and Purification Technology*, 14, 155–162.

Scott, M.J. & Morgan, J.J. (1995) Reactions at oxide surfaces. 1. Oxidation of As(III) by synthetic birnessite. *Environmental Science & Technology*, 29, 1898–1905.

Seidel, A., Waypa, J.J. & Elimech, M. (2001) Role of charge (Donnan) exclusion in removal of arsenic from water by a negatively charged porous nanofiltration membrane. *Environmental Engineering Science*, 18 (2), 105–113.

Shih, M.-C. (2005) An overview of arsenic removal by pressure-driven membrane processes. *Desalination*, 172, 85–97.

Teychene, B., Collet, G., Gallard, H. & Croue, J.-P. (2013) A comparative study of boron and arsenic (III) rejection from brackish water by reverse osmosis membranes. *Desalination*, 310, 109–114.

Toledo, L., Rivas, B.L., Urbano, B.F. & Sánchez, J. (2013) Novel *N*-methyl-D-glucamine-based water-soluble polymer and its potential application in the removal of arsenic. *Separation and Purification Technology*, 103, 1–7.

Urase, T., Oh, J.I. & Yamamoto, K. (1998) Effect of pH on rejection of different species of arsenic by nanofiltration membrane. *Desalination*, 117, 11–18.

Velizarov, S. (2013) Transport of arsenate through anion-exchange membranes in Donnan dialysis. *Journal of Membrane Science*, 425–426, 243–250.

Vrijenhoek, E.M. & Waypa, J. (2000) Arsenic removal from drinking water by a "loose" nanofiltration membrane. *Desalination*, 130, 265–277.

Wang, X.L., Tsuru, T., Nakao, S. & Kimura, S. (1997) The electrostatic and steric-hindrance model for the transport of charged solutes through nanofiltration membranes. *Journal of Membrane Science*, 135, 19–32.

Waypa, J.J., Elimelech, M. & Hering, J.G. (1997) Arsenic removal by RO and NF membranes. *Journal of the American Water Works Associations*, 89 (10), 102–114.

Wickramasinghe, S.R., Han, B., Zimbron, J., Shen, Z. & Karim, M.N. (2004) Arsenic removal by coagulation and filtration: comparison of groundwaters from the United States and Bangladesh. *Desalination*, 169, 231–244.

Xia, S.J., Dong, B.Z., Zhang, Q.L., Xu, B., Gao, N.Y. & Causseranda, C. (2007) Study of arsenic removal by nanofiltration and its application in China. *Desalination*, 204, 374–379.

Xie, Y., Guo, C., Ma, R., Xu, B., Gao, N., Dong, B. & Xia, S. (2013) Effect of dissolved organic matter on arsenic removal by nanofiltration. *Desalination and Water Treating*, 51 (10–12), 2269–2274.

Yeom, C.K., Choi, J.H., Suh, D.S. & Lee, J.M. (2002) Analysis of the permeation and separation of electrolyte solutions through reverse osmosis charged membranes. *Separation Science and Technology*, 37, 1241–1255.

Yoon, Y., Hwang, Y., Ji, M., Jeon, B.-H. & Kang, J.-W. (2011) Ozone/membrane hybrid process for arsenic removal in iron-containing water. *Desalination and Water Treatment*, 31 (1–3), 138–143.

Zhang, T. & Sun, D.D. (2013) Removal of arsenic from water using multifunctional micro-/nano-structured MnO_2 spheres and microfiltration. *Chemical Engineering Journal*, 225, 271–279.

Zhao, B., Zhao, H. & Ni, J. (2010) Arsenate removal by Donnan dialysis: effects of the accompanying components. *Separation and Purification Technology*, 72, 250–255.

Zhao, B., Zhao, H., Dockko, S. & Ni, J. (2012) Arsenate removal from simulated groundwater with a Donnan dialyzer. *Journal of Hazardous Materials*, 215–216, 159–165.

Zheng, Y.-M., Zou, S.-W., Nanayakkara, K.G.N., Matsuura, T. & Chen, J.P. (2011) Adsorptive removal of arsenic from aqueous solution by a PVDF/zirconia blend flat sheet membrane. *Journal of Membrane Science*, 374, 1–11.

CHAPTER 17

Innovative membrane applications for arsenic removal

Alberto Figoli, Priyanka Mondal, Sergio Santoro, Jochen Bundschuh,
Jan Hoinkis & Alessandra Criscuoli

17.1 INTRODUCTION

Arsenic (As) is very toxic to human health and its slow poisoning is observed even in the scalp hair samples of affected people (Uddin *et al.*, 2006). A variety of health hazards is observed due to its long term exposure, including several types of cancer, cardio vascular disease, diabetes, neurological disorders (Abernathy *et al.*, 2003; Kapaj *et al.*, 2006). Therefore, considering the toxicity of As, several national and international agencies including the World Health Organization has reduced the guideline value from 50 to $10\,\mu g\,L^{-1}$. However, a few countries still follow the higher maximum contaminant level (MCL) ($50\,\mu g\,L^{-1}$) (Nriagu *et al.*, 2007). The MCL of the affected countries varies due to lack of appropriate treatment technologies with respect to different socio-politico-economical conditions (Roy *et al.*, 2008). Therefore, an effective treatment technology seems to be essential in order to reduce the As concentration from drinking water. Various treatment technologies have been used till now, including oxidation, coagulation/flocculation, adsorption and membrane based technologies to mitigate the problem (Mondal *et al.*, 2013). A better understanding of these treatment technologies is essential for further research. In this chapter, an overview on traditional technologies (oxidation, coagulation/ flocculation, and adsorption), including membrane pressure driving processes, together with a more detailed discussion on innovative membrane-based technologies, such as forward osmosis (FO) and membrane contactors (MC), is reported.

17.2 TRADITIONAL TECHNOLOGIES FOR TREATMENT OF ARSENIC CONTAMINATED WATER

Among the applied traditional treatment technologies for the treatment of As contaminated water oxidation, coagulation/flocculation and adsorption are widely used together with pressure driven process membranes.

17.2.1 *Oxidation*

This process is often used in combination with other treatment technologies to obtain maximum As removal efficiency. The purpose of oxidation is to convert soluble more toxic As(III) to less toxic As(V). Several oxidants are used to serve the purpose e.g. O_3, pure air and oxygen, H_2O_2, ferrate, Fenton reagent, hypochlorite, permanganate, MnO_2 coated nanostructured capsules, photocatalytic oxidation (TiO_2-based), photochemical oxidation etc. (Criscuoli *et al.*, 2012; Guan *et al.*, 2012; Kim and Nriagu, 2000; Lee *et al.*, 2003; 2011; Vasudevan *et al.*, 2006). Modified oxidants show higher oxidation efficiency than traditional oxidants (air, pure oxygen etc.). Criscuoli *et al.* (2012) observed that MnO_2 coated PEEC-WC nanostructured capsules can achieve over 99% oxidation with initial As concentration of 0.1–$0.3\,mg\,L^{-1}$. Photocatalytic and photochemical oxidation have also been investigated in many studies. UV irradiation helps to increase the oxidation of As(III) by generating hydroxyl radicals through the photolysis of Fe(III) species and in consequence the oxidation rate becomes faster in the presence of both hydroxyl radicals and O_2 (Sharma *et al.*, 2007; Yoon and Lee, 2005). Yoon *et al.* (2008) observed that the oxidation efficiency of As(III) increases in natural water due to the presence of Fe(II) and H_2O_2 when 185 and 254 nm vacuum UV lamps were used. Photocatalytic oxidation, followed by adsorption on TiO_2 was also previously studied by Dutta *et al.* (2004) and Miller *et al.* (2011). Miller and Zimmerman (2010) observed improved adsorption of As during oxidation followed by UV radiation in the presence of TiO_2-impregnated chitosan beads [6400 µg As(III) g^{-1} and 4925 µg As(V) g^{-1}] compared with the absence of UV light [2198 µg As(III) g^{-1} TICB and 2050 µg As(V) g^{-1}]. Therefore, oxidation is usually considered as a pretreatment step. However, the selection of oxidants and their dosage are very important with respect to initial As concentration.

17.2.2   *Coagulation/flocculation*

Coagulation/flocculation is also used for the treatment of As contaminated drinking water. As is removed from water by the addition of coagulant followed by the formation of flocs. Iron and aluminum-based coagulants are widely used for the purpose. Bilici Baskan and Pala (2010) used 42–56 mg L^{-1} of $Al_2(SO_4)_3$ and removed ∼100% of As(V) with initial concentration of 500–1000 µg L^{-1}. Hu *et al.* (2012) also used three aluminum-based coagulants (aluminum chloride and two types of polyaluminum chloride). As was removed below the MCL, when the initial As(V) concentration was 280 µg L^{-1}. Aluminum-based coagulants are less stable, as they can operate only in a very narrow range of pH, compared to iron-based coagulants (Hering *et al.*, 1997). Iron-based coagulants are used by several researchers for the treatment of As contaminated water (Andrianisa *et al.*, 2008; Lacasa *et al.*, 2011; Song *et al.*, 2006). As was removed by these iron-based coagulants due to precipitation/co-precipitation on amorphous iron hydroxide. Lakshmanan *et al.* (2010) used $FeCl_3$ as a coagulant for As removal and compared the efficiency between electro coagulant and chemical coagulant. They observed that iAs(V) removal was efficient at pH 7.5 and 8.5 during electrocoagulation However, the concentration of soluble Fe^{2+} (10–45%) in treated water at pH 7.5 is a major concern. Similarly, Pallier *et al.* (2010) used kaolinite and $FeCl_3$ as coagulant/flocculent for iAs removal. They observed above 90% iAs(V) and 70% iAs(III) removal with 9.2 mg L^{-1} Fe^{3+} and suggested that removal of iAs(V) was related to the zeta potential of colloid suspension, whereas iAs(III) removal was related to the available sites of hydroxide on the coagulant surface. However, formation of a large amount of sludge with high As concentration is a major limitation of this process.

17.2.3   *Adsorption*

Adsorption is most extensively and successfully used among all the conventional processes. Various adsorbents are used to treat As contaminated water, from conventional adsorbents (e.g. activated carbon, zeolite, clay minerals etc.) to new adsorbents, including nanoscale adsorbents (synthesized or modified conventional adsorbents). There are two types of activated carbons that are available commercially, powdered and granular. Powdered activated carbon shows higher efficiency due to its large surface area. However, separation from water is more difficult than with granular activated carbon. Nowadays, modified conventional adsorbents are widely used due to their improved performance. Surface modified clay, manganese-supported activated alumina, zirconium oxide and granular ferric oxide have also been used for removal of As(V) (Altundogan *et al.*, 2002; Badruzzaman *et al.*, 2004; Kunzru and Chaudhuri., 2005; Li *et al.*, 2007; Mann *et al.*, 1999; Su *et al.*, 2011).

Iron-based adsorbents are most popular among all the adsorbents and they can be divided into two groups depending on the remediation chemistry. Iron can either act as a sorbent, co-precipitant or it can be used as a reducing agent to convert contaminants into their lower oxidation state. As(V) and As(III) form strong inner sphere complexes through surface complexation with iron hydroxides e.g. goethite, hydrous ferric oxide, siderite, hematite in the minerals and thus they can potentially remove As via adsorption (Guo *et al.*, 2007; Mohan and Pittman 2007; Smedley and Kinniburgh, 2002). Zero-valent iron (ZVI) can remove As from aqueous solution via precipitation/co-precipitation/adsorption on iron hydroxide precipitation in arid conditions (Leupin and Hug, 2005; Sun *et al.*, 2006). However, in anoxic conditions the reaction mechanism is comparatively slow between lower and higher oxidation states of As [gradually from As(V) to As(III) to As(0)] (Mondal *et al.*, 2014c; Sun *et al.*, 2011). Thus, the reaction mechanism of As with ZVI is different under oxic and anoxic conditions and in the absence of oxygen, ZVI acts as a reducing agent. Researchers are now also focused on the advanced adsorbent materials (e.g. nanomaterials, doped iron-based adsorbents etc.) for As removal. Nanoparticles, especially nZVI, are considered a suitable option, as they can remove contaminants faster and efficiently due to their high specific surface area and reactivity (Bhowmick *et al.*, 2014; Ramos *et al.*, 2009). The major disadvantage of this process is the disposal of the spent adsorbent sorbed with toxic metals.

17.2.4 *Membrane technology*

Arsenic species can also be separated from aqueous solution using membrane-based technologies e.g. microfiltration (MF), ultrafiltration (UF), nanofiltration (NF), reverse osmosis (RO), electro-dialysis (ED) etc. Sato *et al.* (2002) stated that the NF process is more effective than conventional processes, since it is less affected by the chemical composition and pH of the feed solution. However, it is observed that As(V) and As(III) rejection increases with increasing pH.

MF membranes are not very effective for removal of dissolved As, as they are only able to remove the particulate form, which is very rare in natural water (Shih *et al.*, 2005). Similarly, Velizarov *et al.* (2004) reported that UF membranes are also not favorable for As removal. as the membrane pores are not small enough to remove soluble As. Therefore, these membrane technologies alone are not able to remove soluble As from contaminated water. For this reason either a combined process of precipitation and flocculation or a particle size enhancer technique together with a membrane treatment is necessary. Ghurye *et al.* (2004), Wickramasinghe *et al.* (2004) and Han *et al.* (2002) studied coagulation with ferric chloride and ferric sulfate and MF for removal of As. The studies demonstrated that a membrane pore size of around $0.2\,\mu m$ is usually necessary in order to achieve a high degree of As removal, whereas the removal efficiency depends on coagulant dosage, pH and ferric counter-ions.

Iqbal *et al.* (2007) showed that the efficiency of UF membranes can be increased by the addition of micelle enhanced cationic surfactant, where As species bind or adsorb onto these surfactants and are removed from solution via micelle formation.

NF and RO membranes are similar to each other and the only difference is the network structure. NF membranes have a more open network structure than RO and are generally negatively charged in neutral or alkaline solution. Therefore, electrostatic repulsion (Donnan exclusion) between the anionic species and the charge of the membrane is the predominant factor for the separation of negatively charged species (Velizarov *et al.*, 2004). With increase in pH of the solution, $H_2AsO_4^-$, $HAsO_4^{2-}$ and AsO_4^{3-} are the varying predominant species of iAs(V) while for iAs(III), neutral H_3AsO_3 is transformed to anionic $H_2AsO_3^-$. As a result, there is an increase in electrostatic repulsion between the ions and the membrane, resulting in an increase in rejection of both As(III) and As(V). However, the rate of diffusion through the membrane is also another responsible factor for separation (Velizarov *et al.*, 2004). NF are more sensitive than RO with respect to pH and the ionic strength of the solution and similar rejection of contaminants can be achieved as for RO, with high water flux. Regarding As removal, this is only true for the charged species such iAsV at pH around 7. However, for the neutral iAs(III) the rejection of NF is much lower compared to RO membranes (Shih *et al*, 2005). Solution pH is important with respect to the separation of target ions, as the charge of the NF membranes is not fixed and can be varied from positive to negative depending on the pH of the solution. As with NF membranes, RO membranes also have been widely used for water desalination and a very high rejection efficiency (>99%) of the low molecular mass compounds (organic salts or small organic molecules) can be achieved (Velizarov *et al.*, 2004).

Akin *et al.* (2011) and Teychene *et al.* (2012) found that the rejection of metalloids depends on the pH of the solution, the applied transmembrane pressure and characteristics of the RO membranes. They showed that regarding As removal from seawater, membranes are typically more efficient than brackish water membranes. Removal of As(V) is a maximum above pH 4.0, whereas As(III) could be effectively removed only at pH > 9.1 (Akin *et al.*, 2011). So at ambient, neutral pH typically only As(V) shows sufficient removal efficiency. A bench-scale cross-flow flat-sheet filtration system using LFC-1 RO membrane and ESNA and MXO7 NF membranes to remove As(III) and As(V) from model and natural waters, was performed by Yoon *et al.* (2009). While As(III) rejection was low below pH 10 (11–30%), a significant rejection of As(V) was observed (>90%). It was noticed that lessening of total As concentration below the MCL was difficult (efficiency decreases below 50%) when the As(III) concentration was slightly high (>350 μg L^{-1}) (Geucke *et al.*, 2009; Walker *et al.*, 2008). Therefore, the presence of As(III) plays a significant role for removal of total As and a pretreatment step (mostly oxidation) seems essential for obtaining maximum rejection efficiency. Richards *et al.* (2009) studied the efficiency of a photovoltaic powered RO desalination system where UF and RO membranes were combined as a two-stage membrane system to treat Australian groundwater. A decrease in flux, from 24.2 to 22.5 L m^{-2} h^{-1}, was evident due to precipitation on the surface of the membrane, while an As rejection, from 65 to 79%, has been obtained, depending on the type and characteristics of the membrane type used. Therefore, selection of membrane with respect to the presence of aqueous species and contaminants is an important factor.

Removal of As(V) and As(III) by NF was studied by Xia *et al* (2007) from synthetic water at different pHs. They used a spiral NF module with an active polyamide layer for a pilot system and found that As(V) can be totally removed, whereas for As(III), only 5% removal was achieved. However, the rejection was increased with increase in pH for both these species. Similar results were also found by Pérez-Sicairos *et al.* (2009). Recently Saitua *et al.* (2011) used a spiral-wound NF membrane (NF-300, a TFC polyamide membrane) and investigated the importance of ionic composition on the transmission of ions and on rejection of the membrane. The rejection of ions from multi-component solutions was significantly different from that of individual salt solutions and in the presence of divalent co-ions, monovalent co-ion rejection decreased significantly. Thin film composite NF membrane was also very useful and can decrease the concentration of As(V) below the MCL together with total dissolved solids and other contaminants; the rejection can be achieved within 180 min operation of the system (Harisha *et al.*, 2010). However, the separation depends on the pH of the solution and the presence of other ions. In another study, Nguyen *et al.* (2009) investigated the rejection of As(V) and As(III) by varying the initial As concentration, types of electrolytes and As speciation. The authors obtained a higher rejection percentage for As(V) than for As(III) and explained this effect by the dominance of Donnan exclusion over steric exclusion. Similarly, the rejection of As from contaminated groundwater by two commercial NF membranes was studied by Kosutíc *et al.* (2005) and they concluded that the removal of As by the negatively charged membrane occurred due to charge exclusion. On the other hand, Figoli *et al.* (2010) used two commercially available NF membranes and studied the effects of several operating parameters (e.g. transmembrane pressure, pH, temperature, As concentration) on As removal. Maximum rejection of As(V) (91%) was achieved at higher pH and lower operating temperature and feed concentration. However, in the presence of high transmembrane pressure, the rejection of As decreased. On the other hand a pilot scale spiral NF membrane module was tested for removal of As along with other ions (F$^-$ and HCO$_3^-$) by Perez Padilla *et al.* (2010) and 93% removal was achieved at pH 8 and 0.7 MPa pressure. They also used a bicycle pump to produce operational energy (0.2 MPa) and 91.6% As rejection. That can be very useful in the remote areas where the power supply is limited. Although the membrane pressure driving processes can remove As considerably, further improvement is necessary due to some inherent drawbacks of the process, as discussed in details in this paragraph (Figoli *et al.*, 2010b; Mondal *et al.*, 2013). In particular, treatment and disposal of As-laden concentrates pose a problem, especially in remote regions.

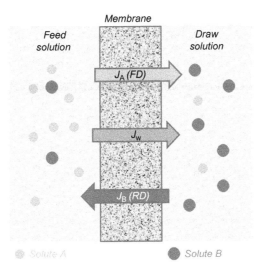

Figure 17.1.   Representation of the FO process. FD: Forward solution diffusion; RD: Reverse solution diffusion.

17.3   NOVEL MEMBRANE TECHNOLOGIES FOR ARSENIC REMOVAL

17.3.1   *Forward osmosis*

Forward osmosis (FO) is a membrane process that has gained considerable interest for sea and brackish water desalination in recent years (Shaffer *et al.*, 2013; Smith and Reynolds, 2014). NF and RO usually use hydraulic pressure, while FO uses, as a driving force across the membrane, an osmotic pressure gradient that is generated between the aqueous feed and a concentrate solution, known as draw solution, to separate water from dissolved solutes. The scheme of the process is shown in Figure 17.1.

The membrane could be usually placed either with its active layer towards the feed solution (AL-FS) or its active layer towards the draw solution (AL-DS), as shown in Figure 17.2, depending on the application.

The FO membrane process has several advantages over pressure driven membrane processes, as documented in many papers (Cath *et al.*, 2006; Macedonio *et al.*, 2012; Mondal *et al.*, 2013). The main advantages can be summarized as follows:

- FO works in the absence of hydraulic pressure and, thus, its energy requirement with respect to the operating pressure is lower. Moreover, it does not require high strength materials (Klaysom *et al.*, 2013). Additionally, if recovery of the draw solution is not required, FO is more energy efficient than RO and can be applied in low electricity accessible areas (Coday *et al.*, 2014);
- It is possible to efficiently reject solutes (including As) and total dissolve solids from complex solutions due to the small mean pore radius (0.25–0.37 nm) of commercial FO membranes (Cath *et al.*, 2006; Coday *et al.*, 2014; Fang *et al.*, 2014; Mondal *et al.*, 2014a);
- FO works in high salt concentration solutions and complex solutions, avoiding any extensive pre-treatment step;
- FO membranes are less susceptible to fouling than RO and NF since the water flux is low (Xie *et al.*, 2014) and easy to clean (Hoover *et al.*, 2011);
- The FO process allows treatment of highly saline streams, at high pressure (>8.3 MPa), which is usually not possible with RO (Hydranautics, 2014);
- The FO technology can be applied for dewatering feed and is much simpler, greener and efficient than other dewatering technologies (Chung *et al.*, 2012).

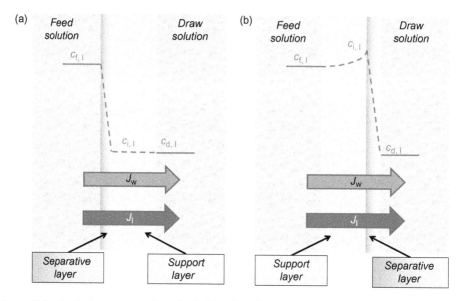

Figure 17.2.   Typical membrane orientation in FO: (a) Active layer towards the feed solution (AL-FS); (b) active layer towards the draw solution (AL-DS).

Therefore, FO can be used for various types of applications. Due to its potentiality, several companies started manufacturing FO membranes mainly in the form of flat sheet. The Company Hydration Technology Innovation (HTI, USA) is the biggest FO manufacturer for cellulose triacetate (CTA) and cellulose acetate (CA) FO membranes. Nanyang Technological University (NTU, Singapore) also developed CA, TFC-1, TFC-2, TFC-3, TFC-4, TFC-5 and nano-fiber composite FO membranes. DOW chemical and Toray produce some TFC RO and CA RO membranes which are applied also in FO applications.

17.3.1.1   *Draw solutions*

Various types of draw solutions have been used for water purification, like organic (glucose, fructose), salts (NaCl, MgSO$_4$, ammonium bicarbonate), volatile solutes (SO$_2$), polymeric hydrogel, magnetic and polymeric nanoparticles etc. The draw solutions can usually be reused after removing the permeated fresh water. Since the draw solution plays an important role in this process, the selection of appropriate draw solutions is very important. In fact, selected draw solutions should not be toxic but easily recyclable and able to generate high osmotic pressure. The diffusivity of the draw solution through the membrane is a significant parameter that needs to be considered (Cornelissen *et al.*, 2008). Careful selection of the draw solution with respect to its application is essential, as mineral scaling on the membrane is possible due to changes in operating parameters (pH, temperature etc.) (Achilli *et al.*, 2010). Recovery of the draw solution from product water is also an important feature for a thermodynamically closed looped system. Thermal recovery, membrane distillation or other pressure driven membrane processes can be used for this purpose. However, it depends on the required recovery rate, type of application and energy consumption of the unit (Lutchmiah *et al.*, 2014). The ideal draw solution does not exist that can fulfill all those important criteria, yet. Therefore, further modification and improvement for the draw solution and also energy evaluation of the integrated systems to be used for the draw solution recovery are required.

17.3.1.2   *FO membrane materials*

An ideal FO membrane should contain an ultrathin, dense, active separating layer for contaminant rejection; the support layer should have high mechanical stability with less ICP; the membrane

Figure 17.3.   SEM image of HTI CTA membrane (adapted from Mondal, 2014c).

should be highly hydrophilic to obtain larger water flux and with a low tendency to fouling. Cellulose triacetate (CTA) FO membrane from HTI is the most popular and widely used for proving the FO potentialities. Figure 17.3 reports a typical cross-section of the FO membrane by scanning electron microscopy. This membrane showed higher water flux and less ICP but it also showed some drawbacks, including less chlorine resistance and low water flux (Lior *et al.*, 2013; Lutchmiah *et al.*, 2014; Mi and Elimelech, 2008).

Klaysom *et al.* (2013) reported that thin film composite (TFC) membranes are much better than CTA membranes, as they are stable over a broader pH range and have also higher permeability. However, hydrophilic cellulose acetate (CA), having higher anti-fouling properties (Zhang *et al.*, 2010), can also be successfully applied in FO applications. These newly synthesized membranes can be doped with pore-forming agents that enhance the water flux and contaminant rejection (Sairam *et al.*, 2011). Several other strong and stable polymers like polybenzimidazole, polyamide-imides, nanoporous polyethersulfones are also used for the synthesis of FO membranes (Setiawan *et al.*, 2011; Wang *et al.*, 2009; Yu *et al.*, 2011). Chemically modified polybenzimidazole membranes (fabricated through cross linking) are made by very strong and temperature-stable thermoplastic polymers that are self-charged in aqueous solution and help to improve salt rejection due to finely tuned pore size with low fouling propensity (Wang *et al.*, 2009). These polymers can increase the contaminant rejection and water flux and decrease the membrane fouling and ICP in FO membranes. Setiawan *et al.* (2011) fabricated positively charged NF-like membranes with polyamide-imides. The positively charged active layer is supported by microporous hollow-fibers. These membranes provide double electric repulsions in the active layer facing the feed solution (ALFS) mode to the salt transfer through the membrane and reduce salt penetration. On the other hand, when the active layer faces the draw solution (ALDS) mode, the same membrane facilitated salt transport due to positive charge. Finally, several researchers employed commercially available RO membranes, which have as their main drawback the relatively thick support layer and thus are not appropriate for FO application (McCutcheon and Elimelech, 2008).

17.3.1.3   *FO membrane development methods*
Usually a membrane can be synthesized using various methods (phase inversion, interfacial polymerization, etc.) but FO membranes are generally synthesized using the traditional phase inversion method. These membranes consist of an asymmetric structure made of a thin, dense active layer supported by a micro-porous layer. However, extensive research is still needed for improving

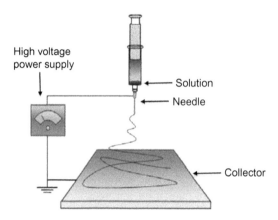

Figure 17.4.   Electrospinning set-up to prepare FO membrane (adapted from Alsvik and Britt Hägg, 2013).

the performance and stability of FO membranes. Recently, Zhang *et al.* (2014) synthesized a hollow-fiber asymmetric FO membrane, employing advanced co-extrusion technology, for better mechanical stability. The membrane consists of a fabricated polyethersulfone support layer and a TFC active layer.

Electro-spinning methods to prepare composite membranes have also gained considerable interest in recent years. A high voltage is induced into the polymer solution that emerges from a needle and is then dried and solidified. The low-cost structured polymer fibers (with diameters in the range 40–2000 nm) are collected on an electrically conducting screen of lower potential (He, 2008; Ramakrishna, 2005) as shown in Figure 17.4.

Bui *et al.* (2011) prepared a different type of flat-sheet polyamide composite membrane supported by a nonwoven web of electrospun nanofibers. This support layer is highly porous and increases the effective membrane area by reducing the area coverage by the support layer. This type of membrane shows two to five times higher water flux than commercial HTI-CTA membrane. However, a proper understanding of polymer chemistry, rheology and electrostatics is necessary. Another innovative PA-coated macro-void-free PES hollow-fiber was synthesized by Sukitpaneenit and Chung (2012). This membrane shows around two to two and half times higher water flux than commercially available HTI-CTA membrane (2 M NaCl was used as draw solution) due to less ICP in the presence of macro-void-free hollow-fiber with a highly sponge-like structure. On the other hand, Tiraferri *et al.* (2011) investigated the effect of TFC composite membrane support layer structure on FO performance. The membranes consisted of a selective PA active layer formed by interfacial polymerization on top of a PS support layer fabricated by phase separation. A rejection of more than 95.5% salt together with high water flux of 4 to 25 $Lm^{-2} h^{-1}$ were obtained when 1 M NaCl draw solution and a deionized water feed solution were used. They confirmed the hypothesis that an optimal FO membrane should contain a mixed-structure support layer, where a thin sponge-like layer sits on top of highly porous macro-voids. However, further development, modification and detailed study is still necessary for making membranes for commercial purposes.

17.3.1.4   *Application of FO for As removal*
FO has been used in several fields for (i) the treatment of industrial wastewater, (ii) seawater desalination, (iii) the treatment of landfill leachate, (iv) the treatment of liquid food and (v) for the water treatment of toxic metal species. Despite the importance of the As removal from contaminated water, only a few studies have been performed to explore the efficiency of FO for this specific application. The effect of organic fouling and membrane orientation on As(III) removal in a laboratory scale FO cross-flow membrane filtration set-up was studied by Jin *et al.*

(2012). It was found that the mode with the active layer (CTA) facing the draw solution (AL-DS) involved severe concentrative internal concentration polarization (ICP) and thus the removal of As species was lower in the AL-DS mode (~50% at pH 6) than with the active layer facing the feed solution (AL-FS) mode (~80% at pH 6). NaCl (5 M) was used as draw solution and obtained 90% rejection of As(III) from aqueous solution at pH ~ 6. The membrane sieving effect also enhanced rejection of relatively large molecular weight arsenious acid (compared with boric acid) in the ALFS mode (from ~78 to 82%) due to the formation of a fouling layer on the membrane active layer. The As(III) rejection between FO and RO operation in both ALFS and ALDS modes was also evaluated. Water flux increases at a faster rate than the flux of solutes permeating across the membrane in FO operation. Therefore, rejection of arsenite increases with increasing water flux due to the enhanced dilution effect in FO operation compared with RO operation (~68% in RO and ~78% in FO operation at pH ~ 6). Cui *et al.* (2014) studied the performance of a novel FO process for the removal of several heavy metal ions (Cr, Pb, Cd, Hg, Cu) from wastewater, including As. TFC FO membrane made from interfacial polymerization on a macro-void-free polyimide support and a novel bulky hydro-acid complex, as draw solution, were used. At room temperature above 99.5% removal of As and other heavy metals (Cr, Pb, Cd, Hg, Cu) and $11 \, L \, m^{-2} \, h^{-1}$ water flux was achieved when 1 M draw solution was used with initial feed concentration of $2000 \, mg \, L^{-1}$. The rejection obtained was higher than in the majority of the NF processes. Moreover, the As rejection above 99.7%, with enhanced water flux to $16 \, L \, m^{-2} \, h^{-1}$, was achieved by increasing the temperature up to 60°C.

Butler *et al.* (2013) performed FO experiments for As(V) removal using a HTI's osmotic water purification (Hydrowell) system. The system was able to remove other toxic metals (e.g. copper, lead, chromium etc.), including As above 88.3%, with a feed concentration of $10 \, mg \, L^{-1}$ during bench-top testing and produced a clean sugar-electrolyte drink from almost all types of water source.

The effect of several physico-chemical parameters (draw solution concentration, feed solution pH, feed solution concentration, membrane orientation) on As removal was recently studied by Mondal *et al.* (2014a). The cellulose triacetate FO membrane (HTI, Scottsdale, AZ) and two different draw solutions (MgSO$_4$ and glucose) were used in the experiments. The removal of As(V) increased, from 93.4 to 96.3%, with increasing draw solution concentration due to the increase in the operating osmotic pressure. Similarly, As(V) rejection increased, from 92.5 to 98.2%, with increasing pH of the solution with glucose draw solution. This is due to the increase in repulsion between negatively charged As species and the negatively charged FO membrane. The water flux remained practically constant, increasing the pH, at a value of about 2, and $1.3 \, L \, m^{-2} \, h^{-1}$, using MgSO$_4$ and glucose, respectively. In the case of As(III), using MgSO$_4$ as draw solution and increasing the pH, the water flux remained constant to a value of about $1.3 \, L \, m^{-2} \, h^{-1}$ while the rejection increased up to 98% at pH = 10 as shown in Figure 17.5.

Finally, increasing the As concentration in the feed solution, from 100 to $500 \, \mu g \, L^{-1}$, the water flux remained practically constant at a value of about 2 and $1.3 \, L \, m^{-2} \, h^{-1}$, using MgSO$_4$ and glucose, respectively. However, the As(V) removal decreased with increasing As(V) feed concentration as As permeation increased, from above 98 to about 94%, through the membrane when glucose was used as draw solution (Mondal *et al.*, 2014a).

The same authors (Mondal *et al.*, 2014b) investigated the effect of several competing and coexisting ions on As removal from aqueous solution for evaluating by FO process. According to their study, silicate did not show any significant effect on As(V) removal, since negatively charged silica species are absent at neutral pH. Vice versa, other ions strongly effect the As removal efficiency. In particular, the following trend has been found: humic acid > bicarbonate > nitrate > fluoride > sulfate > phosphate and similar selectivity was observed for synthetic groundwater. It was also found that humic acid helped to reject As (above 98%) due to the formation of a gel-like foulant layer that improves rejection due to steric hindrance via pore blocking and complexation (Fig. 17.6). On the other hand, P(V) showed a negative effect due to its similar structure with As(V). A hydrated humic acid fouling layer also helped to improve water flux (about $2 \, L \, m^{-2} \, h^{-1}$) by diffusion of water molecules through the membrane. Similarly, As rejection improved in the

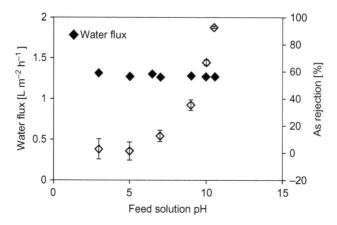

Figure 17.5.   Effect of feed solution pH on permeate flux and removal of As(III) by MgSO₄ (temperature = 22.5°C, pH of draw solution = 7), adapted from Mondal *et al.* (2014a).

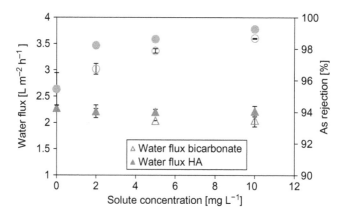

Figure 17.6.   Effect of bicarbonate and humic acid (HA) on permeate flux and removal of As(V) by glucose draw solution for FO membrane module in AL-FS mode (temperature = 22.5 ± 1°C, pH (feed and draw) = 7 ± 0.5, As(V) concentration = 30 ± 0.5 mg L⁻¹), Mondal *et al.*, 2014b.

presence of bicarbonate above 98% and the water flux remained practically constant to a value of about $2\,L\,m^{-2}\,h^{-1}$ (Fig. 17.7). The fraction of negatively charged As species increases in the presence of bicarbonate due to increase of the solution pH by elimination of excess CO_2, which simultaneously helped to reject As by the negative FO membrane (Mondal *et al.*, 2014b).

Although only a few studies are reported on As removal by FO membranes, the potentiality of FO, for removing As from aqueous solutions, has been successfully demonstrated. However, further research is necessary for making this process feasible at the industrial scale.

17.3.2   *Membrane distillation*

Membrane contactors (MC) represent an innovative class of membrane operations developed in the last few decades. MCs are based on the use of a microporous membrane as an inert barrier between two phases, avoiding their mixing but allowing their contact at each pore mouth (Drioli *et al.*, 2006).

Recent studies open up interesting perspectives for the treatment of aqueous solutions containing As by MCs. However the employment of MCs on an industrial scale is still far off.

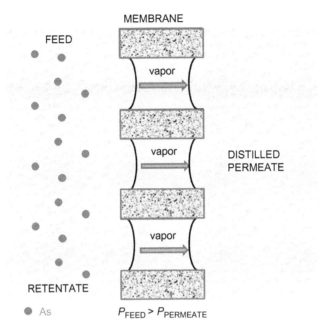

Figure 17.7.   Scheme of direct contact membrane distillation process.

Membrane distillation (MD) is an example of MCs. In the treatment of aqueous solution using MD processes, MCs favor the transport of water vapor and of other volatile compounds through a porous hydrophobic membrane, thanks to a difference of partial pressure that can be created by: (i) sending a colder aqueous stream at the distillate side (direct contact membrane distillation: DCMD); (ii) providing an air gap at the distillate side (air gap membrane distillation: AGMD); in this case the permeating species are condensed over a cold surface inside the membrane module after crossing the air gap; (iii) applying vacuum at the distillate side (vacuum membrane distillation: VMD); (iv) sending a gas stream at the distillate side (sweep gas membrane distillation: SGMD). On the other hand, non-volatile compounds, like salts or metal ions, are rejected. Thus, theoretically speaking, the MD enables the production of pure water purified from As or any kind of non-volatile contaminants. A scheme of direct contact membrane distillation is shown in Figure 17.7.

The hydrophobicity of the membrane avoids the transport of liquid water through the membrane pores at a pressure below the liquid entry pressure (p_{LEP}), which depends on the pore size (r), the surface tension of the water (σ) and the water-membrane contact angle (θ):

$$p_{LEP} = -\frac{2\sigma \cos\theta}{r}$$

MD offers the possibility to operate at relatively low temperature and to provide energy coupling with alternative sources, such as solar and geothermal energy. Furthermore, MD presents the advantage of operating at low pressures (generally below a few hundred kPa), allowing the employment of membranes with moderate mechanical characteristics and minimizing the fouling. It is also able to work efficiently on highly concentrated feeds, overcoming the limits of reverse osmosis. In fact, compared with the common pressure-driven membrane processes, MD is less dependent on the initial concentration of the feed and higher rejections are usually achieved (Lawson et Lloyd, 1997). In general, membranes with pore sizes ranging from 0.05–1 μm should be employed in MD (Manna *et al.*, 2010).

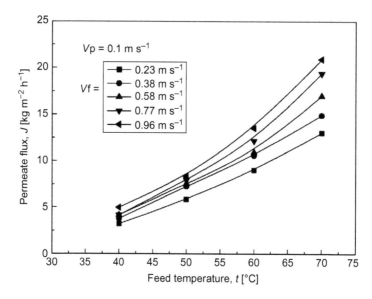

Figure 17.8.   Permeate flux change as a function of the feed temperature and feed velocity (Qu *et al.*, 2009).

The properties required for the preparation of MCs useful in MD are high hydrophobicity and low thermal conductivity. Polytetrafluoroethylene (PTFE), polypropylene (PP), polyethylene (PE), and polyvinylidenefluoride (PVDF) are polymers commonly employed for the preparation of porous hydrophobic membranes in MD.

Among the different membrane configurations, most of the studies published in the literature are performed using the DCMD configuration. However, the research on AGMD and VMD configurations has significantly increased in recent years. In the following, the results obtained with the different configurations, when applied to the treatment of water contaminated by As, are reported and discussed.

Qu *et al.* (2009) reported the removal of arsenite (As(III)) and arsenate (As(V)) by direct contact membrane distillation (DCMD). Self-made hollow-fiber polyvinylidene fluoride (PVDF) membranes, with an average pore size of 0.15 μm and porosity of about 80%, were prepared. A module, containing 50 hydrophobic PVDF membranes, with a total membrane area of $12.56 \times 10^{-4} \, m^2$, was placed in the DCMD system. The feed was pumped in on the lumen-side of the hollow-fibers, while the permeate was collected on the shell side. In Figure 17.8, the results show a typical trend in which the permeate flux increases, from $3–5 \, kg \, m^{-2} \, h^{-1}$ to $12–20 \, kg \, m^{-2} \, h^{-1}$, increasing the feed temperature from 40 to 70°C and the feed velocity from 0.23 to $0.96 \, m \, s^{-1}$, respectively.

As(III) and As(V) in the permeate were below the MCL ($10 \, \mu g \, L^{-1}$) until the feed As(III) and As(V) increased to 40 and $2000 \, mg \, L^{-1}$, respectively. Therefore, the results showed that compared with pressure-driven membrane processes, DCMD has higher removal efficiency 99.95 and 99.99% for both arsenite and arsenate, respectively.

Pal and Manna (2010) investigated the performance of three different commercial membranes (Table 17.1) in As removal from contaminated groundwater by a solar-driven DCMD process, heating the feed water by an evacuated type solar glass panel. In this study, tests were performed using as feed As-contaminated groundwater, coming from Chakdah of south 24 Parganas of West Bengal (India), and containing As in the range of $0.3–0.6 \, mg \, L^{-1}$. Tests on synthetic solutions, in which both trivalent and pentavalent As was dissolved, were also carried out.

The design of the membrane module for minimizing thermal and concentration polarization, together with the choice of the optimal membrane (PTFE), led to complete As removal (100%), and to the achievement of high fluxes ($49.8 \, kg \, m^{-2} \, h^{-1}$).

Table 17.1. Properties of membranes employed in solar-driven DCMD for arsenic removal (modified from Pal and Manna, 2010).

Membrane	Pore size [μm]	Porosity [%]	Thickness active layer [μm]
Composite PTFE/PP	0.22	80	60
Composite PTFE/PET	0.22	80	60
Symmetric PP	0.22	35	160

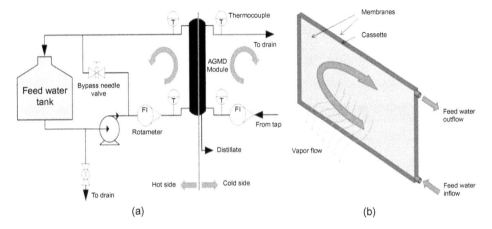

Figure 17.9. (a) AGMD setup; (b) Membrane module containing the PVDF membranes (Khan and Martin, 2014).

Using the same solar-driven DCMD system, Manna *et al.* (2010) tested a flat PVDF membrane, with a nominal pore size of 0.13 μm, thickness of 150 μm and porosity of 70–75%, and were able to remove almost 100% of As from contaminated groundwater with an As content varying from 0.3 to 0.5 mg L^{-1}. Furthermore, the MD system was run continuously, without changing the membranes, for 4 days. By increasing the temperature from 40 to 60°C, the fluxes improved from 74 kg m^{-2} h^{-1} to 95 kg m^{-2} h^{-1} (operative conditions: $Q_{feed} = 120$ L h^{-1} and $Q_{distillate} = 150$ L h^{-1}, $T_{distillate} \approx 20$–22°C, C_{feed} (As) = 396 μg L^{-1}). The performance observed and the efficiency of the solar-driven process were validated by theoretical studies (Pal *et al.*, 2013).

In Bangladesh, an HVR (household water purifier) system, developed by the Swedish HVR Water Purification AB Company (and based on AGMD configuration), was successfully employed for the removal of As from contaminated water containing As, in the range of 0.24–0.33 mg L^{-1} (Islam, 2005). The system includes a membrane module with two flat PTFE membranes, having a thickness of 0.2 mm and porosity of 80%, and a total membrane area of 0.2 m^2. The AGMD system worked with a temperature of 80°C at the feed side, while the purified permeate, containing an amount of As below 1 μg L^{-1}, was condensed on a surface cooled by a stream of water at 29°C. More recently, Khan and Martin (2014) reported again on the use of the same AGMD system (HVR Water Purification AB, Stockholm, subsidiary of Scarab Development AB) as a promising method for small-scale, low cost deployment for rural and remote areas in Bangladesh. In particular, a AGMD commercial prototype, having a nominal capacity of 2 L h^{-1}, has been tested with three different feed-stocks: As-containing groundwater (medium concentration) and As-spiked tap water (medium and high concentrations). Two flat polymeric membranes, allocated in the module inside the AGMD prototype, were also made of polytetrafluoroethylene (PTFE), supplied by Gore, with a porosity of 80% and thickness of 0.2 mm. The operating conditions on the cold side are a flow rate of 1.9 L min^{-1} and a range of coolant inlet temperatures from 15 to 70°C. The AGMD set-up and module are shown in Figure 17.9.

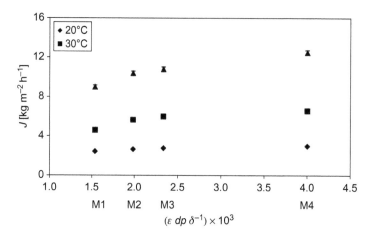

Figure 17.10.   Transmembrane flux (J) of the different membranes used at different temperatures (feed: distilled water, Reynold number: 1700, vacuum pressure: 1 kPa (10 mbar)), (Criscuoli et al., 2013).

By operating at 80°C at the feed side, with a constant feed flow of 3.8 L min^{-1}, the flux obtained ranged between 6 and 30 L m^{-2} h^{-1} by varying the coolant temperature from 70°C to 10°C, respectively. The possibility of integrating AGMD with various innovative thermal systems, such as biomass-derived waste heat, solar thermal, etc., was explored for keeping the temperature level at the hot feed side of about 80°C. Moreover, cold side temperatures were also increased up to 70°C while exhibiting reasonable yields, opening up further possibilities for thermal integration. For all the different feed water types used, it has been proved that the AGMD prototype was able to produce water with As levels well below WHO accepted limits (10 μg L^{-1}), even with the initial As concentrations over 1800 μg L^{-1}.

In another work, VMD was extensively studied by Criscuoli et al. (2013). Flat sheet microporous commercial membranes, made of PP or PVDF, with a membrane area of 180 cm^2, were used. The studies evidenced the key role of the operative conditions (feed temperature and flow-rate) and membrane properties (porosity, pore size and thickness) on the performance of the process, while the vacuum pressure was fixed at 1 kPa (10 mbar). The feed flow-rate was varied with a Reynolds number in the range of 700–1700 while the As concentration was in the range of 0.2–5 mg L^{-1}. VMD is a viable technology for treating water contaminated by As(III) and As(V) at low feed temperatures (20–40°C), thus avoiding the need for the pre-oxidation step to convert As(III) into As(V). In all tests, no As was detected in the permeate and the highest flux ranged between 3 and 12.5 kg h^{-1} m^{-2} at 20°C and 40°C, respectively. In Figure 17.10 the effect of the membrane properties on the trans-membrane flux is shown. The flat membranes employed are: (i) two commercial PP types not supported, (M1) average pore size (dp) of 0.2 micron, membrane thickness of about 90 micron, porosity of about 70% and LEP of about 0.67 MPa; (M2) average pore size of 0.45 micron, membrane thickness of about 170 micron, porosity of about 75%, LEP of about 0.2 MPa; (ii) two self-made PVDF supported, (M3) average pore size (dp) of 0.2 micron, membrane thickness of about 60 micron, porosity of about 70% and LEP of about 0.35 MPa; (M4) average pore size (dp) of 0.2 micron, membrane thickness of about 35 micron, porosity of about 70% and LEP of about 0.35 MPa.

The effect of the porosity, pore size and thickness of the different membranes has been elaborated as reported in the following ratio: $\varepsilon\, dp\, \delta^{-1}$. The trans-membrane flux increases with this ratio, due to the lower membrane resistance offered. However, the effect of membrane properties on the fluxes is lower than that of the feed temperature. The PVDF membrane (M4) showed the highest fluxes, moving from 3 to 12.5 kg h^{-1} m^{-2} for feed temperatures of 20 and 40°C,

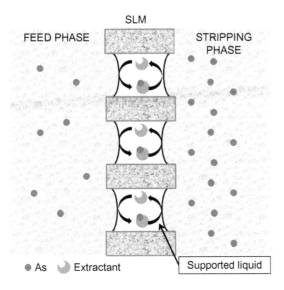

Figure 17.11.   Scheme of extraction of arsenic by supported liquid membranes (SLMs).

respectively, while the PP membrane (M1) registered the lowest fluxes, from $2.3 \, \text{kg h}^{-1} \, \text{m}^{-2}$ (at 20°C) to $9 \, \text{kg h}^{-1} \, \text{m}^{-2}$ (at 40°C). Furthermore, all the tested membranes provided the same performance even after 1 month of continuous use and no wetting phenomena were observed. Macedonio and Drioli (2008) proposed an integrated system consisting of an RO step followed by an MD processing for treating seawater polluted by As(III) and As(V). The MD membrane module is a MD020CP-2N membrane module (Enka Microdyn) containing 40 capillary PP membrane with a nominal pore size of $0.20 \, \mu\text{m}$ (total area $0.1 \, \text{m}^2$). The results showed the positive effect of the temperature on the flux. In fact, the flux increased from 0.36 to $0.72 \, \text{kg m}^{-2} \, \text{h}^{-1}$ by increasing the temperature from 25.3 to 34°C. On the contrary the flux was independent on the As concentration, whereas it slightly increased with the feed flow rate. Moreover, in all the cases the absence of As(V) and As(III) in the treated water was noticed.

17.3.3 *Supported liquid membranes and non-dispersive solvent extractions*

Supported liquid membranes (SLM) consist of organic liquid imbedded in the pores of a polymeric support by capillary forces. If the organic liquid is immiscible with the aqueous feed and strip stream, SLM can act as MC between the two phases. The organic liquid may contain an extractant in order to favor the transport of a target compound from the feed phase to the stripping phase. The SLM scheme is shown in Figure 17.11. Relatively small volumes of organic components in the membrane and simultaneous extraction and re-extraction in one technological step offer the advantages of possible usage of expensive carriers, high separation factors, easy scale-up, low energy requirements, low capital and operating costs (Kocherginsky, 2007). However, the possible loss of organic and the limited life-time of the extractant are typical drawbacks of these systems.

In the last decade, several studies have investigated the potential application of SLM for the As removal from aqueous solution. The SLM combines the extraction process of the species of interest and the subsequent stripping process in a single stage. For instance, As(V) was removed with an efficiency of 94% in 120 min through an SLM from a feed aqueous solution containing H_2SO_4 to a stripping solution containing Na_2SO_4. The membrane utilized was a Millipore GVHP04700 made of PVDF with nominal pore size of $0.22 \, \mu\text{m}$ and 75% porosity. The extractant employed was trioctylphosphine oxide (Cyanex 921) in kerosene as supported liquid. Efforts were made to

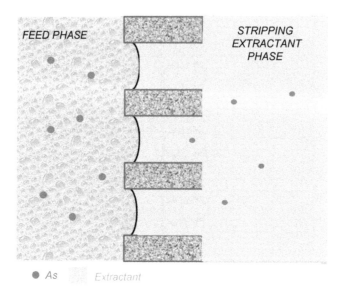

FEED PHASE

STRIPPING
EXTRACTANT
PHASE

● As   Extractant

Figure 17.12.   Scheme of extraction of arsenic by non-dispersive solvent extraction (NDSE).

identify the optimal operative conditions, such as Cyanex 921 and Na_2SO_4 concentrations (0.1 M and 2 M respectively) and the stirring rate (1000 rpm) (Martinez Perez et al., 2007).

Prapasawat et al. (2008) evaluated the separation of As(III) and As(V) ions from sulfate media using LiquiCel Extra-Flow module from CELGARD LLC with PP fiber with a nominal pore size of 30 nm. Again trioctylphosphine oxide, but type Cyanex 923, was used as extractant in toluene. The sulfuric acid concentration positively affected the efficiency of the extraction and the optimal Cyanex 923 was evaluated to be 30 v/v%. The value of As(V) removed was about 45%, whereas 37% of As(III) was extracted.

Another alternative to the classical liquid/liquid separation processes is the non-dispersive solvent extraction (NDSE) carried out in hollow-fiber contactors (HFCs) (Ortiz and Irabien, 2008). The scheme of the process is shown in Figure 17.12. This extraction method overcomes many of the shortcomings associated with liquid-liquid extraction, including solvent loss, emulsion formation, phase separation, and the problems due to flooding and loading. Unlike conventional phase contacting operation, hollow-fiber membranes maintain both of the immiscible phases separated from each other and prevent the intermixing of the organic phase and water stream (containing the species to be extracted) when re-circulated to the module. The NDSE has been successfully tested for the extraction of toxic metals from aqueous solutions of different natures (Choi and Kim, 2003; Juang et al., 2000). The performance of metal extraction by hollow-fiber contactors (HFCs) highly depends on the properties of the organic liquid, which is recirculated on one side of the hollow-fiber, as well as the material type and the structure of the supporting hollow-fiber membranes, which are the core of the treatment. Figure 17.12 shows the use of NDSE for the extraction of As.

Bey et al. (2010) reported the removal of As(V) from an aqueous stream by NDSE, with a hollow-fiber MC, using microporous hydrophobic PVDF hollow-fibers and Aliquat-336 as extractant. PVDF fibers were prepared by the dry/wet spinning technique and presented pore diameters of about 0.2 μm and a porosity of 80%. The study on the effect of operative conditions showed that the extraction was maximized at neutral pH, employing thinner membranes, and increased with the concentration of As in the feed, while it was not affected by the temperature. The membrane contactor HFCs showed a good performance for all the investigated concentrations (20–100 mg L^{-1}) reaching the maximum removal, about 70%, in 6 h of operation for an As content in the feed of 100 mg L^{-1}, as shown in Figure 17.13.

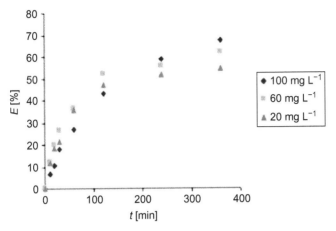

Figure 17.13.   Profile of As(V) extraction against time. Aqueous phase: $[As(V)] = 20–100 \, mg \, L^{-1}$, pH $= 6.98$; flow rate: $0.47 \, mL \, s^{-1}$; Organic phase: $[Aliquat336] = 30\%v/v +$ Kerosene $+$ 4%v/v octanol, flow rate: $1.4 \, mL \, s^{-1}$; Membrane: PVDF hollow-fiber, $T = 25°C$; $\Delta P = 0.03 \, MPa$ (aqueous side) (Bey *et al.*, 2010).

17.4 CONCLUSIONS AND OUTLOOK

A number of remediation methods, including conventional, advanced or hybrid technologies, is usually applied for the treatment of As contaminated water. Since each technique has several drawbacks, researches are still in progress for both improving the performance and developing new processes that can be, eventually, integrated with the existing ones. In this chapter, some of the advanced membrane technologies are reported due to their high potentiality in treating As-contaminated water. The FO process is one of the new advanced membrane technologies, which have found more and more attention in water purification, desalination, food processing, biomedical and energy production. However, the absence of efficient membranes and reverse solute leakage remain the biggest challenge for FO process. Additionally, the choice of a proper draw solution is also an important factor in establishing FO as a successful low-energy demand process, where the draw solution can easily be recovered. However, the potentiality of FO in As removal has been reported and it is expected that, in the years to come, more and more work will focus on improving the actual limitations for producing As-free water. Among the other advanced membrane technologies, MD is the most interesting process, since it is able to completely reject both As(III) and As(V) in a single step without any pretreatment. However, also in this case some drawbacks, as the low trans-membrane fluxes and high energy demand have limited its development and further optimization is needed for a real scale implementation. SLM and NDSE have been also tested and the potentiality of these two membrane technologies for As removal has been illustrated, even if improvements of the obtained removals are needed.

All these innovative membrane technologies are still applied only on the lab scale and the possibility of their scale-up is strictly linked to the development of innovative membrane materials and extractants, to their integration with traditional technologies, as well as to the use of alternative energy sources for reducing the associated energy costs.

REFERENCES

Abernathy, C.O., Thomas, D.J. & Calderon, R. (2003) Health effects and risk assessment of arsenic. *Journal of Nutrition*, 133, 1536–1538.

Achilli, A., Cath, T.Y. & Childress, A.E. (2010) Selection of inorganic based draw solutions for forward osmosis applications. *Journal of Membrane Science*, 364, 233–241.

Akin, I., Arslan, G., Tor, A., Cengeloglu, Y. & Ersoz, M. (2011) Removal of arsenate [As(V)] and arsenite [As(III)] from water by SWHR and BW-30 reverse osmosis. *Desalination*, 281, 88–92.

Alsvik, I. & Hägg, M.-B. (2013) Pressure retarded osmosis and forward osmosis membranes: materials and methods. *Polymers*, 5, 303–327.

Altundoğan, H.S., Altundoğan, S., Tümen, F. & Bildik, M. (2002) Arsenic adsorption from aqueous solutions by activated red mud. *Waste Management*, 22, 357–363.

Badruzzaman, M., Westerhoff, P. & Knappe, D.R.U. (2004) Intraparticle diffusion and adsorption of arsenate onto granular ferric hydroxide (GFH). *Water Research*, 38, 4002–4012.

Bey, S., Criscuoli, A., Figoli, A., Leopold, A., Simone, S., Benamor, M. & Drioli, E. (2010) Removal of As(V) by PVDF hollow fibers membrane contactors using Aliquat-336 as extractant. *Desalination*, 264, 193–200.

Bhowmick, S., Nath, B., Halder, D., Biswas, A., Majumder, S., Mondal, P., Chakraborty, S., Nriagu, J., Bhattacharya, P., Iglesias, M., Roman-Ross, G., Guha Mazumder, D.N., Bundschuh, J. & Chatterjee, D. (2013) Arsenic mobilization in the aquifers of three physiographic settings of West Bengal, India: understanding geogenic and anthropogenic influences. *Journal of Hazardous Materials*, 262, 915–923.

Bhowmick, S., Chakraborty, S., Mondal, P., Van Renterghem, W., Van den Berghe, S., Roman-Ross, G., Chatterjee, D. & Iglesias, M. (2014) Montmorillonite-supported nanoscale zero-valent iron for removal of arsenic from aqueous solution: kinetics and mechanism. *Chemical Engineering Journal*, 243, 14–23.

Bilici Baskan, M. & Pala, A. (2010) A statistical experiment design approach for arsenic removal by coagulation process using aluminum sulfate. *Desalination*, 254, 42–48.

Bui, N.-N., Lind, M.L., Hoek, E.M.V. & McCutcheon, J.R. (2011) Electrospun nanofiber supported thin film composite membranes for engineered osmosis. *Journal of Membrane Science*, 385–386, 10–19.

Butler, E., Silva, A., Horton, K., Rom, Z. & Chwatko, M. (2013) Point of use water treatment with forward osmosis for emergency relief. *Desalination*, 312, 23–30.

Cath, T.Y. (2014) The sweet spot of forward osmosis: treatment of produced water, drilling wastewater, and other complex and difficult liquid streams. *Desalination*, 333, 23–35.

Cath, T.Y., Childress, A.E. & Elimelech, M. (2006) Forward osmosis: principles, applications, and recent developments. *Journal of Membrane Science*, 281, 70–87.

Choi, D.W. & Kim, Y.H. (2003) Cadmium removal using hollow fiber membrane with organic extractant. *Korean Journal of Chemical Engineering*, 20, 768–771.

Chung, T.-S., Zhang, S., Wang, K.Y., Su, J. & Ling, M.M. (2012) Forward osmosis processes: yesterday, today and tomorrow. *Desalination*, 287, 78–81.

Cornelissen, E.R., Harmsen, D., de Korte, K.F., Ruiken, C.J., Qin, J.J., Oo, H. & Wessels, L.P. (2008) Membrane fouling and process performance of forward osmosis membranes on activated sludge. *Journal of Membrane Science*, 319, 158–168.

Criscuoli, A., Majumdar, S., Figoli, A., Sahoo, G.C., Bafaro, P., Bandyopadhyay, S. & Drioli, E. (2012) As(III) oxidation by MnO_2 coated PEEK-WC nanostructured capsules. *Journal of Hazardous Materials*, 211–212, 281–287.

Criscuoli, A., Bafaro, P. & Drioli, E. (2013) Vacuum membrane distillation for purifying waters containing arsenic. *Desalination*, 323, 17–21.

Cui, Y., Ge, Q., Liu, X.-Y. & Chung, T.-S. (2014) Novel forward osmosis process to effectively remove heavy metal ions. *Journal of Membrane Science*, 467, 188–194.

Drioli, E., Criscuoli, A. & Curcio, E. (2006) *Membrane contactors: fundamentals, applications and potentialities*. Elsevier, Amsterdam, The Netherlands.

Dutta, P.K., Ray, A.K., Sharma, V.K. & Millero, F.J. (2004) Adsorption of arsenate and arsenite on titanium dioxide suspensions. *Journal of Colloid and Interface Science*, 278, 270–275.

Fang, Y., Bian, L., Bi, Q., Li, Q. & Wang, X. (2014) Evaluation of the pore size distribution of a forward osmosis membrane in three different ways. *Journal of Membrane Science*, 454, 390–397.

Figoli, A., Cassano, A., Criscuoli, A., Mozumder, M.S.I, Uddin, M.T., Islam, M.A. & Drioli, E. (2010a) Influence of operating parameters on the arsenic removal by nanofiltration. *Water Research*, 44, 97–104.

Figoli, A., Criscuoli, A. & Hoinkis, J. (2010b) Review of membrane processes for arsenic removal from drinking water. In: Kabay N., Bundschuh, J., Hendry, B., Bryjak, M. & Yoshizuka, K. (eds.) *The global arsenic problem: challenges for safe water production*. CRC Press, Boca Raton, FL. pp. 130–145.

Geucke, T., Deowan, S., Hoinkis, J. & Pätzold, C. (2009) Performance of a small-scale RO desalinator for arsenic removal. *Desalination*, 239, 198–206.

Ghurye, G.L., Clifford, D.A. & Trip, A.R. (2004) Iron coagulation and direct microfiltration to remove arsenic from groundwater. *Journal of the American Water Works Association*, 96, 143–152.

Guan, X., Du, J., Meng, X., Sun, Y., Sun, B. & Hu, Q. (2012) Application of titanium dioxide in arsenic removal from water: a review. *Journal of Hazardous Materials*, 215–216, 1–16.

Guo, H., Stüben, D. & Berner, Z. (2007) Removal of arsenic from aqueous solution by natural siderite and hematite. *Applied Geochemistry*, 22, 1039–1051.

Han, B., Runnells, T., Zimbron, J. & Wickramasinghe, R. (2002) Arsenic removal from drinking water by flocculation and microfiltration. *Desalination*, 145, 293–298.

Harisha, R.S., Hosamani, K.M., Keri, R.S., Nataraj, S.K. & Aminabhavi, T.M. (2010) Arsenic removal from drinking water using thin film composite nanofiltration membrane. *Desalination*, 252, 75–80.

He, J.-H. (2008) *Electrospun nanofibers and their applications*. Smithers Rapra, Shrewsbury, UK.

Hering, J.G., Chen, P.Y., Wilkie, J.A. & Elimelech, M. (1997) Arsenic removal from drinking water during coagulation. *Journal of Environmental Engineering*, 123, 800–807.

Hoover, L.A., Schiffman, J.D. & Elimelech, M. (2013) Nanofibers in thin-film composite membrane support layers: enabling expanded application of forward and pressure retarded osmosis. *Desalination*, 308, 73–81.

Hu, C., Liu, H., Chen, G. & Qu, J. (2012) Effect of aluminum speciation on arsenic removal during coagulation process. *Separation and Purification Technology*, 86, 35–40.

Hydranautics (2014) Element Spec Sheets – Hydranautics a Nitto Group Company. Available from: www.membranes.com/index.php?pagename=spec_sheets [accessed July 2015].

IARC Monographs (2004) Some drinking-water disinfectants and contaminants, including arsenic related nitrosamines. *IARC Monographs on the Evaluation of Carcinogenic Risks to Humans* 84.

Iqbal, J., Kim, H.J., Yang, J.S., Baek, K. & Yang, J.W. (2007) Removal of arsenic from groundwater by micellar-enhanced ultrafiltration (MEUF). *Chemosphere*, 66, 970–976.

Islam, A.M. (2005) *Membrane distillation process for pure water and removal of arsenic*. MSc Thesis, Chalmers University of Technology, Gothenburg, Sweden.

Jin, X., She, Q., Ang, X. & Tang, C.Y. (2012) Removal of boron and arsenic by forward osmosis membrane: influence of membrane orientation and organic fouling. *Journal of Membrane Science*, 389, 182–187.

Juang, R.S., Chen, J.D. & Huan, H.C. (2000) Dispersion-free membrane extraction: case studies of metal ion and organic acid extraction. *Journal of Membrane Science*, 165, 59–73.

Kapaj, S., Peterson, H., Liber, K. & Bhattacharya, P. (2006) Human health effects from chronic arsenic poisoning – a review. *Journal of Environmental Science and Health*, Part A, 41, 2399–2428.

Khan, E.U. & Martin, A.R. (2014) Water purification of arsenic contaminated drinking water via airgap membrane distillation (AGMD). *Periodica Polytechnica, Mechanical Engineering*, 58 (1), 47–53.

Kim, M.J. & Nriagu, J. (2000) Oxidation of arsenite in groundwater using ozone and oxygen. *Science of the Total Environment*, 247, 71–79.

Klaysom, C., Cath, T.Y., Depuydt, T. & Vankelecom, I.F.J (2013) Forward and pressure retarded osmosis: potential solutions for global challenges in energy and water supply. *Chemical Society Reviews*, 42, 6959–6989.

Kocherginsky, N.M., Yang, Q. & Seelam, L. (2007) Recent advances in supported liquid membrane technology. *Separation and Purification Technology*, 53, 171–177.

Košutić, K., Furac, L., Sipos, L. & Kunst, B. (2005) Removal of arsenic and pesticides from drinking water by nanofiltration membranes. *Separation and Purification Technology*, 42, 137–144.

Kunzru, S. & Chaudhuri, M. (2005) Manganese amended activated alumina for adsorption/oxidation of arsenic. *Journal of Environmental Engineering*, 131, 1350–1353.

Lacasa, E., Cañizares, P., Sáez, C., Fernández, F.J. & Rodrigo, M.A. (2011) Removal of arsenic by iron and aluminum electrochemically assisted coagulation. *Separation and Purification Technology*, 79, 15–19.

Lakshmanan, D., Clifford, D.A. & Samanta, G. (2010) Comparative study of arsenic removal by iron using electrocoagulation and chemical coagulation. *Water Research*, 44, 5641–5652.

Lawson, K.W. & Lloyd, D.R. (1997) Membrane distillation. *Journal of Membrane Science*, 124, 1–25.

Lee, G., Song, K. & Bae, J. (2011) Permanganate oxidation of arsenic(III): reaction stoichiometry and the characterization of solid product. *Geochimica et Cosmochimica Acta*, 75, 4713–4727.

Lee, Y., Um, I. & Yoon, J. (2003) Arsenic(III) oxidation by iron(VI) (ferrate) and subsequent removal of arsenic(V) by iron(III) coagulation. *Environmental Science & Technology*, 37, 5750–5756.

Leupin, O.X. & Hug, S.J. (2005) Oxidation and removal of arsenic(III) from aerated groundwater by filtration through sand and zero-valent iron. *Water Research*, 39, 1729–1740.

Li, Z., Beachner, R., McManama, Z. & Hanlie, H. (2007) Sorption of arsenic by surfactant modified zeolite and kaolinite. *Microporous and Mesoporous Materials*, 105, 291–297.

Lior, N., Amy, G., Barak, A.Z., Chakraborty, A., Nashar, A.E., El- Sayed, Y., Kennedy, M.D., Kumano, A. & Lattemann, S. (2013) *Advances in water desalination.* Volume 1. John Wiley & Sons Inc., Hoboken, NJ.

Litter, M.I., Morgada, M.E. & Bundschuh, J. (2010) Possible treatments for arsenic removal in Latin American waters for human consumption. *Environmental Pollution,* 158, 1105–1118.

Lutchmiah, K., Verliefde, A.R.D., Roest, K., Rietveld, L.C. & Cornelissen, E.R. (2014) Forward osmosis for application in wastewater treatment: a review. *Water Research,* 58, 179–197.

Macedonio, F. & Drioli, E. (2008) Pressure-driven membrane operations and membrane, distillation technology integration for water purification. *Desalination,* 223, 396–409.

Macedonio, F., Drioli, E., Gusev, A.A., Bardow, A., Semiat, R. & Kurihara, M. (2012) Efficient technologies for worldwide clean water supply. *Chemical Engineering and Processing,* 51, 2–17.

Mann, B.R., Bhat, S.C., Dasgupta, M. & Ghosh, U.C. (1999) Studies on removal of arsenic from water using hydrated zirconium oxide. *Chemical & Environmental Research,* 8, 51–56.

Manna, A.K., Sen, M., Martin, A.R. & Pal, P. (2010) Removal of arsenic from contaminated groundwater by solar-driven membrane distillation. *Environmental Pollution,* 158, 805–811.

Martinez Perez, M.E., Reyes-Aguilera, J.A., Saucedo, T.I., Gonzalez, M.P., Navarro, R. & Avila-Rodriguez, M. (2007) Study of As(V) transfer through a supported liquid membrane impregnated with trioctylphosphine oxide (Cyanex 921). *Journal of Membrane Science,* 302, 119–126.

Matschullat, J. (2000) Arsenic in the geosphere – a review. *Science of the Total Environment,* 249, 297–312.

McCutcheon, J.R. & Elimelech, M. (2008) Influence of membrane support layer hydrophobicity on water flux in osmotically driven membrane processes. *Journal of Membrane Science,* 318, 458–466.

Mi, B. & Elimelech, M. (2008) Chemical and physical aspects of organic fouling of forward osmosis membranes. *Journal of Membrane Science,* 320, 292–302.

Miller, S.M. & Zimmerman, J.B. (2010) Novel, bio-based, photoactive arsenic sorbent: TiO_2-impregnated chitosan bead. *Water Research,* 44, 5722–5729.

Miller, S.M., Spaulding, M.L. & Zimmerman, J.B. (2011) Optimization ofcapacity and kinetics for a novel bio-based arsenic sorbent, TiO_2-impregnated chitosan bead. *Water Research,* 45, 5745–5754.

Mohan, D. & Pittman, C.U. (2007) Arsenic removal from water/wastewater using adsorbents – a critical review. *Journal of Hazardous Materials,* 142, 1–53.

Mondal, P. (2014) *Removal of arsenic from drinking water using physicochemical methods.* Catholic University of Leuven, Leuven, Belgium.

Mondal, P., Bhowmick, S., Chatterjee, D., Figoli, A. & Van der Bruggen, B. (2013) Remediation of inorganic arsenic in groundwater for safe water supply: a critical assessment of technological solutions. *Chemosphere,* 92, 157–170.

Mondal, P., Hermans, N., Tran, A.T.K., Zhang, Y., Fang, Y., Wang, X. & Van der Bruggen, B. (2014a) Effect of physico-chemical parameters on inorganic arsenic removal from aqueous solution using a forward osmosis membrane. *Journal of Environmental Chemical Engineering,* 2, 1309–1316.

Mondal, P., Tran, A.T.K. & Van der Bruggen, B. (2014b) Removal of As (V) from simulated groundwater using forward osmosis: effect of competing and coexisting solutes. *Desalination,* 348, 33–38.

Mondal, P., Bhowmick, S., Jullok, N., Ye, W., Van Renterghem, W., Van den Berghe, S. & Van der Bruggen, B. (2014c) Behaviour of As(V) with ZVI-H_2O system and the reduction to As(0). *Journal of Physical Chemistry, C,* 118 (37), 21,614–21,621.

Nguyen, C.M., Bang, S. Cho, J. & Kim, K.W. (2009) Performance and mechanism of arsenic removal from water by a nanofiltration membrane. *Desalination,* 245, 82–94.

Nriagu, J., Bhattacharya, P., Mukherjee, A., Bundschuh, J., Zevenhoven, R. & Loeppert, R. (2007) Arsenic in soil and groundwater: an overview. In: Bhattacharya, P., Mukherjee, A.B., Bundschuh, J., Zevenhoven, R. & Loeppert, R.H. (eds.) *Arsenic in soil and groundwater environment.* Elsevier, Amsterdam, The Netherlands. pp. 3–60.

Ortiz, I. & Irabien, J.A. (2008) Membrane-assisted solvent extraction for the recovery of metallic pollutants: process modeling and optimization. In: Pabby, A.K., Rizvi, S.S.H. & Sastre Requena, A.M. (eds.) *Handbook of membrane separations: chemical, pharmaceutical, food, and biotechnological applications.* CRC Press, Boca Raton, FL. pp. 1023–1039.

Pal, P. & Manna, A.K. (2010) Removal of arsenic from contaminated groundwater by solar-driven distillation using tree different commercial membranes. *Water Research,* 44, 5750–5760.

Pal, P., Manna, A.K. & Linnanen, L. (2013) Arsenic removal by solar-driven membrane distillation: modeling and experimental investigation with a new flash vaporization module. *Water Environment Research,* 85, 63–76.

Pallier, V., Feuillade-Cathalifaud, G., Serpaud, B. & Bollinger, J.C. (2010) Effect of organic matter on arsenic removal during coagulation/flocculation treatment. *Journal of Colloid and Interface Science*, 342, 26–32.

Perez Padilla, A. & Saitua, H. (2010) Performance of simultaneous arsenic, fluoride and alkalinity (bicarbonate) rejection by pilot-scale nanofiltration. *Desalination*, 257, 16–21.

Pérez-Sicairos, S., Lin, S.W., Félix-Navarro, R.M. & Espinoza-Gómez, H. (2009) Rejection of As(III) and As(V) from arsenic contaminated water via electro-cross-flow negatively charged nanofiltration membrane system. *Desalination*, 249, 458–465.

Polizzotto, M.L., Harvey, C.F., Li, G., Badruzzman, B., Ali, A., Newville, M., Sutton, S. & Fendorf, S. (2006) Solid-phases and desorption processes of arsenic within Bangladesh sediments. *Chemical Geology*, 228, 97–111.

Qu, D., Wang J., Hou, D., Luan, Z., Fan, B. & Zhao, C. (2009) Experimental study of arsenic removal by direct contact membrane distillation. *Journal of Hazardous Materials*, 163, 874–879.

Ramakrishna, S. (2005) *Introduction to electrospinning and nanofibers*. World Scientific Publishing Co., River Edge, NJ.

Ramos, M.A.V., Yan, W., Li, X.-Q., Koel, B.E. & Zhang, W.-X. (2009) Simultaneous oxidation and reduction of arsenic by zero-valent iron nanoparticles: understanding the significance of the core-shell structure. *Journal of Physical Chemistry* C, 113, 14,591–14,594.

Richards, L.A., Richards, B.S., Rossiter, H.M.A. & Schäfer, A.I. (2009) Impact of speciation on fluoride, arsenic and magnesium retention by nanofiltration/reverse osmosis in remote Australian communities. *Desalination*, 248, 177–183.

Roy, M., Nilsson, L. & Pal, P. (2008) Development of groundwater resources in a region with high population density: a study of environmental sustainability. *Journal of Environmental Sciences*, 5, 251–267.

Sairam, M., Sereewatthanawut, E., Li, K., Bismarck, A. & Livingston, A.G. (2011) Method for the preparation of cellulose acetate flat sheet composite membranes for forward osmosis desalination using MgSO$_4$ draw solution. *Desalination*, 273, 299–307.

Saitua, H., Gil, R. & Perez Padilla, A. (2011) Experimental investigation on arsenic removal with a nanofiltration pilot plant from naturally contaminated groundwater. *Desalination*, 274, 1–6.

Sato, Y., Kang, M., Kamei, T. & Magara, Y. (2002) Performance of nanofiltration for arsenic removal. *Water Research*, 36, 3371–3377.

Saxena, V.K., Kumar, S. & Singh, V.S. (2004) Occurrence, behavior and speciation of arsenic in groundwater. *Current Science*, 86, 281–284.

Setiawan, L., Wang, R., Li, K. & Fane, A.G. (2011) Fabrication of novel poly(amide-imide) forward osmosis hollow fiber membranes with a positively charged nanofiltration-like selective layer. *Journal of Membrane Science*, 369, 196–205.

Shaffer, D.L., Arias Chavez, L.H., Ben-Sasson, M., Romero-Vargas Castrillón, S., Yip, N.Y. & Elimelech, M. (2011) Desalination and reuse of high-salinity shale gas produced water: drivers, technologies, and future directions. *Environmental Science & Technology*, 47, 9569–9583.

Sharma, V.K., Dutta, P.K. & Ray, A.K. (2007) Review of kinetics of chemical and photocatalytical oxidation of arsenic(III) as influenced by pH. *Journal of Environmental Science and Health*, Part A, 42, 997–1004.

Shih, M.C. (2005) An overview of arsenic removal by pressure driven membrane processes. *Desalination*, 172, 85–97.

Smedley, P. & Kinniburgh, D. (2002) A review of the source, behavior and distribution of arsenic in natural waters. *Applied Geochemistry*, 17, 517–568.

Smith, M.C. & Reynolds, K.J. (2014) Forward osmosis dialysate production using spiral-wound reverse-osmosis membrane elements. *Journal of Membrane Science*, 469, 95–111.

Song, S., Lopez-Valdivieso, A., Hernandez-Campos, C., Peng, C., Monroy-Fernandez, M.G. & Razo-Soto, I. (2006) Arsenic removal from high-arsenic water by enhanced coagulation with ferric ions and coarse calcite. *Water Research*, 40, 364–374.

Stute, M., Zheng, Y., Schlosser, P., Horneman, A., Dhar, R.K. & Datta, S. (2007) Hydrogeological control of As concentrations in Bangladesh groundwater. *Water Resources Research*, 43, W09417.

Su, J., Huang, H.-G., Jin, X.-Y., Lu, X.-Q. & Chen, Z.-L. (2011) Synthesis, characterization and kinetic of a surfactant-modified bentonite used to remove As(III) and As(V) from aqueous solution. *Journal of Hazardous Materials*, 185, 63–70.

Sukitpaneenit, P. & Chung, T.-S. (2012) High performance thin-film composite forward osmosis hollow fiber membranes with macrovoid-free and highly porous structure for sustainable water production. *Environmental Science & Technology*, 46, 7358–7365.

Sun, F., Osseo-Asare, K.A., Chen, Y. & Dempsey, B.A. (2011) Reduction of As(V) to As(III) by commercial ZVI or As(0) with acid-treated ZVI. *Journal of Hazardous Materials*, 196, 311–317.

Sun, H., Wang, L., Zhang, R., Sui, J. & Xu, G. (2006) Treatment of groundwater polluted by arsenic compounds by zero valent iron. *Journal of Hazardous Materials*, 129, 297–303.

Teychene, B., Collet, G., Gallard, H. & Croue, J.P. (2012) A comparative study of boron and arsenic (III) rejection from brackish water by reverse osmosis membranes. *Desalination*, 310, 109–114.

Tiraferri, A., Yip, N.Y., Phillip, W.A., Schiffman, J.D. & Elimelech, M. (2011) Relating performance of thin-film composite forward osmosis membranes to support layer formation and structure. *Journal of Membrane Science*, 367, 340–352.

Uddin, M., Harun-Ar-Rashid, A.K.M., Hossain, S.M. Hafiz, M.A. Nahar, K. & Mubin, S.H. (2006) Slow arsenic poisoning of the contaminated groundwater users. *Internation Journal of Environmental Science & Technology*, 3, 447–453.

Umita, T. (2008) Biotransformation of arsenic species by activated sludge and removal of bio-oxidized arsenate from waste water by coagulation with ferric chloride. *Water Research*, 42, 4809–4817.

Vasudevan, S., Mohan, S., Sozhan, G., Raghavendran, N.S. & Murugan, C.V. (2006) Studies on the oxidation of As(III) to As(V) by in-situ-generated hypochlorite. *Industrial and Engineering Chemistry Research*, 45, 7729–7732.

Velizarov, S., Crespo, J. & Reis, M. (2004) Removal of inorganic anions from drinking water supplies by membrane bio/processes. *Reviews in Environmental Science and Biotechnology*, 3, 361–380.

Walker, M., Seiler, R.L. & Meinert, M. (2008) Effectiveness of household reverse-osmosis systems in a western US region with high arsenic in groundwater. *Science of the Total Environment*, 389, 245–252.

Wang, K.Y., Yang, Q., Chung, T.S. & Rajagopalan, R. (2009) Enhanced forward osmosis from chemically modified polybenzimidazole (PBI) nanofiltration hollow fiber membranes with a thin wall. *Chemical Engineering Science*, 64, 1577–1584.

Wickramasinghe, S.R., Han, B., Zimbron, J., Shen, Z. & Karim, M.N. (2004) Arsenic removal by coagulation and filtration: comparison of groundwaters from the United States and Bangladesh. *Desalination*, 169, 231–244.

Xia, S., Dong, B., Zhang, Q., Xu, B., Gao, N. & Causseranda, C. (2007) Study of arsenic removal by nanofiltration and its application in China. *Desalination*, 204, 374–379.

Xie, M., Nghiem, L.D., Price, W.E. & Elimelech, M. (2014) Impact of organic and colloidal fouling on trace organic contaminant rejection by forward osmosis: role of initial permeate flux. *Desalination*, 336, 146–152.

Yoon, S.H. & Lee, J.H. (2005) Oxidation mechanism of As(III) in the UV/TiO$_2$ system: evidence for a direct hole oxidation mechanism. *Environmental Science & Technology*, 39, 9695–9701.

Yoon, S.H., Lee, J.H., Oh, S.E. & Yang, J.E. (2008) Photochemical oxidation of As(III) by vacuum-UV lamp irradiation. *Water Research*, 42, 3455–3463.

Yoon, J., Amy, G., Chung, J., Sohn, J. & Yoon, Y. (2009) Removal of toxic ions (chromate, arsenate, and perchlorate) using reverse osmosis, nanofiltration, and ultrafiltration membranes. *Chemosphere*, 77, 228–235.

Yu, Y., Seo, S., Kim, I.-C. & Lee, S. (2011) Nanoporous polyethersulfone (PES) membrane with enhanced flux applied in forward osmosis process. *Journal of Membrane Science*, 375, 63–68.

Zhang, S., Wang, K.Y., Chung, T.S., Chen, H., Jean, Y.C. & Amy, G. (2010) Well-constructed cellulose acetate membranes for forward osmosis: minimized internal concentration polarization with an ultra-thin selective layer. *Journal of Membrane Science*, 360, 522–535.

Zhang, S., Sukitpaneenit, P. & Chung, T.-S. (2014) Design of robust hollow fiber membranes with high power density for osmotic energy production. *Chemical Engineering Journal*, 241, 457–465.

Zhao, S., Zou, L. & Mulcahy, D. (2011) Effects of membrane orientation on process performance in forward osmosis applications. *Fuel Energy Abstracts*, 382, 308–315.

CHAPTER 18

Removal of trace chromate from contaminated water: ion-exchange and redox-active sorption processes

Sudipta Sarkar, Ryan C. Smith, Nicole Blute & Arup K. SenGupta

18.1 INTRODUCTION

Typically, when it comes to contamination of drinking water by Cr(VI), anthropogenic contamination is much more significant compared to natural Cr. Dissolved concentrations of total Cr in groundwater from natural processes are typically below $10\,\mu g\,L^{-1}$ (Richard and Bourg, 1991). In contaminated areas, Cr(VI) concentrations are commonly 300 to $500\,\mu g\,L^{-1}$ (CRWQCB, 2000; Maxwell, 1997) and have been reported to reach $14\,g\,L^{-1}$ (Palmer and Wittbrodt, 1991). A yellow color is imparted to the water at about $1\,mg\,L^{-1}$ Cr(VI) (Palmer and Wittbrodt, 1991). Kharkar *et al.* (1968) reported an average total Cr value in river water of $1.4\,\mu g\,L^{-1}$, while Richard and Bourg (1991) reported total dissolved Cr concentrations from zero to $208\,\mu g\,L^{-1}$ for unpolluted waters, with typical values given of $0.5\,\mu g\,L^{-1}$ for rivers and $1.0\,\mu g\,L^{-1}$ for groundwaters. While it is often the case that Cr(VI) in drinking water is evidence of anthropogenic contamination, over the last decade there have been many reports of naturally occurring Cr(VI) in groundwater and at very high levels. Cr(VI) of natural origin has been found in the groundwater of numerous alluvial basins in Arizona and adjacent parts of California, New Mexico, and Nevada at concentrations of up to $220\,mg\,L^{-1}$ Cr(VI) (Robertson, 1975; 1991). In the remote Cadiz Valley in the Mojave Desert of southeastern California at the location of a proposed water storage project between Metropolitan Water District of southern California (MWD) and Cadiz, Inc., concentrations of 15 to $26\,\mu g\,L^{-1}$ were found in the native groundwater (MWD and Bureau of Land Management, 2001). A combined field and laboratory study of chromite bearing oxidized serpentinite rocks in India indicated the possibility of Cr mobilization from chromite ores to water bodies (Godgul and Sahu, 1995). The authors observed that serpentinization is an intensely oxidizing process that creates alkaline pore waters that would promote oxidation of Cr(III). The study suggested that mining practices enhance the rate and intensity of Cr mobilization (Godgul and Sahu, 1995). Cr(VI) in groundwater has also been documented in the vicinity of UC Davis in the Sacramento Valley (Chung *et al.*, 2001). The Table 18.1 (adapted from AWWA Reports) summarizes the reported instances of naturally occurring Cr(VI) in groundwater.

18.1.1 *Aqueous chemistry of chromium*

The element Cr has an electronic structure of [Ar] $3d^5\,4s^1$. Out of the several oxidation states in which Cr can theoretically exist, only two states, trivalent and hexavalent, are stable enough to occur in the environment. These different types of Cr are key to understanding issues of importance in analytical chemistry, water treatment and distribution. The trivalent oxidation state or Cr(III) is the most stable form as it needs considerable energy to convert it to lower or higher oxidation states (Kotas and Stascika, 2000). Cr(VI), due to its acute toxicity, is the oxidation state of concern. In natural waters, Cr(VI) occurs as its oxyanions or oxyacids; the concentration of total Cr(VI) species as well as the solution pH dictate the ionic form of Cr(VI) that will be prevalent. Figure 18.1 demonstrates the domains of predominance of different species of Cr(VI) at 25°C. Chromic acid (H_2CrO_4) is prevalent at pH < 1 and is a strong acid (Deltombe *et al.*, 1966; Saleh *et al.*, 1989; Sperling *et al.*, 1992). At pH > 1, all the anionic forms of chromic acid

Table 18.1.   Instances of naturally occurring Cr(VI) in groundwater.

Location	Concentration of Cr(VI) [$\mu g\,L^{-1}$]	Reference
Santa Cruz County, CA	4–33	Gonzalez et al. (2005)
Leon Valley, Mexico	12	Robles-Camacho and Armienta (2000)
Cazadero county, CA	12–22	Oze et al. (2004)
Mojave Desert, CA	60	Ball and Izbicki (2004)
Paradise valley, AZ	220	Robertson (1975)
Sao Paulo, Brazil	45	Bourotte et al. (2009)
La Spezia, Italy	5–73	Fantoni et al. (2002)
New Caledonia	700	Becquer et al. (2003)
Yilgarn Craton, Australia	10–430	Gray (2003)
Presidio, SFO, CA	52–98	Steinpress (1998)
Sukinda Valley, India	9–421	Dhakate et al. (2008)

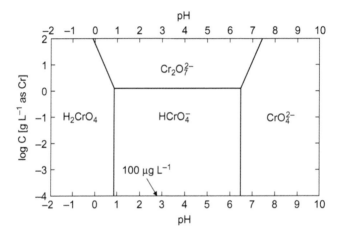

Figure 18.1.   Zone of predominance of different Cr(VI) species in terms of concentration and pH.

prevail; at pH 6.5, the divalent CrO_4^{2-} ions are prevalent in solution throughout the concentration range; in the pH between 1 and 6, $HCrO_4^-$ is the predominant form.

The following are important equilibrium reactions for different Cr(VI) species:

$$H_2CrO_4 \rightleftarrows H^+ + HCrO_4^- \qquad K_1 = 10^{-0.75} \tag{18.1}$$

$$HCrO_4^- \rightleftarrows H^+ + CrO_4^{2-} \qquad K_2 = 10^{-6.45} \tag{18.2}$$

$HCrO_4^-$, however, is the predominant form up to a total Cr(VI) concentration of 10^{-2} M beyond which the $HCrO_4^-$ ions tend to dimerize into dichromate ions ($Cr_2O_7^{2-}$) which have a characteristic orange-red color (Cotton and Wilkinson, 1980; Nieboer and Jusys, 1988).

$$2HCrO_4^- \rightleftarrows Cr_2O_7^{2-} + H_2O \qquad K_3 = 10^{1.52} \tag{18.3}$$

The dichromate ions get protonated to hydrogen dichromate as per the following reaction:

$$HCr_2O_7^- \rightleftarrows Cr_2O_7^{2-} + H^+ \qquad K_4 = 10^{0.07} \tag{18.4}$$

Trichromate ($Cr_3O_{10}^{2-}$) and tetrachromate ($Cr_4O_{13}^{2-}$) species may also be present. However, hydrogen dichromate, trichromate or tetrachromate species are present only when the pH is less than 0 or at total Cr(VI) concentration greater than 1 M.

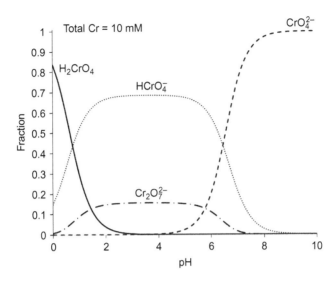

Figure 18.2. Distribution of Cr(VI) species at different pH.

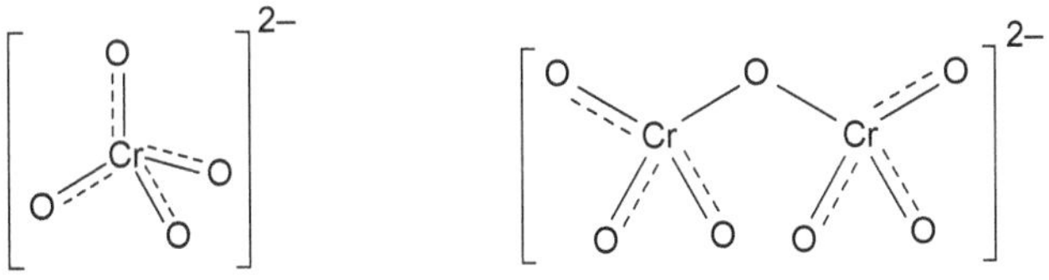

Figure 18.3. Structure of CrO_4^{2-} and $Cr_2O_7^{2-}$ ions.

Therefore, as evident from Figure 18.1, within the normal pH range in natural waters, CrO_4^{2-}, $HCrO_4^-$ and $Cr_2O_7^{2-}$ ions are in the forms of Cr(VI), and at trace concentrations (below $100\ \mu g\ L^{-1}$) dichromate is not expected to occur. The relative distribution of different ionic species of Cr(VI) as a function of pH is illustrated in Figure 18.2 (Tandon *et al.*, 1984).

Figure 18.3 shows the structures of the divalent species of Cr(VI), namely divalent chromate and divalent dichromate.

Cr(VI) is considered to be a strong oxidizing agent. Cr(VI) oxyanions are readily reduced to trivalent forms by electron donors such as organic matter or reduced inorganic species, which are ubiquitous in soil, water and atmospheric systems (Stollenwerk and Grove, 1985).

In acidic solutions where $HCrO_4^-$ is prevalent, the reduction reaction is accompanied by the consumption of H^+ ion, thus causing a decrease in standard potential with decreasing acidity:

$$HCrO_4^- + 7H^+ + 3e^- \rightleftarrows Cr^{3+} + 4H_2O \quad E^0 = 1.35\ V \tag{18.5}$$

$$H_2CrO_4 + 6H^+ + 3e^- \rightleftarrows Cr^{3+} + 4H_2O \quad E^0 = 1.39\ V \tag{18.6}$$

And, in more basic solutions the reduction of Cr(VI) produces OH^- ions as per the following reaction:

$$CrO_4^{2-} + 4H_2O + 3e^- \rightleftarrows Cr(OH)_3 + 5OH^- \quad E^0 = -0.13\ V \tag{18.7}$$

Dichromate ions get reduced to Cr(III) through the following reaction:

$$Cr_2O_7^{2-} + 14H^+ + 6e^- \rightleftarrows 2Cr^{3+} + 7H_2O \quad E^0 = 1.33\ V \tag{18.8}$$

Figure 18.4 shows a pE-pH or predominance diagram illustrating the domain of predominance of different species of Cr(VI) and Cr(III) at different pH as well as redox conditions.

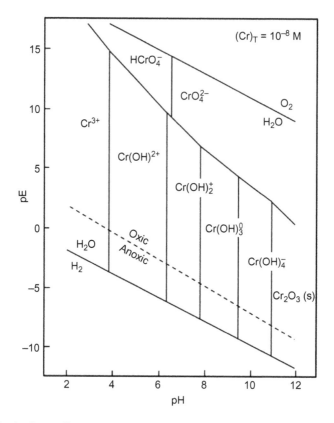

Figure 18.4.   Predominance diagram for Cr(III) and Cr(VI) species.

Cr(III) is not a health concern at levels found in natural waters; in fact at trace level is considered to be essential for human beings as a micronutrient. Cr(III) is a hard acid and shows a strong tendency to form complexes with ligands such as water, ammonia, urea, ethylenediamine, and other organic ligands containing oxygen, nitrogen or sulfur donor atoms (Nakayama et al., 1981a; 1981b; 1981c; Saleh et al., 1989). According to Figure 18.4, at very low concentrations of oxygen or in its absence, Cr(III) is the dominant species, which may remain in cationic (Cr^{3+}, $CrOH^{2+}$, or $Cr(OH)_2^+$) or neutral ($Cr(OH)_3$) form depending on the pH (Hem, 1977; Rai et al., 1987). Cr(III) is highly insoluble between pH 7 and pH 10, with minimum solubility at pH 8 of about $1\,\mu g\,L^{-1}$ (Ball and Nordstrom, 1998; Rai et al., 1987; Richard and Bourg, 1991; Saleh et al., 1989). In a given water sample, Cr(III) can be present as: (i) soluble Cr(III) species, (ii) as suspended particulate such as a precipitate of $Cr(OH)_3$, (iii) adsorbed to the surface of other amphoteric oxides such as $Fe(OH)_3$ and (iv) complexed with naturally occurring organic matter such as humic and fulvic acids (Icopini and Long, 2002). $Cr(OH)_3$ precipitate exhibits amphoteric behavior and at higher pH is transformed into the readily soluble tetrahydroxo complex, $Cr(OH)_4^-$. At more concentrated Cr(III) solutions ($>10^{-6}$M) the polynuclear hydrolytic products, $Cr_2(OH)_4^{2+}$, $Cr_3(OH)_5^{4+}$, $Cr_4(OH)_6^{6+}$ could also be expected (Rai et al., 1987). At high concentrations these ions impart a green color to water. If the complexation from these ligands can be neglected, under redox and pH conditions normally found in natural systems, Cr is removed from the solution as $Cr(OH)_3$, or in the presence of Fe(III), in the form of $(Cr_x,Fe_{1-x})(OH)_3$, (where x is the mole fraction of Cr) (Rai et al., 1987). The mobility of Cr(III) in the aquatic environment is expected to be low because Cr(III) attains its minimum solubility in the pH range of natural waters (pH 7.5–8.5), remaining in the form of sparingly soluble $Cr(OH)_3(s)$ and $(Cr,Fe)(OH)_3(s)$ precipitates or being strongly adsorbed onto other solids under slightly acidic to basic conditions.

The redox potential of the Cr(VI)/Cr(III) couple is high enough so that only a few oxidants present in natural systems are capable of oxidizing Cr(III) to Cr(VI). Without another mediator, oxidation of Cr(III) by only dissolved oxygen has been found to be negligible (Eary and Rai, 1987; Schroeder and Lee, 1975), whereas oxidation by agents such as manganese oxides, elemental chlorine and its derivatives was found to be significant in environmental systems (Bartlett and James, 1979; Johnson and Xyla, 1991; Nakayama *et al.*, 1981d; Saleh *et al.*, 1989; Schroeder and Lee, 1975) and also in drinking water supply systems.

It is worthwhile to mention that mediators such as dissolved oxygen, chlorine, chloramines, ferrous iron and pH can change quite rapidly in drinking water systems, and hence the kinetics of transformation of Cr(VI) and Cr(III) plays a vital role in the relative distribution of these two forms in drinking water. Oxidation by dissolved oxygen is a relatively slow process showing only 3% conversion after 50 days (Eary and Rai, 1987; Schroeder and Lee, 1975); complete conversion occurs over a period of years in oxygenated groundwater, causing the presence of naturally occurring Cr(VI). However, rapid oxidation of Cr(III) to Cr(VI) can occur in the presence of oxidants like dissolved chlorine (HOCl) (Bartlett, 1997; Brandhuber *et al.*, 2004; Clifford and Chau, 1988; Lai and McNeill, 2006; Saputro *et al.*, 2011; Sorg, 1979; Ulmer, 1986), and also by MnO_2 and MnO_4^-:

$$2Cr^{3+} + 3MnO_2 + 2H_2O \rightleftarrows 2CrO_4^{2-} + 3Mn^{2+} + 4H^+ \tag{18.9}$$

$$2Cr^{3+} + 3HOCl + 5H_2O \rightleftarrows 2CrO_4^{2-} + 3Cl^- + 13H^+ \tag{18.10}$$

$$5Cr^{3+} + 3MnO_4^- + 8H_2O \rightleftarrows 5CrO_4^{2-} + 3Mn^{2+} + 16H^+ \tag{18.11}$$

$$2Cr^{3+} + 3NH_2Cl + 8H_2O \rightleftarrows 2CrO_4^{2-} + 3NH_3 + 3Cl^- + 13H^+ \tag{18.12}$$

$$4Cr^{3+} + 3O_2 + 10H_2O \rightleftarrows 4CrO_4^{2-} + 20H^+ \tag{18.13}$$

These rapid conversion reactions are of concern because there is a chance of reformation of Cr(VI) in the disinfection process as well as water distribution systems even if Cr(VI) is completely removed at treatment plants. If these oxidants contact soluble Cr(III) or plumbing surfaces that contain Cr. Chloramine can oxidize Cr(III) to Cr(VI) over a period of hours to days (Brandhuber *et al.*, 2004). Conversely, Izbicki *et al.* (2008) found that reduction of Cr(VI) to Cr(III) occurred when oxygen levels fell below 0.5 mg L^{-1}, even though the reaction is not thermodynamically favored in water with no oxygen.

18.2 OVERVIEW OF CHROMIUM REMOVAL TECHNOLOGIES

According to Sharma *et al.* (2008), Cr(VI) removal technologies can be broadly be classified into five sub-groups, namely: (i) coagulation, precipitation, filtration (including reductive precipitation through chemical/electrochemical processes); (ii) adsorption; (iii) ion exchange (iv) membrane and electrodialysis; and (v) biological removal. However, biological removal of Cr(VI) from drinking water is not practicable for obvious reasons.

The above technologies, singly or in combination, have been successfully used for removal of high concentrations of total Cr present in wastewater or industrial effluents. There have not been many studies for removal of trace concentration of Cr(VI) from drinking water sources due to the absence of strict regulatory standards.

The removal of Cr(VI) using only coagulation and precipitation without prior reduction has showed limited effectiveness. Low Cr(VI) removal rates (<30%) have been observed with iron-based coagulants even with high Fe(III) concentrations (Lee and Hering, 2003). It may be noted from the Pourbaix and predominance zone diagrams above that Cr(VI) does not produce any insoluble species at any pH. On the other hand, the trivalent state of Cr does form insoluble species within the normal pH range. The removal effectiveness is low because Cr(VI) species and complexes are mostly soluble in nature while Cr(III) species are mostly insoluble at normal pH range. Therefore, a prior reduction of Cr(VI) to Cr(III) is necessary before removing the transformed Cr(III) by subsequent coagulation and filtration.

Cr(VI) reduction with various chemicals such as ferrous salts, zero-valent iron, and sodium bisulfite has been researched with special emphasis on reducing Cr(VI) to Cr(III) and then precipitating it as Cr(OH)$_3$or co-precipitation of Cr(III) with Fe(III). The half reactions representing the reduction of various species of Cr(VI) have been already discussed:

$$HCrO_4^- + 7H^+ + 3e^- \rightleftarrows Cr^{3+} + 4H_2O \qquad E^0 = 1.35\,V \qquad (18.14)$$

$$H_2CrO_4 + 6H^+ + 3e^- \rightleftarrows Cr^{3+} + 4H_2O \qquad E^0 = 1.39\,V \qquad (18.15)$$

$$CrO_4^{2-} + 4H_2O + 3e^- \rightleftarrows Cr(OH)_3 + 5OH^- \qquad E^0 = -0.13\,V \qquad (18.16)$$

$$Cr_2O_7^{2-} + 14H^+ + 6e^- \rightleftarrows 2Cr^{+3} + 7H_2O \qquad E^0 = 1.33\,V \qquad (18.17)$$

The other half reaction should be such that there is an electron donor to supply the electrons needed for the reduction reaction. For this purpose, ferrous salts, zero-valent iron, and bisulfite are generally used. The electron donor itself gets oxidized, for instance Fe(0) or Fe(II) oxidizes to Fe(III), bisulfite to sulfate, etc. Two common chemicals, ferrous chloride and ferrous sulfate are commonly used for the reduction of Cr(VI) to the trivalent state for subsequent removal from solution by precipitation of Cr(OH)$_3$ (Faust and Aly 1998; Lee and Hering 2003; Qin et al., 2005). It has been observed that parameters such as pH, temperature, ionic strength and solution composition such as dissolved oxygen, phosphates, or natural organic matter, affect Cr(VI) removal (Pettine et al., 1998; Sedlak and Chan, 1997). A variation of this method is known as electrocoagulation where Fe^{2+} ions are released into the solution from an iron anode and bring forth the reduction of Cr(VI) in the contaminated water (Mukhopadhyay et al., 2007). Removal occurs in two stages: the reduction of Cr(VI) to Cr(III) at the cathode or by the Fe^{2+} ions generated and the subsequent co-precipitation of the Fe(III)/Cr(III) hydroxides. At low pH values, the reduction of Cr(VI) to Cr(III) by Fe^{2+} ions is favored, but under these pH conditions Fe(III)/Cr(III) remain as soluble species. In order to achieve the removal of both iron and chromium the pH needs to be raised beyond 3 at which solubility of hydroxides of the metal species are low giving rise to the precipitation of Fe(III)/Cr(III) hydroxides (Arroyo et al., 2009). The reduction of Cr(VI) can also take place by employing photocatalysis (Qamar et al., 2011; Yang et al., 2010) and Cr-reducing bacteria (Cheung et al., 2007; Mangaiyarkarasi et al., 2011). Laboratory and pilot-scale studies have confirmed that a high degree of treatment leading to a very low levels of residuals ($<5\,\mu g\,L^{-1}$) is easily possible through treatment via reductive coagulation and precipitation (Blute, 2010; Blute and Wu, 2012; Blute et al, 2010; Drago, 2001; Lee and Hering, 2005; McGuire, 2010; McGuire et al., 2006; 2007; Qin et al., 2005). As described by Song et al. (2013), it is possible to remove low level Cr(VI) contamination to sub-$\mu g\,L^{-1}$ levels with this reductive coagulation/filtration process. The downside of this process is the production of voluminous treatment residual as Cr-laden sludge which is difficult to dispose of.

It is already discussed that Cr(VI) oxyanions, being ligands, are effective in getting into Lewis acid-base type of interactions with other Lewis acids by sharing their lone pair of electrons residing in their d-orbital. An abundantly used adsorbent, ferric hydroxide, at pH less than its point of zero charge (pH$_{PZC}$) has positively charged surface functional groups which often act as Lewis acid sites. Thus, adsorbents based on ferric hydroxide should show a good degree of removal of Cr(VI) species from natural waters having low concentration of Cr(VI). Iron-oxide-coated sand (IOCS), produced by coating quartz sand with ferric nitrate, was effective for removal of Cr from wastewater effluent (Bailey et al., 1992; Edwards and Benjamin, 1989).

Activated carbon also possesses affinity for heavy metals, several laboratory-scale studies have been conducted on the effectiveness of activated carbon for the removal of Cr(VI) (Beszedits 1988; Hu et al., 2003; Mohan and Pittman 2006; Selomulya et al., 1999; Sorg 1979). The removal mechanism is believed to be due to two simultaneous reactions: adsorption onto the surface and Cr(VI) reduction to Cr(III) followed by precipitation at the surface. There may be an additional benefit, as reactivation of used up activated carbon may increase adsorption capacity as indicated by Han *et al.* (2000). So far, however, there is no evidence on application and efficiency of activated carbon in removing trace Cr concentrations from drinking water sources (Sharma *et al.*, 2008).

Membrane technology, such as reverse osmosis, although non-selective, removes Cr(VI) efficiently just like the same way it efficiently removes of all other ions (Hafiane *et al.*, 2000; USEPA 2003). Theoretically at high pH, where there is a prevalence of divalent ions, nanofiltration is supposed to effectively remove Cr(VI), though there is not much data available (Hafiane *et al.*, 2000).

Ion exchange has been by far the most practicable technology for removal of trace concentration of Cr(VI) from contaminated water. The main advantages of ion exchange over reduction/coagulation/precipitation are the potential recovery of the metal, higher selectivity, and no sludge production. Ion exchange is considered to be one of the best available technologies recommended by the USEPA for Cr(VI) removal. Several ion-exchange resins including strong base anion-exchange resins (e.g. IRA-900, IRA-458, etc.) and weak base anion-exchange resins (e.g. Amberlite IR 67RF, IRA-94, IRA 68, etc.) have been used for successful removal of trace concentration of Cr from water (Galan *et al.*, 2005; Rees-Novak *et al.*, 2005; Rengaraj *et al.*, 2001; SenGupta *et al.*, 1986a). In this article we shall discuss the mechanisms and potential use of ion exchange and its variants to concentrations of Cr.

18.2.1 *Removal of trace concentration of Cr(VI) from natural waters by ion-exchange process and its variants*

18.2.1.1 *Removal by strong and weak base anion-exchange resins*
Selective removal of trace concentrations of Cr(VI) by anion-exchange resin depends on the solution composition as well as on the characteristics of the anion exchange resins. A considerable amount of work has been done on the use of ion exchange for selective removal of trace Cr from industrial and hazardous wastewaters. However, due to the absence of a low MCL value for the presence of Cr(VI) in drinking water until now, not much work has been done on the removal of trace concentration of Cr(VI) from contaminated drinking water. The concentration of Cr(VI) in the industrial and hazardous waste is typically 10–1000 times higher than the concentration range that is in focus here in this article. However, the relative concentrations of Cr(VI) as compared to other competing species are quite comparable to the groundwater Cr(VI) situation. Before the time when discharge regulations came in force, Cr(VI) was used as a corrosion inhibitor in cooling water systems. It has been demonstrated earlier that at acidic pH some anion-exchange resins can selectively remove and recover chromate from cooling tower blow down water containing 5–20 mg L^{-1} in presence of 2000 mg L^{-1} each of sulfate and chloride. (Höll *et al.*, 2004; Kunin, 1976; Newman and Reed, 1980; Richardson *et al.*, 1968; Sengupta, 1986b; Yamoto *et al.*, 1975). In groundwater too, competing anions such as sulfate and chloride are present in concentrations of 2 to 3 orders of magnitude higher than that of Cr(VI) which is usually present in trace concentration of 10–400 µg L^{-1}.

Natural waters, including groundwater, have a pH range of 6.5 to 8.5. Recognizing that the major species of Cr(VI) are bichromate (HCrO$_4^-$) and chromate (CrO$_4^{2-}$), their relative distribution at different pH in the range of natural water is given by the following expression:

$$pH = 6.45 + \frac{\gamma_1^3 [CrO_4^{2-}]}{[HCrO_4^-]} \tag{18.18}$$

where γ_1 is the monovalent ion activity coefficient, the value of which can be found out from Debye-Hückel theory using the following relationship:

$$-\log \gamma_1 = 0.5 I^{1/2} \tag{18.19}$$

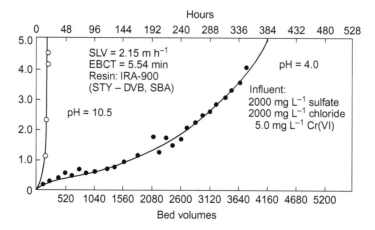

Figure 18.5.   Chromate breakthrough profiles at different pH for an IRA-900 resin with sulfate and chloride as competing species.

where I is the ionic strength of the solution. $HCrO_4^-$ is known to polymerize to form $Cr_2O_7^{2-}$ at higher concentrations; as per Figure 18.1, such polymerization in the aqueous phase largely takes place at pH below pH 6.5 and at a total Cr(VI) concentration of around 1–2 mg L^{-1} as C. So, in natural waters where the total Cr(VI) concentration is usually in the range 10–400 μg L^{-1}, the presence of dichromate is generally ruled out.

When ionic concentration of groundwater is low enough and at pH between 6.5 and 8.5, $[CrO_4^{2-}] > [HCrO_4^-]$. So, at the pH range of groundwater both the species will be present but concentrations of chromate ions dominate over that of bichromate ions. A strong base anion-exchange resin in chloride form will absorb chromate from water according to the following reactions:

$$\overline{R\text{-}Cl^-} + HCrO_4^- \rightleftarrows \overline{R\text{-}HCrO_4^-} + Cl^- \qquad K_{HCr/Cl} \qquad (18.20)$$

$$\overline{2R\text{-}Cl^-} + CrO_4^{2-} \rightleftarrows \overline{R_2\text{-}CrO_4^{2-}} + 2Cl^- \qquad K_{CrO/Cl} \qquad (18.21)$$

where R denotes the functional group on the ion-exchange resin and overbar signifies resin/solid phase. HCr stands for bichromate ion and CrO for chromate ion.

The selectivity coefficients are defined as follows,

$$K_{HCr/Cl} = \frac{[R\text{-}HCrO_4^-][Cl^-]}{[R\text{-}Cl][HCrO_4^-]} \qquad (18.22)$$

$$K_{CrO/Cl} = \frac{[R\text{-}CrO_4^{2-}][Cl^-]^2}{[R\text{-}Cl]^2[HCrO_4^-]} \qquad (18.23)$$

For removal of Cr(VI) from industrial water by ion-exchange, acidic pH has been extensively used for removal of trace concentration of Cr(VI) (Kunin, 1976) because at this pH, in presence of competing sulfate and chloride anions, anion-exchange resins have higher capacity for the pre-dominant species, $HCrO_4^-$, compared to that for CrO_4^{2-}. Figure 18.5 shows Cr(VI) breakthrough history of different column runs performed with influent solutions containing same concentration of Cr(VI) with same background concentrations of sulfate and chloride but at different pH. The tremendous difference in the Cr(VI) removal capacities between acidic and alkaline pH under otherwise similar conditions is evident from the column run histories. Sengupta and Clifford (1986) have proved that such increased chromate removal capacity at acidic pH is attributable to two reasons: (i) removal of greater number of Cr atoms at acidic pH per ion-exchange site of

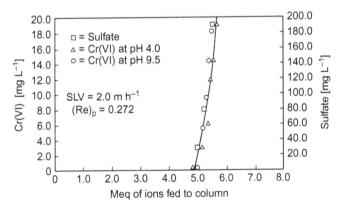

Figure 18.6.   Identical chromate (pH 4.0 and 9.5) and sulfate breakthrough curves for IRA-900 on equivalent capacity basis, one-component feed.

the resin; and (ii) the selectivity reversal effect at the prevailing ionic strength which makes resin prefer $HCrO_4^-$ over CrO_4^{2-}.

Bichromate, $HCrO_4^-$, being monovalent, can attach to a single ion-exchange functional group. Chromate, CrO_4^{2-}, on the other hand, is divalent and hence needs to bind to two ion-exchange functional groups. Thus, $HCrO_4^-$ has one Cr ion per ion-exchange site while CrO_4^{2-} has 0.5 Cr atom per ion-exchange site. However on the basis of milli-equivalents (meq) removed, anion-exchange resin have the same removal capacity for both the ions when they are the sole species present in the influent of a column run. Figure 18.6 shows the breakthrough profiles of Cr(VI) at pH 4 and 9 and also of SO_4^{2-} of such column runs with feed containing only single component, with abscissa calculated in units of meq fed to the column. Exactly same breakthrough profiles for all the three cases prove that ion exchange is the only process responsible for removal of Cr(VI).

It is interesting to note from Figure 18.5 that while $HCrO_4^-$ was the most preferred species among the components in the feed solution, the breakthrough curve, instead of being sharp, is a gradual one. Self-sharpening breakthrough patterns are observed in case of breakthrough of the preferred species. On the other hand, although under competition from chloride and sulfate, there is a sharp breakthrough of Cr(VI) at alkaline pH.

In a binary system (i and j), the selectivity or preference of a component i is indicated by its distribution coefficient (λ_i), which is defined as the ratio of its equivalent fraction in the exchanger phase and the aqueous phase (y_i/x_i); the higher the number is, the greater is the selectivity. y_i and x_i indicate equivalent fractions of component i in the exchanger phase and aqueous phase, respectively, i.e.:

$$y_i = \frac{\overline{C_i}}{Q} \tag{18.24}$$

and

$$x_i = \frac{C_i}{C_T} \tag{18.25}$$

where $\overline{C_i}$ and C_i denote the concentrations of component i in the exchanger phase and aqueous phase, respectively. Q and C_T represent the total exchange capacity of the resin and total aqueous-phase concentration, respectively.

Figures 18.7 and 18.8 show the plot of chromate selectivity versus the resin phase chromate concentration at different concentrations of sulfate and chloride, respectively.

There are two important observations: (i) chromate selectivity, irrespective of the type or concentration of competing ions, goes through a maximum and (ii) at any particular value of y_{Cr} the selectivity is higher at increased sulfate concentration while there is practically no effect on the selectivity with increase in chloride concentration.

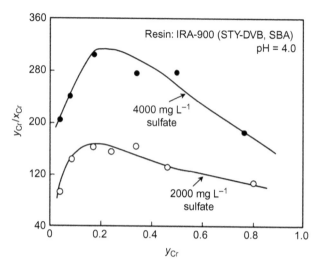

Figure 18.7. Chromate selectivity versus resin phase chromate concentrations at different competing sulfate concentrations.

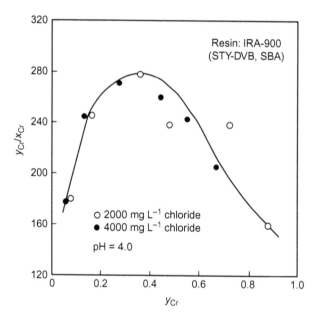

Figure 18.8. Chromate selectivity versus resin phase chromate concentrations at different competing chloride concentrations.

Due to the Donnan exclusion principle, the concentration of counter-ions inside the ion-exchanger phase is order(s) of magnitude greater than that in the bulk aqueous phase (Donnan, 1995). Hence, concentration of Cr(VI) species such as $HCrO_4^-$ inside the ion exchanger is way greater than that in the bulk aqueous phase at acidic pH. Therefore, according to the Figures 18.1 and 18.2, there is a possibility that inside the ion exchanger, both $HCrO_4^-$ and its dimerized species, $Cr_2O_7^{2-}$ are coexisting. Sengupta et al. (1986) have demonstrated that peaks or maxima as observed in Figures 18.6 and 18.7 correspond to the inflection points in the isotherms, and the inflection points are caused by the presence of $Cr_2O_7^{2-}$ in the exchanger phase along with predominant $HCrO_4^-$. Later on, Mustafa *et al.* (1997) provided a direct proof of presence of $Cr_2O_7^{2-}$ through analysis of FTIR spectra of a resin exhausted with Cr(VI) solution at acidic pH.

From the foregoing discussions it is evident that that inside the anion exchangers at this prevailing pH and ionic strength, both monovalent bichromate and dichromate ions are present and they compete with sulfate ions for the available ion-exchange sites. The ion-exchange reactions for exchange of both the Cr(VI) species and sulfate are described by the following reactions:

$$\overline{R_2SO_4} + 2HCrO_4^- \rightleftarrows 2\overline{RHCrO_4} + SO_4^{2-} \, K_{HCr/S} \tag{18.26}$$

$$\overline{R_2SO_4} + Cr_2O_7^{2-} \rightleftarrows \overline{R_2Cr_2O_7} + SO_4^{2-} \, K_{Cr_2/S} \tag{18.27}$$

The equilibrium constants are given by the following relationships,

$$K_{HCr/S} = \frac{\gamma_2[SO_4^{2-}] \, f_{HCr}^2 [\overline{RHCrO_4}]^2}{\gamma_1^2[HCrO_4^-]^2 \, f_S[\overline{R_2SO_4}]} \tag{18.28}$$

and,

$$K_{Cr_2/S} = \frac{[SO_4^{2-}] \, f_{Cr_2}[\overline{R_2Cr_2O_7}]}{[Cr_2O_7^{2-}] \, f_S[\overline{R_2SO_4}]} \tag{18.29}$$

where f and γ are resin and aqueous phase activity coefficients, respectively; overbar denotes resin phase; HCr, Cr_2 and S stand for chromate, dichromate and sulfate, respectively; 1 and 2 denote monovalent and divalent species, respectively.

In the pH range of 3–5, bichromate is the predominant species in the aqueous phase so that the total Cr(VI) balance in aqueous phase is:

$$[Cr(VI)] = 2[Cr_2O_7^{2-}] + [HCrO_4^-] \gg [HCrO_4^-] \tag{18.30}$$

Equation (18.3) and (18.26) yield,

$$[Cr_2O_7^{2-}] = \frac{K_3}{\gamma_1^2}[Cr(VI)]^2 \tag{18.31}$$

Also, according to the Davies equation (Stumm and Morgan, 1996), divalent ion activity coefficients, γ_2, are related to monovalent ion activity coefficients, γ_1, by the relationship:

$$\gamma_2 = \gamma_1^4 \tag{18.32}$$

Equations (18.24) and (18.25) may be rewritten as:

$$[\overline{RHCrO_4}] = \left[\frac{K_{HCr/S} f_S [\overline{R_2SO_4}]}{[SO_4^{2-}]} \right]^{\frac{1}{2}} \frac{[Cr(VI)]}{f_{HCr}\gamma_1} \tag{18.33}$$

$$[\overline{R_2Cr_2O_7}] = \left[\frac{K_3 K_{Cr_2/S} f_S [\overline{R_2SO_4}]}{f_{Cr_2}[SO_4^{2-}]} \right] \frac{[Cr(VI)]^2}{\gamma_1^2} \tag{18.34}$$

The overall chromate exchange isotherm in presence of competing species such as sulfate at the acidic pH is given by:

$$[\overline{Cr(VI)}] = [\overline{RHCrO_4}] + [\overline{R_2Cr_2O_7}]$$

$$= \left[\frac{K_{HCr/S} f_S [\overline{R_2SO_4}]}{[SO_4^{2-}]} \right]^{1/2} \frac{[Cr(VI)]}{f_{HCr}\gamma_1} + \frac{2K_3 K_{Cr_2/S} f_S [\overline{R_2SO_4}]}{f_{Cr_2}[SO_4^{2-}]} \frac{[Cr(VI)]^2}{\gamma_1^2} \tag{18.35}$$

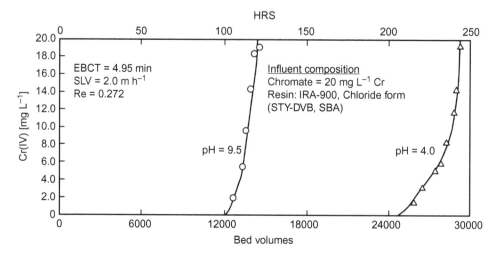

Figure 18.9.   Breakthrough pattern for Cr(VI) species at acidic and alkaline pH.

When Cr(VI) is a trace species and the concentration of the competing species sulfate is constant, all the terms in the right hand side of Equation (18.35) become constant except for [Cr(VI)]. Then, the relationship in Equation (18.35) can be written as:

$$\overline{[Cr(VI)]} = A_1[Cr(VI)] + A_2[Cr(VI)]^2 \tag{18.36}$$

Dividing the both sides of Equation (18.36) by $C_T Q$ we get,

$$\frac{\overline{[Cr(VI)]}}{Q} = \left(\frac{C_T}{Q}A_1\right)\frac{[Cr(VI)]}{C_T} + \left(\frac{C_T^2}{Q}A_2\right)\left[\frac{[Cr(VI)]}{C_T}\right]^2 \tag{18.37}$$

Now, realizing that equivalent weights of $HCrO_4^-$ and $Cr_2O_7^{2-}$ as Cr are the same, the above equation can be converted in terms of equivalent fractions:

$$y_{Cr} = A_3 x_{Cr} + A_4 x_{Cr}^2 \tag{18.38}$$

The constants A_3 and A_4 yield positive values, and hence,

$$\frac{\partial^2 y}{\partial x^2} > 0 \tag{18.39}$$

This means that the isotherm has positive curvature. An isotherm with positive curvature gives rise to an unfavorable (concave upwards) isotherm. For a column run, the breakthrough curve for a species with an unfavorable isotherm ideally should give rise to a gradual breakthrough pattern, even if there is no mass transfer limitation. Figure 18.5 shows the breakthrough curve of Cr(VI) in a column run at pH 4 in the presence of competing sulfate and chloride species. Despite Cr(VI) having a higher selectivity compared to other anions at this pH, the Cr(VI) breakthrough curve has a gradual pattern. SenGupta et al. (1986a) have proved that the gradual nature of Cr(VI) breakthrough from the column in presence of sulfate is not due to any mass transfer limitation; such breakthrough pattern is well predicted by Equations (18.38) and (18.39). However, when the column run was performed with Cr(VI) as the single component, there is a self-sharpening pattern obtained for Cr(VI) breakthrough curve. Figure 18.9 shows breakthrough patterns of Cr(VI) at pH 4 and 9.5 when the influent had only Cr(VI). It may be noted that in both the cases, there is a self-sharpening breakthrough pattern. These curves indicate that Cr(VI), as a single component, has a favorable exchange isotherm and also, there is no mass-transfer limitation in the column.

The presence of $Cr_2O_7^{2-}$ in the ion-exchanger phase causes an early gradual breakthrough of Cr(VI) during column runs, but $HCrO_4^-$ is the prime counter-ion and it dictates the total chromate removal capacity by the resin. Since the removal of chromate is the prime requirement, $HCrO_4^-$ may be considered as the sole exchanging Cr(VI) species. In such a case, Equation (18.34) takes the following form:

$$\overline{[Cr(VI)]} = \left[\frac{K_{HCr/S} f_S \overline{[R_2SO_4]}}{[SO_4^{2-}]} \right]^{1/2} \frac{[Cr(VI)]}{f_{HCr} \gamma_1} \tag{18.40}$$

By dividing both sides of the above equation by $C_T Q$ and combining with Equation (18.10), we can write after algebraic manipulations,

$$y_{Cr} = constant \left(\frac{y_S C_T}{x_S Q} \right)^{0.5} x_{Cr} (10^{0.5 I^{1/2}}) \tag{18.41}$$

Now, for binary exchange reaction, $1 - x_{Cr} = x_S$ and $1 - y_{Cr} = y_S$
As Cr(VI) is a trace species compared to sulfate, $1 - x_{Cr} \approx 1.0$. Thus Equation (18.41) becomes

$$\frac{\left(\frac{y_{Cr}}{x_{Cr}} \right)}{(1 - y_{Cr})^{0.5}} = constant \; C_S^{0.5} (10^{0.5 I^{1/2}}) \tag{18.42}$$

At the same y_{Cr} a decrease in C_S causes an increase in x_{Cr} which means, there will be a decrease in in the chromate selectivity or chromate distribution coefficient (y_{Cr}/x_{Cr}). Thus, the above equation explains the reason as to why the chromate selectivity decreases upon decreasing sulfate concentration. The above equation also takes care of the non-ideality effect in the solution. In the case of binary exchange of chromate and chloride at acidic pH, in the same way it can be shown that

$$\frac{\left(\frac{y_{Cr}}{x_{Cr}} \right)}{1 - x_{Cr}} = constant \tag{18.43}$$

Thus, chromate selectivity remains unaffected by the competing chloride concentration and the experimental data in Figure 18.9 is in agreement with the above deduction.

For chromate removal at alkaline pH, CrO_4^{2-} is the only species of chromate present both in the ion exchanger and in the solution phase. Figure 18.10 shows an isotherm for chromate exchange at alkaline pH. Unlike the isotherms at acidic pH, the isotherm shows a favorable nature. According to the chromatographic theories, chromate ion exchange with a fixed bed column run should give rise to a self-sharpening breakthrough profile, which is evident from Figure 18.9. However, it may be noted that, in acidic pH operation, although there is gradual breakthrough, the chromate removal capacity is much higher. CrO_4^{2-} ions occupy double the number of ion-exchange sites as compared to $HCrO_4^-$ ions and this can be an important reason for lower removal capacity at alkaline pH; the lower capacity for chromate removal is also attributable to strong competition from sulfate ions.

Chromate- sulfate exchange reaction can be explained using the Donnan principle:

$$\{\overline{CrO_4^{2-}}\}\{SO_4^{2-}\} = \{\overline{SO_4^{2-}}\}\{CrO_4^{2-}\} \tag{18.44}$$

where brackets denote activity and overbar denotes resin phase. Introducing activity coefficient terms, we get,

$$\frac{f_{CrO} \gamma_S y_{CrO} x_S}{f_S \gamma_{CrO} y_S x_{CrO}} = 1 \tag{18.45}$$

where f and γ represent the activity coefficients in ion exchanger and the solution phase respectively and the suffixes CrO and S stand for CrO_4^{2-} and SO_4^{2-}, respectively.

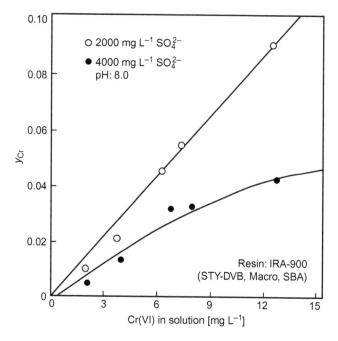

Figure 18.10.   Chromate exchange isotherm at alkaline pH.

At the prevailing ionic strength, the value of the activity coefficients depends on the charges on the ions and hence, for the divalent CrO_4^{2-} as well as SO_4^{2-} ions,

$$\gamma_{CrO} = \gamma_S \qquad (18.46)$$

Therefore, from Equations (18.40) and (18.41), chromate/sulfate selectivity coefficient can be expressed as,

$$K_{CrO/S} = \alpha_{CrO/S} = \frac{y_{CrO}x_S}{x_{CrO}y_S} = \frac{f_S}{f_{CrO}} \qquad (18.47)$$

We can also write,

$$\frac{y_{CrO}}{x_{CrO}} = K_{CrO/S}\frac{y_S}{x_S} \qquad (18.48)$$

When chromate is a trace species, the term y_S/x_S is unity. As both CrO_4^{2-} and SO_4^{2-} are divalent ions, the distribution coefficient y_{Cr}/x_{Cr} is independent of competing sulfate ions at alkaline pH.

The higher the selectivity for chromate over sulfate, the greater would be its chromate removal capacity. From Equation (18.47) it may be inferred that in order to improve the ion exchanger's chromate selectivity over sulfate, the exchanger phase activity coefficient ratio of sulfate and chromate requires to be increased.

CrO_4^{2-} and SO_4^{2-} ions are similar in many ways. Both of them have the same electrical charge and similar tetrahedral structure. CrO_4^{2-} ion has a double tetrahedral structure whereas sulfate has a tetrahedral structure. Due to the structural difference CrO_4^{2-} is less hydrated than SO_4^{2-}. Ion-exchange resins are considered to be a high capacity (>1.0 M) strongly ionized homogeneous electrolyte on the framework of cross-linked polymeric matrix which can have polar (hydrophilic) or non-polar (hydrophobic) characteristics. Due to high concentration of fixed ionic groups (co-ions) inside the resin, a significant fraction of the water inside the resin is oriented towards

Table 18.2.   Influence of individual resin properties on chromate/sulfate selectivity.

Change in resin composition	Chromate/sulfate separation factor	Reason for the change
Matrix changed from polyacrylic to polystyrene	Significantly increases	Caused by increased hydrophobicity and reduced moisture content of the polystyrene matrix
Cross-linking increased	Moderately increases	Caused by reduced moisture content associated with increased cross-linking
Functionality changed from weak base (tertiary amine) to strong base (quaternary ammonium)	Significantly increases	Caused by enhanced hydrophobicity and steric hindrance created by quaternary ammonium functionality
Ion-exchange capacity increased	Decreases	Increased concentration of fixed ionic functional group
Morphology changed from gel to macroporous	No effect	Gel phase of macroporous resins have higher degree of cross-linking which may impart hydrophobicity

these ionic groups so that they are bound with the fixed co-ions. Hence, inside an ion exchanger there is a very small fraction of free water. So, all other conditions remaining the same, the ion exchanger would prefer the less hydrated ion. In this case, CrO_4^{2-} being less hydrated than SO_4^{2-} is the more preferred species. In other words, the selectivity for chromate increases if the resin has less free water available inside it. The resin that tends to contain fewer polar water molecules inside it is termed hydrophobic in comparison to the one which contains more water inside. Therefore, the more hydrophobic the resin, the greater will be the chromate selectivity for the resin. Out of the different matrices available, polyacrylic and polystyrene are the most common ones. Polyacrylic resins have an open chain aliphatic structure containing carbonyl groups and a higher concentration of fixed co-ions as compared to the polystyrene resins. Total ion-exchange capacity remaining constant, polystyrene resins are always more hydrophobic due to less bound water content than the acrylic resins. Thus, polystyrene resins will have higher selectivity towards CrO_4^{2-}. Also, a higher degree of cross-linking in the resin matrix causes higher resistance to swelling and therefore reduces the space available for water molecules inside the resin. Thus, resins with higher degree of cross-linking will prefer CrO_4^{2-} over SO_4^{2-} ions as the latter have higher amount of bound water molecules compared to the former. A summary of the individual resin properties or factors on which chromate/sulfate selectivity shall depend is provided in Table 18.2 (Sengupta et al., 1988).

18.2.1.2   *Regeneration of the exhausted resin*
For removing chromates from water at acidic pH, weak base anion-exchange resins are generally used so that the regeneration process is simpler and more effective using only stoichiometric amount of OH^- ions supplied by NaOH solution (Korngold et al., 2003). The weak base anion-exchange resin's functional group at elevated pH becomes uncharged, thereby losing the electrostatic attraction to the Cr(VI) anion which diffuses out of ion-exchanger phase. For the removal of chromate from tap water, strong base anion exchangers have also been used. From the foregoing discussions, it may be noted that at acidic pH $HCrO_4^-$ is a highly preferred counter-ion as compared to competing chloride and sulfate. So, the efficiency of usual brine regeneration of exhausted ion-exchange resins having strong base quaternary ammonium ion is likely to be poor. However, at alkaline pH, the predominating species is CrO_4^{2-} whose selectivity is much less than $HCrO_4^-$. In order to increase the efficiency of regeneration it is common practice to use sodium hydroxide as the regeneration solution, which transforms the resin from the $HCrO_4^-$ form to the CrO_4^{2-} form. At high brine concentrations, there is a reversal of selectivity between the divalent CrO_4^{2-} ion and the monovalent Cl^- ion. Thus, a highly efficient regeneration of the strong base

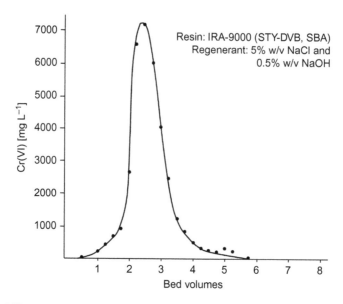

Figure 18.11.   Effluent Cr(VI) history during brine regeneration of an exhausted anion-exchange column at alkaline pH.

anion exchangers can be obtained. The reactions occurring during alkaline regeneration are as follows:

$$2\overline{RHCrO_4} + 2NaOH \longrightarrow \overline{R_2CrO_4} + 2Na^+ + CrO_4^{2-} + 2H_2O \tag{18.49}$$

$$\overline{R_2Cr_2O_7} + 2NaOH \longrightarrow \overline{R_2CrO_4} + 2Na^+ + CrO_4^{2-} + H_2O \tag{18.50}$$

The above reactions 18.45 and 18.46 take place in the ion-exchanger phase in which $HCrO_4^-$ and $Cr_2O_7^{2-}$ produce CrO_4^{2-} at alkaline pH in accordance with the dissociation equilibria as presented in Equations (18.2) and (18.3). These reactions are therefore irreversible reactions and the requirement for alkali is stoichiometric.

At high brine concentration, there is a selectivity reversal between CrO_4^{2-} and Cl^- ion, making Cl^- ions more selective than CrO_4^{2-} according to the following reaction:

$$\overline{R_2CrO_4} + 2NaCl \longrightarrow 2\overline{RCl} + 2Na^+ + CrO_4^{2-} \tag{18.51}$$

The selectivity reversal for CrO_4^{2-}/Cl^- takes place at about 5% brine concentration when Cl-ions become more preferred than CrO_4^{2-}, thus effectively removing CrO_4^{2-} from inside the ion exchanger. Figure 18.11 provides the chromate effluent history of a column during alkaline brine regeneration of a bed of SBA resin exhausted by Cr(VI) at acidic pH. It took only 6 bed volumes for complete regeneration. The sharp profile of front and rear boundaries of chromate elution history indicate that chloride is highly preferred over CrO_4^{2-} under regeneration conditions.

The disadvantage of the process is that the spent regenerant, consisting of a substantial volume of brine with a high concentration of chromate ions, is likely to pose a disposal problem. Korngold et al. (2003), in their studies have used a two-step process for regeneration of the exhausted ion-exchange resin. In the first step, a strong acid solution along with 4% sodium bisulfite was used as regenerant. Sodium bisulfite reduces Cr(VI) to Cr(III) as per the following reactions:

$$\overline{RHCrO_4} + HCl \longrightarrow \overline{RCl} + H_2CrO_4 \tag{18.52}$$

$$2H_2CrO_4 + 3NaHSO_3 + 6HCl \longrightarrow 2Cr^{3+} + 6Cl^- + 3NaHSO_4 + 5H_2O \tag{18.53}$$

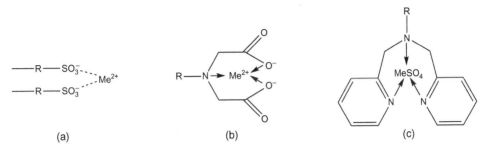

Figure 18.12.   Mechanism of metal ion removal by (a) strong acid cation exchanger, (b) chelating ion exchanger, and (c) DOW-3N. Dashed line shows electrostatic attraction, solid line shows a covalent bond and arrow represents coordinate bonds.

Chromate is thus reduced to Cr(III) cation and gets removed from the ion-exchanger phase due to the Donnan ion exclusion phenomenon. The addition of NaOH until the pH is above 11 makes $Cr(OH)_3$ precipitate out of the solution.

18.2.2   *Removal using a polymeric ligand exchanger (PLE)*

The concept of ligand-exchange was first introduced by Helfferich (1961). A polymeric ligand exchanger (PLE) is composed of a resin phase with functional groups that can firmly bind to a transition metal as well as the immobilized metal ion acting as a Lewis acid site. Ligands are ions, molecules, or functional groups that bind to a central atom to form a coordination complex where the bonding occurs due to donation of one of more ligand's electron pairs. As a result, an exchange of ligands takes place due to concurrent Lewis acid-base (LAB) interactions (metal-ligand complexation) along with electrostatic interactions between the fixed metal ion and the target ligands. Like the ion-exchange resins, ligand exchangers employ transition metal ions as their terminal functional groups. In conventional anion exchangers, the selectivity for various anions is governed by electrostatic interactions only, whereas in case of a PLE it is determined by both the ligand strength (the strength of LAB interaction) as well as the Coulombic attraction of the ligands which depends on the ionic charge. Ideally, strong acid cation-exchange resins as well as chelating ion exchangers have the ability retain the transition metal ions. Strong acid cation exchangers with functional groups such as sulfonic acid retain metal ions through electrostatic attractions while the chelating ion exchangers with iminodiacetate and similar functional groups retain metal ions by Coulombic as well as LAB interactions as shown schematically in Figure 18.12a,b. In such cases, the positive charges of the metal ions are neutralized by the fixed functional groups on the resin. Thus, the central metal ions on the resin-metal framework can act as ligand exchangers by LAB interaction only and the affinity will be limited to uncharged or neutral molecules only. In his pioneering work, Helfferich (1995) prepared PLEs by loading a transition metal (Ni or Cu) onto commercial cation-exchange resins. Because the charges of the loaded metal ions are neutralized by the negative charges of the resin functional groups, the PLEs could only sorb some neutral ligands such as ammonia and diamine. Also, due to the Donnan exclusion principle, the anions are excluded from entering into the ion-exchanger framework. Therefore, PLEs based on the strong acid cation exchangers and chelating ion exchangers cannot effectively sorb anionic ligands. In the case of strong acid cation exchangers, because sorption is due to a weak electrostatic attractive force only, there is a tendency for low metal capacity as well as significant metal bleeding. However, as shown in Figure 18.12c if the chelating group is a neutral ligand such as bis-2-(pyridilmethyl) amine, it forms positively charged chelates with the metal ions, thereby attracting anions to become associated under Coulombic and/or LAB interactions. Numerous industrially and environmentally important trace anionic contaminants in water and wastewater such as arsenate, chromate, phosphate, selenite, cyanide, oxalate, phthalate, and a host of organic derivatives are fairly strong ligands. Conventional sorbents and anion exchangers cannot separate these trace ligands selectively due to the strong competition from anions such as sulfate, carbonate, chloride, and nitrate, which are commonly present in virtually all natural waters and at higher concentrations (Zhao and SenGupta, 1995; 1998a; 1998b). Obviously, these anionic ligands will be more selectively adsorbed than any other anions because the ligands are strongly held due to two reinforcing forces: Coulombic as well as LAB interactions.

The properties of the metal-hosting polymer are of critical importance in preparation of a novel PLE. An ideal metal hosting polymer should possess the following key attributes for its deployment for practical applications such as drinking water treatment: (i) High metal-ion adsorption capacity; (ii) Minimal metal leakage during ligand exchange operations; (iii) The metal-adsorbing functional group on the ion-exchange resin needs to be uncharged to prevent any charge neutralization of the adsorbed metal ions; and (iv) The functional groups of the metal binding polymers should not use up all of the metal's coordination sites only to allow further coordination sites available for the coordinate electron sharing with the aqueous phase ligand.

The properties of the metal that is the central atom on the PLE are also of significant importance. Sorption of metal ions to a neutral chelating resin can be envisioned as a surface complexation process. Therefore, the existing knowledge on solution coordination chemistry is applicable to the PLE's ligand exchange process. To describe the stability of metal-ligand complexes in homogeneous systems, the Irving-Williams order has been often invoked (Stumm and Morgan, 1996). First-row transition metals (Mn^{2+}, Fe^{2+}, Co^{2+}, Ni^{2+}, Cu^{2+} and Zn^{2+}) have partially filled d electron shells, which affect their complex formation (Cotton and Wilkinson, 1988). All first-row transition metals can form high-spin octahedral metal complexes and the general stability sequence can be written as $Mn^{2+} < Fe^{2+} < Co^{2+} < Ni^{2+} < Cu^{2+} > Zn^{2+}$. This is known as the Irving-Williams series which has been found to hold for a wide variety of ligands (Da'na and Sayari, 2012; Irving and Wallace, 1953). More specifically, the ligand field stabilization energy (LFSE) varies with electron occupancy in the d shell (Martell and Hancock, 1996). The above stability sequence suggests that Cu^{2+} may be the best candidate for making high capacity PLEs.

18.2.2.1   Preparation of the PLEs

The chelating resin, DOW 3N (XFS4195), was selected as the metal hosting polymeric resin. The resin is synthesized by attaching bispicolylamine functional groups to a macroporous polystyrene divinylbenzene copolymer. Since the resin contains pyridyl and tertiary-amine groups as functional groups, it is classified as a weak base chelating resin. Extensive studies of this resin were conducted with respect to their sorption behaviors for the removal of heavy metals (Jones and Grinstead, 1977; SenGupta et al., 1991; Zhao, 1997). The following affinity sequence was inferred for DOW 3 N: $Cu^{2+} \gg Ni^{2+} > Fe^{3+} > Cd^{2+} > Zn^{2+} > Co^{2+} = Fe^{2+} > Ca^{2+} > Mg^{2+} = Al^{3+}$. For comparison, a commercial strong base anion exchanger (IRA-900, Rohm and Haas Co., Philadelphia) was also included in the study. IRA-900 was used earlier in a pilot-scale study in Scottsdale, Arizona, to remove chromate from contaminated groundwater. Besides the composition of the functional groups, IRA-900 and the PLE are identical, i.e., they are both macroporous with polystyrene divinylbenzene matrix. DOW 3N was conditioned following the standard procedure of cyclic exhaustion with 1 N HCl and 1 N NaOH. DOW 3N was then converted into Cu(II)-loaded form by passing a 500 mg L^{-1} Cu(II) solution at pH 4.5 through the resin in a column until saturation. The synthesized polymeric ligand exchanger is further designated as DOW 3N-Cu.

All fixed-bed column runs in the study were carried out using glass columns (11 mm diameter), constant-flow stainless steel pumps, and effluent samples were continuously collected using an ISCO fraction collector. To avoid any possible bleeding of copper from the DOW 3N-Cu bed into the exit of the column, a small amount (about 10% of the total bed height) of a virgin chelating ion exchanger with iminodiacetate functionality (IRC-718, Rohm and Haas, Philadelphia) was kept at the bottom of the column. Total dissolved Cr in each sample was analyzed using an atomic absorption spectrophotometer with graphite furnace accessory (Perkin-Elmer Model 6000) while Cr(VI) was determined by the colorimetric technique using diphenylcarbazide. The dissolved

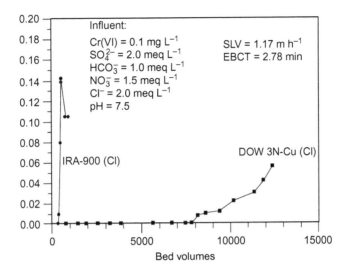

Figure 18.13.   Cr(VI) history in the effluent of columns run with IRA-900 and DOW 3N-Cu, both in Cl form.

Cr(III) constituted the difference between total Cr and Cr(VI). Sulfate, chloride, and nitrate were analyzed using a Dionex ion chromatograph (model DX 120).

18.2.2.2   *Selective chromate removal by PLE*

In order to find out the capability of the PLE and compare it with the commercially available anion-exchange resins we ran two separate fixed-bed column runs using IRA-900 (strong base anion exchanger) and DOW 3N-Cu under otherwise identical conditions. Figure 18.13 shows the chromate effluent histories obtained from the two column runs. In both the column runs, the influent water had Cr(VI) at a concentration of $100\,\mu g\,L^{-1}$ with a background of naturally occurring anions, namely sulfate, bicarbonate, chloride and nitrate, at concentrations in which they are generally found in groundwater. Much higher background concentrations of these anions compared to that of Cr(VI) means that Cr(VI) was a trace species in the test water. In case of IRA 900, Cr(VI) broke through at around 500 bed volumes whereas the column containing DOW 3N-Cu, the Cr(VI) breakthrough occurred at 7500 bed volumes. Analysis of influent and effluent water samples proved that all the Cr present was only in the form of Cr(VI) only, no Cr(III) was present. For the column with IRA-900, the effluent Cr(VI) concentration after the breakthrough reached a peak which is higher than the influent concentration, resembling to a phenomenon known as chromatographic elution. Nitrate broke through the column later than Cr(VI), indicating that nitrate had higher affinity compared to Cr(VI). When nitrate broke through the column, the Cr(VI) concentration in the effluent returned to the inlet concentration.

Figure 18.14 shows effluent breakthrough profiles for all competing anions, namely sulfate, chloride, nitrate and bicarbonate for the same column run with DOW 3N-Cu. All the competing species went into complete breakthrough much before the breakthrough of Cr(VI).

The mechanisms and driving forces underlying the sorption of various ligands, ions, or sorbents can be explained through inspection of various constituents of the overall standard free energy change ($\Delta G^0_{Overall}$) involved in the ligand exchange reactions.

a) In case of removal of Cr(VI) anion by the strong base anion exchanger, sorption takes place only through electrostatic attraction between the Cr(VI) anion and the positively charged quaternary ammonium functional group of the resin. Thus,

$$\Delta G^0_{Overall} = \Delta G^0_{EL} \tag{18.54}$$

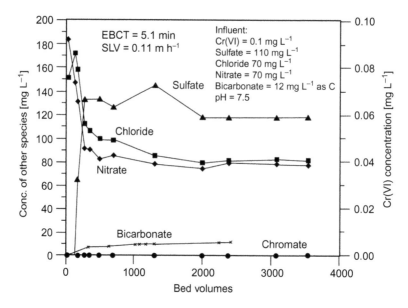

Figure 18.14.   Effluent breakthrough profiles of different competing ions and chromate for a column run using DOW 3N-Cu as the adsorbent.

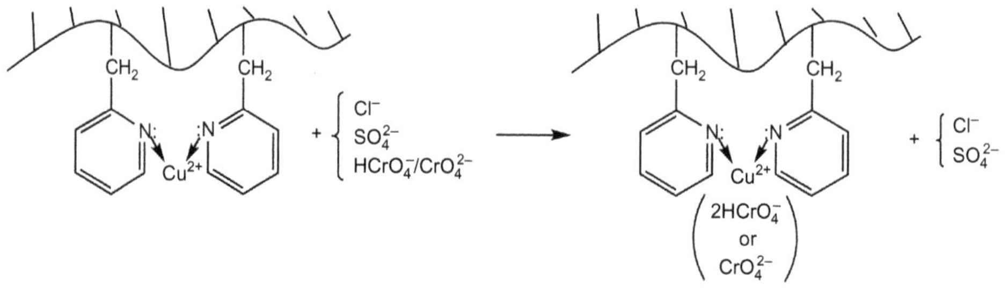

Figure 18.15.   Schematic diagram showing the mechanism for chromate removal by polymeric ligand exchanger.

where ΔG_{EL}^0 is the standard free energy change associated with electrostatic interactions. Noting that, $\Delta G^0 = -RT \ln K$, where K is the equilibrium constant, the overall sorption equilibrium constant ($K_{overall}$) is also equal to the equilibrium constant due to electrical interactions, or

$$K_{overall} = K_{EL} \qquad (18.55)$$

For sorption of chromate species by the PLE or DOW 3N-Cu, the overall driving force of interaction is composed of two additive components, namely electrostatic attractions and Lewis acid-base interactions (ΔG_{LAB}^0). In this case thus,

$$\Delta G_{Overall}^0 = \Delta G_{EL}^0 + \Delta G_{LAB}^0 \qquad (18.56)$$

and therefore,

$$K_{overall} = K_{EL} K_{LAB} \qquad (18.57)$$

As chromate is a preferred species by the anion-exchange sites as well as the PLE, both K_{EL} and K_{LAB} have value higher than 1. Thus, incorporation of Lewis acid-base interaction through PLE acts synergistically with electrostatic interactions interaction, resulting in enhanced selectivity for chromate. Figure 18.15 illustrates the sorption mechanism for chromate by the PLE.

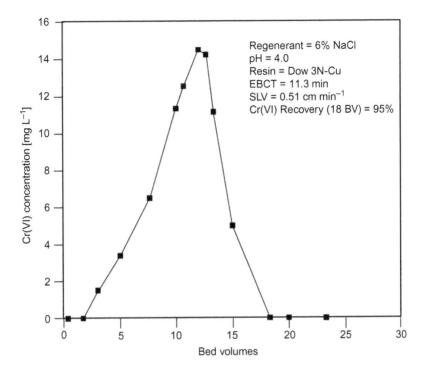

Figure 18.16. Elution history of hexavalent chromium during regeneration of an exhausted column.

18.2.2.3 *Regeneration of the exhausted PLE*

Following experimental trial and error, 6% sodium chloride at pH 4.0 was found to be an efficient regenerant for the exhausted PLE. Figure 18.16 shows the chromate elution during brine regeneration of the exhausted mini column. According to Cr mass balance, more than 95% of total chromate was recovered in less than 18 bed volumes. It is postulated that the selectivity reversal between chloride and chromate at high brine concentration of the regenerant and weaker Lewis acid-base interaction due to competition from H^+ ions are responsible for an efficient chromate desorption.

18.2.3 *Simultaneous removals of heavy metal cations and chromate*

In contaminated groundwaters, chromate may also be present at below-neutral pH with other heavy metal cations, such as copper, zinc, and nickel. PLEs offer a unique solution to the contamination problem in such cases. In PLE, the parent chelating exchanger with nitrogen donor atoms can simultaneously and selectively remove such metal cations and chromate anions. Figure 18.17 shows the effluent history of a column run for a synthetic influent containing both Cu(II) and Cr(VI) along with other competing cations and anions. It may be noted that simultaneous Cu(II) and Cr(VI) removal of about 95% was achieved over 2500 bed volumes. Mechanistically, such a phenomenon can be explained as the following: first, Cu(II) is selectively sorbed onto the chelating polymer in preference to other competing cations; and second, once sorbed onto the chelating polymer, Cu(II) forms positively charged chelates thereby converting the site into a PLE site that act as a selective anion exchanging site for preferential adsorption of chromate over chloride and sulfate.

Other commercially available chelating exchangers with iminodiacetate, thiol, aminophosphonate, and carboxylate functional groups are unable to provide simultaneous removal of heavy metal cations and chromate. The chelating group-metal complex is neutral in such a case and

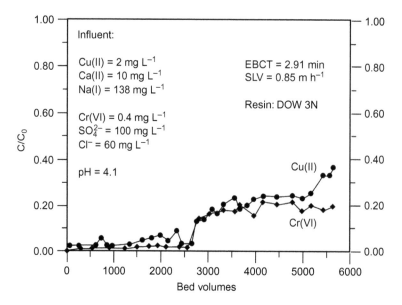

Figure 18.17.   Effluent breakthrough history showing simultaneous removal of Cu and Cr(VI) by PLE.

can have LAB interaction with neutral molecules only. Moreover, as the functional groups on metal selective chelating exchangers are negatively charged, they reject chromate anions through Donnan co-ion exclusion effects.

It is to be noted that both the influent and effluent from the column runs were subjected to the diphenylcarbazide test, specific for determination of Cr(VI). Both water samples showed that all the total Cr, as measured by the AAS, was in the form of Cr(VI) and there was no trace of Cr(III). Testing also confirmed that only Cr(VI) was present in the spent regenerant and Cr(III) was absent altogether. This observation that Cr(VI) was not chemically reduced to Cr(III) within the organic PLE even after a prolonged contact is counterintuitive. Note from Figure 18.17 that copper (II), i.e., Cu^{2+}, is the anion-exchange site for CrO_4^{2-}. Since Cu(II) is the highest oxidation state of elemental copper, the micro-environment in the vicinity of sorbed chromate is not thermodynamically favorable for Cr(VI) reduction although the sorbent is essentially an organic polymer.

18.2.4   *Hexavalent chromium removal through a redox-active anion exchanger*

One recent study involving the removal of trace concentrations of Cr(VI) present in groundwater of Glendale, CA has indicated that two weak base anion exchange (WBA) resins (Duolite A-7 and SIR-700) have shown at least an order of magnitude higher capacity than strong base anion exchange (SBA) resins in presence of significantly high concentrations of competing anions like chloride, sulfate, etc. (McGuire et al., 2007). A similar Cr(VI) removal performance by Duolite A-7 has also been reported elsewhere (Höll et al., 2004). As mentioned before in this article, previous research studies established that ion exchange is the mechanism responsible behind chromate removal by SBA resins. Also, the theoretical exchange capacities of the WBA resins do not significantly differ from those of the SBA resins. It is therefore intriguing to find out the mechanism behind such exceptionally high chromate ion removal capacity shown by these WBA resins during the pilot plant study at Glendale, CA. The field results from pilot plant study suggested presence of one or more mechanisms along with ion exchange. Chromate ion removal by Duolite A-7 was further investigated in the laboratory with an aim to find out the

Table 18.3.   Details of anion-exchange resins used in the study.

Resin	Manufacturer	Type	Type of matrix	Capacity [eq L^{-1}]
A-850	Purolite Co.	SBA	Polyacrylic	1.25
A-600	Purolite Co.	SBA	Styrene-divinyl benzene	1.3
A-400	Purolite Co.	SBA	Styrene-divinyl benzene	1.5
Duolite A-7	Rohm and Haas	WBA	Phenol-formaldehyde	2.1
SIR-700	Resin Tech	WBA	Epoxy-polyamine	2.7

Table 18.4.   General water quality parameters of GS-3 well water at Glendale, CA.

Constituent	Typical concentration
Alkalinity	200 mg L^{-1} as CaCO$_3$
Arsenic (total)	<2 μg L^{-1}
Chromium (total)	35–40 μg L^{-1}
Chromium (hexavalent)	35–40 μg L^{-1}
Conductivity	850 μS cm^{-1}
Copper	20 μg L^{-1}
Hardness	350 mg L^{-1} as CaCO$_3$
Iron (total)	<6 μg L^{-1}
Manganese	<20 μg L^{-1}
Nitrate	7 mg L^{-1} as NO$_3$
pH	6.8
Phosphate	0.3 mg L^{-1} as PO$_4$
Silicate	33 mg L^{-1} as SiO$_2$
Sulfate	100 mg L^{-1} as SO$_4$
Uranium	1.4 μg L^{-1}
Vanadium	7 μg L^{-1}

exact mechanisms through which these ions are removed from the contaminated groundwater by the resin.

The commercially available resins used in this study were A-400, A-600 and A-850 (Purolite Co.), Duolite A-7 (Rohm and Haas Co.), SIR-700 (ResinTech Co.). The properties of the anion-exchange resins and their capacities are indicated in Table 18.3.

In the laboratory fixed bed column runs were carried out at room temperature using epoxy coated ACE glass columns (11 mm in diameter) and constant flow positive displacement pumps. The effluent was collected using an ELDEX fraction collector. The ratio of column diameter to the average diameter of the resins was 20:1. Superficial liquid velocity (SLV) and empty bed contact times were recorded for each column run. All Cr used in the influent was in Cr(VI) form. In order to maintain an oxidizing environment in the influent the reservoir containing the Cr(VI) solution was kept open to atmosphere. In order to examine the behavior of the column under changing conditions, a few column runs were deliberately interrupted for some days, influent pH was changed, or the flow rate was increased.

Synthetic raw water used in the column studies always contained, apart from the contaminant Cr(VI), a background of commonly occurring competing anions like sulfate, chloride and bicarbonate at concentrations representing a typical groundwater. The synthetic water was prepared by mixing together appropriate amounts of reagent grade salts of these ions in deionized water. The actual groundwater used in the column runs was obtained from the well marked GS-3 at Glendale, CA. The composition of the water is indicated in Table 18.4.

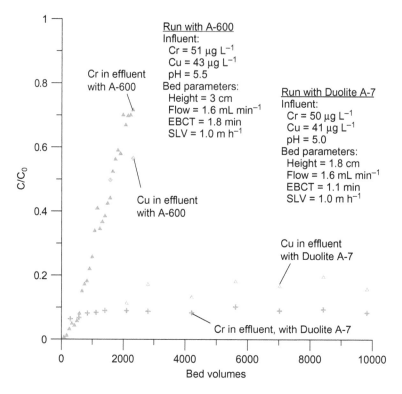

Figure 18.18.   Chromium and copper breakthrough profiles from columns containing strong base anion-exchange resins (A-600) and weak base anion-exchange resins (Duloite A-7).

Total Cr was analyzed using a Perkin-Elmer SIMAA 6000 atomic absorption spectrophotometer with a graphite furnace. The presence of Cr(III) in aqueous samples was analyzed using a Perkin-Elmer UV/VIS spectrophotometer Lambda 2 following the colorimetric diphenylcarbazide method detailed in the Standard Methods (APHA, 1998). X-ray absorption fine structure (XAFS) study was carried out at the advanced photon source facility in the Argonne National Laboratory at Chicago, Illinois.

Figure 18.18 shows effluent histories of two separate fixed-bed column runs with Glendale well water using a strong base anion-exchange resin (A-600) and a weak base anion conditions at acidic pH. In raw water, total Cr was essentially present as Cr(VI) anions in trace conditions compared to other competing anions like sulfate, nitrate and bicarbonate (Table 18.4). For strong base anion-exchange resin, Cr breakthrough took place almost instantaneously at less than 2000 bed volumes, whereas a very tiny concentration of Cr was detected at the effluent of the column even after the column containing weak base anion-exchange resin (Duolite A-7) was run for more than 10,000 bed volumes. Figure 18.18 also includes the breakthrough of copper ions from the columns. It was interesting to note that the Duolite A-7 column was capable of removing a significant amount of copper from the influent groundwater.

The observation of simultaneous Cr and Cu removal by Duloite A-7 was further confirmed by a residuals analysis on the exhausted resin at the pilot plant at Glendale, CA, which revealed the presence of high concentrations of Cu and Cr in the resin phase (average $16,427\,\mu g\,g^{-1}$ and $10,877\,\mu g\,g^{-1}$, respectively). Weak base anion-exchange resins contain primary, secondary and tertiary amine functional groups which, at or near neutral pH, are not protonated. The nitrogen atoms have a free electron pair and can act as Lewis bases. Heavy metal cations like copper act as Lewis acids (electron acceptors) and are preferentially adsorbed at the functional groups through

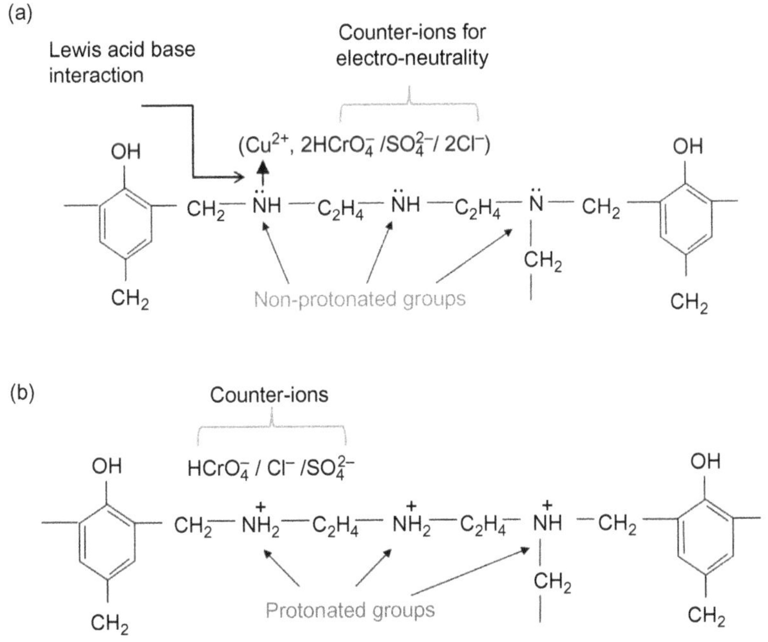

(a)

(b)

Figure 18.19.   Structures of Duolite A-7 anion-exchange resin along with mechanisms of (a) selective adsorption of copper cation through Lewis acid-base interaction with co-adsorption of accompanying anions and (b) regular anion exchange by protonated functional groups at low pH.

Lewis acid-base interaction. Rather than being ion exchange, this is an adsorption phenomenon involving cations; that is why a parallel co-adsorption of equivalent concentrations of accompanying anions should take place in order to maintain electroneutrality. Figure 18.19a describes such an uptake of copper ions by Duolite A-7. The nitrogen atoms of the functional groups, however, prefer H^+ ions over the heavy metal cations. Therefore, with decrease in pH, the functional groups become increasingly protonated and act like other strong base anion-exchange resins. This situation is indicated in Figure 18.19b.

Simultaneous removal of copper and Cr(VI) has earlier been reported by Zhao et al., 1995 using a special chelating resin (DOW, XFS 4195). It has been discussed in the previous section that anion exchangers with the functional groups containing bispicolylamine were responsible for coordinatively binding Cu^{2+} ions which in turn selectively adsorbed ligands like phosphate, selenate, arsenate and chromate in presence of commonly occurring anions like sulfate, chloride, etc. Presence of such a mechanism in Duolite A-7 in addition to the regular ion-exchange phenomenon might be responsible for imparting an extraordinarily high Cr(VI) removal ability through ligand exchange. However, in the present study, Duolite A-7 resins preloaded with copper ions did not show any preference towards arsenate ligand during a column run in laboratory with synthetic influent containing $200 \, \mu g \, L^{-1}$ As along with a background of commonly occurring anions. Hence, the presence of a ligand exchange phenomenon influenced by copper ions adsorbed at the non-protonated functional groups is ruled out.

Figure 18.20 depicts the results of column runs using synthetic water containing Cr(VI) anions along with other competing anions but without any copper ion. There was no change in the breakthrough behaviors of SBA and WBA resin columns run with synthetic water without copper compared to those run with Glendale water containing copper ions. SBA resins, A850 (polyacrylic matrix) and A400 (styrene-divinyl benzene matrix) broke through at the very beginning of column run whereas Duolite A-7 again demonstrated an extraordinarily high Cr(VI) removal capacity.

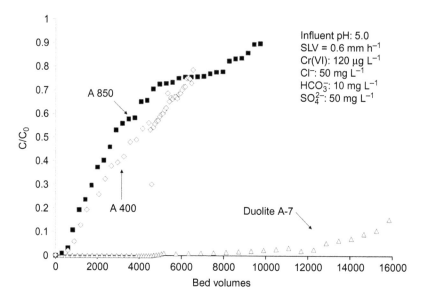

Figure 18.20.   Chromium breakthrough profiles of column runs of strong base anion-exchange resins (A-850 and A-400) and weak base anion-exchange resins (Duolite A-7) with synthetic influent containing Cr(VI) anions but without copper ions.

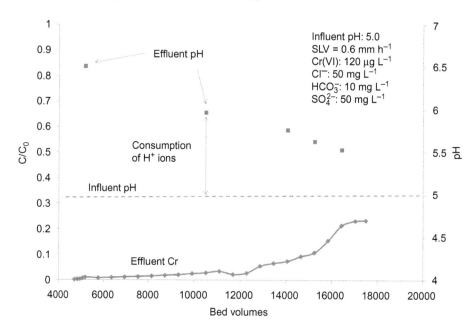

Figure 18.21.   Breakthrough profile of chromium along with effluent pH profile for a column run with Duolite A-7 with synthetic influent containing Cr(VI) anions.

This observation confirms that copper did not play any significant role in removal of chromate ions by Duolite A-7 resin.

Figure 18.21 depicts Cr breakthrough profile along with pH of the effluent for the same column run with Duolite A-7 resin. It may be observed that there was jump in the effluent pH indicating

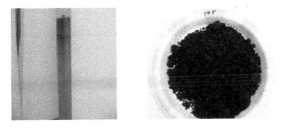

Figure 18.22.   (a) Photograph of a partially progressed column run with Glendale well water (b) Photograph of exhausted Duolite A-7 resin after the pilot plant run at Glendale, CA.

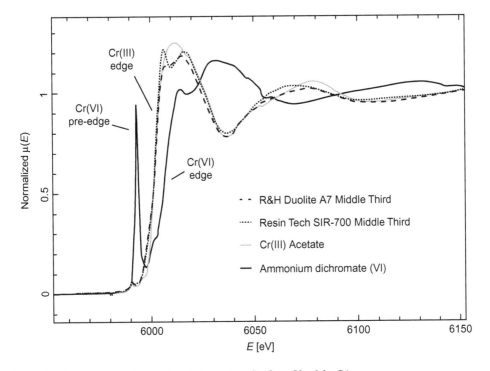

Figure 18.23.   XAFS spectra of exhausted resin samples from Glendale, CA.

a consumption of H$^+$ ions during the removal process. As the column run progressed towards breakthrough there was a continuous drop in the effluent pH. The effluent was tested for the presence of Cr(III). It was found out that all Cr present in the effluent was in Cr(VI) form. There was no presence of Cr(III) in the column effluent.

Figure 18.22a shows photograph of a glass column containing Duolite A-7 at a stage when the column run was partially complete. The change of color of the column from amber to greenish black suggests adsorption of copper ions. Figure 18.22b is a photograph of exhausted Duolite A-7 resin obtained from pilot plant study at Glendale, CA.

The exhausted resins were subjected to two kinds of study. One part of the resin was characterized with X-ray fine structure absorption (XAFS) study to find out the oxidation state of Cr inside the resin. Another part was leached with solutions of sulfuric acid and sodium hydroxide followed by the determination of concentration of different oxidation states of Cr in the leachate. Figure 18.23 shows the result of Cr XAFS spectra for known Cr(III) as well as Cr(VI) compounds

Table 18.5.   Leachate composition of the exhausted resins obtained after pilot plant study at Glendale, CA.

Leaching Solution Resin	6% NaCl & 2% NaOH			2% H_2SO_4		
	Total Cr [mg g^{-1}]	Cr(VI) [mg g^{-1}]	Total Cu [mg g^{-1}]	Total Cr [mg g^{-1}]	Cr(VI) [mg g^{-1}]	Total Cu [mg g^{-1}]
Duolite A-7	0.605	0.006	0.664	2.9	0.00	21.8
SIR-700	0.607	0.012	0.21	0.931	0.00	7.13

along with exhausted resin samples. The edge overlap between spent resin samples and Cr(III) reference compounds and lack of any pre-edge peak with Cr(VI) compound indicated that Cr(III) is the predominant species in both the exhausted resins.

Table 18.5 contains the results of the leaching study for exhausted Duolite A-7 and SIR-700 samples obtained after the field studies at Glendale. It may be noted that Cr(VI) is absent in the eluted solution; This means that, Cr(III) is the only Cr present within Duolite A-7 and SIR-700 that was eluted out regardless of the chemical composition of the regenerant.

Thus, the analysis of treatment residuals confirmed that there is a reduction of Cr(VI) to its trivalent state inside the exhausted resin. Equation (18.58) indicates such a reaction converting Cr(VI) ($HCrO_4^-$) to Cr(III) which forms precipitates of $Cr(OH)_3$ inside the resins near neutral pH range. It may also be noted from the half reaction that there is a consumption of H^+ ions during the reduction process:

$$HCrO_4^- + 4H^+ + 3e^- \rightarrow Cr(OH)_{3(S)} + H_2O \tag{18.58}$$

FTIR spectroscopic analysis of virgin and exhausted Duolite A-7 resin samples as indicated in Figure 18.24 provides evidence of appearance of C-O groups in both the exhausted resin samples. This observation leads to a conclusion that there has been oxidation inside the resin.

Figure 18.25 presents the breakthrough curve for a Duolite A-7 column run with synthetic feedwater. A sharp breakthrough of Cr occurred after 12,000 bed volumes. The column was deliberately stopped for three days at 16,500 bed volumes. The column run was again interrupted at 25,000 bed volumes but this time the stoppage lasted for a longer interval of time. Upon restart of the column after the pause, in both cases, there was a significant drop in effluent Cr concentrations. Also, the slope of the breakthrough curve after the restart was found to become flat compared to that before the interruption. In the inset of Figure 18.25, results of a similar interruption in an A-400 column run are depicted. It showed an opposite breakthrough phenomenon. Unlike the Duolite A-7 column, the A-400 column showed a steeper breakthrough curve after restart than before the interruption, which is typical of intraparticle diffusion controlled process. The interruption allows an evening-out of the particle phase concentration gradient but does not necessarily cause any increase in capacity. That is why, although there is a sudden drop in effluent concentration immediately after restart, the slope of the breakthrough curves become steeper than the interruption because the total exchange capacity remains constant. On the contrary, the breakthrough profile after interruption of the Duolite A-7 column suggested a gain in some of its exhausted capacities during the interruption time and that is why long after restart, the gradient of breakthrough curve remain flatter compared to that before interruption.

Figure 18.26 indicates the change in pH profile at the effluent of the column during the break-through of the column including the interruption and restart events. A few points may be noted. First, the effluent pH was higher than the influent pH. Second, as the column run progressed towards breakthrough, pH in the effluent started dropping towards the influent pH value. Third,

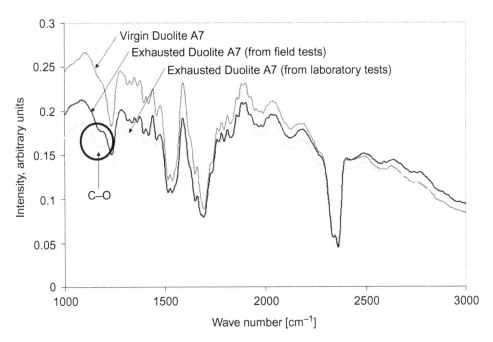

Figure 18.24.   FTIR spectra of Virgin and exhausted Duolite A-7 resins showing appearance of C-O groups in the exhausted samples.

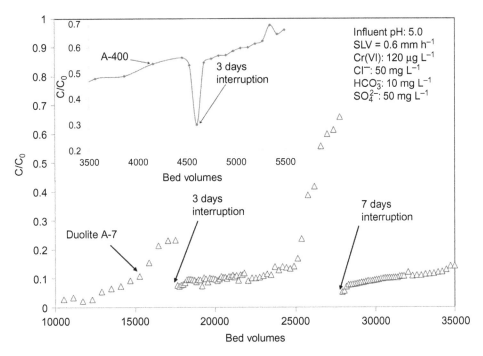

Figure 18.25.   Cr breakthrough profile of an extended column run of Duolite A-7 with periods of interruption.

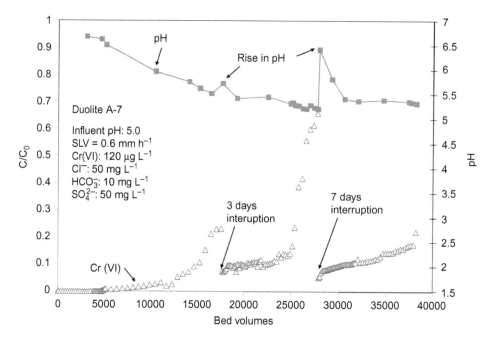

Figure 18.26.   Combined Cr breakthrough and effluent pH profiles during an extended column runs of
Duolite A-7 with periods of interruption.

interruptions in the column run were always associated with a corresponding sharp jump in the
effluent pH upon restart.

These observations suggest that interruption in the column run essentially helped the redox
reaction inside the resin to progress without any further occupation of ion-exchange sites. Redox
reactions are slower compared to ion exchange. As the redox reaction inside anion exchanger
progressed it consumed more H^+ ions from the aqueous phase. The chromate anions bound to
anion-exchange sites were converted to Cr(III) ions and vacated the occupied sites and precipitated
in the form of $Cr(OH)_3$. As some of the occupied sites are vacated as a result of completion of the
redox reaction, the resin seemed to gain the chromate ion removal capacity due to such interruption
process. It can also be concluded that Cr(VI) essentially does not oxidize the functional groups. It
is therefore most probable that it is the phenol-formaldehyde matrix that gets oxidized by Cr(VI).
Other WBA resins with styrene-divinyl benzene matrix did not show any high Cr(VI) removal
capacity. Figure 18.27 shows the mechanism of such ion exchange assisted reduction of Cr(VI) to
Cr(III) leading to a very high capacity for Cr(VI) removal by Duolite A-7 anion-exchange resin.

18.3   CLOSURE

Due to the general absence of stringent regulations covering Cr in drinking water supplies, the
removal of traces of different species of Cr has not received much attention until this time.
Presently, evidence-based studies on ill-effects due to the presence of Cr, specifically Cr(VI),
have revealed the necessity of having stringent regulations for allowable Cr(VI) concentration
in drinking water. There have been some moves by regulating agencies to push forward such
standards for Cr(VI) and it is expected that these moves will increase. Consequently, it is necessary
to develop effective technologies for trace Cr(VI) removal from drinking water to meet new
regulatory standards. Reduction of Cr(VI) to Cr(III) followed by precipitation may prove to be
effective for removal of such trace amounts if the conditions are properly maintained. However,

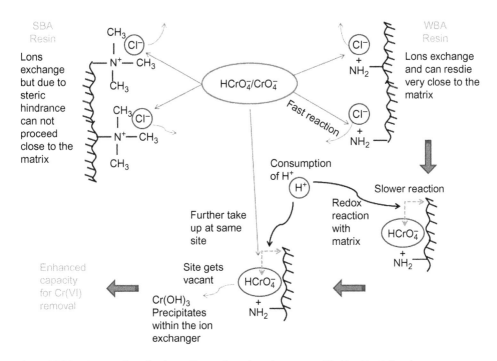

Figure 18.27.   Proposed mechanism of hexavalent chromium removal by Duolite A-7 resin.

issues like the consumption of chemicals, separation of the precipitates and management and disposal of bulky sludge will pose challenges for practical implementation of such technologies. Ion exchange-based technologies provide a practicable solution for selective removal of trace concentration of Cr(VI). In this article we highlighted three such technologies which may offer efficient solutions. While selective removal of Cr(VI) using strong and weak base anion exchange resins has been commercially implemented, only a limited amount of work has been done on the selective removal of trace concentrations of Cr(VI) from drinking water. There is a need to further investigate the role of variables such as pH to achieve ultra-low concentration of Cr(VI) in the product water. The polymeric ligand exchange process has proved to be efficient in the selective removal of Cr(VI). Commercial applications of such processes require further careful investigation of the roles of pH and competing anions that affect the efficiency of the process. The use of redox-active ion exchangers for achieving ultra-low Cr(VI) concentration in the product water has been demonstrated in a pilot study in the field, and the mechanisms of such removal have been elucidated. The process seems effective. However, the study did not include characterization of the treated water with respect to the byproducts of the redox reaction between the polymeric matrix and Cr(VI). Before any attempt is made to commercialize such a process, a study on this particular aspect is considered to be necessary. Also the process was found to be sensitive to pH variation, flow rate, and initial concentration of Cr(VI). Nevertheless, this process opens the door for the design of new class of redox-active agents for trace removal of Cr(VI).

REFERENCES

APHA, AWWA & WEF (1998) *Standard methods of analysis of water and wastewater*. 2nd edition. APHA, AWWA & WEF, New York, NY.

Arroyo, M.G., Pérez-Herranz, V., Montañés, M.T., García-Antón, J. & Guinón, J.L. (2009) Effect of pH and chloride concentration on the removal of hexavalent chromium in a batch electrocoagulation reactor. *Journal of Hazardous Materials*, 169, 1127–1133.

Bailey, R.P., Bennett, T. & Benjamin, M.M. (1992) Sorption onto and recovery of Cr(VI) using iron oxide coated sand. *Water Science and Technology*, 26, 1239–1244.

Ball, J.W. & Izbicki, J.A. (2004) Occurrence of hexavalent chromium in ground water in the western Mojave Desert, California. *Applied Geochemistry*, 19 (7), 1123.

Bartlett, L.B. (1997) *Aqueous Cr(III) oxidation by free chlorine in the presence of model organic ligands: chemistry in the context of environmental regulations*. PhD Thesis, Civil and Environmental Engineering, Duke University, Durham, NC.

Bartlett, R.J. & James, B.R. (1979) Behavior of chromium in soils. III. Oxidation. *Journal of Environmental Quality*, 8, 31–35.

Becquer, T., Quantin, C., Sicot, M. & Boudot, J.P. (2003) Chromium availability in ultramafic soils from New Caledonia. *Science of the Total Environment*, 301, 251–263.

Beszedits, S. (1988) Chromium removal from industrial wastewaters. In: Nriagu, J.O. & Nieboer, E. (eds.) *Chromium in the natural and human environments*. Wiley Series in Advances in Environmental Science & Technology, Volume 20. John Wiley & Sons, New York, NY. pp. 232–263.

Blute, N.K. (2010) Optimization studies to assist in Cr(VI) treatment design. *AWWA Annual Conference and Exposition, 20–24 June 2010, Chicago, IL*.

Blute, N.K. & Wu, Y. (2012) Chromium research effort by the City of Glendale California (interim report). Arcadis U.S. Inc., Los Angeles, CA.

Blute, N.K., Porter, K. & Kuhnel, B. (2010) Cost estimates for two hexavalent chromium treatment processes. *AWWA Annual Conference and Exposition, 20–24 June 2010, Chicago, IL*.

Bourotte, C., Bertolo, R., Almodovar, M. & Hirata, R. (2009) Natural occurrence of hexavalent chromium in a sedimentary aquifer in Urania, State of Sao Paolo, Brazil. *Anais da Academia Brasileira de Ciencias*, 81, 227–236.

Brandhuber, P., Frey, M.M., McGuire, M.J., Chao, P.-F., Seidel, C., Amy, G., Yoon, J., McNeill, L.S. & Banerjee, K. (2004) Low-level hexavalent chromium treatment options: bench-scale evaluation. *AWWARF*. Denver, CO.

California Department of Public Health (2012) Chromium-6 in drinking water: MCL update. Available from: http://www.cdph.ca.gov/certlic/drinkingwater/Pages/Chromium6.aspx [accessed 7th May 2012].

California Department of Public Health (2014) First drinking water standard for hexavalent chromium now final. Available from: http://www.cdph.ca.gov/Pages/NR14-053.aspx [accessed 15th July 2014].

California Office of Environmental Health Hazard Assessment (OEHHA) (2011) Final technical support document on public health goal for hexavalent chromium in drinking water. Available from: http://www.oehha.ca.gov/water/phg/072911Cr6PHG.html [accessed 29th July 2011].

California Regional Water Quality Control Board (CRWQCB) (2000) Waste discharge requirements for in-situ pilot-study for the chemical reduction of Cr, Order No. R1–2000–54.

Cheung, K.H. & Ji-Dong, G. (2007) Mechanism of hexavalent chromium detoxification by microorganisms and bioremediation application potential: a review. *International Biodeterioration & Biodegradation*, 59, 8–15.

Chung, J., Burau, R.G. & Zasoski, R.J. (2001) Chromate generation by chromate depleted subsurface materials. *Water, Air, & Soil Pollution*, 128, 407–417.

Clifford, D. & Chau, J.M. (1988) The fate of chromium (III) in chlorinated water. Project summary, EPA/600/S2-87/100. US Environmental Protection Agency, Cincinnati, OH.

Cotton, F.A. & Wilkinson, G. (1980) Chromium. In: Cotton, F.A. & Wilkinson, G. (eds.) *Advanced inorganic chemistry, a comprehensive text*. 4th edition. John Wiley, New York, NY, pp. 719–736.

Cotton, F.A. & Wilkinson, G. (eds.) (1988) *Advanced inorganic chemistry*. 5th edition. John Wiley & Sons, New York, NY.

Da'na, E. & Sayari, A. (2012) Adsorption of heavy metals on amine-functionalized SBA-15 prepared by co-condensation: applications to real water samples. *Desalination*, 285, 62–67.

Deltombe, E., Zoubov, N. & Pourbaix, M. (1966) Chromium. In: Pourbaix, M. (ed.) *Atlas of electrochemical equilibria in aqueous solutions*. Pergamon Press, Oxford, UK, pp. 256–271.

Dhakate, R. & Singh, V.S. (2008) Heavy metal contamination in groundwater due to mining activities in Sukinda valley, Orissa – a case study. *Journal of Geography and Regional Planning*, 1, 58–67.

Donnan, F.G. (1995) Theory of membrane equilibria and membrane potentials in the presence of non-dialysing electrolytes. A contribution to physical – chemical physiology. *Journal of Membrane Science*, 100, 45–55.

Drago, J.A. (2001) Technology and cost analysis for hexavalent chromium removal from drinking water supplies. In: *AWWA Water Quality and Technology Conference, May 24–27, Nashville, TN*.

Eary, L.E. & Rai, D. (1987) Kinetics of chromium(III) oxidation to chromium(VI) by reaction with manganese dioxide. *Environmental Science & Technology*, 21, 1187–1193.

Edwards, M. & Benjamin, M.M. (1989) Adsorptive filtration using coated sand: a new approach for the treatment of metal bearing wastes. *Journal of Water Pollution Control Federation*, 61, 1523–1533.

Environmental Working Group (2010) *Chromium-6 is widespread in US tap water*. Available from: http://www.ewg.org/chromium6-in-tap-water [accessed 7th July 2011].

Fantoni, D., Brozzo, G., Canepa, M., Cipolli, F., Marini, L., Ottonello, G. & Zuccolini, M.V. (2002) Natural hexavalent chromium in groundwaters interacting with ophiolitic rocks. *Environmental Geology*, 42, 871–882.

Faust, S.D. & Aly, O.M. (1998) *Chemistry of water treatment*. 2nd edition. Ann Arbor Press, Chelsea, MI.

Galan, B., Castaneda, D. & Ortiz, I. (2005) Removal and recovery of Cr(VI) from polluted ground waters: a comparative study of ion-exchange technologies. *Water Research*, 39, 4317–4324.

Godgul, G. & Sahu, K.C. (1995) Chromium contamination from chromite mine. *Environmental Geology*, 25, 251–257.

Gonzalez, A.R., Ndung'u, K. & Flegal, A.R. (2005) Natural occurrence of hexavalent chromium in the aromas red sands aquifer, California. *Environmental Science & Technology*, 39 (15), 5505.

Gray, D.J. (2003) Naturally occurring Cr^{+6} in shallow groundwaters of the Yilgarn Craton, western Australia. *Geochemistry: Exploration, Environment, Analysis*, 3, 359.

Hafiane, A., Lomordant, D. & Dhahbi, M. (2000) Removal of hexavalent chromium by nanofiltration. *Desalination*, 130, 305–312.

Han, I., Schlautman, M.A. & Batchelor, B. (2000) Removal of hexavalent chromium from groundwater by granular activated carbon. *Water Environment Research*, 72, 29–39.

Helfferich, F.G. (1961) Ligand exchange: a novel separation technique. *Nature*, 4769, 1001.

Helfferich, F.G. (1995) *Ion exchange*. Dover Publications, New York, NY.

Hem, J.D. (1977) Reactions of metal ions at surfaces of hydrous iron oxide. *Geochimica et Cosmochimica Acta*, 41, 527.

Höll, W.H., Xuan, Z. & Hagen, K. (2004) Elimination of health relevant heavy metals from raw waters of the drinking water supply in the PR China by means of weakly basic ion exchange resins. *Wissenschaftliche Berichte FZKA*, 6994, A-88.

Hu, Z., Lei, L., Li, Y. & Ni, Y. (2003) Chromium adsorption on high-performance activated carbons from aqueous solution. *Separation and Purification Technology*, 31, 13–18.

IARC (1990) IARC Monographs on the evaluation of carcinogenic risks to humans. Volume 49: Chromium, nickel and welding. International Agency for Research on Cancer, Lyon, France.

Icopini, G.A. & Long, D.T. (2002) Speciation of aqueous chromium by use of solid-phase extractions in the field. *Environmental Science & Technology*, 36, 2994.

Irving, H. & Williams, R.J.P. (1953) The stability of transition-metal complexes. *Journal of the Chemical Society*, 637, 3192–3210.

Izbicki, J.A., Ball, J.W., Bullen, T.D. & Sutley, S.J. (2008) Chromium, chromium isotopes and selected trace elements, western Mojave Desert, USA. *Applied Geochemistry*, 23, 1325–1330.

Johnson, C.A. & Xyla, A.G. (1991) The oxidation of chromium(III) to chromium(VI) on the surface of manganite (gamma-MnOOH). *Geochimica et Cosmochimica Acta*, 55, 2861–2870.

Jones, K.C. & Grinstead, R.R. (1977) Properties and hydrometallurgical applications of two new chelating exchange resins. *Chemistry and Industry*, 3, 637.

Kharkar, D.P., Turekian, K.K. & Bertine, K.K. (1968) Stream supply of dissolved silver, molybdenum, antimony, selenium, chromium, cobalt, rubidium and cesium to the ocean. *Geochimica et Cosmochimica Acta*, 32, 285–298.

Korngold, E., Belayev, N. & Aronov, L. (2003) Removal of chromates from drinking water by anion exchangers. *Separation and Purification Technology*, 33, 179–187.

Kotas, J. & Stascika, Z. (2000) Chromium occurrence in the environment and methods of its speciation. *Environmental Pollution*, 107, 263–287.

Kunin, R. (1976) Amber Hi-Lites No. 151. Rohm and Haas Co., Philadelphia, PA.

Lai, H. & McNeill, L.S. (2006) Chromium redox chemistry in drinking water systems. *Journal of Environmental Engineering*, 132, 842–849.

Lee, G.H. & Hering, J.G. (2005) Oxidative dissolution of chromium(III) hydroxide at pH 9, 3, and 2 with product inhibition at pH 2. *Environmental Science & Technology*, 39, 4921–4930.

Mangaiyarkarasi, M.S.M., Vincent, S., Janarthanan, S., Subba Rao, T. & Tata, B.V.R. (2011) Bioreduction of Cr(VI) by alkaliphilic *Bacillus subtilis* and interaction of the membrane groups. [Saudi] *Journal of Biological Sciences*, 18, 157–167.

Martell, A.E. & Hancock, R.D. (1996) *Metal complexes in aqueous solutions*. Springer, New York, NY.

Maxwell, C.R. (1997) Investigation and remediation of Cr and nitrate groundwater contamination: case study for an industrial facility. *Journal of Soil Contamination*, 6, 733–749.

McGuire, M.J. (2010) Scientific underpinnings: hexavalent chromium treatment basic research and pilot testing. *AWWA Annual Conference and Exposition, 20–24 June 2010, Chicago, IL.*

McGuire, M.J., Blute, N.K., Seidel, C., Qin, G. & Fong, L. (2006) Pilot-scale studies of hexavalent chromium removal from drinking water. *Journal of the American Water Works Association*, 98, 134–140.

McGuire, M.J., Blute, N.K., Qin, G., Kavounas, P., Froelich, D. & Fong, L. (2007) Hexavalent chromium removal using anion exchange and reduction with coagulation and filtration. *AWWARF*, Denver, CO.

Metropolitan Water District of southern California (MWD) and Bureau of Land Management (2001) Cadiz groundwater storage and dry-year supply program, Final EIR/EIS response to comments.

Mohan, O. & Pittman, C.U., Jr. (2006) Activated carbons and low cost adsorbents for remediation of tri- and hexavalent chromium from water. *Journal of Hazardous Materials*, 137, 762–811.

Mukhopadhyay, B., Sundquist, J. & Schmitz, R.J. (2007) Removal of Cr(VI) from Cr-contaminated groundwater through electrochemical addition of Fe(II). *Journal of Environmental Management*, 12, 66–67.

Nakayama, E., Kuwamoto, T., Tokoro, H. & Fujinaga, T. (1981a) Chemical speciation of chromium in sea water. Part 3. The determination of chromium species. *Analytica Chimica Acta*, 131, 247–254.

Nakayama, E., Kuwamoto, T., Tsurubo, S. & Fujinaga, T. (1981b) Chemical speciation of chromium in sea water. Part 2. Effects of manganese oxides and reducible organic materials on the redox processes of chromium. *Analytica Chimica Acta*, 130, 401–404.

Nakayama, E., Kuwamoto, T., Tsurubo, S., Tokoro, H. & Fujinaga, T. (1981c) Chemical speciation of chromium in sea water. Part 1. Effect of naturally occurring organic materials on the complex formation of chromium (III). *Analytica Chimica Acta*, 130, 289–294.

Nakayama, E., Tokoro, H., Kuwamoto, T. & Fujinaga, T. (1981d) Dissolved state of chromium in seawater. *Nature*, 290, 768–770.

National Toxicology Program (2008) NTP technical report on the toxicology and carcinogenesis studies of sodium dichromate dihydrate (Cas No. 7789-12-0) in F344/N Rats and B6C3F1 Mice (Drinking Water Studies): NTP TR 546.

New Jersey Drinking Water Quality Institute (2010) Testing subcommittee meeting minutes. Available from: http://www.state.nj.us/dep/watersupply/pdf/minutes100224.pdf [accessed 7th July 2011].

Newman, J. & Reed, L. (1980) Proceedings, Water-1979. *AIChE Journal*, 197, 76

Nieboer, E. & Jusys, A.A. (1988) Biologic chemistry of chromium. In: Nriagu, J.O. & Nieboer, E. (eds.) *Chromium in natural and human environments*. Wiley Interscience, New York, NY. pp. 21–81.

Oze, C., Fendorf, S., Bird, D.K. & Coleman, R.G. (2004) Chromium geochemistry of serpentine soils. *International Geology Review*, 46, 97–102.

Palmer, C.D. & Wittbrodt, P.R. (1991) Processes affecting the remediation of chromium-contaminated sites. *Environmental Health Perspectives*, 92, 25–40.

Pettine, M., D'Ottone, L., Campanella, L., Millero, F.J. & Passino, R. (1998) The reduction of chromium(VI) by iron (II) in aqueous solutions. *Geochimica et Cosmochimica Acta*, 62, 1509–1512.

Qamar, M., Gondal, M.A. & Yamani, Z.H. (2011) Laser-induced efficient reduction of Cr(VI) catalyzed by ZnO nanoparticles. *Journal of Hazardous Materials*, 187, 258–263.

Qin, G., McGuire, M.J., Blute, N.K., Seidel, C. & Fong, L. (2005) Hexavalent chromium removal by reduction with ferrous sulfate, coagulation, and filtration: a pilot-scale study. *Environmental Science & Technology*, 39, 6321–6328.

Rai, D., Saas, B.M. & Moore, D.A. (1987) Chromium(III) hydrolysis constants and solubility of chromium(III) hydroxide. *Inorganic Chemistry*, 26, 345–352.

Rees-Novak, D., Martson, C. & Gisch, D. (2005) Controlling chromium. *Water & Wastewater International*, 20, 21–29.

Rengaraj, S., Yeon, K.-H. & Moon, S.-H. (2001) Removal of chromium from water and wastewater by ion exchange resin. *Journal of Hazardous Materials*, 87, 273–287.

Richard, F.C. & Bourg, A.C.M. (1991) Aqueous geochemistry of chromium: a review. *Water Research*, 25, 807–816.

Richardson, E., Stobbe, E. & Bernstein, S. (1968) Ion exchange traps chromates for reuse. *Environmental Science & Technology*, 2, 1006–1012.

Robertson, F.N. (1975) Hexavalent chromium in the groundwater in Paradise Valley, Arizona. *Groundwater*, 13, 516–527.

Robertson, F.N. (1991) Geochemistry of grand water in alluvial basins of Arizona and adjacent parts of Nevada, New Mexico, and California. US Geological Survey Professional Paper 1406-C. p. 89.

Robles-Camacho, J. & Armienta, M.A. (2000) Natural chromium contamination of groundwater at León Valley, México. *Journal of Geochemical Exploration*, 68, 167–174.

Saleh, F.Y., Parkerton, T.F., Lewis, R.V., Huang, J.H. & Dickson, K.L. (1989) Kinetics of chromium transformation in the environment. *Science of the Total Environment*, 86, 25–41.

Saputro, S., Yoshimura, K., Takehara, K., Matsuoka, S. & Narsito (2011) Oxidation of chromium(III) by free chlorine in tap water during the chlorination process studied by an improved solid-phase spectrometry. *Analytical Sciences*, 27, 649–655.

Schroeder, D.C. & Lee, G.F. (1975) Potential transformations of chromium in natural waters. *Water, Air, & Soil Pollution*, 4, 355–362.

Sedlak, D.L. & Chan, P.G. (1997) Reduction of hexavalent chromium by ferrous iron. *Geochimica et Cosmochimica Acta*, 61, 2185–2192.

Selomulya, C., Meeyoo, V. & Amal, R. (1999) Mechanism of Cr(VI) removal from water by various types of activated carbons. *Journal of Chemical Technology and Biotechnology*, 74, 111–122.

SenGupta, A.K. & Clifford, D. (1986a) Chromate ion exchange mechanism for cooling water. *Industrial & Engineering Chemistry Fundamentals*, 25, 249–258.

SenGupta, A.K. & Clifford, D. (1986b) Important process variables in chromate ion exchange. *Environmental Science & Technology*, 20, 149–155.

SenGupta, A.K., Clifford, D. & Subramonian, S. (1986) Chromate ion exchange process at alkaline pH. *Water Research*, 20, 1177–1184

SenGupta, A.K., Roy, T. & Jessen, D. (1988) Modified anion exchange resin for improved chromate selectivity and increased efficiency of regeneration. *Reactive Polymers*, 9, 293–299.

SenGupta, A.K., Zhu, Y. & Hauze, D. (1991) Metal ion binding onto chelating exchangers with nitrogen donor atoms. *Environmental Science & Technology*, 25, 481–496.

Sharma, S.K., Petrusevski, B. & Amy, G. (2008) Chromium removal from water: a review. *Journal of Water Supply: Research & Technology-Aqua*, 57, 541–549.

Song, R., Zhu, E., Frit, E. & Brandhuber, P. (2013) *Low level hexavalent chromium reduction*. Louisville Water Company, Madison, WI.

Sorg, T.J. (1979) Treatment technology to meet the interim primary drinking water regulations for organics. Part 4. *Journal of the American Water Works Association*, 71, 454–459.

Sperling, M., Yin, X. & Welz, B. (1992) Differential determination of chromium(VI) and total chromium in natural waters using flow injection online separation and preconcentration electrothermal atomic absorption spectrometry. *Analyst*, 117, 629–636.

Steinpress, M.G., Miller, M., Little, G., Ozbilgin, M., Mabey, R.V., Henderson, R., Handel, B. & Wilkins, D. (1998) Hexavalent chromium in groundwater at the Presidio of San Francisco: anthropogenic or naturally occurring? *Seventh Annual Meeting on California Groundwater Effective and Efficient Usage for the Year 2000 and Beyond, 22–23 October, 1998, Walnut Creek, CA*. California Groundwater Resources Association. p. 23.

Stollenwerk, K.G. & Grove, D.B. (1985) Adsorption and desorption of hexavalent chromium in an alluvial aquifer near Telluride, Colorado. *Journal of Environmental Quality*, 14, 150–155.

Stumm, W. & Morgan, J.J. (1966) *Aquatic chemistry*. Wiley, New York, NY.

Tandon, R.K., Crisp, P.T. & Ellis, J. (1984) Effect of pH chromium(VI) species in solution. *Talanta*, 31, 227–228.

Ulmer, N.S. (1986) Effect of chlorine on chromium speciation in tap water. EPA/600/M- 86/015. USEPA Water, Engineering Research Laboratory, Cincinnati, OH.

USEPA (2003) Water treatment technology feasibility support document for chemical contaminants. In support of EPA six-year review of national primary drinking water regulations. Office of Water, EPA Report 815-R-03-004, June 2003. US Environmental Protection Agency, Washington, DC.

USEPA (2011) Chromium in drinking water. US Environmental Protection Agency, Washington, DC. Available from: http://water.epa.gov/drink/info/chromium/ index.cfm [accessed 24th July 2011].

USEPA, IRIS (1998) Chromium. US Environmental Protection Agency, Washington, DC. Available from: http://www.epa.gov/ncea/iris/subst/0144.htm [accessed 26th July 2011].

USEPA, IRIS (2010) Toxicological review of hexavalent chromium (external review draft). EPA/635/R-10/004A, US Environmental Protection Agency, Washington, DC.

Yang, L., Xiao, Y., Liu, S., Li, Y., Cai, Q., Luo, S. & Zeng, G. (2010) Photocatalytic reduction of Cr(VI) on WO$_3$ doped long TiO$_2$ nanotube arrays in the presence of citric acid. *Applied Catalysis* B: *Environmental*, 94, 142–149.

Zhao, D. (1997) *Polymeric ligand exchange: a new approach toward enhanced separation of environmental contaminants*. PhD Thesis. Civil and Environmental Engineering Department, Lehigh University, Bethlehem, PA.

Zhao, D. & SenGupta, A.K. (1998) Ultimate removal and recovery of phosphate from wastewater using a new class of polymeric exchangers. *Water Research*, 32, 1613–1621.

Zhao, D., SenGupta, A.K. & Zhu, Y. (1995) Trace contaminants sorption through polymeric ligand exchange. *Industrial and Engineering Chemistry Research*, 34, 2676–2680.

Zhao, D., SenGupta, A.K. & Stewart, L. (1998) Selective removal of Cr(VI) oxyanions with a new anion exchanger. *Industrial and Engineering Chemistry Research*, 37, 4383–4389.

CHAPTER 19

Application of membrane technology to chromium elimination from aquatic environments

Michał Bodzek & Krystyna Konieczny

19.1 INTRODUCTION

In the natural environment chromium (Cr) most commonly occurs in the third oxidation state as a cation (Cr^{3+}) or on the sixth oxidation state Cr(VI) in the form of anions. Cr(VI) is a strong oxidant and easily reduces to Cr(III) (Jacukowicz-Sobala, 2009). Chromium (III) naturally present in the environment is an essential nutrient, while Cr(VI) is formed in industrial processes and gets into the environment as anthropogenic pollution. Chromium (VI) and Cr(III) compounds are widely used in many industries due to their durability and aesthetic effects. One can mention the galvanic industry, the production of dyes and pigments, textile and leather articles, tanners and maintenance of wood. Nevertheless, it results in the higher amount of Cr compounds that can be found in surface waters, groundwater or soil. Chromium (VI) compounds are soluble in water and at pH 1–6 they appear as $HCrO_4^-$ and $Cr_2O_7^{2-}$ ions, while at pH $> 6 - CrO_4^{2-}$ ions are formed. Reversible electrolytic dissociation reactions of chromic (VI) acid and anions containing Cr(VI) proceed according to the scheme (Muthukrishnan et al., 2008):

$$H_2CrO_4 \rightleftarrows HCrO_4^- + H^+ \tag{19.1}$$

$$HCrO_4^- \rightleftarrows CrO_4^{2-} + H^+ \tag{19.2}$$

$$2HCrO_4^- \rightleftarrows Cr_2O_7^{2-} + H_2O \tag{19.3}$$

The impact of Cr compounds on living organisms depends on the oxidation state, solubility, and the exposure pathway. Cr(III) is a trace element essential for the proper functioning of plants, animals and humans (Bodzek and Konieczny, 2011a), while Cr(VI) is highly toxic to living organisms. Thus, the permissible concentration of total Cr in drinking water is established at $0.05\,mg\,L^{-1}$, which includes $3\,\mu g\,L^{-1}$ for Cr(VI).

The traditional method used to remove and recover Cr salts from contaminated water and wastewater is the reduction of Cr(VI) to Cr(III), followed by the precipitation of Cr(III) hydroxide and ended with the filtration of the suspension (Owlad et al., 2009). In the case of Cr(III) removal and recovery the reduction stage is omitted. Precipitation methods are the most popular techniques of purification of industrial wastewater from Cr and other heavy metals. Unfortunately, their application results in the formation of large quantities of hazardous sludge significantly contaminated with Cr deposits containing up to a few percent of the element. The produced sludge is usually stored, reused or deposited at secure landfills. However, those sludges undergo oxidation processes, and as a result toxic and mobile forms of Cr (VI), which easily reach soil and groundwater (Jacukowicz-Sobala, 2009), are formed.

There are also several other methods proposed for the removal of Cr, i.e. adsorption and biosorption, ion exchange (used on an industrial scale), solvent extraction and electrochemical methods (Owlad et al., 2009). There is a wide range of natural and synthetic materials, which can be applied as Cr(VI) sorbents, including activated carbon, biological materials, zeolites, chitosan and industrial waste (Owlad et al., 2009). Adsorption on activated carbon appears to be an attractive way to remove Cr due to the exceptionally high surface area of the adsorbent, ranging from 500 to 1500 m^2 g^{-1}, and the existence of a broad spectrum of surface functional groups, such as the carboxylic group (Chingombe *et al.*, 2005). Ion-exchange processes are used mainly for the removal of Cr from galvanic effluents, but also from tannery wastewaters (Jacukowicz-Sobala, 2009). All proposed processes should enable the reuse of metal recovered in industrial processes (e.g. tanners).

Pressure-driven membrane processes, both high (reverse osmosis (RO) and nanofiltration (NF)) and low (ultra- and microfiltration), enhanced with surfactants and polymers, are of great importance in Cr removal and recovery technology (Bodzek and Konieczny, 2011a; Koltuniewicz and Drioli, 2008; Owlad *et al.*, 2009). The application of membrane contactors, especially those with liquid membranes (Hasan *et al.*, 2009; Kozłowski and Walkowiak, 2002; Kulkarni *et al.*, 2007), and processes based on the ion-exchange membranes, including electrodialysis, electrodeionization and membrane electrolysis (Bodzek, 2012), should also be mentioned. Liquid membranes are particularly suitable, because of their selectivity and recovery of Cr in a pure form, even from very dilute solutions (Jacukowicz-Sobala, 2009; Koltuniewicz and Drioli, 2008; Owlad *et al.*, 2009).

19.2   PRESSURE-DRIVEN MEMBRANE PROCESSES

In pressure-driven membrane processes (reverse osmosis, nanofiltration, ultrafiltration, and microfiltration) a pressure exerted on the solution at one side of the membrane serves as a driving force to separate the feed stream into a permeate and a retentate. The permeate is usually pure water, whereas the retentate is a concentrated solution that must be disposed of or treated by other methods. Membranes may be polymeric, organo-mineral, ceramic, or metallic, and filtration techniques differ in pore size, from dense (no pores) to porous membranes. Depending on the type of the technique, salts, small organic molecules, macromolecules, or particles can be retained, and the applied pressure will differ. All pressure-driven membrane processes can be applied to the removal of metals, including Cr.

19.2.1   *Reverse osmosis and nanofiltration*

Reverse osmosis (RO) and nanofiltration (NF) enable the direct separation of Cr compounds from the solutions and find practical application in this field (Bodzek, 2013; Koltuniewicz and Drioli, 2008; Owlad *et al.*, 2009). Both polymeric and inorganic membranes can be used for the removal of Cr(III) and Cr(VI).

Application of reverse osmosis for the treatment of solutions with Cr, e.g. tanning wastewater, is limited by the high concentration of chloride and sulfate ions. The high content of inorganic ions results in the need for the use of high transmembrane pressure. Thus, such a method of removal of Cr becomes economically justifiable only if the salt concentration does not exceed 5 g L^{-1}, and the Cr concentration is around 1 g L^{-1}.

The results of studies on Cr(VI) removal from groundwater with increased Cr content involving RO Osmonics Sepa-S membranes and cellulose acetate (CA) membranes can be found in the literature (Bodzek *et al.*, 2011; Koltuniewicz and Drioli, 2008; Owlad *et al.*, 2009). They showed that the CA membranes retained 96% of Cr(VI) ions, while Osmonics membranes 80–96%, depending on the membrane compactness (Bodzek and Konieczny, 2011b). It was also found that the permeate flux for CA membranes was much smaller than for Osmonics ones.

The application of RO for the removal of heavy metals from solutions can be presented by the example of treatment of wastewaters from the electroplating industry. The wastewater consists mainly of effluent from washing processes of the products after the electroplating coating and from used electroplating baths. The concentration of metal ions in such the wastewater ranges from 0.025 to 1 mg L^{-1} (Bodzek, 1999), and in the case of Cr its concentration can reach up to 2 g L^{-1} (Bodzek, 2012). The reverse osmosis process allows the recovery of water of very high purity (0.0017 mg Cr L^{-1}), which in many cases can be directly returned to the technological

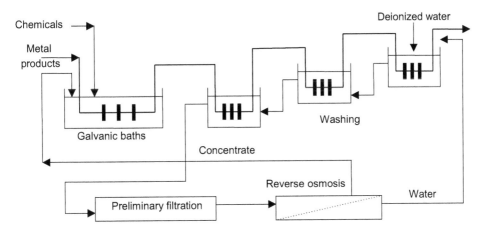

Figure 19.1. The diagram of the electroplating process line integrated with reverse osmosis.

process without additional treatment. A concentrated chromic acid anhydride solution (retentate) containing 5 g Cr L^{-1} (Jacukowicz-Sobala, 2009) may be reused to fill up the electroplating bath (Bodzek, 1999). In Figure 19.1 a diagram of a typical installation operating in a closed cycle applied to such a process is presented (Bodzek, 1999).

In tanning processes, about 20–40% of Cr is usually disposed into sewage system, which adversely affects the environment, especially water. Hafez *et al.* (2002) conducted a study on the removal of Cr from tanning wastewater (1300–2500 mg Cr(VI) L^{-1}) using a pilot scale installation and aimed to recover the metal for further recycling. Low pressure (0.7 MPa) cellulose acetate (RS-Y8-02 type) and medium-pressure (1.6 MPa) Filmtec TFC type (TW 30-40-40) RO membranes were used. The results demonstrated that the RO membrane modules could be operated at medium and low pressure, which decreased the cost, i.e. an economical effect of the separation and recovery of Cr from tannery wastewaters was achieved. Such a use of RO methods results also in the reduction of the amount of Cr deposition into natural waters.

However, nanofiltration seems to be a better solution for the removal of Cr from water (tannery and galvanic wastewaters). The filtrate obtained is then a Cr-free stream, but it contains a significant amount of salts, which can be used for the preparation of new etching baths. The retentate, a concentrated solution of Cr, after further concentration, undergoes hydroxide precipitation, followed by dehydration in sulfuric acid. The final product is a solution which can be directly used in the tanning process (Religa and Gawroński, 2006).

In the case of nanofiltration the retention coefficient increases with pH increase, but the effect is more pronounced for membranes with lower separation capacity (e.g. from 47 to 94.5% for Osmonics membranes) compared to more compact membranes (e.g. from 84 to 99.7% for Osmonics membranes) (Hafiane *et al.*, 2000). The dependence of the retention coefficient on the concentration of Cr in the feed was also observed for NF membranes (Hafiane *et al.*, 2000), however the range of the effect also depended on pH. Higher retention was found for acidic solutions of high Cr concentration, while at pH 6.5–11 the nature of this relationship was opposite, i.e. lower retention was obtained for higher concentrations of Cr. This particular phenomenon, with general importance, is due to the fact that the Cr(VI) changes its ionic form with the change of pH (Fig. 19.2).

In a highly acidic environment, Cr(VI) occurs in the form of non-dissociated chromic acid (H$_2$CrO$_4$) and when the pH increases to 6.5, HCrO$_4^-$ ions are formed and their concentration increases with the parameter increase. Further rising of the pH to above 7 results in the formation of CrO$_4^{2-}$ ions, the concentration of which also depends on pH. Cr$_2$O$_7^{2-}$ ions are also present in the solution and their amount depends on the initial Cr content in the feed as well as on pH. This ion is usually dominant at high concentrations of Cr and in strongly acidic environments (pH 1–7),

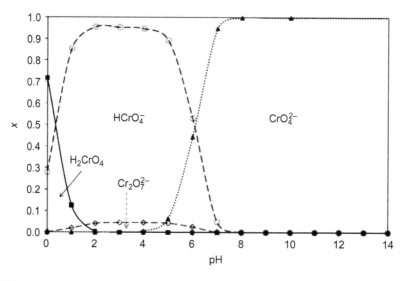

Figure 19.2.   The diagram of the various forms of Cr(VI)–H$_2$O system (100 mg L^{-1}, 25°C), x: mole fraction.

but its concentration decreases with pH increase (Bodzek and Konieczny, 2011a; 2011b; Hafiane *et al.*, 2000).

The nature of the ionic Cr forms has a big impact on the separation using nanofiltration membranes (Kołtuniewicz and Drioli, 2008; Muthukrishnan and Guha, 2008; Taleb-Ahmed *et al.*, 2002). NF membrane has mostly negative charge, which favors the permeation of monovalent ions and higher retention of multivalent ones, depending on their molecular weight. The retention changes with the feed concentration, resulting from the relative changes in the amount of monovalent and multivalent ions in the separation system. This explains the higher retention observed for higher concentrations of Cr at acidic pH, when dissociation of the bivalent ion is higher. However, at basic conditions dissociation takes place to a lesser extent at higher concentration while it is relatively high at low concentrations of Cr in the feed. This shows the opposite phenomenon of separation in relation to changes in the concentration and pH of the feed when conditions are changed from acid to alkaline. The presence of various forms of Cr(VI) ions and the dependence of their separation on pH and concentration of the feed is the main factor affecting the retention of Cr in nanofiltration process at the given environmental conditions.

The separation of Cr(VI) using NF at different pH values was conducted by Hafiane *et al.* (2000). The composite NF membrane made of polyamide (TFCS – Fluid Systems) in the form of flat sheets was applied. Higher retention rates of Cr at basic pH were obtained. The effect of pH on Cr separation can be explained using the system diagram Cr(VI)-H$_2$O, which shows that at pH $= 8$, chromate anions, CrO$_4^{2-}$, are dominant, whereas acidic chromates (HCrO$_4^-$) are the main fraction at pH 2 (Fig. 19.2). With regard to dichromate ion (Cr$_2$O$_7^{2-}$), its concentration in acidic environments depends on the total concentration of Cr according to the following equilibrium:

$$Cr_2O_7^{2-} + H_2O \rightleftarrows 2HCrO_4^- \qquad (19.4)$$

It was found that when total Cr concentration is equal to 1 mmol L^{-1}, the presence of dichromate ions in a solution can be neglected. Chromate ions have a higher retention than HCrO$_4^-$ ions, even at high ionic strength (0.1 mol L^{-1} of NaCl), due to the higher charge density of CrO$_4^{2-}$ ion. The same effect was also observed for sulfate (VI) ions in relation to chlorides (Linde and Jonsson, 1995). Retention measurements with a single salt showed that the Donnan exclusion plays an important role in the process of separation of Cr anions. In the experiments a retention coefficient of Cr of about 60% at pH $= 4$ was obtained, and then gradually increased, reaching a value of

77% at pH = 7. Above pH = 7 the retention rate stabilized at a constant level, due to the change in the relative concentrations of $HCrO_4^-$ and CrO_4^{2-} ions, which resulted from a change in pH. One will notice that the Cr retention in nanofiltration is high over a large range of pH, while such behavior is not observed for other separation techniques such as UF/complexion or adsorption.

The authors of the work (Linde and Jonsson, 1995) identified the impact of ionic strength on the Cr retention using NF membranes. It was found that an increase in ionic strength caused a decrease of the retention coefficient of Cr. It was explained that an increase in the concentration of sodium chloride in solution resulted in the weakening of electrostatic interaction between Cr ions and the membrane and thus the permeation of those ions through the membrane was higher. These observations may be connected with the dependence of a membrane effective charge density (Φ_x) on concentration of electrolytes (c) expressed by the following empirical equation proposed by Wang et al. (1997) for a single electrolyte:

$$\Phi_x = \frac{Ac^{0.5}}{1 + Bc^{0.5}} \tag{19.5}$$

where A and B are constants dependent on electrolyte and membrane properties.

This equation suggests that the charge density of a membrane increases with increase of the salt concentration, and assumes a constant value for a given concentration. Hence the Cr retention decrease results of the smaller membrane electrostatic effect and spherical blocking of the Cr(VI) ions.

Torabian et al. (2010) presented results of experiments on Cr(VI) removal performed on a pilot scale with NF membrane 90_2540 (FilmTec). The effect of transmembrane pressure, the initial concentration of Cr in water, pH, temperature and ionic strength on the efficiency of removal of Cr(VI) in NF process was examined. The obtained results indicated that NF membrane removed hexavalent Cr with the average efficiency of 85%, and pH played an important role in the separation process. Its impact was found to be the most significant among all other parameters, i.e. initial concentration of Cr or ion strength. It was observed that the pressure of 0.8 MPa, at which the Cr retention was 96%, was optimal. Moreover, the retention decreased with the increase of initial concentration of Cr. Nevertheless, the optimal effectiveness of removal was obtained at pH 10.5 at 40°C. It was also shown that the efficiency of Cr(VI) retention decreased in the presence of sulfate and chloride ions.

Bohdziewicz et al. (1995; 1997) also carried out studies on Cr removal from groundwater using NF membranes. Six NF membranes by Osmonics differing in polymers: HG, SX 10, SV, SX 10 01, BQ 01, MX 07 (H – polysulfone, M – polyamide, S – cellulose acetate, B – unknown polymer) were used. The membrane indicated as HG was proven to be the most compact, whereas MX 07 was the one with the most open structure. The obtained rate of Cr(VI) ions removal in the case of cellulose acetate membranes was as follows: SV 10–98%, SX 10–96%, SX 01–90% (the concentration of Cr in the permeate did not exceed 0.01 mg Cr(VI) L^{-1}).

Several studies (Das et al., 2006; Gomes et al., 2010; Ortega et al., 2005; Religa et al., 2011a; 2011b; 2011c) demonstrated very interesting possibilities of the NF process for recovering Cr(III) from mixtures of concentrated salts at low pH. Such solutions are used in the tanning industry. Separation of Cr(III) from a mixture of concentrated salts depends on the feed composition (Religa et al., 2011a; 2011b) and properties of the membranes (Religa et al., 2011b). The presence of mono- and multivalent negative ions and sufficiently high numbers of monovalent positive ions produces a Donnan phenomenon, which can be additionally increased by proper characteristics of the surface layer of the NF membrane. An NF membrane of a negative zeta potential (Religa et al., 2011b) can be especially useful for Cr(III) recovery from mixtures of concentrated salts at low pH. In such conditions the polarization of a membrane is lower, which results in an increase of the permeate flux and a higher Cr(III) concentration factor. However, even with optimum feed composition and the use of membrane with appropriate surface properties, a decrease in process capacity nevertheless results (Religa et al., 2011a).

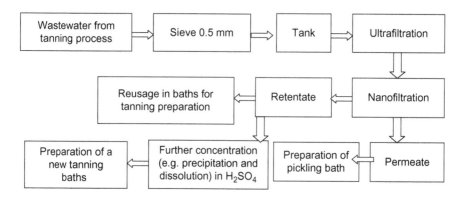

Figure 19.3.   Process scheme for the reuse of chromium compounds in the tanning process.

Cassano *et al.* (1997; 2007) carried out systematic studies on application of pressure-driven membrane techniques for recycling of Cr in the tanning process. They applied various NF, UF and RO membranes by Separem (Italy), i.e.: polysulfone UF (cut-off 3, 15 and 42 kDa) and PVDF (50 kDa), composite polyimide RO (MSCB type 4040) and polyamide NF (MOCD type 4040 N50) membranes. The studies showed that the application of pressure-driven membrane processes integrated with conventional operations e.g. precipitation, enabled the recycling of streams and reduced the loading of the wastewater with Cr compounds. The application of a process comprised of prefiltration, ultrafiltration and nanofiltration separates the substances present in effluents into two valuable streams i.e. Cr enriched retentate and permeate being relatively clean water (Fig. 19.3) (Cassano *et al.*, 2007), which can be reused in the process of tanning and etching operations.

Similar results were achieved by Shaalan *et al.* (2001) who showed that Cr salts recovered by membranes have good properties suitable for the tanning process. The results also indicated that if the process was carried out at optimum conditions it would be efficient and cost effective.

19.2.2   *Ultrafiltration and microfiltration*

Membranes used in the ultrafiltration (UF) and microfiltration (MF) processes do not allow direct concentration of Cr ions in the retentate stream. Thus, low pressure membrane processes are applied for the removal of Cr in integrated systems. Several options of such a solution can be identified (Bodzek, 2012; Koltuniewicz and Drioli, 2008; Owlad *et al.*, 2009):

- The preliminary MF/UF followed by a conventional or secondary high-pressure membrane treatment;
- Modification of UF membrane surface, i.e. pore size reduction or introduction of ion-exchange properties;
- Polymer enhanced ultrafiltration (PEUF) or surfactant/micellar enhanced ultrafiltration (MEUF).

19.2.2.1   *Direct ultrafiltration/microfiltration*
The first method is used in the initial stages of tanning and galvanic wastewater treatment. Direct MF/UF removes suspended solids, fats and emulsions from the aqueous phase (without the use of additional chemicals), easing and improving the performance of subsequent purification (Religa and Gawronski, 2006).

Both polymeric and inorganic UF/NF membranes can be used for direct removal of Cr from water and wastewater, however, inorganic membranes are chemically and thermally more stable (Bodzek *et al.*, 1997). Most of the ceramic and other inorganic material modules are available in the form of tubular modules (Bodzek *et al.*, 1997). Pugazhenthi *et al.* (2005) performed successful

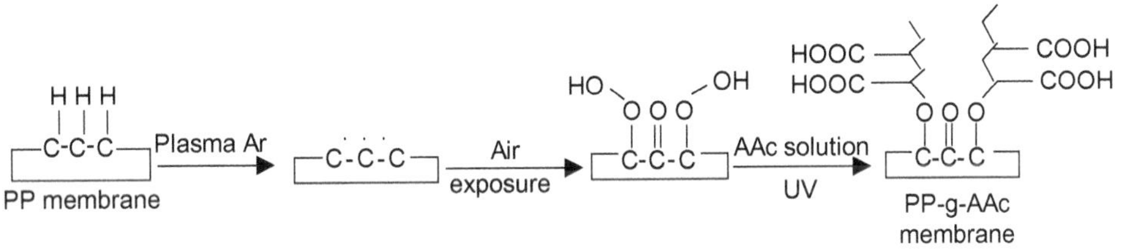

Figure 19.4.   Modification of porous polypropylene membrane with acrylic acid.

studies on Cr removal using modified and unmodified coal membrane. The unmodified UF membranes used in the studies had an effective pore size of 2.0 nm (cut-off 7.5 kDa), 2.8 nm (cut-off 14 kDa) and 3.3 nm (cut-off 17 kDa). Experiments were carried out on Cr(VI) acidic solutions and showed that for unmodified membrane removal of Cr reached 96%, while for membrane modified with gaseous nitrogen oxides – 84%, and for ones modified with hydrazine – 88%. The modified membranes were found to reveal higher capacity than unmodified ones with only 12% in the retention rate.

Introduction of charge to a porous ultrafiltration membrane causes a significant reduction of fouling, improves the efficiency of membrane filtration and increases the life-time of the membrane. The main methods of the charge introduction on the surface and inside the pores of UF membranes are sulfonation of polysulfone membranes and introduction of amino and carboxylic groups. The modification of UF membranes extends their separation capacity by giving them properties of ion-exchange processes (Bryjak, 2001). Chemical modification, including the latest technique i.e. plasma modification, can also be used to introduce appropriate ion-active groups onto a membrane. An exemplary grafting of acrylic acid (AAC) on the surface of polypropylene (PP) membrane is shown on a scheme in Figure 19.4 (Garncarz et al., 1999).

Another method to prepare a porous ion-exchange membrane is to mix polymers before the formation of a membrane (Bodzek and Konieczny, 2005). However, the use of this method is limited due to the lack of polymers with good mixing properties after their dissolution in organic solvent. Mbareck et al. (2006) described the preparation of UF membranes from the mixture of polyacrylic acid (PAA) and polysulfone drained of non-ionic functional groups (PSf). Such membranes were next used for the removal of heavy metals including Cr, from water (Mbareck et al., 2009). The studies showed that the PSf/PAA membranes exhibited high retention of Cr and other heavy metals (lead, cadmium) (even up to 100% at pH > 5.7) which decreased with the pH decrease. This phenomenon is associated with either the presence of carboxylic ion-exchange groups or a complexation process. At low pH, retention of metal ions is very low, because carboxylic groups are present in the acidic form. On the other hand, at high pH, these groups are ionized, which improves the removal of metals. High metal retention at pH > 5.7 allows the conclusion that the complexation between metal ions and carboxylic groups (-COO-) on the inner surface of the membrane matrix pores acts as a barrier against electrostatically free metal ions. High retention rates of metals obtained for membranes with pores larger than the dimensions of ions (hydrated ion radii are 0.426 nm and 0.45 nm, respectively for Cd and Cr) suggest a tortuous way of their permeation through the membrane and therefore transported ions are likely to be "caught" by the carboxylic group located on the pore's walls. It was also observed that the retention of heavy metals decreased with the increase of both their concentration in the feed and of pressure. Some authors (Mbareck et al., 2009) also believed that the hydrophilic groups of acrylic acid were attached to the membrane during the penetration of the material by water molecules (non-solvent) when the membrane was formed (coagulation of membrane forming solution).

19.2.2.2   *Polymer enhanced ultrafiltration*

An interesting solution of the removal of heavy metals from aqueous solutions is ultrafiltration enhanced with polymer (PEUF) (Bodzek et al., 1999; Korus, 2012). It combines UF with metal complexation using water-soluble polymers. The formed complexes have sufficiently large size to be retained by UF membranes. The permeate is deprived of metal ions and the retentate can undergo regeneration in order to recover both the metal and polymer. The usefulness of such a hybrid process was already confirmed for many metals including Cr, Cu, Ni, Zn, Co, Hg and radioisotopes (Molinari *et al.*, 2004; Muslehiddinoglu *et al.*, 1998; Thompson and Jarvinen, 1999; Zakrzewska-Trznadel and Harasimowicz, 2002). Chelating macromolecular compounds as well as ion-exchange polyelectrolytes are mainly proposed as polymers suitable to be used for most metal removal via the PEUF technique. Macromolecular amines, carboxylic and sulfonic acids, amides, aminoacids and alcohols are typical polymers used for binding of metal ions (Aroua *et al.*, 2007; Juang *et al.*, 2000; Llanos *et al.*, 2008; Molinari *et al.*, 2005).

The PEUF method was successfully applied to remove chromate(VI) from groundwater using 1-hexa-decylopyridine chloride, sodium polyacrylate, poly(dimethyldiallylammonium) chloride, chitosan, pectin and other complexing agents (Bodzek and Konieczny, 2011a). Polyethylenimine (PEI) is used for concentration and recovery of Cr(III). One of the most important factors in the metal complexation process using macromolecular compounds is pH, because either protons or hydroxyl anions can compete with metal in polymer-metal bond formation. The presence of protons affects both the equilibrium of the ion-exchange process, as well as complexation. In the case of metal binding with polyethylenimine (PEI), at a lower pH, amino groups as electronodonors adopt a positive charge due to protonation and therefore will not react with other cations. If a number of metal ions form different macromolecular complexes of various stability constants, it is possible to separate them by a change in pH.

Aroua *et al.* (2007) applied three water-soluble polymers, namely, chitosan, polyethylenimine (PEI) and pectin to the removal of Cr from diluted aqueous solutions via PEUF. The studies were carried out using capillary polysulfone membrane of cut-off 500 kDa. The influence of pH and concentration of the polymer on the retention of Cr(III) and Cr(VI) was determined for all types of polymers. In the case of Cr(III) the retention was practically 100% at pH > 7 for all three studied polymers. For chitosan and pectin, retention of Cr(VI) slightly increased with an increase in pH, but did not exceed 50%. For PEI, the retention rate of Cr(VI) was 100% at low pH and decreased with increase of pH (Aroua *et al.*, 2007). Chromium retention rates also depended on the concentration of the polymer. For pectin, retention of Cr(III) was constant and close to 100% regardless of biopolymer dose, which comments on the effective capacity of the metal binding and the formation of macromolecular complexes, even at very low concentrations of the polymer (Aroua *et al.*, 2007). For Cr(VI), the retention rates were lower, and depended on the concentration of the polymer, i.e. the retention initially grew with the concentration of pectins, while above 0.1% it started to decrease. This result confirmed the fact that interactions between Cr(VI) ions and pectins were not a case of complex formation. For PEI, retention increased with increasing polymer concentration, achieved the maximum at a concentration of 0.05%, and above this point slightly decreased. These results are in agreement with the typical behavior of polyelectrolyte, i.e. retention of Cr ions decreases with polymer concentration increases as the solution viscosity is reduced. A study carried out by Aroua *et al.* (2007) showed that the pectin was an excellent polymer for removing Cr from diluted solutions, as it gave even better results of Cr(III) removal than chitosan and PEI. On the other hand, PEI was the most favorable complexing agent of Cr(VI) ions, when compared to other polymers.

Bohdziewicz (2000) examined the removal of Cr(VI) from groundwater by the PEUF method using UF membranes made of polyacrylonitrile (PAN) and hexadecylpyridinium chloride as a complexing polymer. The efficiency of Cr removal depended on the ratio of the complexing agent to Cr(VI) ion concentrations and on pH. The Cr(VI) retention rate was 91–98% at metal to polymer concentration ratios 1:25 – 1:20. It was found that the increase in the concentration of Cr(VI) ions caused an increase in its retention rate. For a range of Cr(VI) concentrations 0.03–0.05 mg Cr L^{-1} the retention was 71–83%, while for concentrations of 0.07–0.2 mg L^{-1} it was constant and kept at the level of 90%. The concentration of Cr(VI) in the permeate was lower than permissible for drinking water and in the range of 0.0007 to 0.00085 mg Cr L^{-1}. The presence of complexing polymer in the permeate was not identified. The retention rate of

Cr(VI) increased with an increase of pH in the range of 0–2 and 5–6, and it was equal to 70% at pH = 2 and 95% at pH = 6. For the remaining pH values retention rates were lower. The highest retention rate was achieved for pH = 6 at the concentration of Cr(VI) 0.01 mg L^{-1}. The results confirmed that the use of PEUF was an effective method of the removal of Cr(VI) ions from groundwater.

A study on the removal of Cr from aqueous solutions by PEUF was also carried out by Korus *et al.* (2009, 2010). The sodium salt of polyacrylic acid (PAA) was proposed for binding Cr(III) ions, and for Cr(VI) polyethylenimine (PEI), a cationic polymer with weak anion-exchange properties, was used. Experiments were conducted with the use of UF membrane MX 50 (Osmonics) made of a modified polyacrylonitrile. The separation efficiency of Cr(III) for PAA was strongly dependent on pH. For pH > 4, in the whole range of polymer concentrations, only small amounts of Cr(III) were retained by UF membrane, which was the result of the weak ion-exchange properties of the PAA. For the pH range of 6–8 the binding conditions of Cr(III) by polyacrylic ions were optimal, and thus the obtained retention coefficient was more than 0.97. For Cr(III), a relatively high ratio of the concentrations of organic to metal should be applied in order to prevent precipitation of chromium hydroxide (III). For Cr(VI) complexation with PEI, good results were obtained in the case of polymer:metal ratio 5:1 at pH 4–6. Lower retention rates observed at pH 2 were probably caused by the protonation of anionic forms of Cr(VI) and the formation of less-dissociated H$_2$CrO$_4$ acid. The decrease of the retention coefficient at higher pH (pH > 6) was related to the competition between OH$^-$ ions and the anionic form of Cr(VI) while binding with the polymer. The results of the concentration experiments showed high efficiency of PEUF methods in Cr removal, particularly for Cr(III). During concentration of a solution containing 5 mg L^{-1} of Cr(III), 50-fold volumetric reduction was achieved, resulting in a 49-times increase in the concentrations of Cr in the retentate, and only less than 2% of the metal passed to the permeate. For Cr(III) concentration 50 mg L^{-1}, a 47-fold increase in the concentration of metal was obtained, which was confirmed by low concentrations of Cr in the permeate and high retention rate (0.99–1) (Korus and Loska, 2009). The reduction of Cr(VI) obtained for PEI (0.8–0.97) indicated a slightly lower efficiency. A 40-fold increase in the concentration of Cr in the retentate was obtained, but the average concentration of Cr(VI) in the permeate was 10 to 20 times larger than that obtained for Cr(III) in similar process conditions.

19.2.2.3   *Micellar enhanced ultrafiltration*

Proposed in the 1980s, by Scamechorn *et al.* (1989), the ultrafiltration process supported by surface active compounds called micellar enhanced ultrafiltration (MEUF) is a promising new method of removal of metal ions and other low-molecular compounds from aqueous waste (Dunn *et al.*, 1985; Gibbs *et al.*, 1987; Huang *et al.*, 2009; Sadaoui *et al.*, 2009; Scamechorn *et al.*, 1989). This is a hybrid process that combines the classical ultrafiltration with the ability of surfactant to solubilize selected components from aqueous solutions. In the process, surfactant is added to the solution containing hydrophobic substances, and its concentration exceeds the critical concentration of micelle formation (CMC), i.e. one at which the free surfactant molecules are in equilibrium with its aggregated form (Dunn *et al.*, 1989; Urbanski *et al.*, 2002). The diameters of the micelles, which are already aggregated with separated compounds, are usually greater than the diameters of the pores of the ultrafiltration membrane. Thus, micelles remain in a retentate, while the permeate can contain non-soluble compounds, ions and small amounts of surfactant monomer, usually not exceeding the concentration of CMC of the surfactant (Aoudia *et al.*, 2003; Juang *et al.*, 2010). Recently, a number of publications on the extraction of metal ions, acid residues and other low-molecular organic compounds with application of the MEUF process have appeared. They showed the possibility of effective removal of heavy metal ions, such as Cr (Staszak *et al.*, 2010) cadmium (Huang *et al.*, 2005) cobalt (Juang *et al.*, 2003), nickel (Yurlowa *et al.*, 2002) and zinc (Zhang *et al.*, 2007) in the presence of anionic surfactants e.g. dodecyl-sulfate (VI) sodium (SDS).

Staszak *et al.*, (2010) presented the results of a study on the possibility of removal of Cr(III) and Cu(II) from aqueous solutions by the MEUF method. The effectiveness of classical UF

Table 19.1.   The influence of CPC:Cr(VI) ratio on retention of CPC and Cr(CI).

| CPC concentration [mmol L^{-1}] | CPC:Cr(VI) ratio | Retention [%] | |
		CPC	Cr(VI)
0.3	1.6		72.6
0.6	3.1	76.0	97.2
0.9	4.7	84.9	100
1.2	6.2	88.8	100
1.5	7.8	91.3	100
2.0	10.4	93.5	100

and MEUF processes was also compared. The impact of salt concentration in the feed, the type of surfactant and membrane material hydrophilicity on the performance of the process and the efficiency of separation was studied. Anionic surfactants SDS, oxyethylene methyl ester of rape oil acid (Rofam) with the general formula of $RCO(OCH_2CH_2)_n OCH_3$ ($R = C_{12}$–C_{24}, $n = 10$) and a mixture of both surfactants, as well as cellulose acetate (CA), PES-(PS) and polyvinylidene difluoride (PVDF) membranes were used. It was shown that for the single-stage UF process the retention coefficient of metal ions was low, less than 25%, and depended on both the membrane material used and the feed concentration. In the MEUF process, Cr(III) retention was higher up to 90–95%, and did not depend on the type of membrane.

Samper et al. (2009) presented the results of the impact of pH, conductivity and concentration of surfactant on the permeate flux and Cr retention during the MEUF process using polyethersulfone membrane (10 kDa) and a cationic active surface substance – cetylpyridinium chloride (CPC). The research led to the following conclusions describing removal of Cr by the MEUF process:

- Retention of Cr(VI) was around 100% in the case of a surfactant to metal (S:M) molar ratio at the level of 4.7 (Table 19.1);
- With increasing electrical conductivity the retention of Cr(VI) decreased from 100% to 30% for the conductivity change in the range of 148 to 10,000 μS cm^{-1};
- Retention of Cr(VI) increased with an increase of pH and at pH 6 > reached 99%.

Huang et al. (1994) also found that metal removal efficiency depended primarily on the molar ratio of concentration of surfactant and surfactant to metal (S:M).

Baek et al. (2004) also used cetylpyridinium chloride (CPC) of concentration above CMC for simultaneous removal of nitrates(V), chromates(VI) and ferrocyanides using the MEUF method. Regenerated cellulose membrane with cut-off of 10 kDa was used. The reduction of Cr(VI) content was approximately 48, 67, 85, 98 and 98%, for the ratio of CPC:chromium equal to 1, 2, 4, 5, and 10, respectively. In turn, Lee et al. (2005), examined the simultaneous removal of trichloroethylene and chromate (VI) by MEUF using CPC as cationic surfactant and Tween 80 as a non-ionic one.

Konopczyńska et al. (2011) presented the applicability of MEUF to the removal of Cr ions from aqueous solutions. The impact of the concentration of Cr(III) in the solution, the dose and the type of surfactant added to the separated system on both the efficiency and the performance of the MEUF process, were analyzed. The obtained results showed that the retention rate of Cr(III) in the MEUF process with SDS surfactant at its concentration of 2.5 and 5 CMC was equal to approximately 90%. The application of a mixture of SDS and Rofam 10 surfactants enabled both further increase of the retention rate of Cr(III) ions and significant reduction of surfactant concentrations added to the separated system. The studies confirmed that MEUF could be an effective method of removal of Cr(III) ions from aqueous solution.

19.3   ELECTROCHEMICAL MEMBRANE PROCESSES

Electrochemical separation techniques are becoming an alternative method of Cr removal from the aquatic environment (Bodzek and Konieczny, 2011a). This toxic metal is present in various streams produced by a number of industrial processes, which also contain other substances (mainly metals), that should be separated from the stream and Cr as well. Unlike other methods of membrane Cr removal from the aquatic environment, in the electrochemical methods ion-exchange membranes, being permeable to recovered cations (impurities) or anions (chromates (VI)), are applied. Therefore, electrochemical technology is applied to recover chromic (VI) acid from baths coating metal parts (large concentration of Cr) or as a method of disposal and recovery of Cr from wastewater produced during washing of these elements. The use of electrochemical technology for chromic (VI) acid recovery has so far been limited mainly due to the weak resistance of the anion-exchange membranes to chromic acid (VI) (Audinos, 1992; Donepudi *et al.*, 2002). However, this was changed when a new generation of membranes was introduced (Dzyazko *et al.*, 2007; Frenzel, 2005; Frenzel *et al.*, 2005a; 2005b). Nowadays, three electrochemical membrane methods are taken into consideration: electrodialysis (ED), electrodeionization (EDI) and electro-electrodialysis (EED). Unlike ED and EDI, EED is based on electrolytic reactions running on the electrodes and on the electrodialysis process (Audinos, 1992; Jacukowicz-Sobala, 2009).

19.3.1   *Electrodialysis and electrodeionization*

Electrodialysis (ED) is a membrane process which separates ions across charged membranes from one solution to another using an electric field as the driving force. ED has also been proven as a promising treatment method in the removal of heavy metals and toxic anions from wastewater. In electrodialysis, the transport of ions is accelerated due to an externally applied electric potential difference. This allows higher fluxes than in e.g. Donnan dialysis (DD), which is also applied to removal of ions. In ED, anion-exchange and cation-exchange membranes are applied alternately, which results in the formation of solutions of different concentrations (diluate and concentrate) (Bodzek and Konieczny, 2011a; Velizarov *et al.*, 2004). The ED systems are usually operated in the so-called electrodialysis reversal mode (EDR) to prevent membrane fouling and scaling. The suitability of ED strongly depends on the ionic composition of the contaminated water. Thus, the process appears to be less applicable to waters of very low salinity (conductivity less than 0.5 mS), for which DD can be a better solution. In cases when removal of ions should be accompanied by the removal of low-molecular weight non-charged compounds, pressure-driven membrane processes may be preferable.

Nataraj *et al.* (2007) investigated the removal of Cr(VI) ions using an ED pilot plant comprised of a set of ion-exchange membranes. In order to check the efficiency of the ED unit, process parameters like electric potential, pH, initial Cr concentration in the feed and flow rates were varied. Significant results were obtained with lower initial concentrations of less than $10\,mg\,L^{-1}$, while the most satisfactory ones were met at maximum contaminate level (MCL) i.e. $0.1\,mg\,L^{-1}$ of Cr. The effect of working parameters on energy consumption using ion-exchange membranes was also investigated. Results of this study are very useful for the design and operation of ED plants of different capacities applied to recover different ions. The ED plant used in the discussed research was found to be sufficient to produce good quality drinking water from the simulated mixture and to remove unwanted ions.

Lambert *et al.* (2006) studied the separation of Cr(III) from sodium ions by ED using modified cation-exchange membranes. Trivalent chromium Cr(III) present in wastewaters produced by leather tanning processes must be treated before its discharge to the environment. The membrane modification consisted of a polyethylenimine (PEI) layer electrodeposited on the membrane surface. Such a layer has a positive charge in acidic conditions and repels multivalent ions, while monovalent ions pass the membrane. The membrane that underwent the modification in this study was Nafion® 324 membrane. The transfer of chromium, sodium, calcium, magnesium, chloride and sulfate ions from a mixture was investigated. The pH had to be regulated in order to avoid

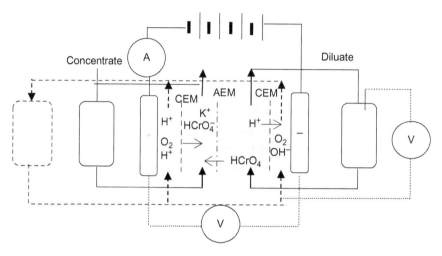

Figure 19.5.   The scheme of the apparatus for electrodialysis/electrodeionization processes (AEM – anion-exchange membrane, CEM – cation-exchange membrane).

Cr hydroxide precipitation in the diluate chamber. The behavior observed for the sulfate-chloride system was unusual as for the AMX membrane and explained by the adsorption of PEI on the membrane surface. The overall current efficiency was close to 96–98% for both cations and anions.

ED is particularly useful and very often applied to treat washery effluents and wastewaters from electroplating plants (Bodzek, 1999). The design of the installation is similar to that in Figure 19.1, but instead of RO, ED is applied. The retentate i.e. a concentrated solution of metal ions, is used to fill up the electroplating bath, whereas the dialysate is returned to the washing installation. Hence, practically the whole quantity of water and salts present in washery effluents can be utilized (Bodzek, 1999). Recently, the application of ED to recovery of metals used in electroplating such as Cr, Au, Pt, Ni, Ag, Pd, Cd, Zn and Sn/Pb from the diluted electroplating wastewaters has gained researchers' attention (Bodzek, 1999).

Both the electrodialysis (ED) and electrodeionization (EDI) processes can be applied to the removal and the separation of metal ions and their mixtures, including Cr. In ED, the electrical resistance in the dialysate chambers increases in time as ions are removed from the diluted solution to the concentrate chamber. This causes an increase of energy consumption and decreases the efficiency of the process. One of the solutions to this problem is the EDI process, in which the dilute solution chamber is filled with an ion-exchange resin (Bodzek and Konieczny, 2011a). The applied voltage improves the migration of ions to the respective electrodes and, thus, to the concentrated stream and causes water dissociation into H^+ and OH^- ions, which regenerate the ion-exchange resin.

Alvarado *et al.* (2009) in his work assessed the feasibility of EDI and ED continuous processes to the removal of Cr(VI) from simulated solutions at pH $= 5$. The ED/EDI installation consisted of electrodes and two acrylic separation plates between which an anion-exchange membrane by Neosepta was placed (Fig. 19.5) (Alvarado *et al.*, 2009).

Two cation-exchange membranes, by the same producer, separated the electrodes from the separation plates. In this way two chambers with diluate and concentrated solution were formed. Simulated wastewater containing 100 mg L^{-1} of Cr(VI) was treated. During the EDI process, the chamber with diluted solution was filled with a mixed ion-exchange resin (1:1), which enabled the removal of 99.8% of Cr(VI) within only 1.3 h of the process at energy consumption equal to 0.167 kWh m^{-3} and the maximum limited current (I_{lim}) of 85%. In the ED process the removal of Cr reached 98% after 6.25 h at energy consumption approximately 1.2 kWh m^{-3}. Therefore, in EDI the time necessary to obtain a high Cr removal rate as well as energy consumption were reduced about 5-fold.

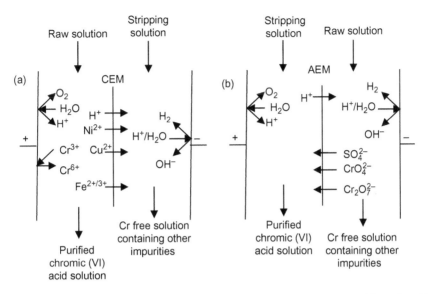

Figure 19.6. The principle of membrane electrolysis with ion-exchange membranes: (a) anode oxidation of Cr(III) and the recovery of metallic impurities using cation-exchange membranes (CEM) and (b) chromic acid recovery using anion-exchange membranes (AEM).

19.3.2 *Electro-electrodialysis*

The electro-electrodialysis (EED) process can be carried out as two- and three-compartment systems. The first is also called membrane electrolysis (ME). Both processes are based on the electrolytic reactions running on electrodes and on the electrodialysis process (Frenzel *et al.*, 2005a).

Membrane electrolysis (ME) is the process which combines electrolysis with electrodialysis and primarily affects Cr(VI). It is characterized by a low consumption of chemicals and energy. An electrolyzer consists of two chambers separated by an anion-exchange membrane or a cation-exchange membrane, to which raw solution is introduced (Fig. 19.6). Depending on the treatment goal and the kind of ion-exchange membrane used there are two procedures of the process performance, in particular the recovery of Cr from plating baths (Frenzel, 2005; Jacukowicz-Sobala, 2009):

- The exhausted plating solution is placed in the anode compartment, where Cr(III) is oxidized, while other cations (impurities) pass from the waste solution (cathode chamber) through the cation-exchange membrane (CEM) (Fig. 19.6a). In the cathode chamber, diluted chromic acid is formed.
- The waste solution containing Cr(VI) anions is introduced to the cathode compartment, and anions migrate towards the anion-exchange membrane (AEM) to the anode compartment. In this compartment chromic acid (VI) is formed due to the presence of protons generated at the anode (Fig. 19.6b), and the acid is then returned to the Cr bath.

During the electrolysis of aqueous solutions at the anode and cathode the following reactions run:

$$\text{Anode:} \quad 2H_2O - 4e \rightleftarrows O_2 + 4H^+ \tag{19.6}$$

$$\text{Cathode:} \quad 4H_2O + 4e \rightleftarrows 2H_2 + 4OH^- \tag{19.7}$$

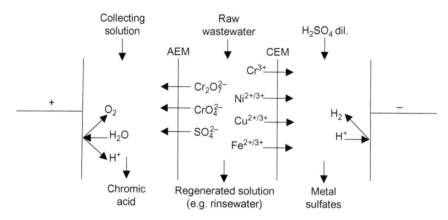

Figure 19.7.   Principles of three-compartment electro-electrodialysis (EED) process for chromic acid
recovery (AEM – anion-exchange membrane, CEM – cation-exchange membrane).

In both cases, in the catholyte, metal cations are present, which in the presence of OH^- ions
formed at cathode, are precipitated in the form of hydroxides. After filtration, the purified solution
can be used as water to rinse the parts coated with metal (Bodzek, 2012; Bodzek *et al.*, 1997;
Dylewski, 2001).

The technology involving anion-exchange membrane for chromic acid recovery was described
by Audinos (1992). A solution of Cr(VI) concentration equal to $280\,g\,L^{-1}$ was obtained and the
consumption of energy was in the range of $10–20\,kWh$ per kg of CrO_3. The membranes were
operated in very adverse redox conditions, and thus they required frequent replacement. The
low stability of anion-exchange membranes significantly limits the use of this technology in
practice. Audran *et al.* (1990; 1992) conducted a study of two-compartment EED using AEM
membranes by Morgane (France). The maximum concentration of anolite (product) obtained was
$310\,g\,CrO_3\,L^{-1}$, using average current yield of 48% at energy consumption of $18\,kWh$ per kg of
CrO_3, temperature 40°C and a current density of $50\,mA\,cm^{-2}$. The membrane worked for 4000
hours, however, its performance characteristic was not intended for such a long research period.
The system of Cr recovery using AEM Ara type 17-10 membrane by Morgan and Ionac 3475
XL membrane by Sybron and Co. was discussed in the work (Frenzel, 2005). Experiments were
carried out on the recovery/treatment of chromic acid present in galvanic solutions ($250\,g\,L^{-1}$ in
60°C), and with a solution containing $8\,g\,CrO_3\,L^{-1}$ and $1\,g\,Fe\,L^{-1}$, which was concentrated to
$323\,g\,L^{-1}$. In the case of more concentrated plating solution, AEM Morgane membrane (Ara type
17-10) showed a higher Cr transport rate and lower power consumption than the Ionac 3475 XL
membrane. However, the authors did not mention membrane stability in the long term studies. A
number of studies describing the properties of anion-exchange membranes operated at chromate
solutions of a low concentration were also carried out (Castillo *et al.*, 2002a; 2002b; Cengeloglu
et al., 2003; Chaudhary *et al.*, 2006; Vallejo *et al.*, 1999; 2000). The following membranes AEM
SB-6407 (Science Gelman), AFN Neosepta and ACM (Tokuyama Soda co.) were used to treat
a solution of $K_2Cr_2O_7$ concentration $0.01\,mol\,L^{-1}$ (Cengeloglu *et al.*, 2003) and Raipore 1030
(RAI Research Co.) membrane was applied to a solution of $0.02\,mol\,L^{-1}$ Cr(VI) content (Castillo
et al., 2002a; 2002b).

A three-compartment EED seems to be the most promising solution for Cr recovery from plating
industry wastewater (Fig. 19.7) (Ann., 2002; Frenzel, 2005; Frenzel *et al.*, 2005b; Jacukowicz-
Sobala, 2009). It can simultaneously manage three different tasks i.e. the removal of impurities, the
recovery of chromic acid and the purification of rinse water. The treated solution feeds the central
chamber of the device, which is separated from the anolyte chamber by the anion-exchange membrane (e.g. Ionac MA-3475) and from the catholyte chamber by the cation-exchange membrane

(e.g. Nafion 324) (Frenzel *et al.*, 2005a). The anolyte chamber is supplied with water, while the catholyte chamber with sulfuric acid. Cr(VI) ions migrate to the anolyte chamber, where they form chromic acid (VI) with protons formed on the anode. In turn, metal cations permeate to the catholyte chamber, where sulfuric acid neutralizes the hydroxide ions formed on the cathode and finally the soluble metal sulfates (VI) are formed. The catholyte can be further treated together with other wastewaters, and dialyzate from the central chamber is returned to rinse items after plating of metal (Jacukowicz-Sobala, 2009).

A comprehensive study with three-compartment EED on recovery of Cr(VI) from galvanic wastewaters using anion-exchange Fumasep®FAP (FuMA-Tech GmbH) and Ionac (MA-3475 and PC-100 D) membranes was conducted by Frenzel (2005) and Frenzel *et al.* (2005a; 2005b). Fumasep®FAP membrane tended to be most effective, because its current performance was much higher than the other two commercial membranes used (Ionac HAS 3475 and PC-100 D). The performance of the process with Fumasep®FAP membrane depended on the concentration gradient between the Cr product (anolyte) and the central chamber (Cr(VI) containing wastewater) and temperature. The obtained results showed that a batch mode was the recommended system and the rate of transport of chromates by AEM membranes increased significantly with the rise of temperature to 50°C (it was the best at 40°C). The results suggested the best process parameters of EED with Fumasep®FAP membrane as: low initial current density (10–20 mA cm^{-2}); high process temperature (between 40–50°C) and high flow rate (above 7 cm s^{-1}). Membrane stability and performance were further evaluated during the pilot testing within a long-term 400 h operation (Frenzel *et al.*, 2005b).

Tor *et al.* (2005) carried out studies on the simultaneous recovery of tri-and Cr(VI) ions applying a three-compartment EED system equipped with ion-exchange membranes (Tokuyama Soda). The impact of current density and the presence of mono- and divalent ions and other metals in the feed on the efficiency of the process were determined. The recovery of Cr ions of both valences was correlated with the data streams of Cr passing through the membrane (Table 19.2) (Tor *et al.*, 2005). The comparison of Cr(III) and Cr(VI) ion streams clearly showed that the values obtained for Cr(III) were about 50% higher than for the Cr(VI), regardless of the conditions of the experiment. This was explained by the fact that Cr(VI) was present in water in the form of various anions ($HCrO_4^-$, CrO_4^{2-}, $Cr_2O_7^{2-}$, $HCr_2O_7^-$), depending on the pH. No significant effect of the presence of mono- and divalent ions in the feed on the value of the Cr(III) ion stream was observed. The opposite was the case for Cr(VI), for which a substantial reduction in effectiveness in the presence of foreign ions was observed, whereby the impact of bivalent ions was larger than of monovalent ones. If the concentration of the Cr ions was decreased from 10^{-2} to 10^{-4} mol L^{-1}, the rate of diffusion of ions through the membrane also decreased, and therefore the values of the Cr(III) and Cr(VI) streams of ions were 12 and 4 times lower, respectively.

Streams of Cr(III) and Cr(VI) ions in the mixture (Table 19.2) were higher in the absence of accompanying mono- and divalent ions than in their presence, and in the latter case, the streams of mono- and divalent ions seemed to be similar. The highest recovery and the Cr ions streams were obtained at the maximum current density. In all cases both oxidation and recovery of Cr gradually increased with current density, in particular for Cr(III). It is clear that with increase in current density it is possible to obtain approximately 100% recovery of the metal. It was shown, therefore, that the use of a three-compartment system EED enabled simultaneous recovery of both tri-and Cr(VI), while the recovery rate increased with current density and decreased in the presence of the accompanying ions in the feed.

19.4   LIQUID MEMBRANES

Liquid membranes were used in the recovery and disposal of metals from various wastewaters, including ones containing Cr. It was shown that Cyanex 923 (phosphoorganic extraction solvent), Aliquat 336 (trioctylmethylammonium chloride) and tri-*n*-octylamine were effective carriers for Cr(VI), and dinonylnafthalenosulfonic acid in a mixture with *o*-xylene and kerosene for Cr(III)

Table 19.2. The value of Cr(III) and Cr(VI) streams for different feed
composition in the EED process.

Feed composition	Stream, $J \times 10^{11}$ [mol cm^{-2} s^{-1}]	
	Cr(III)	Cr(VI)
Cr(III) (10^{-2} mol L^{-1})	300.6	–
Cr(III) + KCl	285.4	–
Cr(III) + CuCl$_2$	296.8	–
Cr(VI) (10^{-2} mol L^{-1})	–	221.6
Cr(VI) + KCl	–	78.1
Cr(VI) + K$_2$SO$_4$	–	128.0
Cr(III) + Cr(VI) (10^{-2} mol L^{-1})	181.1	189.4
Cr(III) + Cr(VI) (10^{-4} mol L^{-1})	13.9	47.8
Cr(III) + Cr(VI) (10^{-4} mol L^{-1} and 1.4 mA cm^{-2})	32.2	83.7
Cr(III) + Cr(VI) + KCl	140.4	148.3
Cr(III) + Cr(VI) + CuSO$_4$	147.9	158.0
Cr(III) + Cr(VI) − 1.4 mA cm^{-2}	331.9	341.6
Cr(III) + Cr(VI) − 5.7 mA cm^{-2}	723.8	1436
Cr(III) + Cr(VI) − 8.5 mA cm^{-2}	1103	1966
Cr(III) + Cr(VI) − 11.3 mA cm^{-2}	1556	2452

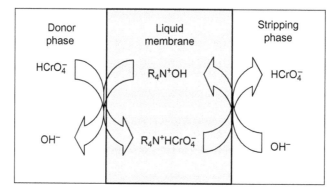

Figure 19.8. The mechanism of Cr(VI) transport with the use of liquid membrane with quaternary ammonium salts.

ions through the liquid membrane (Bodzek and Konieczny, 2011b; Owlad *et al.*, 2009). Elevated temperature, a small concentration of Cr in the effluents and the choice of optimal concentrations of the carrier in the membrane were found to have a beneficial effect on the process performance (Religa and Gawroński, 2006).

The transport mechanism of Cr(VI) anions consists of the complexation of carrier with Cr anions and their decomplexation in an alkaline environment (Fig. 19.8) (Bodzek and Konieczny, 2011b). One of the most important parameters determining the rate and the efficiency of the process is the pH of the donor phase, which affects the ionic form of Cr(VI). The best properties of complexation with most of the carriers used are exhibited by $Cr_2O_7^{2-}$ ions. Optimal transport conditions will therefore occur at pH > 1. Below this value, $HCr_2O_7^-$ ions or $H_2Cr_2O_7$ are formed in a solution, poor dissociation of which slows down the transport process. Too high pH of a solution is also unfavorable, due to the decrease in the concentration of the cationic form of a carrier, caused by a lower availability of H$^+$ ions. Changes in the pH of the solution are directly reflected in changes of the Cr(VI) stream through the membrane.

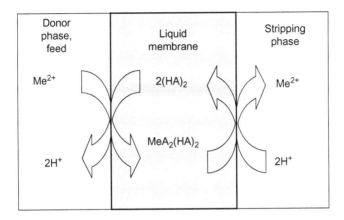

Figure 19.9. The mechanism of metal ion transport in liquid membrane.

Molinari *et al.* (1989), was one of the first who conducted a study on the possibility of direct removal of Cr(III) from aqueous solutions using a supported liquid membrane (SLM). A microporous polypropylene film Accurel with porosity of 17%, thickness of $150\,\mu m$ and with medium size of pores $0.2\,\mu m$ was used as a polymeric support. Dinonylnaphthalenesulfonic acid (DNNS) dissolved in various organic solvents (kerosene, *o*-xylene, hexane, *n*-heptane) played the role of a carrier. Because of the lower viscosity of the aromatic compounds compared to aliphatic ones, the highest stream of Cr(III) was obtained for *o*-xylene (Molinari *et al.*, 1989). The use of mixed solvent, in this case *o*-xylene and kerosene, was also advantageous. The addition of kerosene increased the viscosity of the membrane and thus the stream of substances was reduced, but, on the other hand, it definitely improved the life-time of the membrane. Transport of Cr(III) ions is coupled with transport of H^+ ion in the opposite direction. Thus, it is recommended to maintain the pH of the aqueous feeding phase within the limits of the 4.2–4.5 (at pH > 3 Cr(III) is precipitated as hydroxide), while the stripping phase pH should be kept within the range of 0–0.5 (Molinari *et al.*, 1989). One can also observe a large effect of temperature on the efficiency of the liquid membrane process. In the discussed study the stream of Cr(III) increased 2-fold with increase of the temperature from 25 to 35°C. Simultaneously, this reduced the time in which maximum performance was achieved. A significant impact of the initial concentration of Cr in solution on the efficiency of the process was also found. It was observed that the rate and the efficiency of the process decreased with increase of concentration (Molinari *et al.*, 1989). The transport mechanism of metal cations through liquid membranes is shown in Figure 19.9. The carrier (HA), is usually in the form of a dimer $[HA]_2$ and can be found in the membrane organic phase. Next, it reacts with the metal ion $Me^{2+/3+}$ on the interphase between the aqueous and organic phase and a metal-extractant $MeA_2[HA]_2$ complex (in the case of divalent ions) is created. Then, the complex diffuses through the membrane phase to the interphase of an aqueous solution of the stripping phase, where a decomplexation/separation process takes place. Decomplexation reaction results in both metal ion release to the stripping phase and regeneration of the carrier. The carrier molecules diffuse back through the membrane to membrane – feed interphase and the transport cycle ends.

19.5 FINAL REMARKS

Chromium is an important and widely used element in many branches of industry, especially in steel metallurgy, electroplating and tanning. Industrial activities contribute to the formation of large volumes of wastes and wastewaters containing Cr compounds that enter the groundwater and surface water resources, which are the main source of drinking water. The hexavalent form of Cr seems to be one of the main pollutants from the heavy-metal group, due to the fact that it has high toxicity and reveals carcinogenic and teratogenic properties and causes a serious threat to human health and to the natural environment.

Currently, intensive research on novel techniques of water and wastewater containing Cr treatment is being carried out. Next to the classical processes i.e. reduction, precipitation, filtration, adsorption and ion-exchange, membrane methods are starting to play a significant role. The comparison of different techniques of the removal of Cr(VI) from the aquatic environment is very difficult due to inconsistencies in the presentation of the data, which are provided for different process conditions (pH, initial concentration, temperature, the ratios of various reagents, etc.). In addition, different types of water contaminated with Cr are used as a feed i.e. groundwater, drinking water, wastewaters from actual tannery and electroplating operations as well as simulated solutions.

Membrane filtration is a modern technique for removal of Cr, which is able to treat wastewater with a high concentration of Cr(VI). Depending on the characteristics of the membrane and membrane filtration system it is possible to remove Cr in a wide range of operating conditions. One can apply high pressure membrane processes, reverse osmosis and nanofiltration, as well as ultrafiltration enhanced with polymers or surface-active substances. It seems that nanofiltration is a better solution for the removal of Cr from the aquatic environment. The use of liquid membranes, which are characterized by high selectivity, is also a promising solution. However, operating costs are a major issue of all separation techniques based on membrane separation.

Another group of techniques used to remove the Cr(VI) is that of electrochemical membrane methods. These have relatively low cost and high selectivity. Particularly useful is electro-electrodialysis in the three-compartment system, which additionally allows separation of Cr(VI) ions from other metals.

Most of the research on the methods of Cr removal is limited only to the initial evaluation of the ability of the metal separation, so there is a need to continue the experiments on a pilot scale. In addition, there is still a lot of research work on the mechanisms of Cr removal, which will enable a better understanding of the phenomena accompanying the technological processes and explore new methods for implementation to be done.

REFERENCES

Alvarado, L., Ramírez, A. & Rodríguez-Torres, I. (2009) Cr(VI) removal by continuous electrodeionization: study of its basic technologies. *Desalination*, 249, 423–428.

Anonymous (2002) Poradnik Galwanotechnika [in Polish]. (Electroplating Guide). WNT, Warszawa, Poland.

Aoudia, M., Allal, N., Djennet, A. & Toumi, L. (2003) Dynamic micellar enhanced ultrafiltration: use of anionic (SDS)-nonionic (NPE) system to remove Cr^{3+} at low surfactant concentration. *Journal of Membrane Science*, 217, 181–192.

Aroua, M., Zuki, F.M. & Sulaiman, N.M. (2007) Removal of chromium ions from aqueous solutions by polymer-enhanced ultrafiltration. *Journal of Hazardous Materials*, 147, 752–758.

Audinos, R. (1992) Liquid waste concentration by electrodialysis. In: Li, N.N. & Calo, J.M. (eds.) *Separation and purification technology*. Marcel Dekker, New York, NY. pp. 229–301.

Audran, J., Baticle, P. & Letord, M.-M. (1990) Recycling of chromic acid by electro-electrodialysis. *Proceedings of Vth World Filtration Congress, 5–8 June 1990, Nice, France*. p. 43.

Audran, J., Letord, M.-M. & Ducourroy, A. (1992) *Electrochemical cell, especially for recycling of chromic acid*. Patent EP 048301.

Baek, K. & Yang, J.-W. (2004) Competitive bind of anionic metals with cetylpyridinium chloride micelle in micellar-enhanced ultrafiltration. *Desalination*, 167, 101–110.

Bodzek, M. (1999) Membrane techniques in wastewater treatment. In: Goosen, I.M.F.A. & Shayya, W.H. (eds.) *Water management purification and conservation in arid climates*. Volume 2: *Water purification*. Lancaster-Basel, Technomic Publishing, Lancaster, UK, Basel, Switzerland. pp. 121–184.

Bodzek, M. (2012) Usuwanie metali ze środowiska wodnego za pomocą metod membranowych – stan wiedzy [in Polish]. (Removal of metals from water environment using membrane methods – state of the art). *Monografie Komitetu Inżynierii Środowiska Polskiej Akademii Nauk*, 66, 305–313.

Bodzek, M. (2013) Przegląd możliwości wykorzystania technik membranowych w usuwaniu mikroorganizmów i zanieczyszczeń organicznych ze środowiska wodnego [in Polish]. (An overview of the possibility of membrane techniques application in the removal of microorganisms and organic pollutants from aquatic environment). *Inżynieria i Ochrona Środowiska.* 16 (1), 5–37.

Bodzek, M. & Konieczny, K. (2005) *Wykorzystanie procesów membranowych w uzdatnianiu wody* [in Polish]. (*Application of membrane processes in water treatment*). Oficyna Wydawnicza Projprzem-Eko, Bydgoszcz, Poland.

Bodzek, M. & Konieczny, K. (2011a) *Usuwanie zanieczyszczeń nieorganicznych ze środowiska wodnego metodami membranowymi* [in Polish]. (*Removal of inorganic micropollutants from water environment by means of membrane methods*). Wydawnictwo Seidel-Przywecki, Warszawa, Poland.

Bodzek, M. & Konieczny, K. (2011b) Membrane techniques in the removal of inorganic micropollutants from water environment – state of the art. *Archives of Environmental Protection*, 37 (2), 15–29.

Bodzek, M., Bohdziewicz, J. & Konieczny, K. (1997) *Techniki membranowe w ochronie środowiska* [in Polish]. (*Membrane techniques in environmental protection*). Wydawnictwo Politechniki Śląskiej, Gliwice, Poland.

Bodzek, M., Korus, I. & Loska, K. (1999) Application of the hybrid complexation – ultrafiltration process for removal of metal ions from galvanic wastewater. *Desalination*, 121, 117–121.

Bodzek, M., Konieczny, K. & Kwiecinska, A. (2011) Application of membrane processes in drinking water treatment – state of art. *Desalination and Water Treatment*, 35, 164–184.

Bohdziewicz, J. (2000) Removal of chromium ions (VI) from underground water in the hybrid complexation-ultrafiltration process. *Desalination*, 129, 227–235.

Bohdziewicz, J., Bodzek, M. & Bień, J. (1995) Ocena możliwości usuwania chromu z wody metodą odwróconej osmozy [in Polish]. *Ochrona Środowiska*, 57 (2), 7–10.

Bohdziewicz, J., Bodzek, M. & Bień, J. (1997) Zastosowanie membran osmotycznych i nanofiltracyjnych typu SEPA CF amerykańskiej firmy OSMONICS do usuwania jonów Cr(VI) z wód głębinowych [in Polish]. (Application of osmotic and nanofiltration membranes SEPA CF type by OSMONICS U.S. company to remove Cr(VI) ions from underground waters). *Report of Conference: Nowe technologie w uzdatnianiu wody, oczyszczaniu ścieków i gospodarce osadowej.* Częstochowa-Ustroń 1997, Wydawnictwo Politechniki Częstochowskiej, Seria: Konferencje no. 15, pp. 17–25.

Bryjak, M. (2001) *Procesy separacyjne a polimery. O możliwościach nietypowego wykorzystania syntetycznych polimerów* [in Polish]. (*Separation processes and polymers. about the possibilities of an atypical use of synthetic polymers*). Oficyna Wydawnicza Politechniki Wrocławskiej, Wrocław, Poland.

Cassano, A., Drioli, E., Molinari, R. & Bertolutti, E. (1997) Quality improvement of recycled chromium in the tanning operation by membrane processes. *Desalination*, 108, 193–203.

Cassano, A., Pietra, L.D. & Drioli, E. (2007) Integrated membrane process for the recovery of chromium salts from tannery effluents. *Industrial & Engineering Chemistry Research*, 46, 6825–6830.

Castillo, E., Granados, M. & Cortina, J.L. (2002a) Chemically facilitated chromium(VI) transport throughout an anion-exchange membrane: application to an optical sensor for chromium(VI) monitoring. *Journal of Chromatography*, 963 (1–2), 205–211.

Castillo, E., Granados, M. & Cortina, J.L. (2002b) Chromium(VI) transport through the Raipore 1030 anion exchange membrane. *Analytica Chimica Acta*, 464 (1), 15–23.

Çengeloglu, Y., Tor, A., Kir, E. & Ersöz, M. (2003) Transport of hexavalent chromium through anion-exchange membranes. *Desalination*, 154 (3), 239–246.

Chaudhary, A.J., Ganguli, B. & Grimes, S.M. (2006) Concentrator cell methodology in the regeneration and recycle of chromium etching solutions using membrane technology. *Chemosphere*, 62, 841–846.

Chingombe, P., Saha, B. & Wakeman, R.J. (2005) Surface modification and characterisation of a coal-based activated carbon. *Carbon*, 43 (15), 3132–3143.

Das, C., Patel, P., De, S. & DasGupta, S. (2006) Treatment of tanning effluent using nanofiltration followed by reverse osmosis. *Separation and Purification Technology*, 50, 291–299.

Donepudi, V.S., Khalili, N.R., Kizilel, R. & Selman, J.R. (2002) Purification of hard chromium plating baths – options & challenges. *Plating and Surface Finishing*, 4, 62.

Dunn, R.O., Scamehorn, J.F. & Christian, S.D. (1985) Use the micellar enhanced ultrafiltration to remove dissolved organics from aqueous wastes. *Separation Science and Technology*, 20, 257–284.

Dunn, R.O., Scamehorn, J.F. & Christian, S.D. (1989) Simultaneous removal of dissolved organics and divalent metal cations from water using micellar-enhanced ultrafiltration. *Colloids and Surfaces*, 35, 49–56.

Dylewski, R. (2001) Otrzymywanie kwasu siarkowego i wodorotlenku sodu z odpadowego siarczanu sodu [in Polish]. (Manufacture of sulphuric acid and sodium hydroxide from spent sodium sulphate). *Chemik*, 54, 8–11.

Dzyazko, Y.U.S., Mahmoud, A., Lapicque, F. & Belyakov, V.N. (2007) Cr(VI) transport through ceramic ion-exchange membranes for treatment of industrial wastewaters. *Journal of Applied Electrochemistry*, 37, 209–217.

Frenzel, I. (2005) *Waste minimization in chromium plating industry*. PhD Thesis. The University of Twente, Enschede, The Netherlands.

Frenzel, I., Holdik, H., Stamatialis, D.F., Pourcelly, G. & Wessling, M. (2005a) Chromic acid recovery by electro-electrodialysis. I. Evaluation of anion-exchange membrane. *Journal of Membrane Science*, 261, 49–55.

Frenzel, I., Holdik, H., Stamatialis, D.F., Pourcelly, G. & Wessling, M. (2005b) Chromic acid recovery by electro-electrodialysis. II. Pilot scale process, development, and optimization. *Separation and Purification Technology*, 47, 27–35.

Garncarz, I., Poźniak, G., Bryjak, M. & Frankiewicz, A. (1999) Modification of polysulfone membranes. 2. Plasma grafting and plasma polymerization of acrylic acid. *Acta Polymerica*, 50, 317–326.

Gibbs, L.L., Scamehorn, J.F. & Christian, S.D. (1987) Removal of *n*-alcohols from aqueous stream using micellar enhanced ultrafiltration. *Journal of Membrane Science*, 30, 67–74.

Gomes, S., Cavaco, S.A., Quina, M.J. & Gando-Ferreira, L.M. (2010) Nanofiltration process for separating Cr(III) from acid solutions. Experimental and modelling analysis. *Desalination*, 254, 80–89.

Hafez, A.I., El-Manharawy, M.S. & Khedr, M.A. (2002) RO membrane removal of unreacted chromium from spent tanning effluent. A pilot-scale study. *Desalination*, 144, 237–242.

Hafiane, A., Lemordant, D. & Dhahbi, M. (2000) Removal of hexavalent chromium by nanofiltration. *Desalination*, 130, 305-312.

Hasan, M., Selim, A. & Mohamed, K.M. (2009) Removal of chromium from aqueous waste solution using liquid emulsion membrane. *Journal of Hazardous Materials*, 168 (2–3), 1537–1541.

Ho, W.S., Tarun, H. & Poddar, K. (2001) New membrane technology for removal and recovery of chromium from waste waters. *Environmental Progress*, 20 (1), 44–52.

Huang, J., Zeng, G.M. & Xu, K. (2005) Removal of cadmium ions from aqueous solution via micellar-enhanced ultrafiltration. *Transactions of Nonferrous Metals Society of China*, 15, 184–189.

Huang, J., Zeng, G., Fang, Y., Qu, Y. & Li, X. (2009) Removal of cadmium ions from waste water using micellar-enhanced ultrafiltration with mixed anionic-nonionic surfactants. *Journal of Membrane Science*, 326 (2), 303–309.

Huang, Y.C., Batchelor, B. & Koseoglu, S.S. (1994) Cross-flow surfactant-based ultrafiltration of heavy metals from waste streams. *Separation Science Technology*, 29, 1979–1999.

Jacukowicz-Sobala, I. (2009) Współczesne metody usuwania chromu ze ścieków [in Polish]. (Modern methods of removal of chromium from wastewaters). *Przemysł Chemiczny*, 88 (1), 51–60.

Juang, R., Lin, S. & Peng, L. (2010) Flux decline analysis in micellar-enhanced ultrafiltration of synthetic waste solutions for metal removal. *Chemical Engineering Journal*, 161, 19–26.

Juang, R.S. & Chiou, Ch.H. (2000) Ultrafiltration rejection of dissolved ions using various weakly basic water-soluble polymers. *Journal of Membrane Science*, 177, 207–214.

Juang, R.S., Xu, Y.Y. & Chen, C.L. (2003) Separation and removal of heavy metal ions from dilute solutions using micellar-enhanced ultrafiltration. *Journal of Membrane Science*, 218, 257–267.

Kołtuniewicz, A.B. & Drioli, E. (2008) *Membranes in clean technologies*. Wiley-VCH Verlag GmbH, Weinheim, Germany.

Konopczyńska, B., Staszak, K. & Prochaska, K. (2011) Usuwanie jonów chromu(III) z roztworów wodnych techniką ultrafiltracji miceralnej (MEUF) [in Polish]. (Removal of chromium(III) ions from aqueous solutions by micellar ultrafiltration technique (MEUF)). *Inżynieria i Aparatura Chemiczna*, 50 (5), 58–59.

Korus, I. (2012) *Wykorzystanie ultrafiltracji wspomaganej polimerami do separacji jonów metali ciężkich* [in Polish]. (The application of ultrafiltration enhanced with polymers for the separation of heavy metal ions). Wydawnictwo Politechniki Śląskiej, Gliwice, Poland.

Korus, I. & Bożek, M. (2010) Równoczesna i selektywna separacja Cr(VI) i Zn(II) ze ścieków galwanicznych metodą ultrafiltracji wspomaganej polimerem [in Polish]. (Simultaneous and selective separation of Cr(VI) and Zn(II) from galvanic wastewaters by polymer enhanced ultrafiltration). *Monografie Naukowe Komitetu Inżynierii Żrodowiska PAN*, 65, 233–245.

Korus, I. & Loska, K. (2009) Removal of Cr(III) and Cr(VI) ions from aqueous solutions by means of polyelectrolyte-enhanced ultrafiltration. *Desalination*, 247, 390–395.

Kozłowski, C.A. & Walkowiak, W. (2002) Removal of chromium(VI) from aqueous solutions by polymer inclusion membranes. *Water Research*, 36, 4870–4876.

Kulkarni, P.S., Kalyani, V. & Mahajani, V.V. (2007) Removal of hexavalent chromium by membrane-based hybrid processes. *Industrial and Engineering Chemistry Research*, 46 (24), 8176–8182.

Lambert, J., Avila-Rodriguez, M., Durand, G. & Rakib, M. (2006) Separation of sodium ions from trivalent chromium by electrodialysis using monovalent cation selective membranes. *Journal of Membrane Science*, 280, 219–225.

Lee, J., Yang, J.-S., Kim, H.J., Baek, K. & Yang, J.W. (2005) Simultaneous removal of organic and inorganic contaminants by micellar enhanced ultrafiltration with mixed surfactant. *Desalination*, 184, 395–407.

Linde, K. & Jonsson, A.S. (1995) Nanofiltration of salt solutions and landfill leachate. *Desalination*, 103, 223–232.

Llanos, J., Pérez, Á. & Cañizares, P. (2008) Copper recovery by polymer enhanced ultrafiltration (PEUF) and electrochemical regeneration. *Journal of Membrane Science*, 323, 28–36.

Mbareck, C., Nguyen, Q.T., Alexandre, S. & Zimmerlin, I. (2006) Fabrication of ion-exchange ultrafiltration membranes for water treatment. I. Semi interpenetrating polymer networks of polysulfone and polyacrylic acid. *Journal of Membrane Science*, 278, 10–18.

Mbareck, C., Nguyen, Q.T., Alaoui, O.T. & Barillier, D. (2009) Elaboration, characterization and application of polysulfone and polyacrylic acid blends as ultrafiltration membranes for removal of some heavy metals from water. *Journal of Hazardous Materials*, 171 (1–3), 93–101.

Molinari, R., Drioli, E. & Pantano, G. (1989) Stability and effect of diluents in supported liquid membranes for Cr(III), Cr(VI) and Cd(II) recovery. *Separation Science and Technology*, 24, 1015–1032.

Molinari, R., Gallo, S. & Argurio, P. (2004) Metal ions removal from wastewater or washing water from contaminated soil by ultrafiltration-complexation. *Water Research*, 38, 593–600.

Molinari, R., Poerio, T. & Argurio, P. (2005) Polymer assisted ultrafiltration for copper-citric acid chelate removal from wash solutions of contaminated soil. *Journal of Applied Electrochemistry*, 35, 375–380.

Muslehiddinoglu, J., Uludag, Y., Ozbelge, H.O. & Yilmaz, L. (1998) Effect of operating parameters on selective separation of heavy metals from binary mixtures via polymer enhanced ultrafiltration. *Journal of Membrane Science*, 140, 251–266.

Muthukrishnan, M. & Guha, B.K. (2008) Effect of pH on rejection of hexavalent chromium by nanofiltration. *Desalination*, 219, 171–178.

Nataraj, S.K., Hosamani, K.M. & Aminabhavi, T.M. (2007) Potential application of an electrodialysis pilot plant containing ion-exchange membranes in chromium removal. *Desalination*, 217, 181–190.

Ortega, L.M., Lebrun, R., Noël, I.M. & Hausler, R. (2005) Application of nanofiltration in the recovery of chromium(III) from tannery effluents. *Separation and Purification Technology*, 44, 45–52.

Ortiz, L., Galan, B. & Irabien, A. (1996) Membrane mass transport coefficient for the recovery of Cr(VI) in hollow fiber extraction and back-extraction modules. *Journal of Membrane Science*, 118, 713–221.

Owlad, M., Aroua, M.K., Daud, W.A. & Baroutian, S. (2009) Removal of hexavalent chromium-contaminated water and wastewater: a review. *Water Air, & Soil Pollution*, 200, 59–77.

Pugazhenthi, G., Sachan, S., Kishore, N. & Kumar, A. (2005) Separation of chromium (VI) using modified ultrafiltration charged carbon membrane and its mathematical modeling. *Journal of Membrane Science*, 254, 229–239.

Religa, P. & Gawroński, R. (2006) Oczyszczanie chromowych ścieków garbarskich – procesy membranowe [in Polish]. (Treatment of tanning wastewaters – membrane processes). *Przegląd Włókienniczy – Włókno, Odzież, Skóra*, 12, 41–44.

Religa, P., Kowalik, A. & Gierycz, P. (2011a) A new approach to chromium concentration from salt mixture solution using nanofiltration. *Separation and Purification Technology*, 82, 114–120.

Religa, P., Kowalik, A. & Gierycz, P. (2011b) Application of nanofiltration for chromium concentration in the tannery wastewater. *Journal of Hazardous Materials*, 186, 288–292.

Religa, P., Kowalik, A. & Gierycz, P. (2011c) Effect of membrane properties on chromium(III) recirculation from concentrate salt mixture solution by nanofiltration. *Desalination*, 274, 164–170.

Sadaoui, Z., Hemidouche, S. & Allalou, O. (2009) Removal of hexavalent chromium from aqueous solutions by micellar compounds. *Desalination*, 249, 768–773.

Samper, E., Rodrigues, M., Senatana, I. & Prats, D. (2009) Effects of ionic strength and pH over the removal of chrome by micellar-enhanced ultrafiltration (MEUF) using CPC surfactant. *CD Proceedings of the Conference, Desalination for the Environment: Clean Water and Energy, 17–20 May 2009 Baden-Baden, Germany*.

Scamehorn, J.F., Christian, S.D. & Ellington, R.T. (1989) Use of micellar-enchanced ultrafiltration to remove streams. In: Scamehorn, J.F. & Harwell, J.H (eds.) *Surfactant based separation processes*. Marcell Decker, New York, NY. pp. 29–51.

Shaalan, H.F., Sorour, M.H. & Tewfik, S.R. (2001) Simulation and optimization of a membrane system for chromium recovery from tanning wastes. *Desalination*, 141, 314–324.

Staszak, K., Redutko, B. & Prochaska, K. (2010) Usuwanie jonów metali (Cu(II) I Cr(III)) z roztworów wodnych technika MEUF [in Polish]. *Monografie Naukowe Komitetu Inżynierii Środowiska PAN*, 66, 185–193.

Taleb-Ahmed, M., Taha, R., Maachi, S. & Dorange, G. (2002) The influence of physico-chemistry on the retention of chromium ions during nanofiltration. *Desalination*, 145, 103–108.

Thompson, J.A. & Jarvinen, G. (1999) Using water-soluble polymers to remove dissolved metal ions. *Filtration and Separation*, 36 (5), 28–32.

Tor, A., Büyükerkek, T., Cengeloglu, Y. & Ersöz, M. (2005) Simultaneous recovery of Cr(III) and Cr(VI) from the aqueous phase with ion-exchange membranes. *Desalination*, 171 (3), 233–241.

Torabian, A., Ghadimkhani, A.A, Mohammadpour, A., Mehrabadi, A.R. & Akhtarirad, F. (2010) Removal of hexavalent chromium from potable water by nanofiltration. *International Journal of Chemical Engineering*, 02/2010; 939252.

Urbański, R., Góralska, E., Bart, H.J. & Szymanowski, J. (2002) Ultrafiltration of surfactant solutions. *Journal of Colloid and Interface Science*, 253, 419–426.

Vallejo, M.E., Huguet, P., Innocent, C., Persin, F., Bribes, J.L. & Pourcelly, G. (1999) Contribution of raman spectroscopy to the comprehension of limiting phenomena occurring with a vinylpyridinium anion exchange membrane during the electrolysis of Cr(VI) solutions. *Journal of Physical Chemistry B*, 103 (51), 11,366–11,371.

Vallejo, M.E., Persin, F., Innocent, C., Sistat, P. & Pourcelly, G. (2000) Electrotransport of Cr(VI) through an anion exchange membrane. *Separation and Purification Technology*, 21, 61–69.

Velizarov, S., Crespo, J.G. & Reis, M.A. (2004) Removal of inorganic anions from drinking water supplies by membrane bio/processes. *Reviews in Environmental Science & Biotechnology*, 3, 361–380.

Wang, X.L., Tsuru, T., Nakao, S. & Kimura, S. (1997) The electrostatic and steric-hindrance model for the transport of charged solutes through nanofiltration membranes. *Journal of Membrane Science*, 135, 19–32.

Yurlova, L., Kryvoruchko, A. & Kornilovich, B. (2002) Removal of Ni (II) ions from wastewater by micellar-enhanced ultrafiltration. *Desalination*, 144, 255–260.

Zakrzewska-Trznadel, G. & Harasimowicz, M. (2002) Removal of radionuclides by membrane permeation combined with complexation. *Desalination*, 144, 207–212.

Zhang, Z., Zeng, G.M. & Huang, J.H. (2007) Removal of zinc ions from aqueous solution using micellar-enhanced ultrafiltration at low surfactant concentrations. *Water SA*, 33, 129–136.

CHAPTER 20

Hybrid systems for removal of trace amounts of Cr(VI) and As(V) ions from water

Marek Bryjak & Nalan Kabay

20.1 INTRODUCTION

There is no simple technology for removal of trace amounts of As(V) and Cr(VI) ions from aqueous solutions. Some conventional and advanced treatment methods are reviewed in many published papers. The general comments and specific outcomes are summarized below.

20.1.1 *Removal of As(V)*

In the case of As contaminated water, the following methods are of the highest popularity (Jain and Singh, 2012):

- Oxidation-filtration.
- Biological oxidation where oxidation is performed by microorganisms and then As(V) is removed by sorption on iron and manganese oxides.
- Co-precipitation where oxidation by the following oxidizing agents: chlorine, ozone, permanganate, or hydrogen peroxide, is followed by coagulation and sedimentation.
- Adsorption where such adsorbents as activated alumina, activated carbon, iron-based adsorbents, zero-valent iron or hydrated iron oxide are used.
- Ion exchange through suitable ion-exchange resins.
- Membrane technologies such as reverse osmosis, nanofiltration and electrodialysis are of the highest interest.

Oxidation-filtration based method is designed to remove naturally occurring iron and manganese from water and involves oxidation of the soluble forms of these compounds and formation of insoluble flocks that are removed by simple filtration. When As appears in the treated water, it is removed by adsorption on iron/manganese hydroxides. Hence, the efficiency of the process is strongly related to concentration of iron and As in water; favorable it should reach 20:1 mole ratio.

Biological oxidation by microorganisms is a relatively new method. Groundwater containing the reduced form of As as well as iron and manganese should be aerated. During this process, water is populated by iron and manganese oxidizing bacteria that can oxidize As(III) species. Finally, As(V) is adsorbed on the iron manganese oxides and removed by filtration. It was reported (Katsoyiannis and Zouboulis, 2006) that a faster bio-oxidation of As(III) in comparison to any chemical oxidation process was achieved and remaining As concentration was less than $10\,\mu g\,L^{-1}$.

In case of co-precipitation, water is treated with such coagulants as alum, ferric chloride or ferric sulfate that can form flocks and attract As contaminants. Typically, the process reduces As concentration to $10\,\mu g\,L^{-1}$ and can be used when flocks separation is assisted with removal of the major part of the coagulant. Generally, three modes of operations such as lime softening, gravity coagulation/filtration and coagulation with microfiltration are applied. The first one is used occasionally, mostly where As removal assists water softening, while the second and the third one are more popular.

Table 20.1.   Treatment efficiency and water loss for arsenic removal (reproduced with permission from Jain and Singh, 2012, copyright of Taylor and Francis).

Technology	As removal [%]	Water loss [%]
Oxidation/filtration	50–90	>2
Bio-oxidation/filtration	>95	1–2
Adsorption	90–95	1–2
Ion exchange	95	1–2
Membrane separations	>95	15–50

Adsorption technology is one of the most widely used methods for removal of As from groundwater. The technology typically can reduce As concentration to the permissible level of $10 \, \mu g \, L^{-1}$. However adsorption is sensitive to many interfering ions presented in water and for that reason it should be optimized for each particular case. The most commonly used adsorbents are alumina and sorbents with iron.

Ion exchange is the physical process where ions are swapped between stationary and mobile phases. Typically, the stationary phase is formed by solid material that contains ignitable functional groups. Various types of synthetic ion-exchange resins are used for As removal. When the resin is loaded with As, it should be subjected to regeneration. The time of resin service depends on the quantity of As in the treated water.

Membrane technologies such as nanofiltration, reverse osmosis and electrodialysis can reduce As concentration typically to $50 \, \mu g \, L^{-1}$ and in some cases to $10 \, \mu g \, L^{-1}$. Due to the high operational cost of membrane technologies they are not as popular as precipitation, adsorption and ion-exchange methods for As removal.

The efficiency of the above-mentioned technologies is summarized in Table 20.1.

Table 20.1 clearly shows the reason of low popularity of the membrane technologies. Although their efficiency for As removal is high, the water recovery for membrane separations looks poor.

20.1.2   *Removal of Cr(VI)*

Due to strict regulations for the Cr contents in water, the interest in removal of this element has grown rapidly during the last decades. Wastewaters contaminated with Cr are usually treated with such physicochemical methods as adsorption, solvent extraction and membrane separation. There is no biological-based method as Cr inhibits the growth of any organism (Malaviya and Singh, 2011). Among the separation methods employed, the most popular ones are:

- Precipitation;
- Reduction;
- Flotation;
- Solvent extraction;
- Ion exchange;
- Adsorption;
- Membrane separation.

Precipitation is based on chemical coagulation by addition of some coagulants (lime, sodium hydroxide, sodium carbonate, sodium bicarbonate or calcium hydroxide, magnesium oxide, ammonia or polyelectrolytes) and addition of aluminum salts to flocculate the formed particles.

It is well-known that Cr(III) is less toxic than Cr(VI). Hence one of the frequently used methods is the reduction of Cr(VI) to Cr(III). Various reducing agents such as metallic iron, zinc, aluminum, SO_2 or ascorbic acid were employed as reductants. The reduction is performed with a yield of 50–95%. However, the reducing agents used contaminate the treated wastewater and this makes the reduction method less profitable. Photo-reduction and electro-reduction are not so problematic as chemical reduction. The first is carried out on surfaces coated with titanium oxide mostly while the second employs electrodes covered by conductive polymers.

Flotation is used mostly for separation of solids or dispersed liquids from any liquid phase by means of bubble attachment. Surface active agents assemble around particle and such structures are removed by air bubbles. The air bubbles laden with the metal ions float to the surface and are removed as metal-rich froth. Among few studies on application of flotation for Cr(VI) removal, the most interesting one seems to be the use of hydrotalcite as collector. Lazaridis *et al.* (2001) reported a 95% removal of Cr(VI). Recently, the interest has been shifted from simple flotation to hybrid processes where flotation was combined with adsorption or filtration. In the first case, called sometimes sorptive filtration, Cr(VI) is sorbed on goethite or ferric hydroxide and the obtained flocks are flotated. The second case, hybrid flotation-microfiltration process, employs precipitation and adsorbing colloid flotation. It allows conducting the pre-treatment of contaminated waters which thereby prevents membrane fouling during the microfiltration.

In the case of solvent extraction, liquid ion exchangers are used as solvents for removal of Cr(VI) or Cr(III) ions. High molecular weight amines are used for extraction of Cr(VI) while different acids such as di(2-ethylhexyl)phosphoric acid or carboxylic acids can be employed for Cr(III) removal.

Various inorganic and polymeric ion exchangers were tested for Cr removal. The adsorption of Cr(VI) ions on most of the cation-exchange resins follows first order kinetics and intraparticle diffusion represents the rate-limiting step (Mukherjee *et al.*, 2013). Various ion-exchangers have been tested (Malaviya and Singh, 2011). Their main advantages over the precipitation method were: high selectivity and reduced volume of sludge (Rengaraj *et al.*, 2001).

Among all the physicochemical separations, adsorption is a highly effective, cheap, and easy method to be applied. It is the process when a substance becomes bound by physical and chemical interactions. Adsorption can be an effective and versatile method for Cr removal especially if low cost adsorbents are used.

Membrane techniques involve the separation of metals in the presence of a semipermeable membrane dividing two phases. The membrane restricts mixing of ions and molecules and its separation ability is based on permeant size, differences in diffusion coefficients, electrical charge, and solubility. The following membrane methods have been tested for Cr separation:

- *Ultrafiltration (UF)*: Metal ions are complexed with macroligand or micellar structures and such large bodies are rejected by an UF membrane.
- *Nanofiltration (NF)*: Charged membranes control transportation of water and other ions by diffusion, Donnan exclusion mostly and by convection to a small extent.
- *Reverse osmosis (RO)*: The membrane allows the passing of water molecules and is impermeable to larger molecules.
- *Electrodialysis (ED)*: Transport of ions is driven by electrical potential difference between two electrodes and separation is related to the permeability of membranes towards different ions. Cation-exchange membranes transport the cations preferentially while anion-exchange membranes let the anions pass through them.
- *Liquid membranes (LM)*: Ion transport is driven by the concentration gradient and the presence of suitable carriers in organic phase that divides two aqueous phases. There are two types of LM configurations tested for Cr separation: bulk membranes (contactors) and emulsion liquid membranes.

The summary of methods for treatment of effluents contaminated by Cr is shown in Table 20.2.

Bearing the process performances in mind, it could be considered that all presented technologies are effective and could be used for removal of Cr ions from wastewater. However, as pointed out in the discussion of As removal there is a significant loss of water in the cases of membrane separations. That property pushes membrane techniques to the less profitable group. It seems that the above situation can be improved when membrane techniques are combined with other physicochemical separation. That approach is discussed in the next sections.

Table 20.2. Typical treatment efficiency for chromium removal (adapted from Malaviya and Singh, 2011).

Technology	Cr removal [%]
Precipitation	99–100
Reduction	80–100
Flotation	95–100
Solvent extraction	50–100
Ion exchange	90–100
Adsorption	50–100
Membrane separation	80–100

20.2 HYBRID PROCESSES

Suk and Matsuura (2006) have divided membrane-based hybrid processes into two groups:

- Processes that combine membrane techniques with other conventional methods;
- Processes that combine membrane techniques with other membrane methods.

The first group covers almost all hybrid processes and membrane-enhanced separations, being the topic of this chapter, belongs to it. The second group, called the integrated membrane processes, and deals with those cases when at least two membrane processes are arranged in series. This case is not discussed here for Cr and As separations.

The membrane-enhanced hybrid process can be considered as an alternative for removal of harmful species from their diluted solutions. It combines sorption and membrane separation and merges advantages of both these processes. In the hybrid process, solutes are adsorbed by fine coupling agents while the large complexes are separated by membranes. The main benefit of a membrane-enhanced hybrid process is its efficiency and lower cost in comparison to classical fixed bed and batch sorption methods. In the hybrid systems, coupling agents with relatively small diameter are used. Consequently, it is possible to reduce the amount of sorbent and decrease the cost of the process significantly. The new binding agents comprise organic and inorganic components, chelating and ion-exchange resins, molecularly imprinted polymers, micellar structures and/or coordinating agents. They can appear as solid particles, micelles, colloids or water-soluble polymers. The general principle for their use in the membrane-enhanced separations is their affinity to the target molecules and formation of large-dimension associates that are removed by membrane filtration. In the second step, the separated complexes are decomposed to the binding agent and the target compound. The binding agent is recycled to the next sorption step.

Below is a brief description of typical membrane-enhanced process comprises four steps:

1 Formation of complexes between target molecules and coupling agent;
2 Membrane separation of complexes;
3 Desorption of target molecule from the complexes;
4 Regeneration of coupling agent.

Depending on the kinds of coupling agent, the hybrid systems can be divided into several groups. There is no consensus on the clear nomenclature of the processes and some ambiguity can be found (Kabay and Bryjak, 2013). The hybrid process is based on the formation of complexes of small target molecules with larger structures of binding agent and separation of such complexes by membrane. The type of membrane used depends on the size of separated complexes. Hence, it could be said that the mechanisms of complexation have secondary importance to the nature of the process and the clue to the process is membrane filtration. The target molecules can employ ionic forces, hydrophobic interactions, chelating or other forms of affinity to bind to the coupling agent while the membrane-enhanced process will not change its nature. Hence, taking into account the size of complexing agent the membrane-enhanced processes useful for removal of Cr and As ions can be classified as follows:

Membrane-enhanced process	Size of binding agent
Polymer-enhanced ultrafiltration (PEUF)	several nanometers
Micellar-enhanced ultrafiltration (MEUF)	several nanometers
Suspension-enhanced microfiltration (SEMF)	several micrometers

20.2.1 *Polymer-enhanced ultrafiltration (PEUF)*

The PEUF process belongs to the class of complexation-enhanced filtration processes. It has various names in the literature – some authors call it *liquid-phase polymer-based retention* (LPR) (Rivas *et al.*, 2011) while some others refer to it as *polyelectrolyte-enhanced ultrafiltration* (Canizares *et al.*, 2008). The process is based on separation of target ions by binding them to water-soluble polymers and separation of the formed complexes on an UF membrane. The process takes place in a homogeneous media and avoids the phenomenon of mass transfer or diffusion restrictions. The PEUF process can be run in batch, semicontinuous and continuous modes of operations (Palencia *et al.*, 2009a; Sabate *et al.*, 2006). Taking into account the economy of the process, polymer regeneration seems to be of special concern. There are two methods for regeneration of water-soluble polymers: by altering solution pH or by electroregeneration (Llanos *et al.*, 2008; Palencia *et al.*, 2009b). In the first case, the regeneration is carried out by changing the solution pH. It is a simple process and can be run at low cost. However, low concentration of target species in brine and high consumption of water are the negative features of such kinds of regeneration. The second case, electrochemical regeneration, is environmentally friendly as it does not produce large volumes of effluents but its cost seems to be a strong barrier in some cases. Membrane fouling by water-soluble polymers is one of the main disadvantages and this hinders the popularity of the PEUF hybrid process. Fouling has a direct impact on UF cost as its effect would be overcome by increasing the pressure and periodical cleaning of the membrane. It is accepted generally that fouling is affected by the character of membrane material, character of feed, and process parameters. To make the membrane fouling resistant, its surface is sometimes coated with hydrophilic polymers (Jana *et al.*, 2011).

20.2.1.1 *Removal of As(V) by PEUF*
Recently, the ability of some polymers to remove As species in solution by the PEUF method has been studied intensively (Rivas and Aguirre 2007; Rivas *et al.*, 2007a; 2007b; Sanchez *et al.*, 2013). The water-soluble polymers used to separate As(V) ions are shown in Figure 20.1.

When the presented water-soluble polymers were used with the molar ration to As as 20:1, it was possible to remove 80–100% of arsenates at pH range of 6 to 9. When the process was carried out at pH of 3, the separation efficiency was not higher than 60%. The studies pointed out that reduction of membrane fouling was possible by the use of hydrophilic membranes and that some water-soluble polymers could be efficiently regenerated and used several times.

20.2.1.2 *Removal of Cr(VI) by PEUF*
P(ClAPTA), P(ClVBTA) and P(SAETA) were applied as extracting agents in PEUF technique. The results showed highest retention capacity of Cr(VI) for all applied water-soluble polymers when solution pH was 9 or larger and molar ratio of polymer to Cr(VI) was 10:1 (Sanchez and Rivas, 2011). Doganay *et al.* (2010) have used polyglycidyl methacrylate modified with hydroxyethylamino glycerol and poly(*N*,*N*-diallyl morpholinium bromide) for removal of chromate.

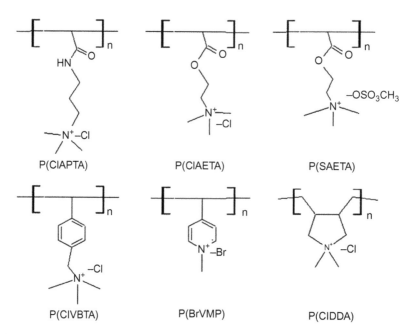

Figure 20.1.   Chemical structure of polyelectrolytes used for PEUF separation of As(V) (reproduced with
permission from Rivas *et al.*, 2011, copyright of Elsevier).

20.2.2   *Micellar-enhanced ultrafiltration (MEUF)*

The sizes of binding agents in MEUF are of order of several nanometers and are similar to PEUF
case. In micellar-enhanced ultrafiltration process, some specific surfactants are added at a concentration higher than their critical micelle concentrations. That allows them to create self-organized
aggregates called *micelles* to which the target ions are bound. It is accepted that ionic species are
adsorbed on the oppositely charged micelles via electrostatic interactions. Finally, the micellar
complexes are separated by UF membranes. The MEUF is counted as a low energy consuming
process with other benefits such as small footprint of the unit plant, no phase change during
the process, low operation pressure and high permeate flux. To counterbalance these plusses,
some drawbacks of MEUF have to be shown. Among them, the possibility to contaminate the
permeate with surfactant molecules and membrane fouling are the most important disadvantages.
To overcome the problem of permeate contamination, appropriate selection of surfactant as well
as micellar reinforcement by functional polymers (Bryjak *et al.*, 2010) could be applied. The
following process parameters have to be taken into consideration in the design of MEUF separation: transmembrane pressure, surfactant concentration, temperature, salt concentration, feed
flow rate, residence time, and type of UF membrane (Mungray *et al.*, 2013). When membrane
and micelle have the same net charge, the repulsion forces do not allow deposition of a fouling
layer on the membrane surface (Pozniak *et al.*, 2006).

20.2.2.1   *Removal of As(V) by MEUF*

The most frequently used surfactants for As removal are cetylpyridinum chloride, cetyl tri-methyl
ammonium bromide or octa-decyl amine acetate. The percent rejection of As is reported as 80–
98% (Beolchini *et al.*, 2007; Iqbal *et al.*, 2007; Mohsen-Nia *et al.*, 2007; Sadaoui *et al.*, 1989).
The studies showed the possibility of using MEUF process with highly permeable membranes
and low surfactant concentration. Hence, low membrane area and low surfactant consumption
improved overall process economics.

20.2.2.2 Removal of Cr(VI) by MEUF

The cationic surfactants used mainly for MEUF removal of Cr(VI) are cetyltrimetylammonium-bromide, cetylpyridinum chloride, di-decyl di-methyl ammonium bromide (Danis and Keskinler, 2009; Kamble and Marathe, 2005; Liu, 2004; Zeng *et al.*, 2008). The conducted studies were focused on the effect of process parameters on the separation efficiency with special attention paid to membrane fouling. It was noted that the rejection of Cr was 92–99%.

20.2.3 Suspension-enhanced microfiltration (SEMF)

This membrane-enhanced filtration system employs binding agents in the form of polymer particles that offer versatile properties. They can appear as hydrophobic sorbents, ion-exchange resins, chelating sorbents, carriers for solvent-impregnated resins or molecularly imprinted polymer sorbents. The functionalization of polymer matrix is a well-known process and can be easy adapted to fine particles. However, it is not possible to find the fine particles on the market. On the other hand, the monodisperse small polymer particles were prepared at laboratory scale following several different routes (Bryjak and Wolska, 2011; Samatya *et al.*, 2010; 2012). Selective removal of trace amounts of harmful metals in the presence of interfering ions such as Ca, Na, and Mg is a difficult task when commercially available ion-exchange resins are used. Thus, chelating polymers are produced by incorporating specific organic ligands into the polymer matrix to improve the selectivity of ion-exchange resins towards target ions (Sengupta and Sengupta, 2002). Slow kinetics of sorption can be controlled by selecting morphology of particles, charge density, nature of polymer used or its hydrophobicity. However, the kinetic characteristics do not decelerate the growing interest in the use of ion-exchange materials in SEMF systems. Abdulgader *et al.* (2013) rationalized that fact by enhancement of filtration performances of the system; possibility to prepare a high quality water; significant decrease of discharge waste streams and reduction of cost.

The SEMF process, evaluated by Kabay and Bryjak (2013), allows correlating the process efficiency with some process parameters. The evaluation was based on the following assumptions:

- The system is fed with a stream J_f of the target species at concentration of C_f;
- Suspension of sorbent is added to the system with volume flux of J_s^i, sorbent concentration, X^i, that is loaded with q^i of target species;
- The system outlets are the stream of permeate (flux J_p and species concentration of C_p) and the stream of loaded suspension (J_s^o, X^o, and q^o, respectively).

For mass balance of target species and for the steady state regime, $J_f = J_p$, $X^i = X^o$ and $J_s^o = J_s^i$, it is possible to relate concentration of species in permeate with process parameters $C_p = f(C_f, q^o, X^i, J_s^i, J_f)$ according to the following formula:

$$C_p = C_f - \frac{J_s^i}{J_f} X^i (q^o - q^i) \tag{20.1}$$

Hence, to improve efficiency of separation (or to reduce C_p), it is profitable to increase the stream of suspension and concentration of sorbent. It is profitable also to use sorbent with high uptake (q^o) that is fully regenerated (q^i as low as possible).

20.2.3.1 Removal of Cr(VI) and As(V) by SEMF

According to our best knowledge there is no report on the use of SEMF process for removal of neither As nor Cr pollutants. Of course, some specific sorbents for these species have been prepared recently, but they have not been applied for the SEMF hybrid system. As an example, synthesis of aminated poly(glycidylmethacrylate) microspheres (Li *et al.*, 2012) or poly(aniline) modified silica particles (Karthik and Meenakshi, 2014) should be considered. The sorbents carrying *N*-methyl-D-glucamine ligands were used for separation of Cr(VI) by Santander *et al.* (2014) and the obtained results can be good references for finding new binding agents in SEMF.

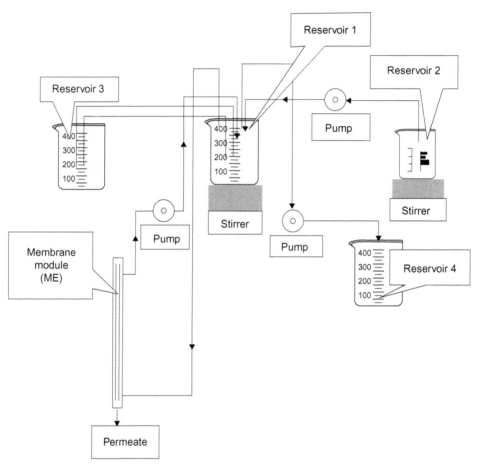

Figure 20.2. Set-up of the SEMF system (reproduced with permission from Wolska and Bryjak (2014), copyright of Elsevier).

In the case of As(V) removal, it is possible to find a lot of data for preparation of specific sorbents: polymer reinforced nanoparticles (Badruddoza at al., 2013; Cumbal *et al.*, 2003), activated carbon (Zhu *et al.*, 2009) or titanium dioxide (Guana *et al.*, 2012). Thus, there is the chance that fine particles of sorbent will available soon.

The correctness of $C_p = f(C_f, q^o, X^i, J_s^i, J_f)$ equation discussed above for removal of Cr(VI) ions by fine particles was evaluated in the frame of CHILTURPOL 2 project. The core structure of sorbent was formed by fine, average diameter 36 μm, particles of Dowex XUS 43594.00 resin that were coated with polyethyleneimine (MW = 60 kDa) and crosslinked with glutaraldehyde. The system employed for this study was shown in Figure 20.2.

The hybrid system was loaded with 0.25, 0.50, 0.75, 1.0 and 1.5 g L^{-1} suspension of sorbent and tested for 3 h period. Feed concentration was kept at 0.51 mmol L^{-1} of Cr(VI) and permeate flux at 0.67 mL min^{-1}. To the system 0.17 mL min^{-1} of fresh suspension was added. The Cr(VI) concentrations in the permeate, given as mmol L^{-1}, are shown in Table 20.3.

When concentration of sorbent increases, the permeate contains less and less harmful Cr(VI). To illustrate that phenomenon, it is enough to compare the concentration of chromate in permeate after 3 h of the process. Hence, a significant reduction of Cr(VI) concentration can be noted when

Table 20.3.   Cr(VI) concentration in permeate in the runs of SEMF system loaded with different amounts of sorbent.

Time [min]		0	30	60	90	120	150	180
Amount of sorbent	$0.00\,g\,L^{-1}$	0.51	0.44	0.44	0.43	0.42	0.41	0.40
	$0.25\,g\,L^{-1}$	0.51	0.40	0.33	0.32	0.32	0.32	0.32
	$0.50\,g\,L^{-1}$	0.51	0.36	0.29	0.280	0.28	0.27	0.27
	$0.75\,g\,L^{-1}$	0.51	0.26	0.25	0.24	0.23	0.23	0.23
	$1.00\,g\,L^{-1}$	0.51	0.23	0.22	0.22	0.20	0.20	0.21
	$1.50\,g\,L^{-1}$	0.51	0.20	0.19	0.18	0.18	0.18	0.18

$1.5\,g\,L^{-1}$ suspension was added. That shows the unique feature of the evaluated hybrid system with a very fast reduction of Cr(VI) ions using relatively small amount of sorbent.

20.3   CONCLUSIONS

Some studies carried out on application of submerged membrane systems for removal of boron (Blahusiak and Schlosser, 2009; Blahusiak et al., 2009; Guler et al., 2011) have exhibited that SEMF method may be considered as an efficient alternative for separations. The particles of sorbent did not attach to the membrane surface due to a running process under the sub-critical flux conditions and using air bubbles to scrape the membrane surface gently. Hence, the use of constructions similar to the ones applied in wastewater treatment plants could be of great interest for process engineers involved in removal of Cr(VI) and As(V). To implement such systems for removal of both ions two problems need to be solved first: to find fine sorbents, not too expensive to be applied in a large scale, and to establish a cheap regeneration method. These elements form the bottleneck of the SEMF process (Abdulgader et al., 2013).

ACKNOWLEDGEMENT

The authors acknowledge the CHILTURPOL2 grant (MCA, 7FP Project no. 269153) for the support.

REFERENCES

Abdulgader, H.A., Kochkodan, V. & Hilal, N. (2013) Hybrid ion exchange-pressure driven membrane processes in water treatment: a review. *Separation and Purification Technology*, 116, 253–264.

Badruddoza, A.Z.M., Shawon, Z.B.Z., Rahman, M.T., Hao, K.W., Hidajat, K. & Uddin, M.S. (2013) Ionically modified magnetic nanomaterials for arsenic and chromium removal from water. *Chemical Engineering Journal*, 225, 607–615.

Beolchini, F., Pagnanelli, F., De Michelis, I. & Veglio, F. (2007) Treatment of concentrated arsenic(V) solutions by micellar enhanced ultrafiltration with high molecular weight cut-off membrane. *Journal of Hazardous Materials*, 148, 116–121.

Blahusiak, M. & Schlosser, S. (2009) Simulation of the adsorption-microfiltration process for boron removal from RO permeate. *Desalination*, 241, 156–166.

Blahusiak, M., Onderkova, B., Schlosser, S. & Annus, J. (2009) Microfiltration of microparticulate boron adsorbent suspensions in submerged hollow fibre and capillary modules. *Desalination*, 241, 138–147.

Bryjak, M., Duraj, I. & Pozniak, G. (2010) Colloid-enhanced ultrafiltration in removal of trace amounts of borates from water. *Environmental Geochemistry and Health*, 32, 275–277.

Canizares, P., Perez, A., Llanos, J. & Rubio, G. (2008) Preliminary design and optimization of PEUF process for Cr(VI) removal. *Desalination*, 223, 229–237.

Cumbal, L., Greenleaf, J., Leun, D. & SenGupta, A.K. (2003) Polymer supported inorganic nanoparticles: characterization and environmental applications. *Reactive and Functional Polymer*, 54, 167–180.

Danis, U. & Keskinler, B. (2009) Chromate removal from wastewater using micellar enhanced crossflow filtration: effect of transmembrane pressure and crossflow velocity. *Desalination*, 249, 1356–1364.

Doganay, C.D., Ozbelgea, H.O., Yilmaz, L. & BiCcak, N. (2006) Removal and recovery of metal anions via functional polymer based PEUF. *Desalination*, 200, 286–287

Guana, X., Dub, J., Meng, X., Sun, Y., Sun, B. & Hu, Q. (2009) Application of titanium dioxide in arsenic removal from water: a review. *Journal of Hazardous Materials*, 172, 1591–1596.

Guler, E., Kabay, N., Yuksel, M., Yigit, N.O., Kitis, M. & Bryjak, M. (2011) Integrated solution for boron removal from seawater using RO process and sorption-membrane filtration hybrid method. *Journal of Membrane Science*, 375, 249–257.

Iqbal, J., Kim, H., Yang, J., Baek, K. & Yang, J. (2007) Removal of arsenic from groundwater by micellar-enhanced ultrafiltration (MEUF). *Chemosphere*, 66, 970–977.

Jain, C.K. & Singh, R.D. (2012) Technological options for the removal of arsenic with special reference to South East Asia. *Journal of Environmental Management*, 107, 1–18.

Jana, S., Saikia, A., Purkait, M.K. & Mohanty, K. (2011) Chitosan based ceramic ultrafiltration membrane: preparation, characterization and application to remove Hg(II) and As(III) using polymer enhanced ultrafiltration. *Chemical Engineering Journal*, 170, 209–219.

Kabay, N. & Bryjak, M. (2013) Hybrid processes combining sorption and membrane filtration In: Hoek, E.M.V. & Tarabara, V.V. (eds.) *Encyclopedia of membrane science and technology*. Wiley & Sons, Chichester, UK.

Kamble, S.B. & Marathe, K.V. (2005) Membrane characteristics and fouling study in MEUF for the removal of chromate anions from aqueous streams. *Separation Science and Technology*, 40, 3051–3070.

Karthik, R. & Meenakshi, S. (2014) Removal of hexavalent chromium ions using polyaniline/silica gel composite. *Journal of Water Process Engineering*, 1, 37–45.

Katsoyiannis, I.A. & Zouboulis, A.I. (2006) Use of iron- and manganese-oxidizing bacteria for the combined removal of iron, manganese and arsenic from contaminated groundwater. *Water Quality Research Journal of Canada*, 41, 117–129.

Lazaridis, N.K., Matis, K.A. & Webb, M. (2001) Flotation of metal-loaded clay anion exchangers. Part I: The case of chromate. *Chemosphere*, 42, 373–378.

Li, P., Yang, L., He, X., Wang, J., Kong, P., Xing, H. & Liu, H. (2012) Synthesis of PGMA microspheres with amino groups for high-capacity adsorption of Cr(VI) by cerium initiated graft polymerization. *Chinese Journal of Chemical Engineering*, 20, 95–104.

Liu, C., Li, C. & Lin, C. (2004) Micellar-enhanced ultrafiltration process (MEUF) for removing copper from synthetic wastewater containing ligands. *Chemosphere*, 57, 629–634.

Llanos, J., Perez, A. & Canizares, P. (2008) Copper recovery by polymer enhanced ultrafiltration (PEUF) and electrochemical regeneration. *Journal of Membrane Science*, 323, 28–36.

Malaviya, P. & Singh, A. (2011) Physicochemical technologies for remediation of chromium-containing waters and wastewaters. *Critical Reviews in Environmental Science & Technology*, 41, 1111–1172.

Mohsen-Nia, M., Montazeri, P. & Modarress, H. (2007) Removal of Cu^{2+} and Ni^{2+} from wastewater with a chelating agent and reverse osmosis processes. *Desalination*, 217, 276–281.

Mukherjee, K., Saha, R., Ghosh, A. & Saha, B. (2013) Chromium removal technologies. *Research on Chemical Intermediaries*, 39, 2267–2286.

Mungray, A.A., Kulkarni, S.V. & Mungray, A.K. (2012) Removal of heavy metals from wastewater using micellar enhanced ultrafiltration technique: a review. *Central European Journal of Chemistry*, 10 (1), 27–46.

Palencia, M., Rivas, B.L., Pereira, E.D., Hernandez, A. & Pradanos, P. (2009a) Study of polymer-metal ion-membrane interactions in liquid-phase polymer-based retention (LPR) by continuous diafiltration. *Journal of Membrane Science*, 336, 128–139.

Palencia, M., Rivas, B.L. & Pereira, E.D. (2009b) Metal ion recovery by polymer enhanced ultrafiltration using poly(vinyl sulfonic acid). Fouling description and membrane-metal ion interaction. *Journal of Membrane Science*, 345, 191–200.

Pozniak, G., Gancarz, I. & Tylus, W. (2006) Modified poly(phenylene oxide) membranes in ultrafiltration and micellar-enhanced ultrafiltration of organic compounds. *Desalination*, 198, 215–224.

Rengaraj, S., Yeon, K. & Moon, S. (2001) Removal of chromium from water and wastewater by ion exchange resins. *Journal of Hazardous Materials*, 87, 273–287.

Rivas, B.L. & Aguirre, M.C. (2007) Arsenite retention properties of watersoluble metal-polymers. *Journal of Applied Polymer Science*, 106, 1889–1894.

Rivas, B.L., Aguirre, M.C. & Pereira, E. (2007a) Cationic water-soluble polymers with the ability to remove arsenate through ultrafiltration technique. *Journal of Applied Polymer Science*, 106, 89–94.

Rivas, B.L., Aguirre, M.C., Pereira, E., Moutet, J.C. & Saint-Aman, E. (2007b) Capability of cationic water-soluble polymers in conjunction with ultrafiltration membranes to remove arsenate ions. *Polymer Engineering Science*, 47, 1256–1261.

Rivas, B.L., Pereira, E.D., Palencia, M. & Sánchez, J. (2011) Water-soluble functional polymers in conjunction with membranes to remove pollutant ions from aqueous solutions. *Progress in Polymer Science*, 36, 294–322.

Sabate, J., Pujola, M. & Llorens, J. (2006) Simulation of continuous metal separation process by polymer enhanced ultrafiltration. *Journal of Membrane Science*, 268, 37–47.

Sadaoui, Z., Azong, C., Charbit, G. & Charbit, F. (1998) Surfactants for separation processes: enhanced ultrafiltration. *Journal of Environmental Engineering*, 124, 695–700.

Samatya, S., Orhan, E., Kabay, N. & Tuncel, A. (2010) Comparative boron removal performance of monodisperse porous particles with molecular brushes via "click chemistry" and direct coupling. *Colloids and Surfaces* A: *Physicochemical Engineering Aspects*, 372, 102–106.

Samatya, S., Kabay, N. & Tuncel, A. (2012) Monodisperse porous *N*-methyl-D-glucamine functionalized poly(vinylbenzyl chloride-co-divinylbenzene) beads as boron selective sorbent. *Journal of Applied Polymer Science*, 126 (4), 1475–1483.

Sánchez, J. & Rivas, B.L. (2011) Cationic hydrophilic polymers coupled to ultrafiltration membranes to remove chromium (VI) from aqueous solution. *Desalination*, 279, 338–343.

Sánchez, J., Rivas, B.L., Bastrzyk, A., Bryjak, M. & Kabay, N. (2013) Removal of As(V) using liquid-phase-polymer-based retention (LPR) technique with regenerated cellulose membrane as a filter. *Polymer Bulletin*, 70, 2633–2644.

Santander, P., Rivas, B.L., Urbano, B., Leiton, L., Yilmaz-Ipek, I., Yuksel, M., Kabay, N. & Bryjak, M. (2014) Removal of Cr(VI) by a chelating resin containing *N*-methyl-D-glucamine. *Polymer Bulletin*, 71, 1813–1825.

Sengupta, S. & Sengupta, A.K. (2002) Trace heavy metal separation by chelating ion exchangers. In: Sengupta, A.K. (ed.). *Environmental separation of heavy metals – engineering processes*. CRC Press LLC, Boca Raton, FL. pp. 45–96.

Suk, D.E. & Matsuura, T. (2006) Membrane-based hybrid processes: a review. *Separation Science and Technology*, 41, 595–626.

Wolska, J. & Bryjak, M. (2014) Removal of bisphenol A from aqueous solutions by molecularly imprinted polymers. *Separation Science and Technology*, 49, 1643–1653.

Zeng, G.M., Xu, K., Huang, J.H., Li, X., Fang, Y.Y. & Qu, Y.H. (2008) Micellar enhanced ultrafiltration of phenol in synthetic wastewater using polysulfone spiral membrane. *Journal of Membrane Science*, 310, 149–160.

Zhu, H., Jia, Y., Wu, X. & Wang, H. (2009) Removal of arsenic from water by supported nano zero-valent iron on activated carbon. *Journal of Hazardous Materials*, 172, 1591–1596.

CHAPTER 21

Removal of arsenic and chromium ions by dialysis and electrodialysis

Stanisław Koter

21.1 INTRODUCTION

There are many techniques used for removal of As and Cr. To remove As from water many varied treatment methods have been developed, like ion exchange, activated alumina, lime softening, oxidation/filtration, co-precipitation, coagulation/microfiltration, adsorption, membrane separation (e.g. reverse osmosis, NF, electrodialysis) (Jain and Singh, 2012; Katsoyiannis and Zouboulis, 2005; Loo and Wen, 2005; O'Connor, 2002; Pickard and Bari, 2004; Wang and Zhao, 2008). Generally, it should be stated that only a combination of different separation methods can result in an efficient removal/recovery treatment.

In this chapter, the application of dialytic and electrodialytic methods in As and Cr removal is reviewed. These methods are based on the semipermeable anion- and cation-exchange membranes which are permeable to anions, cations, respectively. Therefore, the As and Cr species in the solutions should be in an ionic form. This aspect is briefly discussed in Section 1.2 (Chapter 1) where the basic speciation of As and Cr (dependence on pH) is presented. The literature on the removal of As by Donnan dialysis (DD) is quite abundant. It is divided into the following subjects: modeling of As removal, test of membranes, influence of the presence of other ions and of pH, coupling of DD with coagulation, and household Donnan dialyzers intended for rural areas. Only a few papers have been published on Cr removal. Here two aspects are distinguished: the influence of metal cations on Cr(III) transport and simultaneous recovery of Cr(III) and Cr(VI) from their mixture. The literature on the electrodialytic removal of As and Cr refers not only to aqueous solutions but also to contaminated solid wastes, like soil, ash, and wood. A few preliminary papers on the Cr(VI) removal by the electrodeionization method are also shortly reviewed.

21.2 DONNAN DIALYSIS

Donnan dialysis (DD) is a membrane process used for concentrating valuable ions or for removing undesirable ions from dilute solutions to a high degree (Pyrzynska, 2006; Sionkowski and Wódzki, 1995; Wódzki et al., 1996). It is a slow process, because the driving force is the concentration difference across the membrane. However, this is also a reason for its simplicity – no additional devices like DC power supplies (ED) or high pressure pumps (NF, RO) with their instrumentation are needed. Apart from that, in the diffusional processes the tendency to membrane fouling is minimized. Thus, the process does not need so rigorous control as in the case of ED or filtration techniques.

21.2.1 Basic principles

In DD an ion-exchange membrane separates two solutions – a feed solution containing an ion to be removed and a strip solution into which the ion will be moved. If the ion to be concentrated/removed is positively charged, then a cation-exchange membrane (CM) is used. If it is anionic,

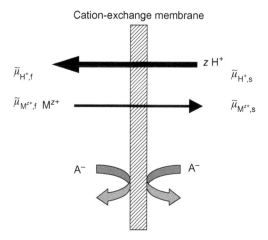

Figure 21.1.   Donnan dialysis with a cation-exchange membrane; the transport of cations M^{z+} is forced by the transport of H^+ against the concentration difference of M^{z+}; anions A^- do not cross the membrane.

then an anion-exchange membrane is used. The principle of DD operation is illustrated in Figure 21.1.

On the left side of CM there is a dilute solution of the M^{z+} cations which should be removed (feed solution, donor phase), on the right side – a concentrated solution of strong acid (receiving solution, acceptor phase, strip phase). CM is ideally selective, i.e. anions cannot pass through the membrane. Because of different concentrations interdiffusion of M^{z+} and H^+ takes place until their electrochemical potentials on both sides of CM become equal:

$$\tilde{\mu}_{i,f} = \tilde{\mu}_{i,s} \quad i = M^{z+}, H^+ \tag{21.1}$$

Here, H^+ is called a stripping or driving ion, because it forces the permeation of M^{z+} even against the M^{z+} concentration difference across the membrane. Substituting the electrochemical potential of ith ion:

$$\tilde{\mu}_i = \mu_i^\ominus(T, p) + RT \ln a_i + z_i F \varphi \tag{21.2}$$

into Equation (21.1) the electric potential difference, $\varphi_s - \varphi_f$ (isobaric conditions) is obtained:

$$\varphi_s - \varphi_f = \frac{RT}{F} \ln \left(\frac{a_{i,f}}{a_{i,s}} \right)^{1/z_i} = \frac{RT}{F} \ln \left(\frac{a_{k,f}}{a_{k,s}} \right)^{1/z_k} \tag{21.3}$$

In the above equations a_i, z_i, φ, F, R denote activity, charge number of ith ion, electric potential, absolute temperature, Faraday and gas constants, respectively; $\mu_i^\ominus(T, p)$ is the part of chemical potential dependent only on temperature, T, and pressure, p. From Equation (21.3) one gets the basic equation of DD which sets the theromodynamic limit of removal/recovery of ions 2, 3, ..., n:

$$\left(\frac{a_{1,s}}{a_{1,f}} \right)^{1/z_1} = \left(\frac{a_{2,s}}{a_{2,f}} \right)^{1/z_2} = \cdots = \left(\frac{a_{n,s}}{a_{n,f}} \right)^{1/z_n} \tag{21.4}$$

The activity of ith ion, a_i, is related to its concentration, c_i, by $a_i = c_i y_i / c^\ominus$, where $c^\ominus = 1 \, \text{mol dm}^{-3}$, y_i is the activity coefficient of ith ion, n is the number of ions of the same sign in the solution. According to Equation (21.4) the highest concentration ratio of the ion to be removed is obtained for the monovalent driving ion. E.g., having initially on the left side 10 L of 0.001 M M^{2+} solution and on the right side 1 L of 0.1 M HCl, at equilibrium we get on the left

6×10^{-6} M M^{2+} and 0.002 M HCl, whereas on the right 0.00994 M M^{2+} and 0.08 M HCl. It was calculated from Equation (21.4) simplified to the concentration ratio:

$$\frac{c_{1,s,0} - 2x/V_s}{c_{1,f,0} + 2x/V_f} = \left(\frac{c_{2,s,0} + x/V_s}{c_{2,f,0} - x/V_f}\right)^{1/2} \tag{21.5}$$

where $c_{i,0}$ is the initial concentration of ith ion, x is the number of moles of M^{2+} transferred through the membrane, which is calculated from this equation.

For monovalent cations DD is much less effective – the concentration of M^+ decreases only to 90×10^{-6} M. These results are ideal ones – in reality the effect will not be so good because of nonideal membrane selectivity (diffusion of electrolytes, osmosis); also the process time is usually shorter than that needed to reach the equilibrium. Similar considerations are valid for anions and an anion-exchange membrane.

Adding to a receiving solution a chemical reagent which selectively binds a target ion it is possible to increase the permeation rate of that ion (its concentration difference is higher because the ion concentration in the receiving solution is always very low – depending on the equilibrium constant of the reaction between that ion and the chemical agent) and to obtain a higher degree of recovery. The condition is that the reagent and the product of reagent-ion reaction should not cross the membrane to the feed side (easily). The Donnan dialysis with the help of such a chemical agent is called chemically facilitated Donnan dialysis (CFDD) (Castillo *et al.*, 2002; Lin and Burgess, 1994). The use of complexing agents in DD was briefly reviewed by Sionkowski and Wódzki (1995). Regarding As, it can be also coagulated (Oehmen *et al.*, 2011) as will be described later. Instead of chemicals the use of sulfate-reducing bacteria in the strip compartment was proposed by Oehmen *et al.* (2006). The role of bacteria is to convert arsenate into As_2S_3 which is recovered as a precipitate.

The fluxes of ions, J_i, in DD are coupled by the zero-current condition:

$$\sum_{i=1}^{n} z_i J_i = 0 \tag{21.6}$$

In the case of ideally selective ion-exchange membranes (AM or CM), across which only counter-ions can permeate, the electroneutrality condition inside the membrane (no sorption of coions) is:

$$\sum_{i} |z_i| \bar{c}_i = |z_m| \bar{c}_m \tag{21.7}$$

where $\bar{c}_m$ is concentration of ion-exchange groups, z_m – their charge number ($z_m > 0$, $z_m < 0$ for anion-, cation-exchange membrane, respectively). Taking the Nernst-Planck equation for the description of ion transport inside the membrane pores:

$$J_i = -\bar{D}_i \left(\frac{d\bar{c}_i}{dx} + z_i \bar{c}_i \frac{F}{RT} \frac{d\bar{\varphi}}{dx}\right) \tag{21.8}$$

we get the general expression for the flux of ion 2 from the feed solution to the receiving one:

$$J_2 = \frac{-\bar{D}_2}{\sum_{i=1}^{n} \bar{D}_i z_i^2 \bar{c}_i} \left(\left(\bar{D}_1 z_2^2 \bar{c}_2 + \sum_{\substack{i=1, \\ i \neq 2}}^{n} \bar{D}_i z_i^2 \bar{c}_i\right) \frac{d\bar{c}_2}{dx} + z_2 \bar{c}_2 \sum_{i=3}^{n} (\bar{D}_1 - \bar{D}_i) z_i \frac{d\bar{c}_i}{dx}\right) \tag{21.9}$$

Here, the stripping ion is denoted by subscript 1, other ions (of the same sign) are numbered $3, \ldots, n$. Generally, according to the electroneutrality condition and the Donnan equilibrium (equality of electrochemical potentials of ions on the boundary solution|membrane)

$$\left(\frac{\bar{a}_1}{a_1}\right)^{1/z_1} = \left(\frac{\bar{a}_2}{a_2}\right)^{1/z_2} = \cdots = \left(\frac{\bar{a}_n}{a_n}\right)^{1/z_n} \tag{21.10}$$

where $\bar{a}_i$ is the activity of ion i inside the membrane, the presence of other ions will decrease the concentration of ion 2 and thus will diminish its flow. From Equation (21.9) it also results that if the concentration gradients of ion 2 and the ion i are of the opposite sign and $\overline{D}_1 > \overline{D}_i$, or vice versa, then the term referring to ion i will even more decrease the transport of ion 2. However, the thermodynamic limit of ion 2 recovery will remain the same.

For two ions Equation (21.9) reduces to:

$$J_2 = -\left(\frac{\bar{c}_1 + k_z^2\bar{c}_2}{\bar{c}_1 + k_D k_z^2\bar{c}_2}\right)\overline{D}_2\frac{d\bar{c}_2}{dx} \tag{21.11}$$

where $k_D = \overline{D}_2/\overline{D}_1$, $k_z = z_2/z_1$, and $\bar{c}_1$ and $\bar{c}_2$ are related by Equation (21.7). It can be noticed that for $\bar{c}_2 << \bar{c}_1$ Equation (21.11) takes the form of Fick's law.

At steady state ($J_2 =$ const.), after the separation of variables x and $\bar{c}_2$, Equation (21.11) can be integrated from the left to the right side of membrane; the final formula for J_2 is:

$$\begin{aligned}J_2 = {} & \frac{\overline{D}_2}{l_p(1 - k_D k_z)^2}\Big((1 - (1 + k_D)k_z + k_D k_z^2)(\bar{c}_{2,f} - \bar{c}_{2,s}) \\ & + \frac{z_m\bar{c}_m}{z_1}(1 - k_D)\ln\left(\frac{\bar{c}_{1,f} + k_D k_z^2\bar{c}_{2,f}}{\bar{c}_{1,s} + k_D k_z^2\bar{c}_{2,s}}\right)\Big)\end{aligned} \tag{21.12}$$

where l_p is the length of membrane pore. For a monovalent driving ion ($|z_1| = 1$) k_z becomes $|z_2|$. J_2 should be multiplied by the surface fraction of membrane pores to obtain the flux per unit membrane area. Taking this equation and the Donnan equation (21.10) with the activities replaced by concentrations one can check that only a monovalent stripping ion ensures much higher flux of ion 2 than a divalent one. In the case of As(V) it was experimentally verified that when Cl^- as the driving ion is used, the arsenate anion transport is much higher than when SO_4^{2-} is the driving ion (Zhao et al., 2012). One can also check that the higher the charge number of ion 2, the lower is its flux (at the same k_D and external concentrations).

21.2.2   Modeling of As removal

A simple model for As(V) removal, based on the Nernst-Planck equation, is proposed by Zhao et al. (2010b). It is developed under the following assumptions:

• the diffusion boundary layers on both sides of a membrane are neglected;
• the As concentration is much smaller than that of accompanying electrolyte (NaCl) – in that case the interdiffusion coefficient $\overline{D}_{As\text{-}Cl}$ reduces to the diffusion coefficient of arsenate $\overline{D}_{As}$;
• the ion flux is the same along the flow path through the membrane;
• the initial volumes of the feed and stripping solutions are the same – thus the sum of the As(V) concentrations in the feed and stripping phases after reaching the steady state is constant: $c_{As,s} + c_{As,f} = c_{As,t_0}$.

According to the above assumptions the As flux through the membrane takes the form:

$$J_{As} = -\overline{D}_{As}\frac{K_s c_{As,s} - K_f c_{As,f}}{l_m} \tag{21.13}$$

where $c_{As,f}$, $c_{As,s}$ are the concentrations of As in the feed, strip solution, respectively, l_m is the membrane thickness, K_f and K_s are the partition coefficients of As on the feed and strip side of the membrane, respectively: $K = \bar{c}_{As}^{(m)}/c_{As}$, $\bar{c}_{As}^{(m)}$ is the As concentration per unit volume of membrane. Substituting Equation (21.13) into:

$$J_{As} = -\frac{V_f}{S_m}\frac{dc_{As,f}}{dt} \tag{21.14}$$

then separating variables and performing integration the final model equation is obtained:

$$c_{As,f,t} = \frac{K_s}{K_f + K_s}\left(c_{As,t_0} - \left(c_{As,t_0} - \left(\frac{K_f}{K_s} + 1\right)c_{As,f,t_0}\right)\exp\left(-\frac{(K_f + K_s)\overline{D}_{As}S}{V_f l_m}(t - t_0)\right)\right)$$

(21.15)

where $c_{As,f,t}$ is the concentration of As in the feed solution at time t, t_0 is the time at which steady state is reached. The model was checked using 0.013 M Na$_2$HAsO$_4$ + 0.01 M NaCl and 0.10 M NaCl as feed and strip phase, respectively. The pH of both solutions was the same and adjusted to the values 4.5, 7.0 and 9.2. At these pH values As(V) is in the form H$_2$AsO$_4^-$, H$_2$AsO$_4^-$/HAsO$_4^{2-}$, and HAsO$_4^{2-}$, respectively (see Section 1.2.1). The As concentrations calculated from the model were slightly higher than the experimental ones. The probable reason was the accumulation of As(V) inside the membrane. Unfortunately, the authors performed calculations assuming $t_0 = 0$, although they experimentally found that t_0 was quite high (ca. 2–8% of the experiment time).

More advanced models of the DD process, taking into account also the boundary diffusion layers at the membrane surfaces, were proposed by Ktari *et al.* (1993), Nwal *et al.* (2004) or Prado-Rubio *et al.* (2010). Analogous models can be developed for As or Cr removal. However, there is no space to present them here.

21.2.3 Arsenic removal

21.2.3.1 Test of membranes
A proper choice of membrane in the membrane techniques is of prime importance. Tests of commercial membranes in the DD process of As removal were published by Oehmen *et al.* (2011) and Velizarov (2013). Three commercial anion-exchange membranes – AR204-UZRA (Ionics), FTAM (Fumatech) and PC-SA (PCA-Polymerchemie Altmeier GmbH) – were tested by Oehmen *et al.* (2011) using 250 µg L^{-1} As solution (prepared from Na$_2$HAsO$_4 \cdot$ 7H$_2$O) as a feed and 23.4 mg L^{-1} NaCl solution as the receiver solution. To prepare both solutions tap water was used. It turned out that the fluxes through these membranes depended on the form of As(V). When the dominant form of As(V) was divalent arsenate HAsO$_4^{2-}$ (pH of both solutions was 7.8–8) the highest flux – ca. 3.4 mg m^{-2} h^{-1} – was observed for AR204-UZRA, the lowest – ca. 1.9 mg m^{-2} h^{-1} – for PC-SA, although its thickness (0.11 mm) was five times smaller. At pH $= 5.3$–6.4 (the dominant form is H$_2$AsO$_4^-$) the As(V) fluxes were similar (the data were not shown). However, taking the thickness into account the H$_2$AsO$_4^-$ permeability through AR204-UZRA was five times higher than that through PC-SA. The reason for these differences was a higher water content of AR204-UZRA (58%) and higher IEC (2.8 meq g^{-1}) than those of PC-SA (31%, 1.1 meq g^{-1}). It should be noted that a high initial As flow through a membrane does not mean that the As recovery will be high. In another series of experiments it was found that the As recovery using Ionics AR204-UZRA was only 26% whereas using PC-SA – 94% (Velizarov 2013).

A series of various membranes: Ionac MA3475, Fumasep FTAM, Ionics AR204-UZRA, Ionics AR103-QDP, PCA membranes (PC-SA, PC 100D, PC Acid 60, PC Acid 100), and monovalent-anion permselective Neosepta ACS were tested by Velizarov (2013). The DD dialyzer was the same as used by Oehmen *et al.* (2011). A 0.1 mmol L^{-1} solution of arsenate (prepared from Na$_2$HAsO$_4 \cdot$ 7H$_2$O) and 10 mmol L^{-1} NaCl were used as the feed and stripping phases, respectively; the volumes of both solutions were the same. Contrary to Oehmen *et al.* (2011), Velizarov (2013) used deionized water as a solvent. The initial pH of the feed was ca. 6.7 which corresponds to ca. 72% of H$_2$AsO$_4^-$ and 28% of HAsO$_4^{2-}$; the initial pH of the strip solution was 5.9. After 24 h of batch process the feed pH decreased to 4.5–5.6, the strip pH increased to 6.0–6.8, depending on the membrane – at those pH values still the dominant form is monovalent H$_2$AsO$_4^-$. Similarly as in the previous work (Oehmen *et al.*, 2011), the removal rate of As(V) in that form was comparable for PC-SA and for AR204-UZRA, as was shown by Oehmen *et al.* (2011). A high flow through PC-SA was explained by its high porosity and a thin functional layer – 20 µm thick (Zhang *et al.*, 2010). Although the prevailing form in the external solutions was H$_2$AsO$_4^-$, the

As(V) flow through the monovalent-anion permselective membrane (Neosepta ACS) was the lowest one. This fact can be explained by the presence of highly cross-linked polymeric layer on the ACS surface which is responsible for the monovalent-anion selectivity. From the changes of the As concentration in the feed it results that for each membrane except ACS the removal exceeded 96%. However, the increase of As(V) concentration in the strip solution was significantly lower, indicating that some amount of As(V) remained inside the membranes (Velizarov, 2013). The highest ratio $c_{As,s}/c_{As,f,0}$ was obtained for PC-SA – 0.74, PC-100 – 0.72 at 6 h, then it decreased to 0.68, similar values were obtained for PC Acid 60, PC Acid 100, and Ionac MA3475 (Velizarov, 2013). For ACS $c_{As,s}$ was increasing with time which means that the maximum removal had not been reached yet.

21.2.3.2 Influence of other ions

In groundwater the As concentration is usually much lower than that of chloride, sulfate, bicarbonate, nitrate and other ions. Their influence on As removal was investigated by Zhao et al. (2010a) and Güell et al. (2011). Zhao carried out the experiments in a batch Donnan dialyzer equipped with the homogeneous anion-exchange membrane JAM (Huanyld, China). The feed solutions contained $500 \mu g L^{-1}$ As(V), $(10 - x)$ mmol L^{-1} NaCl, and x mmol L^{-1} of other salt: $x = 0.1$ (Na_2SiO_3), 2.5 ($NaNO_3$, Na_2SO_4, Na_2HPO_4, $NaHCO_3$). The stripping solution was 1.0 M NaCl. The pH of all the solutions was 7, adjusted by adding 0.1 M HCl or NaOH before the experiment. At that pH value As(V) is only in the ionic form ($H_2AsO_4^-$ and $HAsO_4^{2-}$). It was found that the As(V) removal rate from the feed is slower when a part of NaCl is substituted by an equimolar amount of other salt, especially of Na_2HPO_4. After 2 h of experiment the As(V) removal was ca. 43, 37, 28, 32% for Cl^-, NO_3^-, SO_4^{2-}, HPO_4^-, respectively; after 24 h (practically the end of process) it was 97, 94, 94, 85%. Except for SO_4^{2-}, this is in accordance with the measured uptake of As(V) into the membrane – in the presence of Cl^- the uptake was highest, in the presence of sulfates – lowest: $Cl^- > NO_3^- > HPO_4^- > SO_4^{2-}$. It is not clear why the flow of As(V) was lowest in the presence of phosphate ions. Probably in contact with these ions the anion-exchange membrane loses its selectivity. This is indicated by a low degree of removal of As(V), significantly lower than in the presence of other anions. The loss of selectivity of anion-exchange membranes was also observed in the electrodialytic experiments with a mixture of sulfuric and phosphoric acids (Koter and Kultys, 2010).

Contrary to the above mentioned anions the presence of hydrogen carbonate increased the As(V) flow. The reason was the increase of pH to ca. 7.7–8.2 due to the release of CO_2 which resulted from the pH adjustment (HCl was added). The increased pH resulted in the increased content of $HAsO_4^{2-}$ which, according to the Donnan equilibrium, is more sorbed than H_2AsO_4. Thus, the total flux of As(V) was increased. The presence of silicate, undissociated at pH 7 (the dissociation constant of the silicic acid is $pK_{a,1} = 9.9$), did not show any significant effect on the permeation of As(V).

The effect of anions: NO_3^-, SO_4^{2-}, HCO_3^-, $H_2PO_4^-$ on the As(V) transport through the anion-exchange membrane PC-SA (PCA) was studied by Güell et al. (2011). The concentration of the species $A^{z_A}/|z_A|$ in the feed for all those anions was the same – 2 mmol L^{-1}, whereas that of As(V) – 10 mg L^{-1} As(V) ($\approx$0.13 mmol L^{-1}); pH was 7, as in (Zhao et al., 2010a). The stripping phase was 0.1 M NaCl. After 24 h of the process practically 100% of As(V) was removed. Contrary to the results of Zhao et al. (2010a), no significant difference in the removal of As with and without anions was found. A probable reason for that discrepancy could be the 20 times lower concentration of As(V) and 10 times higher concentration of the stripping agent used by Zhao, at which the selectivity of anion-exchange membrane is lower.

21.2.3.3 Influence of pH

Studies on the influence of pH on the effectiveness of the DD removal of As were published by Güell et al. (2011). The chosen pH values were: 5, 7, 10, and 13. At these values As(V) is in the following forms: $H_2AsO_4^-$, $H_2AsO_4^-$ and $HAsO_4^{2-}$ (ratio ca. 1:1), $HAsO_4^{2-}$, AsO_4^{3-}, respectively). The highest As(V) permeability during the first 6 h of the process was obtained at pH $= 7$ – after

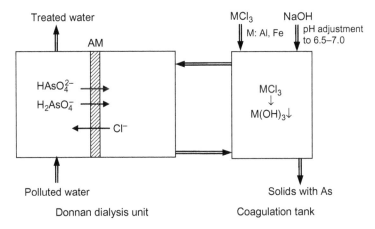

Figure 21.2. Coupling of DD with coagulation; AM – anion-exchange membrane, M – Al or Fe (Oehmen *et al.*, 2011).

6 h 89% of As(V) was removed from the feed side, whereas at pH $= 5$, 10 and 13 – 78, 42 and 12%, respectively (Güell *et al.*, 2011). After 24 h the highest removal, nearly 100%, was observed at pH $= 5$, whereas at pH $= 7$, 10 and 13 – 95, 92 and 30%, respectively. However, these data are not consistent with the data shown in Figure 3 in the same work (Güell *et al.*, 2011) where after 24 h at pH $= 7$ almost 100% of As(V) is removed and at pH $= 5$ – ca. 95%. The smallest removal value at pH $= 13$ (30%) seems to be surprising, because according to the Donnan equilibria, a better removal should be obtained for an anion of a higher absolute value of charge number, i.e. for AsO_4^{3-}. One can conclude that after 24 h the equilibrium was not reached, because the AsO_4^{3-} flux through the membrane is lower than that of mono- or divalent As(V) anions. A smaller flux results from the decreased AsO_4^{3-} mobility inside the membrane and it can be also deduced from Equation (21.12). It can be also supposed that the trivalent anion is more strongly bound with the positive fixed charges of the membrane, which makes the membrane less selective or even may reverse its sign.

Regarding As(III), the authors performed only one experiment at pH $= 7$. In neutral solution only uncharged H_3PO_3 is present which freely diffuses through the membrane. Indeed, even in the long-term experiment (160 h) the As(III) concentration in the strip solution did not exceed that in the feed (the volumes of both solutions were the same).

21.2.3.4 *Coupling of DD with coagulation*
To enhance As removal from drinking water and to avoid its contamination by coagulants and pH controlling reagents, the coupling of DD with the coagulation process was proposed by Oehmen *et al.* (2011). The system is shown in Figure 21.2.

Among three tested membranes (AR204-UZRA – Ionics, FTAM – Fumatech, and PC SA – PCA) the AR204-UZRA membrane of the highest divalent arsenate permeation was chosen for the long term experiments. The feed was a tap water spiked with $100\,\mu g\,L^{-1}$ of As(V) (added as $Na_2HAsO_4 \cdot 7H_2O$). This was continuously supplied to the dialyzer (single-pass mode). The receiver solution was a $10\,g\,L^{-1}$ (0.17 M) NaCl solution with the periodic addition of coagulant (0.1 g every 12 h during the first 21 days of operation). It was circulated through the chamber with the volume flow, F_v, per membrane area $F_v/A = 2.4\,L\,m^{-2}\,h^{-1}$. Two coagulants were tested – $AlCl_3$ and $FeCl_3$. It was found that during the $AlCl_3$ dosing period (first 21 days) and during the next 22–96 days of experiment without the $AlCl_3$ addition the As concentration in the effluent was below $10\,\mu g\,L^{-1}$. Regarding the coagulants, in the case of $AlCl_3$ the dependence of the As flow through the membrane on F_v/A was linear in the range 4.8–37 $L\,m^{-2}\,h^{-1}$ and was slightly lower than the flow needed to reach the $10\,\mu g\,L^{-1}$ limit. For $FeCl_3$ it was observed that at $F_v/A = 37\,L\,m^{-2}\,h^{-1}$

the As flux reached almost its maximum value, ca. $2 \, mg \, m^{-2} \, h^{-1}$, which was 31% lower than the flux in the presence of $AlCl_3$. Apart from that $AlCl_3$ was more effective than $FeCl_3$ because of a lower membrane scaling. This was explained by a formation of positively charged particles of precipitate, which were repelled from the AM surface, also positively charged.

Compared to the coagulation/filtration process the quality of the drinking water from the coupled DD-coagulation process is better because no secondary contamination of the water by coagulants and the pH-controlling agents occurs here.

21.2.3.5 Household Donnan dialyzer

In areas without any central system of water supply (e.g. rural environments), household devices for cleaning groundwater would be useful. For such purposes, a household Donnan dialyzer for removal of As(V) from groundwater was proposed and investigated by Zhao *et al.* (2012). As in their previous work (Zhao *et al.*, 2010), the dialyzer worked in the batch mode and it was equipped with a homogeneous strongly basic anion-exchange membrane. The treatment capacity was 35 L per batch. The solutions were stirred by an air pump. To prepare the feed and stripping solutions an As-free groundwater was used. The As(V) concentration in the feed solution was $500 \, \mu g \, L^{-1}$, the stripping solution contained 12 g of commercial table salt per liter (ca. 0.2 M NaCl). One run lasted 1 day, after each run the treated water was removed and a new portion of feed solution was added; the stripping solution remained the same. One day was enough to decrease the As concentration in the treated water below $50 \, \mu g \, L^{-1}$ (maximum allowable level). After 13 days the As concentration slightly exceeded $50 \, \mu g \, L^{-1}$ (Zhao *et al.*, 2012) and the As(V) concentration in the strip phase was ca. 120 times higher than in the feed. On 16th day of operation, the As(V) content in the feed water was decreased from 500 to $250 \, \mu g \, L^{-1}$ and during the next 9 days the As concentration in the treated water was still around $50 \, \mu g \, L^{-1}$. On 24th day, there was $9135 \, \mu g \, L^{-1}$ As(V) in the strip phase. During the process a small leakage of Na^+ ions to the feed was observed, whereas Cl^- concentration in the feed significantly increased due to the counter transport of sulfates and hydrogen carbonates from the feed to the stripping phase. However, the treated water fulfilled the drinking water standards. The content of chosen ions, including the As species, before/after treatment was (in mg L^{-1}, data taken from Table 3, Zhao *et al.*, 2012):

	As	Na^+	Cl^-	SO_4^{2-}	NO_3^-	HCO_3^-
before:	0.25–0.54	21–25	19–24	59–67	7–11	207–222
after:	0.013–0.065	44–55	183–276	0.6–8.5	0.5–6.0	12–126

The content of K^+, Mg^{+2}, Ca^{+2} was practically unchanged. During the experiments the effect of membrane fouling was not observed – the As(V) removal rates for the membrane after 26 batches and for the new one were practically the same. In those studies 770 L of purified water and 35 L of the stripping solution containing ca. $9 \, g \, L^{-1}$ As(V) were produced. The authors evaluated the operating cost to be 0.087 US$cent L^{-1}. They concluded that although the DD method of As removal was not so fast as other methods it could be used for the effective As removal from water in rural areas. However, still the question remains – how to utilize the stripping solution and how much it would cost.

21.2.4 Chromium removal

Only a few papers have been published on Cr removal by the Donnan dialysis method. In one of them the influence of the presence of metal cations of various charge numbers on the Cr(III) transport was investigated (Tor *et al.*, 2004b). For the experiments the membranes based on polysulfone with sulfonic groups (SA3S and SA3T) were chosen. The feed was a mixture of 0.005 M $CrCl_3$ and 0.005 M salt (NaCl, $ZnCl_2$ or $AlCl_3$), the pH of the feed solution was adjusted to 2.5, the strip phase was a HCl solution (0.05, 0.1 and 1 M); the volumes of both solutions were the same. At first, the influence of the HCl concentration on the Cr(III) removal rate was

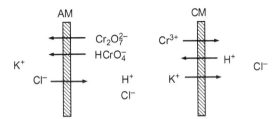

Figure 21.3. Simultaneous removal of Cr(III) and Cr(VI) by the DD method, as proposed by Tor *et al.* (2004a).

determined. It was found that although at 1 M HCl the highest initial Cr(III) flow was observed, with time the flow substantially decreased, as is indicated by the time dependence of the recovery fraction, RF (Tor *et al.*, 2004b). Unfortunately, the experiments were stopped long before the near equilibrium state was reached (RF < 40%). It should be expected that at higher HCl concentration the membrane selectivity is lower and, consequently, a lower value of RF will be obtained. In further experiments with additional salt a 0.1 M HCl solution was chosen as a stripping solution. It turned out that the presence of a metal cation decreased the Cr(III) permeation through the SA3S membrane in the order: Na^+ (16) > Zn^{2+} (29) > Al^{3+} (43), where in parentheses the percent decrease of the Cr(III) flow compared to that in the absence of salt is given. This observation is in accordance with Equation (21.10) which predicts the decrease of the Cr^{3+} sorption with the increase of the charge number of the accompanying cation.

In another paper (Sardohan *et al.*, 2010), the effectiveness of the plasma modified membranes (treated by electron cyclotron resonance plasma) in the Donnan dialysis of Cr(III) and Cr(VI) solutions was studied. Only a small increase of the recovery factor was found (5–10% compared to the unmodified membranes) during the initial stage of the process. Unfortunately, no longer experiments were performed. Therefore it is difficult to judge the usefulness of these membranes in the removal/recovery of Cr.

A simultaneous recovery of Cr(III) and Cr(VI) from their mixture containing other ions was investigated by Tor *et al.* (2004a). Apart from DD, electrodialytic experiments were also performed. In both types of experiments the same 3-compartment module was used (Fig. 21.3). The middle compartment was filled with a feed solution containing Cr(III) and Cr(VI) (0.01 M) prepared from $CrCl_3 \cdot 6H_2O$ and/or $K_2Cr_2O_7$. The compartment on the left of AM was filled with 0.1 M KCl, on the right of CM – with 0.1 M KCl. Thus, in the DD experiments two interdiffusion processes took place simultaneously. The Cr(III) cation (and also K^+) was driven through CM by H^+ and the Cr(VI) anion was driven through AM by Cl^-. Unfortunately, only short-term experiments were performed and the recovery factor of Cr(III) and Cr(VI) was low – 5 and 3%, respectively. It should be noted here that one cannot expect a high RF of Cr(VI) and Cr(III). Regarding Cr(VI) the reason is that protons enter the feed through CM and decrease the selectivity of AM thus reducing the efficiency of Cr(VI) removal. Regarding Cr(III) the K^+ ions (from $K_2Cr_2O_7$) are also driven through CM by protons thus reducing the maximal RF of Cr(III).

As expected, in the electrodialytic experiments (the same duration as that of DD), RF increased significantly: Cr(III) – from 5% ($j=0$) to ca. 40%, Cr(VI) – from 2.5% ($j=0$) to ca. 30% (Cr(III)-Cr(VI) mixture without additional salt, $j=11.4\,mA\,cm^{-2}$). Although the authors did not comment on the current efficiency, using their data it was possible to estimate it. Assuming that Cr(III) is a trivalent cation and Cr(VI) – divalent anion, the current efficiency was ca. 40 and 20%, respectively. These low values result from the presence of substantial amounts of ions K^+, Cl^- ($[K^+]=[Cr(VI)]$), $[Cl^-]=3[Cr(III)]$) and also of H_3O^+ (hydrolysis of Cr(III) – reaction (1.11, Chapter 1)) which transfer a significant part of the electric charge.

The presence of an additional salt in the feed would further decrease the recovery factor of Cr in both types of the recovery process.

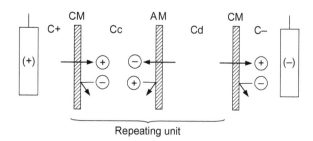

Repeating unit

Figure 21.4. Principle of electrodialysis – in the external electric field the cation, CM, and anion-exchange, AM, membranes allow for the transport of cations, anions, respectively; thus, between CM and AM an electrolyte concentration increases (compartment Cc), whereas between AM and CM – decreases (compartment Cd); C+, C– are the anode, cathode compartments, respectively.

21.3   ELECTRODIALYSIS

21.3.1   *General remarks*

Electrodialysis, ED, is a membrane separation process driven by an external electric field. It is used for the treatment of seawater, brackish water, industrial effluents, in the food industry, etc. (Koltuniewicz and Drioli, 2008; Koter, 2011; Strathmann, 2004). In the conventional ED a membrane module consists of cation- and anion-exchange membranes alternately arranged, separated by thin spacers. The principle of ED is shown in Figure 21.4.

The ED module shown schematically in Figure 21.4 can be shortly written as:

$$(+) \, C+ \, \{CM \, Cc \, AM \, Cd\}_n \, CM \, C- \, (-) \tag{21.16}$$

where C+, Cc, Cd, C– denote the anode, concentrate, diluate, cathode compartment, respectively, n is the number of repeating unit consisting of the pair of compartments.

Generally, when the ion mixture is electrodialyzed, the removal of a given ion is proportional to its transport number in the membrane, which in turn is proportional to the ion concentration and mobility inside the membrane. The ED process cannot be taken to the very end, i.e. to the total removal of ions because the electric resistance of the dilute solution would increase too much. Usually in the ED desalination of e.g. well waters the product water salinity is ca. 500 mg L^{-1} (<0.01 M NaCl) which still provides a good conductivity and, consequently, a low energy consumption. The concentration value 0.01 M would correspond to 750 mg L^{-1} As or 520 mg L^{-1} Cr – it is far above the critical limits. In the case of high As/Cr concentrations ED can be treated as an intermediate step in As/Cr removal. If As or Cr is a minor contamination of a solution to be desalted then during ED desalination also the content of As/Cr decreases, however not to the same degree as the dominant ions. An example of this case is Schoeman's work (Schoeman, 2008) where the ED treatment of a hazardous leachate (ca. 0.6 M NaCl with small amounts of As, Cr, Ba, Sr, Mn, Ni) is reported. After ED, the content of Cl, As and Cr in the product water decreased by 98% (21,000 → 345 mg L^{-1}), 60% (3.41 → 1.37), and 60% (0.191 → 0.076), respectively. Unfortunately, nothing is said about the oxidation state of As. At the feed pH 7.1 (product water – 6.6, brine – 7.8) As(V) and Cr(VI) would be in the form of divalent anions – such anions need 2 times more electric charge for their transport than monovalent chloride anions. Therefore their removal should be also ca. 2 times lower. As(III) would not be removed because at that pH value it would be mainly in the undissociated form H$_3$AsO$_3$.

When the ED process proceeds at constant voltage, at the final stage of the process the electric current becomes lower and lower because the electric resistance of the dilute solution significantly increases. At that stage the electrolyte back diffusion through the nonideally selective membranes becomes more and more significant, which reduces the current efficiency. On the other side too high current density cannot be applied because it would decrease the ion concentration at the membrane surfaces to zero (at the AM, CM surface from the cathode, anode, respectively). A lack of ions to be transported would result in the water splitting at these surfaces – in this way the electric energy would be wasted. Apart from that, OH^- anions originated from the water splitting on the AM surface would migrate on the anode side of AM where the hydroxides of multivalent cations (if present) would precipitate and block the membrane (Strathmann, 2004). It also should be mentioned that during ED substantial changes of solution volumes are observed due to the electroosmotic flow of water caused by the movement of counter-ions. As it is directed from the dilute solution to the concentrated one, it decreases the efficiency of the desalination/concentration process.

Quantitatively, the negative influence of membrane nonideality, expressed by the transport number of coions, $\bar{t}_2$, electroosmotic number of water, $\bar{t}_0$, diffusion flux, J_s, osmotic flux of water, J_0, on the current efficiency can be expressed by the equation (Koter and Narebska, 1989/90) (1:1 electrolyte):

$$CE \equiv \frac{\Delta(\Delta c_s)}{\Delta(\Delta c_s)_{id}} = 1 - \bar{t}_2 - 0.018\,\hat{m}\bar{t}_0 - (J_s - 0.018\,\hat{m}\,J_0)\frac{F}{j} \qquad (21.17)$$

In Equation (21.7) $\hat{m}$ is the mean molality of these solutions, j – current density. Because the desalination/concentration effect is the most important, the current efficiency is defined as the ratio of the increase of concentration difference across a real membrane, $\Delta(\Delta c_s)$, to that obtained for the ideally selective membrane, $\Delta(\Delta c_s)_{id}$. The negative contribution of $\bar{t}_2$ and $\bar{t}_0$ to CE is more or less constant, the contribution of J_s and J_0 (J_0 is usually directed to the more concentrated solution, in that case it has a negative sign) is roughly proportional to the concentration difference and inversely proportional to the current density.

During the ED process a fouling of membranes can occur, e.g. by large ions which cannot pass the membranes, by precipitation of low soluble salts, etc. To reduce membrane fouling electrode polarity reversal (a couple of times per hour) can be applied. This kind of ED is called electrodialysis reversal (EDR). As dilute and concentrate compartments are also reversed, an EDR setup is equipped with additional valves redirecting the solution flows to the proper tanks.

21.3.2 *Removal of As and Cr from contaminated waters*

There are only a few papers on As/Cr removal from wastewaters by the ED method (Cox and Alborzfar, 2007; Nataraj *et al.*, 2007; Peng *et al.*, 2004).

Cox and Alborzfar (2007) tested an electrodialysis reversal (EDR) pilot unit (Aquamite III EDR Pilot) on groundwater of high concentration of silica, fluoride and As at Fort Irwin in California (USA). The unit was equipped with the Ionics membranes AR204SXZL and CR67HMR, the polarity reversal frequency was 4 times per hour. The 75 days tests demonstrated the usefulness of the EDR process in producing water of good quality. Regarding As, it was possible to reduce the As concentration from 21–43 $\mu g\,L^{-1}$ to below 1.8 $\mu g\,L^{-1}$. Using feed water of higher As content (66 $\mu g\,L^{-1}$) the As concentration in the product water increased to 2.7 $\mu g\,L^{-1}$. In both cases it was much below the 10 $\mu g\,L^{-1}$ limit set by WHO. Unfortunately, neither cost nor efficiency of the process was estimated.

The electrodialysis of very dilute solutions of Cr(VI) (10, 20, 50 mg L^{-1}) was investigated by Nataraj *et al.* (2007). They used the ED unit (21.16) comprising 10 repeating units. The Cr(VI) solutions were obtained by dissolving $K_2Cr_2O_7$ in deionized water. In, the electrode compartments ca. 0.18 M Na_2SO_4 was circulated. The experiments were conducted at constant voltage (40, 50 and 60 V). The results are summarized in Table 21.1, where the removal of Cr(VI) after 165 min of the process is given.

It is seen that in the case of the lowest initial Cr(VI) concentration (10 mg L^{-1}) the removal is almost complete (the final concentration is 0.22, 0.08 and 0.04 mg L^{-1} for 40, 50, 60 V, respectively). In the case of a high Cr(VI) concentration (50 mg L^{-1}) the removal hardly reaches 80%

Table 21.1.   Removal of Cr(VI) [in %] after 165 min of the ED process
(calculated using the data from Nataraj et al., 2007).

Voltage [V]	Initial Cr(VI) concentration [mg L^{-1}]		
	10	20	50
40	97.8	82.2	69.4
50	99.2	94.5	79.2
60	99.6	96.0	79.6

indicating that the ED process should last longer. The energy consumption was estimated to be 2–4 kWh m^{-3} for the final Cr(VI) concentration 0.08 mg L^{-1}. Unfortunately, the authors did not mention what was the Cr(VI) concentration in the concentrate and how to utilize that solution.

For the elimination of Cr(VI) from an electroplating wastewater a combination of electrodialysis and chemical precipitation (ChP) was proposed by Peng et al. (2004). The ED process at constant current was carried out using the module containing two concentrate and two dilute compartments (scheme (21.16), $n = 2$). In the preliminary experiments the dependence of the Cr(VI) removal on the Cr(VI) concentration in the feed was determined. It was found that within the investigated concentration range 1–8.5 mg L^{-1} the Cr(VI) removal reaches maximum (ca. 93%) at 3.5–5.5 mg L^{-1}, and above that value it strongly decreases. This dependence was explained by the formation of polychromate anions at a higher Cr(VI) concentration, which compared to the monochromate ones have a smaller mobility inside the membrane pores. However, the authors did not say if the amount of electric charge passed through the system in the experiments was always the same. To remove effectively Cr(VI) from wastewater containing 19 mg L^{-1} Cr(VI), the authors decided at first to reduce the Cr(VI) concentration to the value corresponding to the removal maximum range (3.5–5.5 mg L^{-1}) and then to perform ED. The Cr(VI) content was diminished by chemical precipitation (ChP) which was carried out in two steps: (i) reduction of Cr(VI) to Cr(III) (using a mixture FeCl$_2$:Na$_2$S = 10:1) and (ii) precipitation with NaOH. After ChP the Cr content decreased to 3.62 mg L^{-1}, however, the concentration of Cl$^-$ and Na$^+$ increased to 111 and 44.7 mg L^{-1}, respectively. After applying ED to the ChP-treated water the removal of Cr reached 99% (0.18 mg L^{-1}), Na and Cl – ca. 70%. The concentration of other metals (Cu, Zn) decreased to practically zero. The product water of such characteristics can be used again in the electroplating process, thus reducing the operation cost and environmental impact. The concentrate can be purified using the ChP-ED combination.

21.3.3   Removal of As and Cr from solid wastes

Remediation of solid wastes, like industrially polluted soil, CCA-treated wood or ashes, is investigated mainly by the researchers from the Technical University of Denmark, Lyngby. Usually it is conducted in a 3-compartment cell:

$$(+) C+ AM \ Cd \ CM \ C- (-) \tag{21.18}$$

The medium (suspension of soil, wood, ash) to be remediated is placed in the middle compartment, Cd, where desalination takes place. Generally, it should be stated that the removal of As, Cr, and other heavy metals is not effective – the problem is their tight binding with the solid particle matrix. No soaking agent tested till now was efficient enough.

Below the results of such investigations are briefly described. The content is divided according to the type of the remediated material.

Table 21.2.   Results of the ED treatment of sludge after soil washing (compilation from Jensen *et al.*, 2012).

Sludge no.	Soil contaminated by	Mass loss [%]	pH initial	pH after ED	As [mg kg⁻¹ DW] initial	As [mg kg⁻¹ DW] after ED	Cr [mg kg⁻¹ DW] initial	Cr [mg kg⁻¹ DW] after ED
1	Unknown	25	7	1.9	178	not given	80	116
2	Metal foundry	22	6.85	1.6	24	not given	97	127
3	Wood preservation	12	6.15	3.7	9260	2140	2310	1040
4	Wood preservation	11	4.4	2.2	3030	1900	1680	1270
5	Chlor-alkali process	17	4.1	1.5	<0.1		196	182

21.3.3.1   *Soil*

The ED remediation of soils contaminated by different sources (sludges no. 1–5 in Table 21.2) was investigated by Jensen *et al.* (2012) using a 3-compartment cell equipped with the Ionics membranes (AM – AR204SZRA, CM – CR67 HVY HMR427). The soil suspension was prepared by adding the solid to distilled water by weight 1 to 10. The removal of the elements As, Cr, Cu, Cd, Ni, Zn and Pb was investigated. The electric charge passed through the system was one order higher than that needed to remove all the investigated ions, assuming ideality of the membrane. The chosen results are summarized in Table 21.2.

It is seen that the Cr concentration after ED treatment can even be increased (sludges no. 1 and 2) which is caused by the loss of sludge mass due to the dissolution of carbonates and other soil components. Only in the case of the soils contaminated by the wood preservation process (sludges no. 3 and 4) the Cr removal reaches 24–50%. Since Cr was found mainly in the cathode compartment, it was Cr(III) not Cr(VI). Compared to Cr As is more efficiently removed – its concentration after the ED treatment is reduced by 77 and 37% for sludges no. 3 and 4, respectively. Here the dominating species was As(V) not As(III) which at pH < 7 would occur in the electrically neutral form of H_3PO_3. By increasing the pH the As removal would be increased, however, hydroxides of heavy metals would precipitate. Summarizing – the obtained values of As and Cr concentrations are far above the limits that are obligatory in Denmark – 20 for As and 500 mg kg⁻¹ for Cr(III).

The ED remediation of the soil contaminated by tannery effluents in the past was studied by Nieto Castillo *et al.* (2012). During a period of years more mobile fractions of metals were washed out by the rain and mainly the metals bound to the soil particles remained. The soil contained Cr (3570 mg kg⁻¹, 99.9% Cr(III), 0.1% Cr(VI)), Zn (183), Pb (138), and Cu (64). Soil particles of the size < 2 mm were used for the tests. According to the sequential extraction analysis (BCR) ca. 90% of Cr(III) was in the least mobile fraction whereas in the most mobile one – only 0.3%. Thus, as expected, after the 7-day ED run the Cr removal was very poor – only 1.0 mg of Cr per kg of soil was recovered, mainly in a catholyte solution (0.8 mg kg⁻¹). To improve the metal removal, a 0.05 M KNO_3 solution was replaced by 0.05 M sodium persulfate in the cathode compartment. The role of the oxidant was to mobilize the metals by their oxidation or by oxidation of the organic matter/sulfide complexes. In the presence of oxidant the Cr removal increased to almost 20% of Cr. This is still a low value, however much higher than that obtained by 4-day soil washing with a persulfate solution (2.8%).

Hansen *et al.* (1997) remediated electrodialytically different types of polluted soil, among them loamy sand polluted by 50 years activity of a wood preservation plant. The concentration of Cr and Cu in the sand was 330 and 1400 mg kg⁻¹, respectively. A 3-compartment cell (scheme (21.18)) with SYBRON membranes (MA3470 and MC3475) was used. The middle compartment of length 15 cm was filled with the contaminated sand, not with its suspension in water. It should be noted that the compactness of the sand sample has an effect on its transport properties. Unfortunately, the authors did not mention if the sample was squeezed or not. The experiments were performed at a constant current 0.3 mA cm⁻² and lasted 54 days. From the direction of the Cr migration (to the cathode) it could be stated that it was Cr(III) cation. During the process, in the half of the sand sample from the anode side the Cr concentration decreased to ca. 22–36% of the initial value, i.e. to ca. $100 \, \text{mg kg}^{-1}$, whereas on the cathode side it practically remained unchanged. Compared to Cr, the Cu removal was much more efficient – the concentration in the soil slice close to CM (cathode side) was ca. 25% of the initial value, i.e. ca. $350 \, \text{mg L}^{-1}$, and towards AM it decreased gradually to zero. The average value $200 \, \text{mg kg}^{-1}$ Cu was below the limit set by the Danish Environmental Protection Agency.

The removal of As and Cu from the soil of different granular size was investigated by Sun et al. (2012). Two fractions of particles – fine particles of size $<63 \, \mu\text{m}$ and the original soil (particle size $<2 \, \text{mm}$) – were obtained from the same soil by sieving or wet-sieving (fine particles). The fine particles contained much more As (ca. $4600 \, \text{mg kg}^{-1}$) and Cu (ca. $2050 \, \text{mg kg}^{-1}$) than the original soil (1180 and $570 \, \text{mg kg}^{-1}$ of As and Cu, respectively) indicating that As and Cu were located at the particle surfaces. The experiments were carried out at different current densities ($j = 0.05, 0.1, 0.2 \, \text{mA cm}^{-2}$) using suspensions of particles (distilled water to solid ratio $L/S = 3.5$ and 7). In the electrode compartments a $0.01 \, \text{M NaNO}_3$ solution, adjusted to $\text{pH} = 2$ with HNO_3, was circulated. More As and Cu could be removed from the fine particles because of a higher concentration of these elements. However, the removal efficiency from the fine particles or the original soil in similar experimental conditions was practically the same. The highest removal – 96% of Cu and 64% of As – was obtained in the longest experiment (22 days), at $L/S = 3.5$ and $j = 0.1 \, \text{mA cm}^{-2}$. Still, the quantity of As remaining in the fine particles ($1715 \, \text{mg kg}^{-1}$) was much higher than the Danish limit for land use ($20 \, \text{mg kg}^{-1}$). The Cu concentration – $107 \, \text{mg kg}^{-1}$ – was below the limit value $500 \, \text{mg kg}^{-1}$. It was also found that a higher current density ($j = 0.2 \, \text{mA cm}^{-2}$) led to a higher energy consumption – $2.5 \, \text{Wh g}^{-1}$ soil, whereas for $j = 0.1 \, \text{mA cm}^{-2}$ it was $1.0 \, \text{Wh g}^{-1}$ soil. Using the data from Sun (Sun et al., 2012) one can calculate that for the removal of 1 g As and Cu ca. 21 Wh was consumed. It should be noted that the use of a more dilute suspension ($L/S > 3.5$) or a higher current density is not justified because of too low ion concentration in the suspension. To obtain the optimal results the suspension dilution should be minimized and the current density should be adequate for that dilution.

Simultaneous removal of Cu ($830 \, \text{mg kg}^{-1}$) and As(III) ($900 \, \text{mg kg}^{-1}$, As(V) is not present) from soil polluted by wood preservation activity was studied by Ottosen et al. (2000). To remove simultaneously Cu and As both elements must be in an ionic form. As(III) occurs as H_2AsO_3^- at $\text{pH} \geq 9$, whereas at that pH Cu(OH)_2 would precipitate. Thus a suitable agent should be chosen to keep As and Cu in the ionic forms. The authors decided to use ammonia which alkalizes the soil (As(III) $\rightarrow \text{H}_2\text{AsO}_3^-$) and with Cu^{2+} forms the charged complex $\text{Cu(NH}_3)_4^{2+}$. The process was carried out in the cell of configuration:

$$(+) \, \text{C}+ \, \text{AM Cii FP Ciii CM Civ AM C}- \, (-) \qquad (21.19)$$

In C+ and Cii a 2.5% NH_3 solution was circulated, in Civ and C– – a mixture of 0.01 M $\text{NaNO}_3 + 0.01\text{M HNO}_3$, whereas the sample of soil was in Ciii. On the anode side AM reduced the migration of H^+ (produced in the anode reaction) from C+ to Cii. Ammonia was easily transported through a filter paper, FP, to the soil in Ciii forming there the complex cation $\text{Cu(NH}_3)_4^{2+}$. To prevent the Cu precipitation on the cathode, it was separated from CM by AM. Thus $\text{Cu(NH}_3)_4^{2+}$ was gathered in Civ. CM prevented the useless migration of NO_3^- from Civ to Ciii. The lowest As concentration obtained was $90 \, \text{mg kg}^{-1}$, i.e. 10 times less than the initial As concentration. In the case of Cu 45% decrease of concentration was obtained, i.e. the final concentration ($370 \, \text{mg kg}^{-1}$) was below the Danish limit ($500 \, \text{mg kg}^{-1}$). From the acidified soil the Cu removal was much better – the level $< 50 \, \text{mg kg}^{-1}$ was reached, however, As(III) in the neutral form H_3AsO_3 could not be removed. It should be added that the current efficiency of the As and Cu removal was very low – below 3%.

21.3.3.2 Ash

Various assisting agents (0.25 M ammonium citrate/1.25% $\text{NH}_{3,\text{aq}}$, 0.25 M sodium citrate, 2.5% $\text{NH}_{3,\text{aq}}$, deionized water) were tested by Pedersen (2002) in the electrodialytic removal of heavy metals (Cd, Pb, Zn, Cu and Cr) from municipal solid waste incineration (MSWI) fly ash. The content of heavy metals was as follows: Cd – 241, Pb – 8070, Zn – 17140, Cu – 1570, and Cr – 285 mg kg^{-1} of dry ash. For the tests the following cell was used:

$$(+) \, C- \, CM \, Cii \, AM \, Ciii \, CM \, Civ \, AM \, C- \, (-) \tag{21.20}$$

Here Ciii is the compartment with a stirred ash slurry (L/S ratio ca. 6.5). In Cii and Civ the ions from Ciii are gathered, from those solutions the metal ions can be extracted by other methods (e.g. electroplating, ion exchange, precipitation). The experiments (constant current, $j = 0.8 \, \text{mA cm}^{-2}$) lasted 2 weeks. It was found that for the removal of all five metals simultaneously the best agent was 0.25 M ammonium citrate/1.25% NH$_3$ solution. The highest value of Cr removal was only 20%, obtained for that agent. Cr migrated mainly towards the anode which means that Cr was in the form of chromate (VI) or a negative Cr(III)-citrate complex (Gabriel and Salifoglou, 2005). Pedersen *et al.* (2005) performed remediation experiments of variable duration (from 5 to 70 days) with the same ash (Cr – 285 mg kg^{-1}) using ammonia citrate. In the 14-days experiment the same level of Cr removal as previously (20%) was obtained, in the 35-days run it increased to ca. 31%, and after 70 days it practically remained the same (calculated using the data in Pedersen *et al.*, 2005). It should be noticed that although in each experiment the same mass of ash was taken, the total mass of Cr (and other metals) determined in each experiment differed significantly (from 15 to >30 mg). During the experiments the mass of ash decreased due to dissolution of soluble mineral salts. Thus the Cr content in the ash could be even higher than before the remediation. For the 70-days run it was even 1.6 times higher than the initial concentration, the lowest content was obtained for the 21-day run – ca. 60% of the initial value, the Cr removal for that run exceeded 30%.

A significant dissolution of ash during the ED remediation was reported by Lima *et al.* (2009; 2010). Two kinds of ash were studied there – ST ash (13.5 mg kg^{-1} Cr) generated during the combustion of straw and straw pellets and CW ash (185 mg kg^{-1} Cr) from the co-combustion of wood pellets/fuel oil/natural gas. In the 14-days process only ca. 5% of the ST ash was recovered with the Cr content slightly lower (13%) than the initial value. The high solubility of that ash resulted from a high content of chlorides. The CW ash (almost no chlorides) was much less soluble – after the same remediation time 70% of CW was recovered with the Cr content ca. 26% higher than that before remediation. From the mass balance it results that in the case of the ST ash ca. 95% of the initial Cr moved to a solution, whereas in the case of the CW ash it was only 12%.

Also unsuccessful was the Cr removal from the ash obtained by a combustion of CCA-treated wood waste (Pedersen and Ottosen, 2006). The concentrations of As, Cr, and Cu in the ash was high: approximately 35, 62, and 69 g kg^{-1}, respectively. The ED treatment of a suspension of ash in 0.5 M H$_2$SO$_4$ (S/L = 5.9) was performed in a 3-compartment cell (25). The ash recovery after the process was 92%. The results obtained are ambiguous. According to Pedersen (Pedersen and Ottosen 2006) only 3% of total Cr was removed from the ash whereas the Cr concentration determined in the treated ash was reduced by 27%. The removal of Cu was also poor (ca. 30%, the decrease of concentration in the ash – 7%). Both Cr and Cu were still found incorporated in the silica-based matrix particles and also in the incompletely combusted wood particles. The removal of As was much better – ca. 8% was left in the ash or 15% in concentration. According to Pedersen and Ottosen (2006) and other works cited there, in the ash As is mainly in the form of Ca-arsenates which dissolved during the ED treatment. However, As was found not only in the anolyte (32%) but also in the catholyte (ca. 12%) and at the cathode (ca. 31%) where it electro-precipitated together with Cu. The rest of the As (17%) remained in the aqueous part of the suspension. It is not excluded that at low pH (0.5 M H$_2$SO$_4$) AsO$^+$ was present which moved towards cathode.

21.3.3.3   *CCA-treated wood*

The electrodialytic remediation of CCA-treated wood in a pilot plant containing up to 2 m^3 waste wood chips was investigated by Pedersen *et al.* (2005) and Christensen *et al.* (2006). The plant had the configuration:

$$(-) \, C- \, \{CM \, Cd \, AM \, Cc\}_n \, CM \, Cd \, AM \, C+ \, (+) \tag{21.21}$$

The number of the repeating unit, n, was varied from 0 to 4. The wood with a soaking agent was placed in the Cd compartment, in Cc an electrolyte solution circulated and gathered ions coming from Cd. The wood was chipped and sorted into three fractions – fine (<2 cm), medium (2–4 cm), and large (>4 cm) by sieving. The best results were obtained for a 3-weeks experiment with the medium wood fraction and the mixture of 0.5 M H_3PO_4 and 5% oxalic acid as a soaking agent. The applied voltage was 30–57 V, the current density – 2–5 A m^{-2}. The removal efficiency obtained for As, Cr, Cu was >96, 82, 88%, respectively; this corresponds to the final concentrations: ca. 30 (near detection limit), 252, and 163 mg L^{-1}, respectively. As these values are close to the nonhazardous limits, it would be possible to use remediated wood in MSWI plants. The above results obtained for the pilot plant do not differ too much from those obtained using a laboratory cell (batch mode, 2.5% w/w oxalic acid as a soaking agent, duration 14 days). In those experiments the maximum removal efficiency for As, Cr, Cu was 95, 87, 84%, respectively (Velizarova et al., 2002).

A series of papers devoted to the ED removal of As, Cr, Cu from an 8-year out-of-service CCA-treated *Pinus pinaster* Aiton pole was published by Ribeiro et al. (2000; 2007) and Moreira et al. (2005). Sawdust (particles of size 20 mesh diameter) prepared from that pole was used by Ribeiro and Moreira. The concentration of As, Cr, and Cu in the sawdust was ca. 8110, 8280, and 3150 mg kg^{-1}. The tested soaking agent was: distilled water, oxalic acid solutions of different concentrations (2.5, 5.0, and 7.5% w/w), 2.5% formic and 2.5% citric acid solutions. In all the experiments the same current density was applied (0.2 mA cm^{-2}), the duration was 30 days. In each electrode compartment 0.01 M $NaNO_3$ was circulated, the catholyte pH = 2 was maintained with HNO_3. The highest removal was obtained for the dust saturated with 2.5% oxalic acid: As – 98.7, Cr – 94.8, Cu – 93.1% (Ribeiro et al., 2000). Most of the removed As and Cr was found in the anolyte – 93, 84%, respectively. This does not mean that Cr was only in an anionic form of Cr(VI). In an acidic solution oxalate can reduce chromate to Cr(III) which forms a negatively charged complex with oxalate, $CrOx_3^{3-}$. A cationic form of Cr(III) was also present – 9% of the Cr was found on the cathode side. Arsenic moved mainly to the anode. If the pH were ≤ 7 (unfortunately pH was not measured), then As(V) would move as $H_2AsO_4^-$; As(III) being in the undissociated form H_3AsO_3 would not be mobile. A small amount of As found in the catholyte (4%) could migrate there as As(III) in the form of cations AsO^+.

A model of electrodialytic removal of Cr, Cu, As from CCA-treated wood chips in the presence of 2.5% oxalic acid solution was presented by Ribeiro et al. (2007). The model is too extensive to be described here. Generally, it is based on the mass conservation equation and the Nernst-Planck equation. The membranes were treated as ideally selective with one exception – as well as anions protons could pass through the AM. Regarding As and Cr, the following chemical equilibria were taken into account: dissociation of H_3AsO_4, H_2CrO_4, precipitation-dissolution of $Cr(OH)_3$, and the formation of the complex $Cr(C_2O_4)_3^{3-}$. The model satisfactorily described the removal of Cr and As, only the Cu removal was substantially underestimated.

21.4   ELECTRODEIONIZATION

Electrodeionization (EDI) is a process similar to electrodialysis shown in Figure 21.4. In EDI the dilute compartments are filled with a mixture of cation- and anion-exchanger particles which largely increases the conductivity of these compartments. Consequently, salty water can be desalted to a much higher degree and much higher current density can be applied compared to the classical ED. The role of ion-exchangers is twofold. Firstly, their particles absorb the ions from the water which migrate along the bed to the membranes and leave the compartment. Secondly, near the compartment outlet, where the ion concentration in the water is very low, in the contact places of anion- and cation-exchange material (exchanger, membrane) a splitting of water takes place – the hydronium and hydroxide ions produced enhance ionization and, consequently, removal of weakly ionized species (e.g. CO_2, SiO_2, H_3BO_3, NH_3). Improvements of EDI were reviewed by Wang et al. (2011).

Although EDI is better suited for the removal of trace amounts of As and Cr than ED, till now only a few papers have appeared on that subject (Alvarado *et al.*, 2009; 2013). Both these papers concern the EDI removal of Cr(VI) from its 100 mg L^{-1} solution using the same module:

$$(+) \text{ C+ CM Cc AM Cd CM C− (−)} \tag{21.22}$$

but different modes of the process: batch – (Alvarado *et al.*, 2009), continuous mode – (Alvarado *et al.*, 2013). The authors report 99.8 and 98.5% removal of Cr(VI) for those processes, respectively. However, the current efficiency, *CE*, calculated from their data, exceeds the upper limit 1. E.g. for the continuous mode *CE* is higher than 2:

$$CE = |z_{Cr(VI)}| \frac{F_v([\text{Cr(VI)}]_{in} - [\text{Cr(VI)}]_{out})}{Aj/F} = 2.08 \tag{21.23}$$

even on the assumption that only monovalent anions of Cr(VI) ($|z_{Cr(VI)}| = 1$) migrates through AM. In Equation (21.23) the following values taken from Alvarado *et al.* (2013) were substituted: $[\text{Cr(VI)}]_{in} = 100$ mg L$^{-1} \approx 1.92$ mol m^{-3}, $[\text{Cr(VI)}]_{out} = 1.5$ mg L$^{-1} \approx 0.029$ mol m^{-3}, $F_v = 4.5$ cm^3 min^{-1}, $A = 40$ cm^2, $j = 0.165$ mA cm^{-2}. The most probable explanation of such a high value of *CE* is sorption of a significant part of Cr(VI) by the membranes and the ion-exchange resin. The same remark concerns the work of Alvarado *et al.* (2009). Thus further tests are needed to evaluate the usefulness of the EDI process in Cr(VI) removal.

21.5 SUMMARY

Donnan dialysis (DD) and electrodeionization (EDI) are the methods suitable for the removal of ions from dilute solutions. The existing literature has demonstrated that using the DD method As can be successfully removed from groundwaters. The important factor is the pH of the treated solutions which influences the concentration of ionic forms of As. The concentration of stripping electrolyte (ion) should be appropriately chosen; too high concentration decreases the selectivity of the anion-exchange membrane, too low – decreases the interdiffusion rate of ions. A coupling of DD with a coagulation process can increase the As removal efficiency. In areas with no central system of water supply a household Donnan dialyzer for cleaning groundwater was tested with good results. However, no suggestion has been given on how to utilize the exhausted stripping solution with a high As content.

Investigations on Cr removal by the DD or EDI methods are scarce and incomplete. Removal of Cr(VI) anions may be problematic because the selectivity of anion-exchange membranes in those solutions is low. No work on As removal by EDI has been found.

Using the electrodialysis reversal (EDR) method it is possible to reduce the As content in groundwater much below the WHO limit 10 μg L^{-1}. Unfortunately, no information on the cost of the process was given. With the ED method it is possible to reduce the Cr(VI) concentration in electroplating wastewater below 0.2 mg L^{-1}. As the maximal efficiency of ED strongly decreases with the Cr(VI) concentration, before applying ED it should be reduced to the optimal value (ca. 4 mg L^{-1}) by e.g. chemical precipitation.

Many papers have been published on the ED removal of As, Cr and other heavy metals from contaminated soils and ashes. As these elements are located at the soil particle surfaces, their content is higher in the particles of smaller size. However, the removal efficiency in similar experimental conditions is practically the same for fine and thick particles. Unfortunately, because of strong binding with the soil particles or because of forming insoluble compounds it is not possible to reduce the As and Cr content below the obligatory limits (e.g. in Denmark – 20 mg kg^{-1} for As and 500 mg kg^{-1} for Cr(III)). The use of an oxidant or various soaking agents slightly increases the Cr removal efficiency.

A significant part of ash can be dissolved during the electrodialytic treatment. Depending on the ash origin its recovery can vary from only 5 to 92%. Because of that and because of a strong association with the ash particles the Cr content after the ED treatment can be even higher than before. Arsenic is much more removable from the ash than Cr because in the ash As is mainly in the form of Ca-arsenates soluble in water.

NOMENCLATURE

A	membrane area [m^2]
a_i	activity of ith species [–]
CE	current efficiency [–]
c_i	concentration of ith ion [mol m^{-3}]
D_i	diffusion coefficient [m^2 s^{-1}]
F	Faraday constant [96485 C mol^{-1}]
F_v	volume flow [m^{3}s^{-1}]
IEC	ion-exchange capacity [mol kg^{-1}]
J_i	flux of ith $i$on [mol m^{-2}s^{-1}]
j	electric current density [A m^{-2}]
K	thermodynamic constant [–]
K	partition coefficient [–]
K_c	concentration quotient [–]
l_m	membrane thickness [m]
p	pressure [Pa]
R	gas constant [8.314 J K^{-1} mol^{-1}]
T	absolute temperature [K]
t	time [s]
$\bar{t}_0$	electroosmotic transport number of water [–]
$\bar{t}_i$	transport number of ith ion [–]
V	volume [m^3]
y_i	activity coefficient of ith ion [–]
z_i	charge number of ith ion [–]
φ	electric potential [V]
μ_i	chemical potential [J mol^{-1}]
$\tilde{\mu}_i$	electrochemical potential [J mol^{-1}]

SUBSCRIPTS

f	feed
m	membrane
s	strip

ABBREVIATIONS

AM	anion-exchange membrane
C	cell compartment
CCA	chromated copper arsenate
CM	cation-exchange membrane
ChP	chemical precipitation
DD	Donnan dialysis
EDR	electrodialysis reversal
ED	electrodialysis
EDI	electrodeionization
L/S	liquid to solid ratio
MSWI	municipal solid waste incineration
RF	recovery factor, $RF = 100(1 - c_{f,t}/c_{f,0})$, where $c_{f,0}$, $c_{f,t}$ are the concentrations of an ion removed from the feed at the beginning of experiment, and at time t, respectively.

REFERENCES

Alvarado, L., Ramírez, A. & Rodríguez-Torres, I. (2009) Cr(VI) removal by continuous electrodeionization: study of its basic technologies. *Desalination*, 249, 423–428.

Alvarado, L., Torres, I.R. & Chen, A. (2013) Integration of ion exchange and electrodeionization as a new approach for the continuous treatment of hexavalent chromium wastewater. *Separation and Purification Technology*, 105, 55–62.

Castillo, E., Granados, M. & Cortina, J.L. (2002) Chromium(VI) transport through the Raipore 1030 anion exchange membrane. *Analytica Chimica Acta*, 464, 15–23.

Christensen, I.V., Pedersen, A.J., Ottosen, L.M. & Ribeiro, A.B. (2006) Electrodialytic remediation of CCA-treated waste wood in a 2 m³ pilot plant. *Science of the Total Environment*, 364, 45–54.

Cox, E.E. & Alborzfar, M. (2007) Pilot testing of edr membrane process for fluoride and arsenic reduction at Fort Irwin, California. *World Environmental and Water Resources Congress.* pp. 1–10. Available from: 10.1061/40927(243)453.

Gabriel, K. & Salifoglou, A. (2005) A chromium(III)-citrate complex from aqueous solutions. *Agroalimentary Processes and Technologies*, XI, 57–60.

Güell, R., Fontàs, C., Anticó, E., Salvadó, V., Crespo, J.G. & Velizarov, S. (2011) Transport and separation of arsenate and arsenite from aqueous media by supported liquid and anion-exchange membranes. *Separation and Purification Technology*, 80, 428–434.

Hansen, H.K., Ottosen, L.M., Kliem, B.K. & Villumsen, A. (1997) Electrodialytic remediation of soils polluted with Cu, Cr, Hg, Pb and Zn. *Journal of Chemical Technology and Biotechnology*, 70, 67–73.

Jain, C.K. & Singh, R.D. (2012) Technological options for the removal of arsenic with special reference to South East Asia. *Journal of Environmental Management*, 107, 1–18.

Jensen, P.E., Ottosen, L.M. & Allard, B. (2012) Electrodialytic versus acid extraction of heavy metals from soil washing residue. *Electrochimica Acta*, 86, 115–123.

Katsoyiannis, I. & Zouboulis, A. (2005) Technologies for arsenic removal from contaminated water sources. In: Lehr, J.H. & Keeley, J. (eds.) *Water encyclopedia. Domestic, municipal, and industrial water supply and waste disposal.* John Wiley & Sons, Inc., New York, NY. pp. 636–639.

Koltuniewicz, A.B. & Drioli, E. (2008) *Membranes in clean technologies.* Wiley-VCH Verlag GmbH & Co. KGaA, Weinheim, Germany.

Koter, S. (2011) Ion-exchange membranes for electrodialysis. A patents review. *Recent Patents on Chemical Engineering*, 4, 141–160.

Koter, S. & Kultys, M. (2010) Modeling the electric transport of sulfuric and phosphoric acids through anion-exchange membranes. *Separation and Purification Technology*, 73, 219–229.

Koter, S. & Narebska, A. (1989/1990) Current efficiency and transport phenomena in systems with charged membranes. *Separation Science and Technology*, 24, 1337–1354.

Ktari, T., Larchet, C. & Auclair, B. (1993) Mass transfer characterization in Donnan dialysis. *Journal of Membrane Science*, 84, 53–60.

Lima, A.T., Ottosen, L.M. & Ribeiro, A.B. (2009) Electroremediation of straw and co-combustion ash under acidic conditions. *Journal of Hazardous Materials*, 161, 1003–1009.

Lima, A.T., Rodrigues, P.C. & Mexia, J.T. (2010) Heavy metal migration during electroremediation of fly ash from different wastes-modelling. *Journal of Hazardous Materials*, 175, 366–371.

Lin, Z. & Burgess, L.W. (1994) Chemically facilitated Donnan dialysis and its application in a fiber optic heavy metal sensor. *Analytical Chemistry*, 66, 2544–2551.

Loo, W.W. & Wen, J.J. (2005) Treatment of arsenic, chromium, and biofouling in water supply wells. In: Lehr, J.H. & Keeley, J. (eds.) *Water encyclopedia. Ground water.* John Wiley & Sons, Inc., New York, NY. pp. 22–28.

Moreira, E.E., Ribeiro, A.B., Mateus, E.P., Mexia, J.T. & Ottosen, L.M. (2005) Regressional modeling of electrodialytic removal of Cu, Cr and As from CCA treated timber waste: application to sawdust. *Wood Science and Technology*, 39, 291–309.

Nataraj, S.K., Hosamani, K.M. & Aminabhavi, T.M. (2007) Potential application of an electrodialysis pilot plant containing ion-exchange membranes in chromium removal. *Desalination*, 217, 181–190.

Nieto Castillo, A., García-Delgado, R.A. & Cala Rivero, V. (2012) Electrokinetic treatment of soils contaminated by tannery waste. *Electrochimica Acta*, 86, 110–114.

Nwal Amang, D., Alexandrova, S. & Schaetzel, P. (2004) Mass transfer characterization of Donnan dialysis in a bi-ionic chloride-nitrate system. *Chemical Engineering Journal*, 99, 69–76.

O'Connor, J.T. (2002) Arsenic in drinking water. Part 4: Arsenic removal methods. *Water Engineering and Management*, 149, 20–25.

Oehmen, A., Viegas, R., Velizarov, S., Reis, M.A.M. & Crespo, J.G. (2006) Removal of heavy metals from drinking water supplies through the ion exchange membrane bioreactor. *Desalination*, 199, 405–407.

Oehmen, A., Valerio, R., Llanos, J., Fradinho, J., Serra, S., Reis, M.A.M., Crespo, J.G. & Velizarov, S. (2011) Arsenic removal from drinking water through a hybrid ion exchange membrane coagulation process. *Separation and Purification Technology*, 83, 137–143.

Ottosen, L.M., Hansen, H.K., Bech-Nielsen, G. & Villumsen, A. (2000) Electrodialytic remediation of an arsenic and copper polluted soil – continuous addition of ammonia during the process. *Environmental Technology*, 21, 1421–1428.

Pedersen, A.J. (2002) Evaluation of assisting agents for electrodialytic removal of Cd, Pb, Zn, Cu and Cr from MSWI fly ash. *Journal of Hazardous Materials*, B95, 185–198.

Pedersen, A.J. & Ottosen, L.M. (2006) Elemental analysis of ash residue from combustion of CCA treated wood waste before and after electrodialytic extraction. *Chemosphere*, 65, 110–116.

Pedersen, A.J., Kristensen, I.V., Ottosen, L.M., Ribeiro, A.B. & Villumsen, A. (2005a) Electrodialytic remediation of CCA-treated waste wood in pilot scale. *Engineering Geology*, 77, 331–338.

Pedersen, A.J., Ottosen, L.M. & Villumsen, A. (2005b) Electrodialytic removal of heavy metals from municipal solid waste incineration fly ash using ammonium citrate as assisting agent. *Journal of Hazardous Materials*, B122, 103–109.

Peng, C., Meng, H., Song, S., Lu, S. & Lopez-Vaidivieso, A. (2004) Elimination of Cr(VI) from electroplating wastewater by electrodialysis following chemical precipitation. *Separation Science and Technology*, 39, 1501–1517.

Pickard, B. & Bari, M. (2004) Feasibility of water treatment technologies for arsenic and fluoride removal from groundwater. Report, U.S. Army Center for Health Promotion and Preventive Medicine Water Supply Management Program Aberdeen Proving Ground, Maryland, Presented at the *AWWA Water Quality Technology Conference*, 14–18 November, 2004, San Antonio, TX. Available from: http://www.dtic.mil, 2013-10-16 [accessed July 2015].

Prado-Rubio, O.A., Møllerhøj, M., Jørgensen, S.B. & Jonsson, G. (2010) Modeling Donnan dialysis separation for carboxylic anion recovery. *Computers & Chemical Engineering*, 34, 1567–1579.

Pyrzynska, K. (2006) Preconcentration and recovery of metal ions by Donnan dialysis. *Microchimica Acta*, 153, 117–126.

Ribeiro, A.B., Mateus, E.P., Ottosen, L.M. & Bech-Nielsen, G. (2000) Electrodialytic removal of Cu, Cr, and As from chromated copper arsenate-treated timber waste. *Environmental Science & Technology*, 34, 784–788.

Ribeiro, A.B., Rodríguez-Maroto, J.M., Mateus, E.P., Velizarova, E. & Ottosen, L.M. (2007) Modeling of electrodialytic and dialytic removal of Cr, Cu and As from CCA-treated wood chips. *Chemosphere*, 66, 1716–1726.

Sardohan, T., Kir, E., Gulec, A. & Cengeloglu, Y. (2010) Removal of Cr(III) and Cr(VI) through the plasma modified and unmodified ion-exchange membranes. *Separation and Purification Technology*, 74, 14–20.

Schoeman, J.J. (2008) Evaluation of electrodialysis for the treatment of a hazardous leachate. *Desalination*, 224, 178–182.

Sionkowski, G. & Wódzki, R. (1995) Recovery and concentration of metal ions. I. Donnan dialysis. *Separation Science and Technology*, 30, 805.

Strathmann, H. (2004) Ion-exchange membrane separation processes. *Membrane Science and Technology Series*, Volume 9. Elsevier, Amsterdam, The Netherlands.

Sun, T.R., Ottosen, L.M., Jensen, P.E. & Kirkelund, G.M. (2012) Electrodialytic remediation of suspended soil – comparison of two different soil fractions. *Journal of Hazardous Materials*, 203–204, 229–235.

Tor, A., Büyükerkek, T., Çengeloglu, Y. & Ersöz, M. (2004a) Simultaneous recovery of Cr(III) and Cr(VI) from the aqueous phase with ion-exchange membranes. *Desalination*, 171, 233–241.

Tor, A., Çengelolu, Y., Ersöv, M. & Arslan, G. (2004b) Transport of chromium through cation-exchange membranes by Donnan dialysis in the presence of some metals of different valences. *Desalination*, 170, 151–159.

Velizarov, S. (2013) Transport of arsenate through anion-exchange membranes in Donnan dialysis. *Journal of Membrane Science*, 425–426, 243–250.

Velizarova, E., Ribeiro, A.B. & Ottosen, L.M. (2002) A comparative study on Cu, Cr and As removal from CCA-treated wood waste by dialytic and electrodialytic processes. *Journal of Hazardous Materials*, 94, 147–160.

Wang, J., Fan, G., Dong, H., Fei, Z., Chen, W. & Lu, H. (2011) Recent patents review on electrodeionization. *Recent Patents on Chemical Engineering*, 4, 183–198.

Wang, S. & Zhao, X. (2008) A review on advanced treatment methods for arsenic contaminated soils and water. *Journal of ASTM International*, 5 (10).

Wodzki, R., Sionkowski, G. & Hudzik-Pieta, T. (1996) Recovery of metal ions from electroplating rinse solutions using the Donnan dialysis technique. *Polish Journal of Environmental Studies*, 5, 45–50.

Zhang, Y., Ghyselbrecht, K., Meesschaert, B., Pinoy, L. & van der Bruggen, B. (2010) Electrodialysis on RO concentrate to improve water recovery in wastewater reclamation. *Journal of Membrane Science*, 378, 101–110.

Zhao, B., Zhao, H. & Ni, J. (2010a) Arsenate removal by Donnan dialysis: effects of the accompanying components. *Separation and Purification Technology*, 72, 250–255.

Zhao, B., Zhao, H. & Ni, J. (2010b) Modeling of the Donnan dialysis process for arsenate removal. *Chemical Engineering Journal*, 160, 170–175.

Zhao, B., Zhao, H., Dockko, S. & Ni, J. (2012) Arsenate removal from simulated groundwater with a Donnan dialyzer. *Journal of Hazardous Materials*, 215–216, 159–165.

Part IV
Case studies: Latin America and
the United States of America examples

CHAPTER 22

Latin American experiences in arsenic removal from drinking water and mining effluents

José Luis Cortina, Marta I. Litter, Oriol Gibert, Cesar Valderrama,
Ana María Sancha, Sofía Garrido & Virginia S.T. Ciminelli

22.1 INTRODUCTION

The presence of As in water resources used for human consumption has been identified as a growing problem in Latin America (LA), causing high impact in both developed areas and the poorest regions (Bundschuh *et al.*, 2010a; 2012; Litter *et al.*, 2010a; 2014). In the last decade, several countries in LA have adopted the World Health Organization (WHO) allowable limit of As in drinking water of $10\,\mu g\,L^{-1}$ (WHO, 2011), with the exception of some countries such as Mexico, where the recommended value is still $25\,\mu g\,L^{-1}$. In Argentina, for the regions with soils with high As content, a period of five years was set to reach the $10\,\mu g\,L^{-1}$ limit (Código Alimentario Argentino, 2007).

In most of the cases, such as in Chile and Argentina, the origin of As in groundwater is natural and the As problem has been known since several decades ago, affecting large and medium cities (defined as those with more than 10,000 inhabitants) and rural and periurban areas lacking drinking water supply systems (Litter *et al.*, 2012). In the cases where the drinking water supply contains unsafe levels of arsenic (As), the immediate concern has been to find a safe source of drinking water. Basically, two options have been applied: (i) to find a new safe source to be used directly or to be blended with the contaminated resource as in the case of Mexico, or (ii) to remove As from the contaminated source by an adequate technology, as in the case of Chile, Argentina, Peru or Guatemala.

As is known, most As removal processes are based on a simple sequence of physicochemical treatments that can be applied alone, combined or in a sequence: oxidation/reduction, coagulation/filtration, precipitation, adsorption, solid/liquid separation and membrane technologies, among others (Bundschuh *et al.*, 2010b; Höll and Litter, 2010b; Litter *et al.*, 2010a; 2010b; Newcombe and Möller, 2008; Ravenscroft *et al.*, 2009; Sancha, 2010a). In a scenario such as the case of LA, the choice of the right technology to solve contamination problems in drinking water must be based on one hand on technical factors such as physicochemical (oxidation state – As(III) or As(V), conductivity, pH, redox potential – E_h, level of dissolved organic matter or dissolved oxygen) and microbiological characteristics of the water, capacity of the water treatment plant, requirements of final As concentration in the treated water, type of source, and others (Litter *et al.*, 2010a; 2010b; Sancha, 2010b). On the other hand, the sociological aspects of the population to be supplied, such as poverty, malnutrition, incidence of chronic illnesses, educational skills and others are important factors in the selection of the right technology.

Another important point to be addressed is the increasing effort for removal and disposal of As in the mining (especially copper and gold) and metallurgical industry, because both activities are relevant in LA and important sources of As residues. Due to its toxicity, environmental regulations are becoming increasingly stringent regarding the disposal of As containing wastes. In many Latin American areas, the processing of gold, copper and other sulfide ores containing arsenopyrite (FeAsS) results in solid residues and aqueous solutions containing As. The latter requires As removal and stabilization before its disposal in landfills. Adequate management of large amounts of mining waste is mandatory. It is therefore important that As residues are disposed of in stable phases in order to avoid its further release into the environment (Pantuzzo and Ciminelli, 2010).

22.1.1   *Technological approaches for arsenic removal*

Typical As treatment processes are a combination of an oxidative pretreatment step followed by precipitation, co-precipitation and adsorption onto coagulated flocs, lime treatment, adsorption onto suitable surfaces, use of ion-exchange resins or membrane processes (Bundschuh *et al.*, 2010a; 2010b; Kabay *et al.*, 2010; Litter *et al.*, 2010a; Mondal *et al.*, 2013; Newcombe and Moller, 2008; Ravenscroft *et al.*, 2009; Sharma and Sohn, 2009). Due to the reduction of the maximum allowed As concentration defined by the WHO ($10\,\mu g\,L^{-1}$), technologies such as the membrane-based processes have been introduced, especially in areas where water shortage is a severe problem and when the pressure or volume of water in industry is a limiting aspect.

Most of the precipitation and coagulation methods take advantage of the insolubility of certain arsenical inorganic compounds as well as the adsorption of soluble arsenate and arsenite on metal oxy-hydroxides. By adding cations (e.g. iron (III), calcium, magnesium, aluminum, manganese (II)) to As(V) solutions, As-containing solids are formed, which can be removed by sedimentation or filtration. The stability of the solid residues produced by each of these methods should be taken into consideration in the case of direct disposal.

22.1.1.1   *Preconditioning processes*

Oxidation is in many cases a required step to transform As(III) species to more easily removable, more stable and less toxic As(V). Because the oxidation of As(III) by direct aeration is not kinetically favored (Bissen and Frimmel, 2003), typical oxidizing reagents, such as sodium hypochlorite, hydrogen peroxide and potassium permanganate aqueous solutions, including gases such as chlorine and ozone, have been used (Caetano *et al.*, 2009; Kartinen and Martin, 1995). Chlorine and chlorine dioxide provide a rapid and effective oxidation, but they may react with organic matter, producing harmful trihalomethanes as byproducts. Potassium permanganate effectively oxidizes arsenite, and it has been used as a widely available, relatively low cost reagent, but with some restrictions related to the presence of soluble manganese species in the effluent.

22.1.1.2   *Coagulation and filtration*

The most common technologies used on the medium and large scale to remove As are based on coagulation and filtration. The coagulation/flocculation process is a standardized technology used for turbidity removal (i.e., organic and inorganic species in the form of suspended or colloidal particles). In natural waters (pH 6–9), suspended colloidal particles are stabilized by negative electrical charges on their surface, a process that hinders collisions and formation of higher agglomerates that can be separated from water (flocks formation) (Bilici Baskam and Pala, 2010). Addition of coagulants and flocculants causes destabilization of the colloids by neutralization of the electrical charges (coagulation stage), with subsequent agglomeration and formation of flocs (flocculation stage). The initially small flocs form bigger aggregates that settle easily. Negatively charged, soluble As species in the pentavalent form, can be attracted to the coagulated flocs by adsorption processes or formation of complexes, and the flocs can later be removed by sedimentation or filtration. The most used coagulants for As removal are aluminum sulfate, ferric chloride and ferrous sulfate. Iron salts are generally better removal agents. $FeCl_3$ generates relatively large flocs, while smaller ones are formed with $FeSO_4$. Coagulation with ferric salts works better at pH below 8, while aluminum salts have a narrower effective pH range, ca. 6–7. If the water pH is above 7, removal may be improved by acid addition. In general, the higher the coagulant dose, the better the As removal. Typical doses are $5–30\,mg\,L^{-1}$ ferric salts or $10–50\,mg\,L^{-1}$ alum. Figure 22.1 shows the effect of the various stages of a classical treatment with a coagulation process. As can be seen, a pre-oxidation step using chlorine, as well as a post-oxidation step on the residual As content is needed (Sancha and Frenz, 2000).

The efficiency of the coagulation/flocculation process depends on factors such as the water source, pH, coagulant dose, and duration of the filtration step or interval between filter washings. All these factors can be controlled and adjusted during the operation of the plant. Other factors such as the mixing time and the flocculation gradient should be considered in the initial design of a plant. As removal by coagulation/flocculation generates residues (sludge) with a content of As depending on the treated water, water used in washing filter and As removal efficiency. This sludge must be submitted to a leaching TCLP test (http://www.epa.gov/waste/hazard/testmethods/index.htm) to decide on its potential danger and the final disposal. In general, the sludge generated could be classified as not toxic according to the TLCP protocol.

After this first coagulation step, filtration is performed e.g. by sand filters. Without the filtration step, arsenate removal is around 30%, whereas its use increases removal rates up to more than 95% (Madiec *et al.*, 2000; Sancha, 1999; 2000).

The coagulation/filtration technology is simple; it requires only common chemicals, the installation costs are small and the system can work at high flow rates. As a drawback, the technology generates large volumes of As-containing sludge, normally disposed of in landfills, creating a potential contamination source (Vogels and Johnson, 1998).

22.1.1.3 *Membrane processes*
Membrane processes are considered an attractive technology for removing As from groundwater for drinking water production. Nanofiltration (NF) or reverse osmosis (RO) membranes are appropriate to remove anionic As species over a wide pH range (3–11) (Akin *et al.*, 2011; Seidel *et al.*, 2001; Uddin *et al.*, 2007). However, selection of pH is very important with respect to the target ions separated. Uddin *et al.* (2007) demonstrated that the rejection of As(V) was better than As(III) by nanofiltration membranes and the required minimum concentration level (MCL) could not be obtained for solutions containing $50 \,\mu g \, L^{-1}$. Then, oxidation of As(III) to As(V) is an essential pre-treatment step.

Operation and maintenance requirements for membranes are less: no chemicals are needed, and maintenance consists only of ensuring a constant pressure and periodical cleaning of the membranes. The main disadvantages, especially for RO, are high electricity consumption, relatively high capital and operating costs, the risk of membrane fouling and scaling and the need to treat the generated brines containing As..

However, as the reduction of operating costs has been already accomplished and it has led to expansion of the application of the technology, even for the production of industrial water, high concentrations of suspended solids and colloids (e.g. silica) and dissolved organic matter promote membrane fouling and biofouling, while hardness, sulfates and phosphates, among other factors, cause scaling problems interfering with the overall performance. Discharge of rejected water (20–25% of the influent) or brine is also a concern; therefore, the technology is not useful in areas where water is scarce.

RO provides high As(V) and low As(III) removal, and oxidation is difficult because residual oxidants can damage the membranes (Saitua *et al.*, 2005; USEPA, 2006). In addition, RO eliminates not only As but other ions, altering the chemical composition or organoleptic properties of drinking waters. In the case of water resources with high salinity content, RO is the most suitable technological solution (D'Ambrosio, 2005; Sen *et al.*, 2010; Walker *et al.*, 2008).

22.1.1.4 *Ion-exchange resins*
Strong anionic ion-exchange (IX) resins of polystyrene/divinylbenzene have been used for arsenate removal to produce water with less than $1 \,\mu g \, L^{-1}$ As. Uncharged As(III) species are not removed; thus, a previous oxidation step is necessary. Typical oxidizing reagents such as those previously described are used; however, residual amounts of these oxidants should not affect the stability of the resins. Commonly, resins are in the chloride form. Arsenate removal is relatively independent of pH and influent concentration. $HAsO_4^{2-}$, being more charged, has higher adsorption ability than $H_2AsO_4^-$. Competing anions, especially sulfate, fluoride and nitrate, interfere strongly and can affect the operation capacity. Suspended solids, precipitated iron and aluminum

oxyhydroxides can cause clogging (Wang *et al.*, 2000). In the last decade, efforts of the manufacturers of ion exchangers (e.g., Lanxess, Purolite, Dow Chem, Rohm and Haas) have accelerated the development of new tailored anionic exchangers to attain As values below $10 \mu g L^{-1}$ (An *et al.*, 2010). However, the aforementioned limitations and the relatively high cost in comparison to conventional IX resins have limited the application of these tailor-made resins to the separation of As species (e.g., As(III) and As(V)) for analytical purposes (Francesconi and Kuehnelt, 2004).

22.1.1.5 *Adsorbents-based on iron and aluminum oxides*

Aluminum oxides (activated alumina), iron oxides/hydroxides, titanium dioxide, cerium oxide, or reduced metals can be used as adsorbents (Litter *et al.*, 2010a; Mohan and Pittman, 2007; Ravenscroft *et al.*, 2009; Yamani *et al.*, 2012). Granular activated alumina ($Al_2O_3/Al(OH)_3$) is a commercially available, porous oxide adsorbent, successfully applied at slightly acid pH (5–7), giving efficiencies higher than 95% for both As(V) and As(III) (Pirnie, 2000); above pH 7, removal efficiency drops sharply. The technology is very simple, as it does not require addition of other chemicals. Activated alumina is used in packed beds, and it needs longer empty bed contact times (about 5 to 8 minutes) than ion-exchange resins. Activated alumina beds have usually much longer run times than ion-exchange resins and, typically, several tens of thousands of beds can be treated before As breakthrough. Concentrations of other solutes in the water matrix have a relatively small effect on As removal. For better results, if raw water contains arsenite, it should be oxidized before the treatment. Saturated activated alumina can be regenerated by flushing with strong base followed by strong acid. Regenerated media lose some volume and eventually they must be replaced. Like ion-exchange resins, activated alumina beds can be clogged by precipitation of iron (Ravenscroft *et al.*, 2009).

Granular iron hydroxide (Siemens AG, GFH), a synthetic akaganeite, proved able to retain As(V) and As(III) (Driehaus, 2002), as demonstrated also by granular iron oxide (Bayoxide, GFO) containing 70% Fe_2O_3 (Severn Trent Services, 2007). Commercial titanium dioxide (Bang *et al.*, 2005; Dow Chem, 2005) and manganese dioxide (Driehaus *et al.*, 1995) also proved to be effective. Among As adsorbents, iron-based adsorption has extensively used for As removal, especially from groundwater. Depending on the removal processes involved, the different technologies can be classified as those where iron acts as sorbent, co-precipitant or contaminant immobilizing agent and those where iron behaves as a reductant. Metallic zero-valent iron (ZVI) has been used for the treatment of groundwater as reactive media in permeable reactive barrier systems (Bang *et al.*, 2005). ZVI oxidizes spontaneously in water to form a number of Fe(II) and Fe(III)oxides/hydroxides, and As(III) and As(V) form inner sphere complexes with the corrosion products (Gibert *et al.*, 2010; Manning *et al.*, 2002) or the ZVI is precipitated and co-precipitated with the corrosion products (Farrell *et al.*, 2001).

As was described for the case of hydrated iron oxide-impregnated ion-exchange resins, the relative cost in comparison with standard removal technologies is still a limitation for their application in medium to large-scale plants, and most of the work in Latin America has been performed at laboratory and pilot research level.

22.1.2   *Treatment and disposal of As in industrial effluents from mining industries*

It is well recognized that the feasibility of mining projects is increasingly dependent on their environmental performance. Arsenic is one of the contaminants of concern associated with mining and metallurgical operations, in LA and elsewhere. Processing of ores containing As minerals, the most abundant being arsenopyrite, FeAsS, results in As contaminated solid residues and aqueous solutions. The latter require As removal and fixation before disposal. An adequate management of large amount of mining wastes (such as As sulfides or As residues produced in the treatment of aqueous effluents) is mandatory. Due to its toxicity, environmental regulations are becoming increasingly more stringent regarding the disposal of As containing wastes. It is therefore important that As residues are disposed of in stable phases in order to avoid further release of As into the environment (Pantuzzo and Ciminelli, 2010).

A former survey of industrial practice for removal and disposal of As from process solutions, covering the main copper and gold plants in Latin America, indicated the predominance of co-precipitation methods using iron salts (Harris, 2003). Iron compounds, either as amorphous arsenical ferrihydrite (a mixture of ferric arsenates and As adsorbed onto iron oxyhydroxides produced in ambient conditions) or as crystalline arsenates (produced under hydrothermal conditions) are the most environmentally accepted forms for As disposal under oxidizing conditions. Precipitation with lime or lime combined with ferric salts follows in importance as common industrial practices.

A series of metal (e.g., ferric, calcium, aluminum and others) arsenates, ferric arsenates being the most common ones, are found in mining residues. Calcium arsenate should be avoided due to its relatively high solubility in water and the transformation to $CaCO_3(s)$, by atmospheric CO_2 under alkaline conditions (Pantuzzo and Ciminelli, 2010). The use of chemically or biologically produced sulfide sources as precipitation agents, Na_2S for instance, has been investigated as an alternative solution to the usual metal arsenate co-precipitation. A two-stage process is often used, whereby lime neutralization causes sulfide precipitation.

Precipitation techniques for As removal from aqueous effluents often result in large amounts of residues with low As content, thus implying large disposal costs. For example, the arsenical ferrihydrite process described above generates large volumes of As-containing wastes (e.g. 3–8%). Thus, large areas for the final disposal of these wastes are needed, which in turn results in additional operational costs. The precipitation of crystalline ferric arsenate (scorodite) under atmospheric conditions has been investigated with the aim of increasing As content, tailings stability and to decrease disposal area (Caetano *et al.*, 2009; Demopoulos *et al.*, 1994). Another option is to precipitate ferric arsenate, which is then transformed hydrothermally to scorodite (Ruanala *et al.*, 2014). An industrial aplication of the atmospheric precipitation of scorodite (10,000 t/y of As as scorodite) from copper-bearing arsenic ores has been recently commissioned in Chile.

Fe-As compounds usually show incongruent dissolution behavior – As dissolves preferentially over Fe (Carageorgos and Monhemius, 1993) and Al (Pantuzzo *et al.*, 2014). If redissolution of arseniferous wastes takes place and As compounds are free to move downward into the soil matrix, the selection of the type of soil, which in turn will influence As uptake and fixation, plays a key role in preventing groundwater contamination.

Examples of As treatment plants in copper and gold mining operations in Chile and Brazil are discussed in this chapter.

22.2 IMPLEMENTED TECHNOLOGICAL SOLUTIONS FOR ARSENIC REMOVAL IN LATIN AMERICA

In Latin America, interesting experiences on As removal at large and medium plants have been reported, some of which have been working for several decades. Chile was a pioneer in facing the problem of As in water supply systems in the north of the country. In Argentina, as reported by Bundschuh *et al.* (2012), vast areas are affected, but fortunately many plants have been operating since the 1990s. The most widely used technologies in these cases have been those based on coagulation/adsorption/filtration processes and membrane processes such as reverse osmosis. Other procedures were developed in Guatemala, Peru and Mexico and all of them will be summarized below, in order to describe and provide solutions to be applied in similar situations; both successes and deficiencies perceived will be also highlighted.

22.2.1 *Chile*

22.2.1.1 *Treatment plants for As removal on the large and medium scale*
Chile successfully faced the challenge of As removal from water resources used for human consumption in the north of the country in the 1970s (Sancha and Ruiz, 1984). In these regions,

Table 22.1.   Arsenic concentration [$\mu g\,L^{-1}$] in some Chilean water resources (Sancha and Ruiz, 1984; Sancha, 1999; 2000; 2003).

Site	As concentration range [$\mu g\,L^{-1}$]
Northern zone	
San José	50–100
Lluta	600–700
Toconce	600–900
Lequena	150–350
Colana	70–90
Siloli	20–50
Inacaliri	80–90
Central zone	
Elqui	50–60
Blanco	20–30
Aconcagua	10–20
Maipo	10–20
Mapocho	10–20
Southern zone	
Bío-Bío	<5
Valdivia	<5

with a population of approximately 2 million inhabitants, As is present in all the ecosystems, originating either from natural (geological or volcanic) sources or caused by the active mining industry. A further problem is that water scarcity limits the feasibility of most of the As-removing technologies in this region.

Table 22.1 provides background information on the As levels in some water resources of the country. Studies on As speciation suggested that the element is mainly present as As(V) soluble species (Sancha *et al.*, 1992).

The first As removal projects were developed in the second half of the 20th century. During the 60s, it became clear that the consumption of water from the Toconce river, containing As concentrations in the range 600–900 $\mu g\,L^{-1}$, was causing serious health damage to the population of the cities of Antofagasta and Calama (Borgoño and Greiber, 1971; Borgoño and Vicent, 1977; Ferreccio and Sancha, 2006; Rosenberg, 1974; Zaldivar, 1980). This impact is still under evaluation (Borgoño and Vicent, 1977; Cerda *et al.*, 1999; Rivara *et al.*, 1997; Smith *et al.*, 1998).

This critical situation led to the Chilean Government at that time to commission a study on As removal from the German company Berkefeld Filter in collaboration with Chilean researchers (Latorre, 1966). This study demonstrated that a potential technology based on coagulation/flocculation was the most suitable to reduce the As content below the required values, according to the regulations at that time (120 $\mu g\,L^{-1}$) (NCh 409 of. 1970). The role and the critical factors affecting the removal efficiency of the various stages involved were studied, allowing the optimization of the design parameters. These factors were used afterwards to build a number of plants, constituting a good solution for the problem of As in small and medium cities (Barahona and Gonzalez, 1987; Sancha, 1999; 2000; 2003; 2006; 2010c; Sancha and Fuentealba, 2009; Sancha and Ruiz, 1984). The first plant, which used $Al_2(SO_4)_3$ and then $FeCl_3$ as coagulants, was installed in 1970 in Antofagasta (Salar del Carmen plant). The As removal plants of Calama (Cerro Topater), Chuquicamata and Taltal also use $FeCl_3$ as the coagulant (Sancha and Fuentealba, 2009; Sancha, 2000).

In the case of the Chilean plants, due to the low turbidity of raw surface water of northern Chile, coagulation/flocculation processes removed mainly As. The concentration of the other dissolved ions remained almost unchanged. The low turbidity of this raw water is caused by the absence of rainfall and because the water originates from aquifers recharged at the slopes of the Andes,

Table 22.2. Details of As removal plants in Chile (Sancha, 1999; 2000; 2003; Sancha and Ruiz, 1984).

Plant	Start	Capacity [L s^{-1}]	Source	As range [μg L^{-1}]
Complejo Salar del Carmen*			Mixture (1)	400–450
Old plant	1970	500		
New plant	1978	520		
Cerro Topater*	1978	500	Mixture (2)	400–450
Chuquicamata*	1989	210	Colana	70–90
			Inacaliri	80–90
Taltal**	1998	32	Agua Verde	60–80

*surface water
**groundwater
Mixture (1): mixed waters of Toconce, Lequena, Quinchamale – Siloli Polapi plants
Mixture (2): mixed waters of the Toconce, Lequena – Quinchamale plants

Table 22.3. Chilean plants operation (Salar del Carmen, Cerro Topater and Taltal) (Sancha, 1999; 2000; 2003; Sancha and Ruiz, 1984).

Operation	Salar del Carmen (Antofagasta)	Cerro Topater (Calama)	Taltal
Raw water As concentration [μg L^{-1}]	400–450	400–450	60–80
Chemical dose [mg L^{-1}] oxidant (Cl$_2$)	1.0	1.0	1.0
Coagulant [mg L^{-1} FeCl$_3$]	40.5	40.5	8.0
Sedimentation rate [m^3 m^{-2} d^{-1}]	70–75	70–75	–
Filtration rate [m^3 m^{-2} d^{-1}]	143	143	150
Sludge generation [kg As day^{-1}]	25–30	20–30	–
Outlet water As concentration [μg L^{-1}]	10	10	10

which latter emerge as rivers on the west. For As removal from groundwater, where the turbidity is even lower and the volume of particulate formed is small, further studies allowed simplification of this process, removing the sedimentation step (Barahona and Gonzalez, 1987; Fuentealba 2003; Sancha, 2006; Sancha and Ruiz, 1984; Sancha et al., 2000).

Currently, there are four plants in operation using classical coagulation with a total capacity of approximately 1800 L s^{-1} and one plant with the simplified process of coagulation/filtration. Table 22.2 collects details of these plants (year of construction, treatment capacity and As content in the treated influent water). In the case of the Complejo Salar del Carmen plant, data are provided for the first plant built in 1970 and for the plant rebuilt in 1978 after the Chilean standard for As level was moved from 120 to 50 μg L^{-1}. Figure 22.1 shows a general outline of the plants currently operative in Chile for surface and groundwater, respectively. Table 22.3 shows operative conditions of the Salar del Carmen (Antofagasta), Cerro Topater (Calama) and Taltal plants. Operational parameters are similar for the three plants, since the preliminary experience gained on the Salar del Carmen plant was transferred to the others. A preoxidation step to oxidize As(III) to As(V) is accomplished by dosing of 1 mg L^{-1} of chlorine, while the coagulant dose (FeCl$_3$) is 40 mg L^{-1} for the raw water with the highest As contents (400–450 μg L^{-1}) and 8 mg L^{-1} for the raw water with the lowest As contents (60–80 μg L^{-1}) as it is in the case of the Taltal plant. Sedimentation rates used were 70–75 (m^3 m^{-2} d^{-1}) and the filtration rate was fixed at 140–145 (m^3 m^{-2} d^{-1}). In both cases, the As removal efficiency provides final As concentrations around 10 μg L^{-1}. In the case of the Salar del Carmen (Antofagasta) and Cerro Topater (Calama) plants, the sludge production ranged between 20–30 kg As d^{-1}.

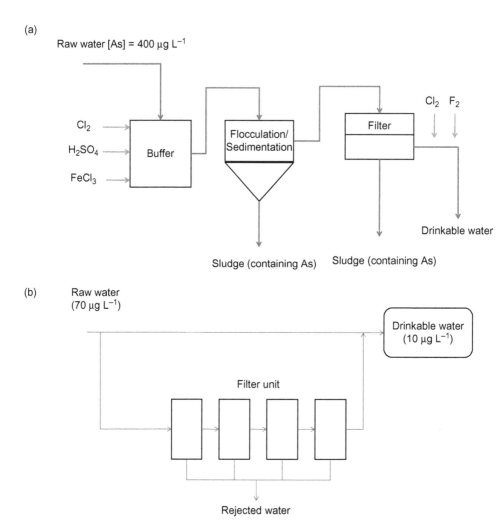

(a)

Raw water [As] = 400 µg L^{-1}

Cl$_2$

H$_2$SO$_4$

FeCl$_3$

Buffer

Flocculation/Sedimentation

Filter

Cl$_2$ F$_2$

Drinkable water

Sludge (containing As)   Sludge (containing As)

(b)   Raw water
(70 µg L^{-1})

Drinkable water
(10 µg L^{-1})

Filter unit

Rejected water

Figure 22.1.   (a) Simplified flow-sheet of the surface water treatment plant of Salar del Carmen, Chile (adapted from Sancha and Ruiz, 1984) and (b) Groundwater treatment scheme of Taltal plant (Chile) (adapted from Sancha and Fuentealba, 2009; Sancha, 2006).

The Chilean experience in the field of coagulation-based As removal plants shows that coagulation is very effective, reaching the recommended standards (10 µg L^{-1}, WHO), in a cost-effective way. The process is able to deal with medium to high flows (30 to 520 L s^{-1}). The modified coagulation process can be performed with low-turbidity groundwater where, as usual, As speciation and water matrix are the limiting factors: As(V) is more efficiently removed than As(III). Removal efficiencies of 80–100% were obtained for As(V) with FeCl$_3$, while only 40–70% was achieved for As(III). Experiences with Al$_2$(SO$_4$)$_3$ showed removal efficiencies lower than with FeCl$_3$ of 75% for As(V) and only 5–20% for As(III). Thus, it is recommended to oxidize As(III) to As(V) before the treatment (Sancha et al., 1992).

Because the maximum admissible As concentration values changed from 120 to 50 µg L^{-1} in 1984 and finally to 10 µg L^{-1} since 2005 (NCh 409/1, 2005), studies by Chilean researchers have attempted to improve the coagulation process of the Complejo Salar del Carmen plant by adding sulfuric acid (Granada et al., 2003). The Chilean Norm (NCh409) is only applicable to plants

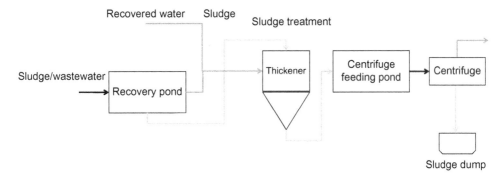

Figure 22.2.   Schematic flow-sheet of the sludge treatment plants in Chile (adapted from Sancha, 2003).

constructed since 2005. Plants built before this year had a period of ten years to reach a threshold of 30 $\mu g\,L^{-1}$ and a maximum of ten years to reach 10 $\mu g\,L^{-1}$.

The decrease of pH affects the acid-base properties of the $Fe(OH)_3$ particles increasing the positive charge of the flocs, favoring then the removal of the As(V) anionic species (Chwirka *et al.*, 2000). Coagulant dose also affects the volume of sludge generated. Since 1997, sludge is left in a disposal site especially designed for As containing wastes. A schematic flow sheet of the sludge treatment plants is described in Figure 22.2. Sludge produced in the coagulation/sedimentation step, after being separated in the settling units, is sent to a dewatering and thickening stage to increase the solids content from 1–2% up to 30–40%. To achieve a higher solid content in the sludge, coagulants could be used; however, as this process consumes chemicals, the sludge is dried using piles before being sent for disposal onto land fields.

As described, in Chile the problem has been partially solved by installation of plants in some large and medium cities; however, the problem persists in the north of the country, typically in small rural communities. Solutions for these areas are proposed elsewhere (Bundschuh *et al.*, 2010b; Litter *et al.*, 2008; 2010a; 2010b; 2012).

22.2.1.2   *Treatment of mining effluents in Chile*

Chile is the largest copper producer in the world and hosts about 30% of the known copper resources in the globe, accounting for over 35% of global copper production. Within the copper metallurgical processes and due to the presence of As in the processed ores, concentrated effluents from the refining stage of copper have to be treated. In 2008, a plant for the treatment of mining effluents was designed (SGA, 2008) to treat 1250 $m^3\,d^{-1}$ of effluents. Figure 22.3 illustrates the scorodite process proposed for the plant. The ferric sulfate needed for the stabilization of As in the effluent generated by leaching of smelter dusts is produced in the plant. Byproducts of the iron mining industry, such as magnetite, hematite, hydroxides of iron and/or foundry slag, can be also used. A stable waste (10 $t\,d^{-1}$) containing As and antimony would be generated when working at the maximum plant capacity.

The need to process As-rich copper concentrates combined with the increased penalties and lower acceptable amounts of As in copper concentrates have motivated a number of investigations and technological strategies. One of these approaches is controlled oxidation roasting to remove As from the copper concentrate. This treatment generates highly toxic As dusts, such as arsenic trioxide, which then require further processing prior to disposal (Van der Meer, 2011). Codelco and its subsidiary EcoMetales (ECL) inaugurated an industrial plant to treat As dusts from three copper smelters and to stabilize As as scorodite in 2012. Magnetite and limestone are used to precipitate arsenic and antimony. Claimed as the first of its kind in the world, the unit was designed with the capacity to treat up to 10,000 tons of As per year, to be disposed of in a deposit with the

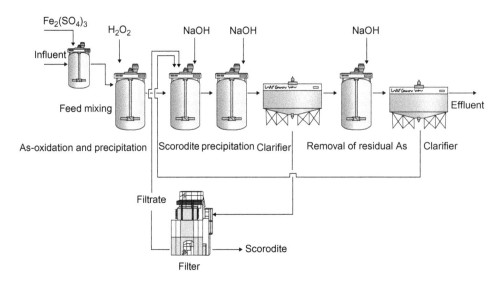

Figure 22.3.   Flow-sheet of a treatment plant for arsenic mining effluents in Chile (SGA, 2008).

capacity of 1.1 million tons of waste over a 15-year period. The concentrated copper solution, free from the impurities, is send back to the metallurgical plant (Superneau, 2012).

22.2.2   *Argentina*

In Argentina, many regions suffer from the presence of As in groundwater. It is estimated that under the new regulations approved by the WHO ($10 \, \mu g \, L^{-1}$), approximately 4 million people are at risk. Moreover, some surface water sources are also affected. Although the number of As removal plants is unknown, the use of membranes is the most used alternative, whereas coagulation/adsorption/filtration processes come second (D'Ambrosio, 2005). Reverse osmosis is the membrane-based process for As removal used in Argentina, and is the favorite option because they allow the treatment of waters where other pollutants, such as organic micropollutants, chlorides, sulfates, nitrates and heavy metals are present. Membrane technologies are applicable to highly mineralized water, which cannot be treated by other methods. To achieve high production ratios of water, membrane units incorporate treatments having more than one pass, where the rejected water of the first pass is treated by a second membrane unit and so on. Current available membranes and membrane-based technologies are more expensive than other As removal options, but they have been selected for use in municipal waterworks of Argentina, where very low As levels are required.

22.2.2.1   *Coagulation/adsorption/filtration process for groundwater treatment*
The development of plants using coagulation/adsorption/filtration processes, followed by a double filtration, was carried out by the *Universidad Nacional de Rosario* (Fernández and Ingallinella, 2010; Ingallinella *et al.*, 2003a; 2003b; Litter *et al.*, 2008). In 2000, given the conditions existing in the Santa Fe province, it was necessary to develop an As removal process that also eliminated fluoride. The province of Santa Fe is located in the *Pampa Húmeda* (central Argentina) with 3,200,000 inhabitants of which not more than 85% have access to safe drinking water. It has 334 water services, 96% of which originate from groundwater. Much of this groundwater contains As, in concentrations ranging between 50 and 300 $\mu g \, L^{-1}$, with 30% of As(III) in some plants. In many cases, high fluoride concentrations (2.0–2.5 mg L^{-1}), above the amount officially allowed in the country (Código Alimentario Argentino, 2007), are also found in these waters. In some

Table 22.4. Groundwater quality parameters in Santa Fe (Argentina) (Ingallinella *et al.*, 2003a; 2003b).

Parameter	Unit	Mean value
Color	UH	2
Turbidity	UNT	0.10
pH		7.85
TDS	$mg\,L^{-1}$	749
Conductivity	$\mu S\,cm^{-1}$	1160
Total hardness	$mg\,L^{-1}\,CaCO_3$	45
Ca^{2+}	$mg\,L^{-1}\,Ca^{2+}$	7
Mg	$mg\,L^{-1}\,Mg^{2+}$	7
Alkalinity	$mg\,L^{-1}\,CaCO_3$	470
Cl^-	$mg\,L^{-1}\,Cl^-$	48
SO_4^{-2}	$mg\,L^{-1}\,SO_4^{2-}$	40
NO_3^-	$mg\,L^{-1}\,NO_3^-$	10
NO_2^-	$mg\,L^{-1}\,NO_2^-$	<0.005
NH_3	$mg\,L^{-1}\,NH_4^+$	<0.05
Total iron	$mg\,L^{-1}\,Fe^{3+}$	<0.1
As	$\mu g\,L^{-1}\,As$	150
F^-	$mg\,L^{-1}\,F^-$	1.7
SiO_2	$mg\,L^{-1}\,SiO_2$	55

localities, the total salinity is low (TDS < 1000 mg L^{-1}), but in others the total salt content is over the regional regulation (TDS < 1500 mg L^{-1}). Equally remarkable, the groundwater contains 55–90 mg L^{-1} of dissolved silica in the form of silicates, which compete for adsorption sites during coagulation processes. Silicates also cause scaling problems in RO processes: they precipitate, disabling the membrane, and force the use of antiscalants. In the province of Santa Fe, there are more than 80 RO plants. Thirty-two of them treat flows between 10 and 100 $m^3\,h^{-1}$, which are supplied to the distribution network, and 50 are small plants (0.2–5 $m^3\,h^{-1}$), where the water is bottled. However, the water provided by the water treatment system does not meet the conditions of drinkable water. RO plants present some additional problems, which will be commented on further. In addition, there are two plants operating with granular iron hydroxide as adsorbent and seven coagulation/adsorption/filtration plants, four of which are based on the ArCIS process, described.

The *ArCIS-UNR* process abates As and fluoride simultaneously using an up-flow filtration system, followed by a rapid filtration step to separate the flocs generated by the coagulant (Azbar and Turkman, 2000; Ingallinella, 2003a). The process was implemented for the city of Villa Cañás, located in the south of the Santa Fe province, to supply water to 11,000 inhabitants. The main physicochemical characteristics of this groundwater are shown in Table 22.4. In order to determine the optimal coagulant, its dose and the working pH, several tests were carried out. PAC, an inorganic aluminum polymer of growing application recently in water treatment processes, was chosen as the tested coagulant. Laboratory-scale tests showed that As concentrations lower than 50 $\mu g\,L^{-1}$ could be obtained in a pH range 6.5–8 (the best results were reached at pH 7.3). Fluorine removal increased as pH diminished. With PAC doses of 100 mg L^{-1}, at pH 6, a F^- concentration below 1.5 mg L^{-1} could be achieved. Higher PAC doses were evaluated as not economical; on the other hand, pH is a critical parameter and it needs careful monitoring if As concentrations below 10 $\mu g\,L^{-1}$ need to be reached (Fernández *et al.*, 2009).

The pilot-plant tests were carried out, as is shown in Figure 22.4, using gravel up-flow filters (6 m h^{-1}) and a conventional rapid filter showed: (i) the largest As removal was produced in the prefilter; (ii) although Si removal and turbidity generated during the treatment by the rapid filter seem to be unrelated, reduction of turbidity can be taken as both a good indicator of As removal and of the treatment efficiency. The final As, F^- and Al concentrations and turbidity levels were below the limits established by the current regulations (Fernández et al., 2009). A scaled treatment plant was designed for the city of Villa Cañás to supply water with the quality specified in Table 22.4. The plant was built on reinforced concrete and consisted of two modules with a capacity of $600\,m^3\,d^{-1}$ each. The water consumption for backwashing of rapid filters ($10\,min$, $60\,m\,h^{-1}$ backwash velocity) was around 6–7% of the treated water ($1200\,m^3\,d^{-1}$). As concentration was reduced from $200\,\mu g\,L^{-1}$ (Viña Cañás) or $70\,\mu g\,L^{-1}$ (López) to $20\,\mu g\,L^{-1}$. Fluoride was removed from a raw water concentration of $2.3\,mg\,L^{-1}$ down to $1.5\,mg\,L^{-1}$. The accumulated sludge volume of each washing operation for both modules was around $50\,m^3$, with 99% moisture. The sludge was disposed of in an evaporation lagoon where As was analyzed from a leachate sample. Arsenic content in this waste was $30\,\mu g\,L^{-1}$, well below the Argentine regulation for hazardous wastes (below $1000\,\mu g\,L^{-1}$). After the extraction from the lagoons, sludge was disposed of together with other solid urban wastes. The dry sludge production was approximately $0.8\,m^3$ per month.

The ArCIS-UNR process was used to design the water treatment plants of López (Santa Fe) with $150\,m^3\,d^{-1}$, Santa Isabel (Santa Fe) with $150\,m^3\,d^{-1}$, Andino (Santa Fe) with $380\,m^3\,d^{-1}$ and Lezama (Buenos Aires) with $850\,m^3\,d^{-1}$. Figure 22.5 shows the installations of the Villa Cañás and López plants, respectively. The first constructed plants were intended to provide As levels below $50\,\mu g\,L^{-1}$ for the treated flows. However, the new regulations that set a maximum concentration of $10\,\mu L^{-1}$ of As forced an optimization of the developed technology. For the new plants, the prefilter granulometry was optimized and drying beds were designed to generate sludge dehydration. The operating costs of the process are approximately 0.3 US$ m^{-3} treated water. The ArCIS-UNR technology proved to achieve 80–90% As removal in the plants installed in the Santa Fe province.

Recently, the ArCIS-UNR process was improved to treat groundwater containing phosphates by adding the PAC coagulant in two stages (González et al., 2014).

22.2.2.2 *Coagulation/filtration process for surface water treatment*

Jáchal treatment plant: Water of the department of Jáchal (11,000 inhabitants (2001)) (San Juan) is supplied from the river of the same name. Its quality parameters are shown in Table 22.5. Arsenic levels in the raw water ranged from 70 to $110\,\mu g\,L^{-1}$, a silica content of $15\,mg\,L^{-1}$ and high contents of chlorides (270–$300\,mg\,L^{-1}$) and sulfates (290–$420\,mg\,L^{-1}$) as can be seen in Table 22.5. Water is treated in a conventional plant ($50\,m^3\,h^{-1}$), previously designed to remove turbidity with values ranging from 2 to 70 NTU. Originally, the plant consisted of five lagoons as equalization vessels followed by settlers and slow sand filters. In a second step, the corresponding doses of aluminum sulfate as coagulant and an organic flocculant were added at two points before the lagoon inlet (OSE, 2006). $Al_2(SO_4)_3$ doses of $60\,mg\,L^{-1}$ led to As concentrations below $10\,\mu g\,L^{-1}$ without previous acidification.

San Antonio de los Cobres treatment plant: Water for the 5000 inhabitants of the town of San Antonio de los Cobres (Salta) is supplied from a channel; the quality of this water is shown in Table 22.6. Arsenic levels in the raw water ranged from 120 to $250\,\mu g\,L^{-1}$ and there were lower levels of chlorides (107–$127\,mg\,L^{-1}$) and sulfates (25–$44\,mg\,L^{-1}$) compared with the Jáchal waterworks. A dose of $65\,mg\ FeCl_3\ L^{-1}$ with pH modifications (up to 7) was found optimal to achieve As concentrations below $10\,\mu g\,L^{-1}$.

22.2.2.3 *Water treatment plant integrating reverse osmosis membrane processes*

Several RO plants have been installed recently in Argentina, (e.g. in the provinces of Santa Fe, Córdoba and La Pampa) (D'Ambrosio, 2005; Fernández and Ingallinella, 2010). After some time of operation of RO plants in Santa Fe and in other parts of Argentina, it has been observed that, in some cases, the inlet water contained a large amount of solids due to deficient design of perforation or due to geological conditions of the aquifers. In those cases, it was necessary to introduce some pretreatment step such as sand filters.

Figure 22.4.   Images of the As removal units of the water treatment of (a), (b) Villa Cañas and (c) López, both in the Santa Fe province (Argentina) (from Ingallinella *et al.*, 2003b).

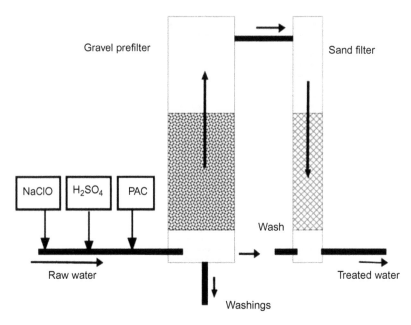

Figure 22.5.   Flow sheet of the arsenic and fluoride removal plant of Villa Cañás in the province of Santa Fe (Argentina) (adapted from Ingallinella *et al.*, 2003a).

Table 22.5.   Quality of inlet water of San José de Chácal (Santa Fe Argentina) (Ingallinella, 2003a; OSE, 2006).

Parameter	Unit	Value range
Turbidity	UNT	2.0–72
Color	UC	5
Alkalinity (CaCO$_3$)	mg L^{-1}	75–200
pH		8.1–8.4
Total hardness (as CaCO$_3$)	mg L^{-1}	350–415
Dry residue 105°C	mg L^{-1}	970–1170
Cl$^-$	mg L^{-1}	270–300
SO$_4^{2-}$	mg L^{-1}	290–420
Fe	mg L^{-1}	0.01–0.04
As	μg L^{-1}	70–110

In other cases, the high silica contents obliged the use of antiscalants, with the subsequent cost increase up to 20%. Since water rejection in normal RO is around 50%, this is an important problem in dry regions. Frequently, the useful life of the membranes is too short due to deficient pretreatment design. Many existing plants are installed without previous studies and they require trained personnel, which is not always available. Many plants work under their capacity, or are not used because service providers cannot afford to cover the operating costs. Distribution networks are old and have consequently many leaks, forcing an increase of treatment plant capacity with subsequent increasing costs. In addition, there is no safe disposal of sludge. The technology is not applicable to low-income communities because they cannot afford the costs. Many plants have been installed without performing previous studies and the operation experiences indicate the need of improving the design of the extraction wells.

Table 22.6. Quality of inlet water of San Antonio de los Cobres (Salta province, Argentina) (Ingallinella, 2003a; OSE, 2006).

Parameter	Unit	Value range
Turbidity	UNT	0.3–1.8
Color	UC	<1
Alkalinity ($CaCO_3$)	mg L^{-1}	129–139
pH		7.8–8.1
Conductivity	mg L^{-1}	645–684
Total hardness ($CaCO_3$)	mg L^{-1}	137–161
Dry residue (105°C)	mg L^{-1}	461–479
NH_4^+	mg L^{-1}	<0.03
Cl^-	mg L^{-1}	107–127
SO_4^{2-}	mg L^{-1}	25–44
Fe	mg L^{-1}	<0.06
NO_2^-	mg L^{-1}	<0.01
Mg	mg L^{-1}	<0.04
As	mg L^{-1}	0.12–0.25
B	mg L^{-1}	4.1-6.4

The treatment and safe disposal of wastes generated in the conventional and RO treatment plants are still a not totally solved problem. The most used disposal options frequently affect either the soil (groundwater) or other surface water sources. Moreover, there are no systematic sludge environmental impact studies.

Several RO plants have been installed in the Buenos Aires province to treat waters of medium-size localities such as Suipacha, but there is no reliable documentation about the results of these plants, particularly in relation to their maintenance and sustainability. The only documented results are those from AySA (Aguas y Saneamientos Argentinos), which has installed the largest potabilizing plant by RO in South America to treat borehole waters coming from González Catán and Virrey del Pino, in the southern periphery of Buenos Aires City, benefiting 400,000 inhabitants (AySA 2014a). A similar plant is being installed at present in the district of Ezeiza (AySA 2014b).

Another interesting work is the installation of an industrial scale water treatment plant in Venado Tuerto city (Santa Fe province, Argentina), which uses RO processes with special membranes Ultra-Osmosis (URO) developed by Dow Chemical. These present very high As rejection and reduced retention of SiO_2 which is an important membrane fouling source. The plant has been successfully running for more than a year with approximately 4800 m^3 day^{-1} of water intake and can reduce the As levels in the wells from around 325 to less than 10 µg L^{-1}. The plant is currently expanding to overcome the average consumption by connection, which is continuously growing. New modules are planned to be installed with the RO-URO system (Armas *et al.*, 2014).

22.2.3 *Guatemala*

In Guatemala, an interesting example of a plant installed to supply As-free water to the Naranjo County (Mixco, close to the capital city of Guatemala), with 3000 inhabitants, has been recently reported (Cardoso *et al.*, 2010; Garrido and Aviles, 2008). The raw water is supplied from two wells 370 m depth, with maximum flows of 400 and 530 L s^{-1}, respectively. The concentration of As in the wells ranged between 140 and 170 µg L^{-1}, 14 and 17 times over the value of the Guatemala COGUANOR 29 NGO 001:99 and WHO standards (10 µg L^{-1}). Another parameter that did not comply with the Guatemala standard was pH, with a value of 6.0 in both wells and below the lower limit of 7.5 (Guatemala, 2001).

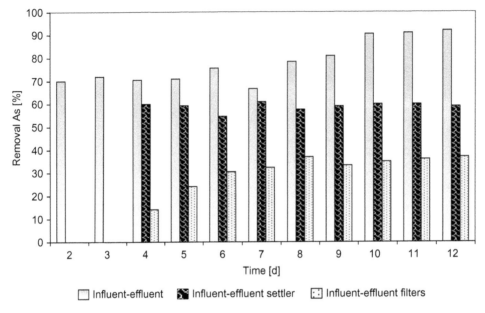

Figure 22.6.   Evolution of arsenic removal efficiency with time of the full scale treatment plant installed in Mixco, Guatemala (Cardoso *et al.*, 2010; Garrido *et al.*, 2009).

Pilot tests, basic engineering design and the development of a full scale As removal plant based on conventional processes using coagulation/co-precipitation with $FeCl_3$ were initiated in 2007, with the objective of achieving the limit of $10 \mu g L^{-1}$ of As in the treated water. The As removal plant consisted of the following units: a pre-oxidation, pH regulation, fast mixing unit (velocity gradient: $1.2 s^{-1}$), four flocculation basins, two sedimentation basins (surface loading rate: $72 m^3 m^{-2} d^{-1}$), four pressure filter units (0.15 MPa) with a ceramic medium (Microlite-Kinetivo® filtration rate of 25 m h^{-1}) and a disinfection unit. Dosage of 12 mg $FeCl_3 L^{-1}$, 1 mg L^{-1} Poliflocal-CH, 1 mg NaClO L^{-1}, and pH adjustment at 7.5 with sodium hydroxide were found to be the optimal operational conditions (Archer and Elmore, 2010), providing an As removal efficiency higher than 90% (Garrido and Avilés, 2008).

Once the plant was operating, the efficiency of the treatment was evaluated. The maximum elimination efficiencies were: 100% for turbidity, 90% for color and 92% for As, with values below $10 \mu g L^{-1}$ in the filter effluent (Fig. 22.6). On the other hand, an increase of the color and iron of 100% in the filter influent was observed (80 UPt-Co and 1.6 mg L^{-1}, respectively), due to the ferric chloride dosage. The Microlite-KINETICO® filters removed this color and iron content to a large extent (79% and 85%, respectively), achieving final values in the effluent of 3 UPt-Co for color and 0.32 mg L^{-1} for iron.

A full-scale treatment plant, built in March 2008 (Río Azul Plant) with a Macrolite-KINETICO® filter media, gave excellent results: an effluent with As concentration below $10 \mu g L^{-1}$ was obtained, in accordance with the water quality standard of Guatemala. The operating and maintenance cost is 0.2 US$ m^{-3} treated drinking water. This cost does not include the management of the As-containing sludge generated, which, after being concentrated, is typically dried up to more than 80–90%, and then stored in landfills.

22.2.4   *Peru*

The treatment plant in the city of Ilo was built in 1982 and designed to eliminate As and turbidity. The treatment used initially massive doses of 90% lime (CaO), but many difficulties were encountered and it was necessary to use new methods in order to reduce costs, increase the efficiency and improve the quality of the treated water. After laboratory and pilot trials, the solution taken was based on the use of ferric chloride or ferric hydroxide and sulfuric acid, removing As at high pH through coagulation and flocculation with natural $Mg(OH)_2$ (Castro de Esparza, 2006).

22.2.5 *Mexico*

Mexican law through the 2000 Amendment to Mexican Official Standard, NOM 127 SSA1-1994 established a permissible limit of $25\,mg\,L^{-1}$ of As in drinking water (Diario Oficial de la Federación, 2000) to be fulfilled in 2005. In Mexico, there is not currently any estimation of the population exposed to the ingestion of As in drinking water; however, in some localities of the states of Chihuahua, Coahuila, Durango, San Luis Potosí, Guanajuato, Jalisco, Morelos, Hidalgo and Guerrero, water sources are contaminated with As (Cebrian *et al.*, 1994; Armienta *et al.*, 2001; Armienta and Segovia, 2008). In most cases, no treatment processes have yet been implemented due to the cost of the initial investment, the cost of the operation and maintenance of the plant or processing equipment, among other reasons. Efficient processes for As removal, such as adsorption on activated alumina or ferric hydroxides, ion-exchange resins and reverse osmosis, have limitations for application in Mexican groundwater. The main causes are the requirement of expensive chemicals, their correct management, and importation of supplies, equipment or components. Among other causes, the loss of 20–40% of the total water flow disposed through rejection in RO processes and the generation of considerable amounts of sludge highly concentrated in As coming from the regeneration of adsorbents or ion exchangers can be cited. Then, the technological solution typically adopted has been to find new safe sources of water, or to mix waters to decrease As concentration by dilution.

In recent years greater importance has been attached to As removal by filtration. (Garrido, *et al.*, 2013a; 2013b), have been studying the behavior of two filter media (manganese greensand and sand-anthracite) to remove As from groundwater in Torreon (Coahuila), with As concentrations averaging from 0.070 to $0.085\,mg\,L^{-1}$. Pilot tests conducted in Torreon wells, at $0.6\,L\,s^{-1}$ flow rate and 4, 7 and $10\,m\,h^{-1}$ linear velocity to evaluate the efficiency of As removal, concluded that reduction of the size of the grain media favors the As removal capacity. Deriving from this study, the Federal and State Governments of the states of Coahuila and Durango and Metropolitan Fund have invested in the installation of ten filtration plants in the wells situated in the Comarca Lagunera, to comply with Mexican regulations regarding As from $0.025\,mg\,L^{-1}$.

Additionally, a case study was presented to demonstrate the feasibility of the ultrafiltration (UF) technology in a municipal water treatment plant in Celaya, a small city in the state of Guanajuato, Mexico, with a population of 340,400 inhabitants (de Carvalho *et al.*, 2014). The Celaya plant has a production capacity of $108\,m^3\,h^{-1}$, based on 32 ultrafiltration modules (2 skids with 16 modules each), a system recovery higher than 98% and design flux of $66\,L\,m^{-2}\,h^{-1}$. The UF technology uses pressurized vertical modules ($51\,m^2$ active area) with hydrophilic PVDF polymeric hollow-fiber membranes providing high strength and chemical resistance, $0.03\,\mu m$ nominal pore diameter for removal of bacteria, viruses, and particulates, outside-in flow configuration for high tolerance to feed solids, and dead-end flow that offers higher recoveries. The application of UF proved to be adequate to produce high quality water appropriate to be distributed to the community. The UF pore size allows the removal of the As flocks providing high quality water and reducing the amount of coagulant needed in the process. Additionally, the method is capable of removing bacteria and virus without using a secondary step. The use of ultrafiltration decreases the amount of waste to be disposed of, this being more environmentally friendly. Furthermore, the water used in the backwash is recycled and the slurry is sent to adequate disposal. Future studies should seek to increase the As removal, making UF technology become more widespread.

22.2.6 *Brazil*

The Iron Quadrangle (IQ) is a worldwide known iron-gold geo-province located in the south-center part of the Minas Gerais state, in Brazil. Among the mineral deposits existing in this area, gold is one of the oldest exploited. Gold mineralization occurs in hydrothermal veins rich in As-sulfide minerals. There is a direct correlation between gold prospects and As-anomalies and, therefore, high As concentrations are found in soils and sediments related both to lithology and to mining activities. The estimated 500 tons gold production from the Morro Velho mine is associated with 40 million metric tons of tailings containing 0.8–8% As (Matschullat *et al.*, 2007). However, despite the important As anomalies and long-lasting mining activities in the IQ, the enriched iron environment creates favorable conditions for a natural mitigation process that has contributed to reducing the potential impact of As release from tailings and residues. Arsenic fixation in the environment was shown to take place via co-precipitation and adsorption onto iron, aluminum and manganese oxyhydroxides (Deschamps *et al.*, 2003; Ladeira and Ciminelli, 2004; Ladeira *et al.*, 2002; Vasconcelos *et al.*, 2004).

Industrial practice of As fixation and disposal related to gold processing in the IQ region has involved three main processes: (i) high-lime (HL), (ii) ferrihydrite precipitation (FH) and (iii) hydrothermal precipitation. Precipitation with lime or lime combined with ferric salts leading to the formation of calcium arsenates, widely used in the past, has been replaced by precipitation with iron salts due to the well-known instability of calcium arsenates.

Gold ores from the Cuiaba mine have been processed in the Nova Lima district since the middle 80s. Gold is found along shear zones, hosted in oxide-facies and carbonate-facies of banded iron formations-BIFs and associated mainly with pyrite or arsenopyrite (Barcellos, 1992; Lobato *et al.*, 2001; Magalhaes, 2002; Vieira, 1991). Gold extraction from the Cuiaba deposit requires a stage of roasting prior to cyanidation. Minor arsenopyrite (FeAsS) present in the sulfide concentrate is volatized in the form of arsenic trioxide (As_2O_3) during roasting, absorbed in water in the gas washing-towers and, finally, removed by co-precipitation/adsorption with addition of iron salts and lime. The resulting residues (arsenical slimes) are disposed of in lined pits located in the area under influence of the metallurgical operation. The drained water is recycled to the industrial unit (Pantuzzo and Ciminelli, 2010; Pantuzzo *et al.*, 2007).

In the HL process, used from 1987 to 1994 (Fig. 22.7) the sulfate rich arsenical effluent (pH 5, 200–300 mg As L^{-1}) leaving the roasting washing tower was neutralized with lime suspension in the two consecutive Pachuca (air stirred) tanks (pH $\sim$ 10, and pH $\sim$ 13, respectively). Iron solution was simultaneously added ($(Fe/As)_{molar} \sim 0.4$) in the first Pachuca tank. After thickening, the resultant precipitate was directed to pits of 13,500 m^3 capacity. A portion of the suspension was recirculated as seeds. Note that the As:Fe ratio was not sufficient to guarantee the formation of Fe-As co-precipitates. Part of the As was likely removed as calcium arsenate compounds.

Since 1994, the residues have been generated by means of the FH process. In this process, As is firstly oxidized and co-precipitates as ferric arsenate ($(Fe/As)_{molar} \approx 1.3$, pH ≈ 5). The suspension goes to a second Pachuca tank to be lime-neutralized until a pH of approximately 8. In the ferrihydrite process, an oxidant (sodium peroxide) is added to the first tank in order to assure the oxidation of arsenic and ferrous ions and the formation of ferric arsenate. A second stage of polishing is required to further reduce As concentration. In this stage, As is mainly removed by adsorption.

Long-term stability of As residues produced by the HL and FH processes was investigated by determining As phases remaining in mining residues after two decades of impoundment (Pantuzzo and Ciminelli, 2010). As(V) was found in both cases, predominantly in the form of amorphous iron arsenate (55–75% As_{total}) and sorbed onto amorphous iron oxi-hydroxides (20–37% As_{total}). There was a clear passive enrichment of iron in the HL residues. This finding, combined with the relatively low concentration of calcium, sulfur and As, compared to those of FH residues, indicated the remobilization of soluble Ca-arsenate (e.g. CaHAsO$_4 \cdot$ H$_2$O), gypsum and As(III), during 16–23 years of residue disposal (Bothe and Brown, 1999).

The precipitation of arsenical ferrihydrite at room temperature generates large volumes of waste with low As content (e.g. 3–8%). The precipitation as crystalline ferric arsenate offers the advantages of combining relatively high As content and relatively low release of As in aqueous solutions. Crystalline ferric arsenates are readily produced under hydrothermal conditions (temperatures of 150°C or higher). However, the relatively high capital and operational costs associated with hydrothermal operations make this option economically attractive only if it can

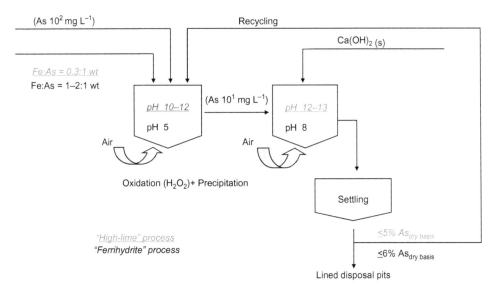

Figure 22.7.   Treatment plant for arsenic stabilization using lime and ferrihydrite (adapted from Pantuzzo *et al.*, 2007).

be coupled with the extraction of a valuable product, such as gold. The Sao Bento refractory gold ores consist of sulfide replacements in iron formation where gold is mainly associated with arsenopyrite. The processing of Sao Bento refractory gold ores required arsenopyrite oxidation prior to cyanidation. The oxidation was carried out by bioxidation and pressure oxidation. Most of the As was fixed within the autoclave by precipitation as scorodite ($FeAsO_4 \cdot 2H_2O$) and more complex compounds ($Fe(AsO_4)_x(SO_4)_y(OH)_z)$). These compounds are considered some of the most stable As compounds (Carageorgos and Monhemius, 1993). The final residue contained approximately 4% As and 30% ferric oxide.

As an original example of industrial management of As tailings in mining areas, oxisol liners have been used in tanks providing containment for sulfide tailings (ca. 15% S, as pyrite and arsenopyrite, 100 mg L^{-1} of dissolved As) generated in gold cyanidation of flotation concentrates (Esper *et al.*, 2007). Over 15 years of groundwater monitoring has proved that the enriched Fe- and Al-oxisol liner is an efficient system for the retention and attenuation of tailings dam seepage at this mining site (Bundschuh *et al.*, 2009, 2010b). The highest As(III) and As(V) uptakes were correlated with the highest content of aluminum and iron oxihydroxides in the oxisol samples (gibbsite, goethite with minor hematite) and the highest specific surface area of sample (Ladeira and Ciminelli, 2004; Ladeira *et al.*, 2002). Zero-valent iron has also been used to remove low As concentrations from mining waters (Pantuzzo *et al.*, 2014).

22.3   CONCLUSIONS

Conventional technologies (coagulation/ co-precipitation, adsorption and reverse osmosis) have been applied for As removal from centralized water supplies on the large and medium scale in Chile, Argentina, Guatemala and Peru. Technologically complex techniques such as adsorption using tailored adsorbents or membrane processes may be too expensive for populations with low economic resources such as those in Latin America. Sometimes, reluctance of consumers in accepting changes in the organoleptic properties of waters they have been drinking for years is another problem for the implementation of a removal procedure. In addition, As treatment units require very sensitive monitoring and maintenance facilities; this is not always feasible in some areas in LA. Furthermore, the volume, handling and final disposal of the wastes generated during the treatment processes is becoming a new problem due to the recent regulations, as has been reported in Chile and in Argentina (Código Alimentario Argentino 2007; Sancha, 2003).

After decades of experience in Chile and Argentina in the field of coagulation-based As removal plants, it was shown that coagulation is a very efficient procedure for removing As from water in the tested cases, reaching the recommended WHO and national standards ($10 \, \mu g \, L^{-1}$) (WHO, 2011). This is a cost-effective technology dealing with low to medium production capacity plants (30 to $520 \, L \, s^{-1}$), which requires moderate capital investment (reactive dosifiers, mixers, filters, etc.), when compared with other technologies like RO. The modified coagulation process (C/F) is a simplified procedure that can be applied to groundwater with low turbidity, with a good correlation between residual turbidity (flocs with adsorbed As) and As concentration in the treated water. The composition of water and the As oxidation state are the critical factors that regulate the application of the process, and water has to be preoxidized to assure the total oxidation of As(III). The coagulation/adsorption/filtration processes are low-cost technologies that do not require extremely trained personnel or expensive chemicals.

In the particular case of the Chilean plants, the optimization of the coagulation step by control of the pH and the coagulant dose allowed the attainment of a residual concentration of As of $10 \, \mu g \, L^{-1}$, achieving the national regulations since 2008. Although plants built before 2005 had a period to reach the legislation requirement, all the existing plants in Chile began to produce water with As levels of $10 \, \mu g \, L^{-1}$ since 2004, demonstrating that the coagulation technology was suitable to reach these requirements.

In the case of Argentina, plants developed with the ArCIS-UNR process (coagulation/adsorption/filtration process with PAC or ferric chloride, followed by double filtration), were found efficient in removing both As and F^-. The particular use of PAC provides removal capacities allowing to reach the value recommended by WHO in a pH range of 6.5 to 8.

On the other hand, in the last decades several RO plants have been installed in Argentina to remove As and salinity. However, the high levels of solids, including high silica content obliged the use of antiscaling reagents. More importantly, water rejection is around 50%, a detrimental factor for desert zones, and the lifetime of the membranes used so far in most of the cases is low due to a deficient pretreatment design. The experience in Argentina shows that RO has elevated operating costs, being not affordable by poor communities. Moreover, there are no systematic sludge environmental impact studies. However, recently, an improved RO plant has been installed in the periphery of Buenos Aires.

Other examples of As removal plants have been reported in Peru and in Guatemala. In the first case, a technology using coagulation/co-precipitation with $FeCl_3$ has been used, whereas in the second case, $FeCl_3$, a flocculant plus ceramic-based filtering media were used.

Concerning removal and disposal of As from industrial effluents containing high levels of the element in solution (copper and gold mining and metallurgical plants of Chile and Brazil), co-precipitation methods with iron salts either as amorphous (arsenical ferrihydrite) or crystalline arsenates are the predominant technologies.

In spite of all the difficulties and problems that had to be faced, the Latin American experience gives valuable information that could be used to solve this problem in other regions of the world, especially in countries of Asia where the first option is to find other water sources not contaminated with As.

REFERENCES

Akin, I., Arslan, G., Tor, A., Cengeloglu, Y. & Ersoz, M. (2011) Removal of arsenate [As(V)] and arsenite [As(III)] from water by SWHR and BW-30 reverse osmosis. *Desalination*, 281, 88–92.
An, B., Fu, Z., Xiong, Z., Zhao, D. & SenGupta, A.K. (2010) Synthesis and characterization of a new class of polymeric ligand exchangers for selective removal of arsenate from drinking water. *Reactive and Functional Polymers*, 70, 497–507.

Archer, A. & Elmore, A.C. (2010) Use of ceramic pot filters for drinking water disinfection in Guatemala. *Proceedings of the World Environmental and Water Resources Congress, 16–20 May, Providence, Rhode Island*. pp. 545–558.

Armas, A., Olivieri, V., Mauricci, J.J. & Silva, V. (2014) Reduction of arsenic from groundwater by using a coupling of RO and processes. In: Litter, M.I., Nicolli, H.B., Meichtry, J.M., Quici, N., Bundschuh, J., Bhattacharya, P. & Naidu, R. (eds.) *One Century of the Discovery of Arsenicosis in Latin America (1914–2014): Proceedings of the 5th International Congress on Arsenic in the Environment, 11–16 May 2014, Buenos Aires, Argentina*. CRC Press, Boca Raton, FL. pp. 759–761.

Armienta, M.A. & Segovia, N. (2008) Arsenic and fluoride in the groundwater of Mexico. *Environmental Geochemistry and Health*, 30, 345–353.

Armienta, M.A., Villaseñor, G., Rodríguez, R., Ongley, L.K. & Mango, H. (2001) The role of arsenic bearing rocks in groundwater pollution at Zimapán Valley, México. *Environmental Geology*, 40, 571–581.

AySA (2014a) Planta potabilizadora por el sistema de ósmosis inversa – construcción de la Planta en La Matanza. Available from: http://www.aysa.com.ar/index.php?id_seccion=562 [accessed June 2014].

AySA (2014b) Resumen ejecutivo – estudio de impacto ambiental: expansión del sistema de producción y distribución de agua potable Batería Ezeiza, Batería La Unión y redes secundarias. Available from: http://www.aysa.com.ar/index.php?id_contenido=1550&id_seccion=570 [accessed June 2014].

Azbar, N. & Turkman, A. (2000) Defluoridation in drinking waters. *IWA Water Science and Technology*, 42, 403–407.

Bang, S., Patel, M., Lippincott, L. & Meng, X. (2005) Removal of arsenic from groundwater by granular titanium dioxide adsorbent. *Chemosphere*, 60, 389–397.

Barahona, J. & González, Z. (1987) Estudio a nivel de planta piloto sobre abatimiento de arsénico del agua de Taltal. *Proceedings VII Congreso Chileno de Ingeniería Sanitaria y Ambiental Copiapó, Chile*. pp. 880–896.

Barcellos, C. (1992) Arsenic contamination in a coastal environment affected by a zinc semelting plant (Sepetiba Bay, Brazil). In: Sancha, A.M. (ed.) *Arsenic in the environment and its incidence on health. International Seminar at University of Chile, May 1992, Santiago de Chile, Chile*. pp. 59–62.

Bilici Baskan, M. & Pala, A. (2010) A statistical experiment design approach for arsenic removal by coagulation process using aluminum sulfate. *Desalination*, 254, 42–48.

Bissen, M. & Frimmel, F.H. (2003) Arsenic – a review. Part II: Oxidation of arsenic and its removal in water treatment. *Acta Hydrochimica et Hydrobiologica*, 31, 97–107.

Borgoño, J. & Greiber, R. (1971) Epidemiologic study of arsenic poisoning in the city of Antofagasta. *Revista Medica Chile*, 99, 702–707.

Borgoño, J. & Vicent, O. (1977) Arsenic in the drinking water of the city of Antofagasta: epidemiological and clinical study before and after the installation of a treatment plant. *Environmental Health Perspectives*, 19, 103–105.

Bothe, J.V. & Brown, P. (1999) Arsenic immobilization by calcium arsenate formation. *Environmental Science & Technology*, 33, 3806–3811.

Bundschuh, J., García, M.E., Birkle, P., Cumbal, L.H., Bhattacharya, P. & Matschullat, J. (2009) Occurrence, health effects and remediation of arsenic in groundwaters of Latin America. In: Bundschuh, J., Armienta, M.A., Birkle, P., Bhattacharya, P., Matschullat, J. & Mukherjee, A.B. (eds.) *Natural arsenic in groundwater of Latin America*. CRC Press, Boca Raton, FL. pp. 3–15.

Bundschuh, J., Bhattacharya, P., Hoinkis, J., Kabay, N., Jean, J. & Litter, M. (2010a) Groundwater arsenic: from genesis to sustainable remediation. *Water Research*, 44, 5511–5523.

Bundschuh, J., Litter, M., Ciminelli, V., Morgada, M.E., Cornejo, L., Garrido Hoyos, S., Hoinkis, J., Alarcón-Herrera, M.T., Armienta, M.A. & Bhattacharya, P. (2010b) Emerging mitigation needs and sustainable options for solving the arsenic problems of rural and isolated urban areas in Iberoamerica – a critical analysis. *Water Research*, 44, 5828–5845.

Bundschuh, J., Litter, M.-I., Parvez, F., Román-Ross, G., Nicolli, H.B, Jiin-Shuh, J., Chen-Wuing, L., López, D., Armienta, M.A., Guilherme, L.R.G., Gomez Cuevas, A., Cornejo, L., Cumbal, L. & Toujaguez, R. (2012) One century of arsenic exposure in Latin America: a review of history and occurrence from 14 countries. *Science of the Total Environment*, 429, 2–35.

Caetano, M.L., Ciminelli, V.S.T., Rocha, S.D.F., Spitale, M.C. & Caldeira, C.L. (2009) Batch and continuous precipitation of scorodite from diluted industrial solutions. *Hydrometallurgy*, 95, 44–52.

Carageorgos, T. & Monhemius, A.J. (1993) Iron arsenic compounds formed during acidic pressure oxidation of arsenopyrite. In: Harris, B. & Krause, E. (eds.) *Proceedings Impurity Control and Disposal in Hydrometallurgical Processes*. CIM, Toronto, Canada. pp. 101–124.

Cardoso, S., Grajeda, C., Argueta, S. & Garrido, S. (2010) Experiencia satisfactoria para la remoción de arsénico en Mixco, Guatemala. In: Litter, M.I., Sancha, A.M. & Ingallinella, A.M. (eds.) *Tecnologías económicas para el abatimiento de arsénico en aguas*. Editorial Programa Iberoamericano de Ciencia y Tecnología para el Desarrollo, Buenos Aires, Argentina. pp. 179–189.

Castro de Esparza, M.L. (2006) Removal of arsenic from drinking water and soil bioremediation. *Natural arsenic in groundwaters of Latin America: International Congress, 20–24 June, Mexico-City, Mexico*.

Cebrián, M.E., Albores, A., García-Vergas, G. & Del Razo, L.M. (1994) Chronic arsenic poisoning in humans: the case of Mexico. In: Nriagu, J.O. (ed.) *Arsenic in the environment. Part II: Human health and ecosystem effects*. John Wiley & Sons Inc., New York, NY. pp. 93–107.

Cerda, W., Gatica, R. & Veneros, M. (1999) Sistema de recuperación de aguas de descarte y disposición final de lodos arsenicados. *AIDIS-Chile*, 24, 26–31.

Chwirka, J., Colvin, C., Gómez, J. & Mueller, P. (2000) Arsenic removal from drinking water using the coagulation/microfiltration processes. *Journal of American Water Works Association*, 96 (3), 106–114.

Código Alimentario Argentino (CAA) (2007) Secretaría de Políticas, Regulación y Relaciones Sanitarias y Secretaría de Agricultura, Ganadería, Pesca y Alimentos Resolución Conjunta 68/2007 y 196/2007 Modificación. Buenos Aires, Argentina.

Comisión Nacional del Agua (CNA) (2007) National Water Program 2007–2012. Secretaría de Medio Ambiente y Recursos Naturales Boulevard Adolfo Ruiz Cortines No. 4209, Col. Jardines en la Montaña, C.P. 14210, Tlalpan, Mexico, D.F.

D'Ambrosio, C. (2005) Evaluación y selección de tecnologías disponibles para remoción de arsénico. In: Galindo, G., Fernández Turiel, J.L., Parada, M.A., Gimeno Torrente, D. (eds.) *Arsénico en aguas: origen movilidad y tratamiento. Taller II Seminario Hispano-Latinoamericano sobre temas actuales de hidrología subterránea-IV Congreso Hidrogeológico Argentino, 25–28 October 2005, Río Cuarto, Argentina*. pp. 28–38.

De Carvalho, F., Masetto, N., Lima, A., Arbelaez, J., Ramos, R.G. & Avalos, C. (2014) Arsenic removal enhanced by the use of ultrafiltration. In: Litter, M.I., Nicolli, H.B., Meichtry, J.M., Quici, N., Bundschuh, J., Bhattacharya, P. & Naidu, R. (eds.) *One Century of the Discovery of Arsenicosis in Latin America (1914–2014): Proceedings of the 5th International Congress on Arsenic in the Environment, 11–16 May 2014, Buenos Aires, Argentina*. CRC Press, Boca Raton, FL. pp. 775–777.

Demopolous, G.P., Droppert, D.J. & Van Weert, G. (1994) Options for the immobilisation of arsenic as crystalline scorodite. In: Harris, B. & Krause, E. (eds.) *Proceedings Impurity Control and Disposal in Hydrometallurgical Processes, 21–24 August 1994, Toronto, ON*. Canadian Institute of Mining, Metallurgy and Petroleum, CIM, Montreal, Quebec. pp. 57–69.

Deschamps, E., Ciminelli, V.S.T., Weidler, P. & Ramos, A.Y. (2003) Arsenic sorption onto soils enriched in Mn and Fe minerals. *Clays and Clay Minerals*, 51, 197–204.

Diario Oficial de la Federación (2000) Wednesday, 22 November 2000. Modificación a la Norma Oficial Mexicana NOM-127-SSA1-1994, Salud ambiental. Agua para uso y consumo humano. Límites permisibles de calidad y tratamiento a que debe someterse el agua para su potabilización. Available from: http://www.salud.gob.mx/unidades/cdi/nom/127ssa14.html [accessed July 2015].

Driehaus, W. (2002) Arsenic removal-experience with the GEH process in Germany. *Water Science and Technology: Water Supply*, 2, 276–280.

Driehaus, W., Seith, R. & Jekel, M. (1995) Oxidation of arsenate (III) with manganese oxides in water treatment. *Water Research*, 29, 297–305.

Esper, J.A.M.M., Amaral, R.D. & Ciminelli, V.S.T. (2007) Cover design performance at a Kinross gold mine in Brazil. XXII Encontro Nacional de Tratamento de Minérios e Metalurgia Extrativa (ed.) *Proceedings VII Meeting of the Southern Hemisphere on Mineral Technology, 20–23 November 2007, Ouro Preto, Brazil*. Volume II, pp. 607–612.

Farrell, J., Wang, J., O'Day, P. & Conklin, M. (2001) Electrochemical and spectroscopic study of arsenate removal from water using zero-valent iron media. *Environmental Science & Technology*, 35, 2026–2032.

Fernández, R.G. & Ingallinella, A.M. (2010) Experiencia Argentina en la remoción de arsénico por diversas tecnologías, In: Litter, M.I., Sancha, A.M. & Ingallinella, A.M. (eds.) *Tecnologías económicas para el abatimiento de arsénico en aguas*. Buenos Aires, Argentina, Editorial Programa Iberoamericano de Ciencia y Tecnología para el Desarrollo. pp. 155–167.

Fernández, R.G., Ingallinella, A.M. & Stecca, L.M. (2009) Arsenic removal from groundwater by coagulation with polyaluminum chloride and double filtration In: Bundschuh, J., Armienta, M.A., Birkle, P., Bhattacharya, P., Matschullat, J. & Mukherjee, A.B. (eds.) *Natural arsenic in groundwaters of Latin-America*. CRC Press, Boca Raton, FL. pp. 665–679.

Ferreccio, C. & Sancha, A.M. (2006) Arsenic exposure and its impacts on health in Chile. *Journal of Health, Population and Nutrition*, 24, 164–175.

Francesconi, K.A. & Kuehnelt, D. (2004) Determination of arsenic species: a critical review of methods and applications, 2000–2003. *Analyst*, 129, 373–395.

Fuentealba, C. (2003) *Planta piloto para remover arsénico en una fuente subterránea de agua potable.* Santiago de Chile, Faculty of Physical and Mathematical Sciences, University of Chile, Santiago de Chile, Chile.

Garrido, S. & Avilés, M. (2008) *Análisis de la información técnica y evaluación del funcionamiento de la planta para remoción de arsénico, Mixco, Guatemala.* Final Report. Mexican Institute of Water Technology (IMTA), Jiutepec, Mexico.

Garrido, S., Avilés, M., Ramírez, A., Calderón, C., Ramírez-Orozco, A., Nieto, A., Shelp, G., Seed, L., Cebrián, M. & Vera, E. (2009) Arsenic removal from water of Hualtla, Morelos, Mexico using a capacitive deionization. In: Bundschuh, J., Armienta, M.A, Birkle, P., Bhattacharya, P., Matschullat, J. & Mukherjee, A.B. (eds.) *Natural arsenic in groundwaters of Latin-America.* CRC Press, Boca Raton, FL. pp. 665–679.

Garrido, S., Avilés, M., Grajeda, C., Cardozo, S., Velásquez, H. & Ramírez, A. (2013a) Comparing two operating configurations in a full-scale arsenic removal plant. Case study Guatemala. *Water*, 5, 834–851.

Garrido, S., Piña, M., López, I., De La O, D. & Rodríguez, R. (2013b) Behavior of two filter media for to remove arsenic from drinking water. The Filtration Society. *International Journal for Filtration and Separation*, 13, 21–26.

Gibert, O., Pablo, J., Cortina, J.-L. & Ayora, C. (2010) In situ removal of arsenic from groundwater by using permeable reactive barriers of organic matter/limestone/ zero-valent iron mixtures. *Environmental Geochemistry and Health*, 32, 373–378.

González, A., Ingallinella, A.M., Sanguinetti, G.S., Pacini, V.A., Fernández, R.G. & Quevedo, H. (2014) Enhancing arsenic removal by means of coagulation-adsorption-filtration processes from groundwater containing phosphates. In: Litter, M.I., Nicolli, H.B., Meichtry, J.M., Quici, N., Bundschuh, J., Bhattacharya, P. & Naidu, R. (eds.) *One Century of the Discovery of Arsenicosis in Latin America (1914–2014): Proceedings of the 5th International Congress on Arsenic in the Environment, 11–16 May 2014, Buenos Aires, Argentina.* CRC Press, Boca Raton, FL. pp. 705–707.

Granada, J., Cerda, W.E. & Godoy, D.A. (2003) El camino para reducir notoriamente el arsénico en agua potable. *AIDIS-Chile*, 34, 44–49.

Guatemala, Ministerio de Salud Pública y Asistencia Social (2001) Unidad Ejecutora del Programa de Acueductos Rurales. Norma Guatemalteca obligatoria, agua potable (especificaciones), CONGUANOR, NGO, 29.001. UNEPAR Guatemala.

Harris, B. (2003) The removal of arsenic from process solutions: theory and industrial practice. In: The Minerals, Metals and Materials Society *Hydrometallurgy*. Volume 2. pp. 1889–1902.

Höll, W. & Litter, M.I. (2010) Ocurrencia y química del arsénico en aguas. Sumario de tecnologías de remoción de arsénico de aguas. In: Litter, M.I., Sancha, A.M., Ingallinella, A.M. (eds.) *Tecnologías económicas para el abatimiento de arsénico en aguas.* Editorial Programa Iberoamericano de Ciencia y Tecnología para el Desarrollo, Buenos Aires, Argentina. pp. 17–27.

Ingallinella, A.M., Fernández, R.G. & Stecca, L.M. (2003a) Proceso ARCIS-UNR para la remoción de As y flúor en aguas subterráneas: una experiencia de aplicación. Part 1. *Revista Ingeniería Sanitaria y Ambiental*, *AIDIS Argentina*, 66, 36–39.

Ingallinella, A.M., Fernández, R.G. & Stecca, L.M. (2003b) Proceso ARCIS-UNR para la remoción de As y flúor en aguas subterráneas: una experiencia de aplicación. Part 2. *Revista Ingeniería Sanitaria y Ambiental*, *AIDIS Argentina*, 67, 61–65.

Kabay, N., Bundschuh, J., Hendry, B., Bryjak, M., Yoshizuka, K., Bhattacharya, P. & Anaç, S. (eds.) (2010) *The global arsenic problem: challenges for safe water production.* CRC Press, Boca Raton, FL.

Kartinen, E.O., Jr. & Martin, C.J. (1995) An overview of arsenic removal processes. *Desalination*, 103, 79–88.

Ladeira, A.C.Q. & Ciminelli, V.M.T. (2004) Adsorption and desorption of arsenic on an oxisol and its constituents. *Water Research*, 38, 2087–2094.

Ladeira, A.C.Q., Ciminelli, V.S.T. & Nepomuceno, A.L. (2002) Soil selection for arsenic immobilization (in Portuguese). *REM*, 55, 215–221.

Latorre, C. (1966) Estudio físico-químico para la remoción del arsénico en el río Toronce. *AIDIS Argentina*, 12–21.

Litter, M.I., Fernández, R.G., Cáceres, R.E., Grande Cobián, D., Cicerone, D. & Fernández Cirelli, A. (2008) Tecnologías de bajo costo para el tratamiento de arsénico a pequeña y mediana escala. *Revista Ingeniería Sanitaria y Ambiental*, *AIDIS Argentina*, 100, 41–50.

Litter, M.I., Morgada, M.E. & Bundschuh, J. (2010a) Possible treatments for arsenic removal in Latin American waters for human consumption. *Environmental Pollution*, 158, 1105–1118.

Litter, M.I., Sancha, A.M. & Ingallinella, A.M. (eds.) (2010b) *Tecnologías económicas para el abatimiento de arsénico en aguas*. Editorial Programa Iberoamericano de Ciencia y Tecnología para el Desarrollo, Buenos Aires, Argentina.

Litter, M.I., Alarcón-Herrera, M.T., Arenas, M.J., Armienta, M.A., Avilés, M., Cáceres, R.E., Cipriani, H.N., Cornejo, L., Dias, L.E., Fernández Cirelli, A., Farfán, E.M., Garrido, S., Lorenzo, L., Morgada, M.E., Olmos-Márquez, M.A. & Pérez-Carrera, A. (2012) Small-scale and household methods to remove arsenic from water for drinking purposes in Latin America. *Science of the Total Environment*, 429, 106–121.

Litter, M.I., Nicolli, H.B., Meichtry, J.M., Quici, N., Bundschuh, J., Bhattacharya, P. & Naidu, R. (eds.) (2014) *One Century of the Discovery of Arsenicosis in Latin America (1914–2014): Proceedings of the 5th International Congress on Arsenic in the Environment, 11–16 May 2014, Buenos Aires, Argentina*. CRC Press, Boca Raton, FL.

Lobato, L., Rodrigues, L.R. & Vieira, F. (2001) Brazil's premier gold province. Part II: Geology and genesis of gold deposits in the Archean Rio das Velhas greenstone belt, Quadrilátero Ferrífero. *Mineralium Deposita*, 36, 249–277.

Madiec, H., Cepero, E. & Mozziconacci, D. (2000) Treatment of arsenic by filter coagulation: a South American advanced technology. *International Water Association, Water Supply*, 18, 613–618.

Magalhaes, M.C.F. (2002) Arsenic. An environmental problem limited by solubility. *IUPAC*, 74, 1843–1850.

Manning, B.A., Hunt, M.L. & Amrhein, C. (2002) Arsenic(III) and arsenic(V) reactions with zerovalent iron corrosion products. *Environmental Science & Technology*, 36, 5455–5461.

Matschullat, J., Birmann, K., Borba, R.P., Ciminelli, V., Deschamps, M.E., Figueiredo, B.R., Gabrio, T., Hassler, S., Hilscher, A., Junghanel, I., Oliveira, N., Rassbach, K., Schmidt, H., Schwenk, M., Villena, M.J.O. & Weidner, U. (2007) Long-term environmental impact of As-dispersion in Minas Gerais, Brazil. Arsenic in soil and groundwater environment. In: Bhattacharya, P., Mukherjee, A.B., Bundschuh, J., Zevenhoven, R. & Loeppert, R.H. (eds.) *Arsenic in soil and groundwater environment*. Volume 9. *Trace metals and other contaminants in the environment*. Elsevier. pp. 365–382.

Mohan, D. & Pittman, C.U. (2007) Arsenic removal from water/wastewater using adsorbents – a critical review. *Journal of Hazardous Materials*, 147, 1–53.

Mondal, P., Bhowmick, S., Chatterjee, D., Figoli, A. & Van der Bruggen, B. (2013) Remediation of inorganic arsenic in groundwater for safe water supply: a critical assessment of technological solutions. *Chemosphere*, 92, 157–170.

NCh 409/1 Instituto Nacional de Normalización, INN-Chile (2005) Drinking Water. Part 1, Requirements.

Newcombe, R.L. & Möller, G. (2008) Arsenic removal from drinking water: a review. Available from: http://www.blueh2o.net/docs/asreview%20080305.pdf [accessed July 2015].

OSE (Obras Sanitarias del Estado) (2006) Provincia de San Juan, Argentina, (Personal communication).

Pantuzzo, F.L. & Ciminelli, V.S.T. (2010) Arsenic association and stability in long-term disposed arsenic residues. *Water Research*, 44, 5631–5640.

Pantuzzo, F.L., Ciminelli, V.S.T. & Braga, I. (2007) Arsenic speciation in arsenic slimes. *Proceedings 7th Meeting of the Southern Hemisphere on Mineral Technology, 20–24 November 2007, Ouro Preto*. Volume 1. pp. 107–114.

Pantuzzo, F.L., Santos, L.R.G. & Ciminelli, V.S.T. (2014) Solubility-product constant of an amorphous aluminum-arsenate phase ($AlAsO_4 \cdot 3.5H_2O$) at 25°C. *Hydrometallurgy*, 144–145, 63–68.

Pirnie, M. (2000) Technologies and costs for removal of arsenic from drinking water. USEPA Report 815-R-00–028.

Ravenscroft, P., Brammer, H. & Richards, K. (2009) *Arsenic pollution: a global synthesis*. Wiley-Blackwell, Oxford, UK.

Rivara, M.I., Cebrian, M., Corey, G., Hernandez, M. & Romieu, I. (1997) Cancer risk in an arsenic-contaminated area of Chile. *Toxicology and Industrial Health*, 13, 321–338.

Rosenberg, H. (1974) Systemic arterial disease and chronic arsenicism in infants. *Archives of Pathology*, 97, 360–365.

Ruanala, M., Leppinen, J. & Miettinrn, V. (2014) Method for removing arsenic as scorodite. US 8790516 B2.

Saitua, H., Campderros, M., Cerutti, S. & Padilla, A.P. (2005) Effect of operating conditions in removal of arsenic from water by nanofiltration membrane. *Desalination*, 172, 173–180.

Sancha, A. (1999) Full-scale application of coagulation processes for Arsenic removal in Chile: a successful case study. In: Chappell, W., Abernathy, C., Calderon, R., (eds.) *Arsenic exposure and health effects*. Elsevier, Amsterdam, The Netherlands. pp. 373–378.

Sancha, A. (2000) Removal of arsenic from drinking water supplies: Chile experience. *Water Supply*, 18, 621–625.

Sancha, A.M. (2003) Removing arsenic from drinking water: a brief review of some lessons learned and gaps arisen in Chilean water utilities. In: Chappell, W.R., Abernathy, C.O., Calderon, R.L. & Thomas, D.J. (eds.) *Arsenic exposure and health effects*. Elsevier, Amsterdam, The Netherlands. pp. 471–481.

Sancha, A.M. (2006) Review of coagulation technology for removal of arsenic: case of Chile. *Journal of Health, Population and Nutrition*, 24, 267–272.

Sancha, A.M. (2010a) Remoción de arsénico por coagulación y precipitación. In: Litter, M.I., Sancha, A.M., Ingallinella, A.M. (eds.) *Tecnologías económicas para el abatimiento de arsénico en aguas*. Editorial Programa Iberoamericano de Ciencia y Tecnología para el Desarrollo, Buenos Aires, Argentina. pp. 33–41.

Sancha, A.M. (2010b) Importancia de la matriz de agua a tratar en la selección de las tecnologías de abatimiento de arsénico. In: Litter, M.I., Sancha, A.M. & Ingallinella, A.M. (eds.) *Tecnologías económicas para el abatimiento de arsénico en aguas*. Editorial Programa Iberoamericano de Ciencia y Tecnología para el Desarrollo, Buenos Aires, Argentina. pp. 145–153.

Sancha, A.M. (2010c) Experiencia Chilena en la remoción de arsénico a escala de planta de Tratamiento, In: Litter, M.I., Sancha, A.M., Ingallinella, A.M. (eds.) *Tecnologías económicas para el abatimiento de arsénico en aguas*. Editorial Programa Iberoamericano de Ciencia y Tecnología para el Desarrollo, Buenos Aires, Argentina. pp. 169–178.

Sancha, A.M. & Frenz, P. (2000) Estimate of the current exposure of the urban population of northern Chile to arsenic. In: *Interdisciplinary perspectives on drinking water risk assessment and management*. IAHS Publication. Volume 260. pp. 3–8.

Sancha, A.M. & Fuentealba, C. (2009) Application of coagulation-filtration processes to remove arsenic from low-turbidity waters. In: Bundschuh, J., Armienta, M.A., Birkle, P., Bhattacharya, P., Matschullat, J. & Mukherjee, A.B. (eds.) *Natural arsenic in groundwaters of Latin America*. CRC Press, Boca Raton, FL. pp. 581–588.

Sancha, A.M. & Ruiz, G. (1984) Estudio del proceso de remoción de arsénico de fuentes de agua potable empleando sales de aluminio. *Proceedings XIX Congreso Interamericano de Ingeniería Sanitaria y Ambiental*. AIDIS Chile. pp. 380–410.

Sancha, A.M., Vega, F. & Fuentes, S. (1992) Speciation of arsenic present in water inflowing to the Salar del Carmen treatment plant in Antofagasta, Chile, and its incidence on the removal process. *International Seminar Proceedings. Arsenic in the environment and its incidence on health*. Universidad de Chile, Santiago de Chile, Chile. pp. 183–186.

Sancha, A.M., O'Ryan, R. & Pérez, O. (2000) The removal of arsenic from drinking water and associated costs: the Chilean case. *Interdisciplinary perspectives on drinking water risk assessment and management*. IAHS Publication. Volume 260. pp. 17–25.

Seidel, A., Waypa, J.J. & Elimelech, M. (2001) Role of charge (Donnan) exclusion in removal of arsenic from water by a negatively charged porous nanofiltration membrane. *Environmental Engineering Science*, 18, 105–113.

Sen, M., Manna, A. & Pal, P. (2010) Removal of arsenic from contaminated groundwater by membrane-integrated hybrid treatment system. *Journal of Membrane Science*, 354, 108–113.

Severn Trent Services (2007–2009). Available from: http://severntrentservices.com/en_us/Arsenic_Removal_Media/Bayoxide__E33_prod_103.aspx.

SGA Estrategias Ambientales (2008) Planta de abatimiento de arsénico y antimonio para el tratamiento de polvos y efluentes de refinería de cobre. Available from: http://www.e-seia.cl/archivos/DIA_Planta_AAA_v12_05_08.pdf [accessed July 2015].

Sharma, V.K. & Sohn, M. (2009) Aquatic arsenic: toxicity, speciation, transformations, and remediation. *Environment International*, 35, 743–759.

Smith, A., Goycolea, M., Haque, R. & Biggs M. (1998) Marked increase in bladder and lung cancer mortality in a region of northern Chile due to arsenic in drinking water. *American Journal of Epidemiology*, 147, 660–669.

Superneau, L. (2012) Codelco, EcoMetals launch arsenic treatment plant. Available from: http://www.bnamericas.com/news/mining/codelco-ecometals-launch-arsenic-treatment-plant [accessed January 2015].

Uddin, M.T., Mozumder, S.I., Figoli A., Islam, A. & Drioli E. (2007) Arsenic eemoval by conventional and membrane technology: an overview. *Indian Journal of Chemical Technology*, 14 (5), 441–450.

USEPA (2006) Arsenic in drinking water. Washington DC, Office of Groundwater and Drinking Water US Environmental Protection Agency. Available from: http://www.epa.gov/ogwdw/arsenic/basicinformation.html [accessed July 2015].

Van der Meer, T. (2011) Arsenic removal from effluents. Outotec reports 2011. Available from: http://www.outotec.com/imagevaultfiles/id_552/cf_2/arsenic_-_sources-_pathways_and_treatment_of_minin.pdf [accessed July 2015].

Vasconcelos, F.M., Ciminelli, V.S.T., Oliveira, R.P. & Silva, R.J. (2004) Determination of chemical speciation and the potential of As remobilization in mining sites. *Geochimica Brasiliensis*, 18, 115–120.

Vieira, F.W.R. (1991) Textures and processes of hydrothermal alteration and mineralization in the Nova Lima Group, Minas Gerais, Brazil. In: Ladeira, E.A. (ed.) *Brazil gold '91, the economics, geology, geochemistry and genesis of gold deposits*. Balkema, Rotterdam, The Netherlands. pp. 319–325.

Vogels, C.M. & Johnson, M.D. (1998) Arsenic remediation in drinking waters using ferrate and ferrous ions. Technical Completion Report 307, Account No. 01-4-23922. New Mexico Water Resources Research Institute, Las Cruces, NM.

Walker, M., Seiler, R.L. & Meinert, M. (2008) Effectiveness of household reverse-osmosis systems in a western US region with high arsenic in groundwater. *Science of the Total Environment*, 389, 245–252.

Wang, L., Chen, A. & Fields, K. (2000) *Arsenic removal from drinking water by ion exchange and activated alumina plants*. USEPA Report 600-R-00-088.

WHO (2011) *Guidelines for drinking-water quality: arsenic in drinking water*. Fact sheet No. 210, World Health Organization, Geneva, Switzerland.

Yamani, J.S., Miller, S.M., Spaulding, M.L. & Zimmerman, J.B. (2012) Enhanced arsenic removal using mixed metal oxide impregnated chitosan beads. *Water Research*, 46, 4427–4434.

Zaldivar, R. (1980) A morbid condition involving cardio-vascular, broncopulmonary, digestive and neural lesions in children and young infants after dietary arsenic exposure. *Zentralblatt für Bakteriologie. 1. Abt. Originale B, Hygiene, Krankenhaushygiene, Betriebshygiene, präventive Medizin*, 170, 44–56.

CHAPTER 23

The removal of arsenic and chromate (Cr(VI)) from drinking water in the United States of America

Paul Sylvester

23.1   INTRODUCTION

As has been discussed previously in this book, the consumption of drinking water containing even relatively low levels of arsenic (As) over long periods of time has been shown to be deleterious to human health. A range of diseases have been linked to As exposure via drinking water, including cancers, dermal lesions, and skin diseases (Kapaj *et al.*, 2006; Smith *et al.*, 1999; Steinmaus *et al.*, 2000) This link between As exposure due to the consumption of contaminated water and human health prompted the United States Environmental Protection Agency (EPA) to lower the Maximum Contaminant Level (MCL) for As in drinking water from $50 \,\mu g \, L^{-1}$ to $10 \,\mu g \, L^{-1}$, effective January 2006. There are large geographic areas of the United States where As occurs naturally in the groundwaters, including parts of New England, the South West, the Mid West and Southern Texas. In the majority of these areas, the As concentration in municipal water supplies rarely exceeded the old arsenic MCL of $50 \,\mu g \, L^{-1}$ so there were relatively few large scale As removal plants in operation. Consequently, when the MCL was reduced to $10 \,\mu g \, L^{-1}$, bringing the United States into line with the World Health Organization (WHO) recommended limit, a large number of water providers were obliged to treat the water to reduce the levels of As. This necessitated the construction and operation of many new water treatment plants specifically designed for As removal.

The regulation of chromate (Cr(VI)) in drinking water has yet to be enforced in the USA and as such there are currently no full scale plants dedicated to Cr(VI) removal in operation to the author's knowledge. At present, chromate is included in the total chromium MCL, which in the United States has been set by the EPA at $0.1 \, mg \, L^{-1}$. The California Environmental Protection Agency (CA-EPA) has a more stringent MCL for total Cr of $50 \,\mu g \, L^{-1}$ and also recently decided to regulate chromate separately. In July, 2014 California implemented an MCL for chromate in drinking water of $10 \,\mu g \, L^{-1}$ (CA EPA – Cr-6, 2014). The USEPA is also considering setting a separate nationwide MCL for chromate following studies that have shown that low levels of chromate are present in the drinking water of numerous cities in the USA (Sutton, 2010). Several pilot scale trials have been undertaken in an effort to identify optimum chromate removal methodologies that can be successfully and economically utilized on a commercial scale.

23.2   AQUEOUS CHEMISTRY OF ARSENIC

The aqueous chemistry of As has been discussed in detail in previous chapters so it is not necessary to go into details when discussing the full scale implementation of As removal technologies. However, it is essential to know the oxidation state of the As species present in the source water prior to making any treatment decisions since this will have a large impact on the plant design and the technology selected to remove As. In oxygenated waters, As exists primarily in the $+5$ oxidation state, as either the anion $HAsO_4^{2-}$ or $H_2AsO_4^{-}$ or a combination of the two depending upon the pH. In relatively anaerobic waters, As in the $+3$ oxidation state predominates and is primarily present as the uncharged species H_3AsO_3 at drinking water pH values. Studies have demonstrated that there are other organic forms of As present in some groundwaters but the concentrations are generally insignificant compared to the amounts of inorganic species present and thus do not influence the selection of a technology. The transition between the As(III) and As(V) oxidation states is generally slow in the absence of a strong oxidant so it is therefore not uncommon to find both As(III) and As(V) species present simultaneously in a water source. Since As adsorbents all have a much lower capacity for As(III) compared to As(V), it is usual to add an oxidant if significant As(III) is present to ensure all of the As(III) is oxidized to As(V) to maximize media life and therefore improve the economics of the water treatment. Also, because As(III) is predominantly present as an uncharged species, ion exchange is an ineffective technique unless the As(III) is first oxidized to As(V).

23.3　ARSENIC REMOVAL TECHNOLOGIES

A variety of methods have been successfully utilized on a municipal scale for the removal of As from drinking water. These include anion exchange, coagulation/filtration and adsorption onto fixed beds of As-selective media. (Reverse osmosis, although effective for As removal, is generally not viable at an industrial scale due to high costs compared to other treatment methods and is thus not used as a primary method for As remediation.) The decision on which type of plant to use is governed by a number of factors including ease of waste disposal, available footprint for the plant, water chemistry and size of the water treatment system.

23.3.1　*Ion-exchange*

In the United States, the use of anion exchange for As removal is generally not favored because of the large amounts of waste brine generated because of the higher affinity of ion-exchange resins for sulfate over arsenate. Discharge of waste brines to the municipal sewers may not be allowed, or access to sewers may not be available, resulting in large disposal fees being incurred to ship waste brine containing As off site for disposal at an appropriate facility. For As removal, a strong base anion-exchange resin is generally used (Berdal *et al.*, 2000) which has the following selectivity in groundwaters:

$$SO_4^{2-} > HAsO_4^{2-} > NO_3^- > Cl^- > H_2AsO_4^-, HCO_3^- >> Si(OH)_4, H_3AsO_4 \qquad (23.1)$$

Since the sulfate anion is present in most waters at concentrations several orders of magnitude greater than the As, this limits the run length which makes As removal by ion exchange economically unattractive when the sulfate levels are over $150 \, mg \, L^{-1}$ (Wang *et al.*, 2007). This problem is particularly acute in some of the more arid regions (e.g. New Mexico) in the United States where elevated sulfate concentrations regularly occur (Aragon *et al.*, 2007).

An additional disadvantage of using ion exchange for As removal is that much of the alkalinity of the water is initially removed which can lead to a drop in pH and a change in the aesthetic properties (i.e. taste) of the finished water. In the author's opinion, ion exchange is best suited for the treatment of groundwaters containing other contaminants (e.g. nitrate) in addition to As which are also amenable to removal by anion exchange.

23.3.2　*Coagulation/filtration*

Coagulation/filtration (CF) plants have been widely employed for the removal of As from municipal drinking water. This method uses the *in situ* generation of a ferric (iron (III)) hydroxide floc to remove As from potable water by a combination of precipitation and adsorption. This approach is particularly popular if there is already significant soluble iron present in the raw

Figure 23.1.   A CF plant at Casa Grande, Arizona designed to treat a flow rate of up to 1350 gpm (307 m^3 h^{-1}).

water because this can be utilized to help to remove the As thus lowering the amount of additional iron required. The steps involved in a CF plant typically comprise the following:

- Iron (usually in the form of a solution of iron (III) chloride) is added to the water. This is often accompanied by the addition of an oxidant to oxidize any iron(II) naturally present to iron (III). The oxidant will also oxidize As(III) to As(V), which is the most amenable oxidation state of As for removal via adsorption or precipitation methods.
- The added iron is allowed to react and form a precipitate of ferric hydroxide which adsorbs As from the water. This reaction is rapid and is essentially complete in seconds.
- The solid iron hydroxide precipitate and adsorbed As is then filtered through a sand/media filter or ultrafiltration membrane. The filter is backwashed, when required, and the iron hydroxide sludge and adsorbed As stored prior to off-site disposal. (In some areas, direct discharge of the iron hydroxide sludge to the municipal sewers is permitted, though this is not common).

The efficiency of CF plants is affected by a number of factors including pH, and the concentration of phosphate, silica and other competing species (Meng *et al.*, 2000). Consequently, pH control is often utilized for high pH waters to reduce the pH to between 7.0 and 7.5 which minimizes the amount of iron addition required, thus reducing the volumes of sludge produced. Adjusting the water to pH values below 7.0 is usually avoided due to the possibility of pipe corrosion downstream from the treatment plant which could lead to elevated levels of copper, lead and other trace elements in the finished water.

A photograph showing a typical CF plant is shown in Figure 23.1.

The illustration shows the six media pressure filtration tanks used to remove the iron hydroxide precipitate from the water and the large settling tank used to settle the sludge from the filter backwash. Each pressure filter is approximately 77.25 inches (approximately 196 cm) in diameter and contains filter media with a depth of 36 inches (91.4 cm). The settling tank, situated behind the media tanks, is fitted with a conical bottom and the sludge is allowed to settle under gravity. After a period of approximately 4 hours, the supernatant water from the sludge settling tank is decanted and pumped to the front of the plant. The settled sludge is then drawn off the bottom of the settling tank and pumped to the dewatering vessel situated to the right of the filters. Supernatant from the dewatering vessel is also reclaimed and the residual sludge is periodically transported off site for disposal at a licensed facility. Typically, the waste sludge contains about 0.8% solids by weight, though this number may vary according to the water chemistry and rate of ferric chloride dosing. Additional concentration of the sludge can be performed if desired using either polymer addition or a filter press to reduce waste volumes for disposal. The backwash cycle is triggered by the pressure differential across the media beds and thus the frequency will vary according to the rate of ferric hydroxide generation.

The major operating expenses associated with the CF plants are:

- Hydrochloric acid (for pH adjustment)
- Sodium hydroxide (for pH adjustment)
- Sodium hypochlorite (for As and Fe(II) oxidation and finished water disinfection)
- Ferric chloride for iron hydroxide floc formation and subsequent As removal
- Sludge disposal
- Labor to periodically monitor chemical feed tanks, pH, finished water quality, filter performance, etc.

Once a plant has been commissioned, the labor required is usually minimal and just consists of periodic inspections to confirm that the system is operating according to design parameters and that all equipment is functioning properly. It is worth noting that in general terms, the footprint and complexity of a CF plant is greater than that of a standard adsorption plant of a similar design capacity.

23.3.3   Adsorption

Adsorption plants rely on the relatively simple process of passing the groundwater through a fixed bed of an As-selective adsorbent media. As a consequence, the plant design is relatively simple compared to a CF plant or a standard IX plant because there is no sludge generation or filtration equipment required and the media is not subject to a regular regeneration process. Some granular As adsorbents require periodic backwashing to maintain a reasonable pressure drop across the beds which leads to the generation of relatively low volumes of backwash waste which needs to be stored on site and disposed of appropriately. The hybrid As adsorbents (described later) do not require backwashing thus simplifying the system design, and reducing the plant footprint and capital costs. Arsenic is removed by the adsorbent with minimal interference from other common anions found in drinking water (e.g. sulfate, bicarbonate, chloride, etc.) leading to a much longer bed life than traditional anion-exchange resins. These As adsorbents are all adversely affected by high levels of phosphate anions, which directly compete with As for the available adsorption sites. High concentrations of silica can also drastically reduce the performance of the media and this is especially true when the pH of the water is above 7.5 (Davis et al., 2002; Möller and Sylvester, 2008; Smith and Edwards, 2005). Arsenic-selective adsorbents work best when the pH is between 6.5 and 8 and the operating capacity drops off significantly at pH values greater than 8.5. Consequently, pH reduction needs to be considered for the treatment of high pH groundwaters. There are a large number of media commercially available which have been successfully used in the United States in full scale treatment plants. The bulk of these media are based upon hydrous iron oxides but there are also products which use oxides of zirconium, titanium and lanthanum (Aragon et al., 2007). Most of the media are supplied in the form of granules and are used on a once through basis and disposed of when they are saturated with As. However, there are also several media on the market that consist of hydrous iron oxide particles immobilized in a polymeric anion-exchange resin which are designed to be regenerated and reused, thus reducing waste volumes and media replacement costs. Regeneration is not done on site but is performed at a central location where the hazardous As-laden regeneration solutions can be handled and disposed of in a safe manner.

Figure 23.2.   The arsenic treatment system installed at El Mirage, Arizona. The picture shows the two arsenic adsorption beds, each of which contains 225 ft^3 (6.4 m^3) of ArsenX^{np} media.

An example of an As adsorption system is shown in Figure 23.2. This system contained the hybrid As adsorbent, ArsenX^{np}, one of the first hybrid media, and was installed at El Mirage, Arizona in 2007. ArsenX^{np} was based upon the initial research of SenGupta at Lehigh University (SenGupta and Cumbal, 2007) and differs from standard granular adsorption media used for As removal from potable water in that it combines the As affinity of hydrous iron oxide with the durability of a conventional polymeric ion-exchange resin and thus has a greater physical integrity and does not require backwashing (Sylvester *et al.*, 2007).

As can be seen from the photograph, the system at El Mirage is very simple and consists of two seven foot (213 cm) diameter media vessels in a lead-lag configuration. The tanks are engineered so that the flow direction can be adjusted so that either tank can be in the lead or lag position. The system is designed to operate at a flow rate of 263 gpm (60 m^3 h^{-1}), with a maximum flow rate of 425 gpm (96.5 m^3 h^{-1}), which would equate to a minimum empty bed contact time (EBCT) of approximately 4 minutes per vessel. Other than pre-filtration of the groundwater through a 25–50 μm filter, there is no other treatment to the raw water. At this specific location, the life of the media varied from 11 to 18 months (depending upon the volume of water treated) before As breakthrough above the MCL occurred and the media required regenerating.

23.4   CHROMATE REMOVAL TECHNOLOGIES

As mentioned at the beginning of this chapter, the removal of chromate from drinking water is only just being regulated and, as a consequence, the technologies utilized are still effectively in their infancy. However, there are two major approaches to chromate removal that have demonstrated to be effective in field trials and are potentially suitable for full-scale implementation. These techniques are ion exchange and a reduction-coagulation/filtration technology based upon the addition of a ferrous (Fe(II)) salt. A description of these technologies is given below, though there is very limited information to allow relative costs or effectiveness to be assessed. Other technologies have been assessed but the results have tended to suggest they are of limited effectiveness or economic viability (McGuire *et al.*, 2006; Sharma *et al.*, 2008).

23.4.1   *Ion-exchange for chromate removal*

Chromate has been shown to be effectively removed by strong base anion-exchange resins but the effective bed life was short making it unlikely to be cost effective for full scale applications where Cr(VI) was the only contaminant. However, weak base anion (WBA) resins have proven to be far more effective with operational lives measured in excess of 38,000 BVs in field trials which suggest they could potentially be used without regeneration in a 'once through' basis. However, the uptake of chromate is highly pH dependent and the incoming water must be adjusted to have a pH of <6 to be effective. The mechanism of chromate removal by the WBA resins is not certain and the detection of Cr(III) in the effluent indicates that there may be reduction of Cr(VI) to Cr(III) occurring on the resin. This observation is of critical importance for potential applications of WBA resins in California because the reaction between amine groups on the WBA resin and adsorbed chromate (a relatively strong oxidant) has the potential to lead to the formation of N-nitrosodimethylamine (NDMA) which is a potent carcinogen. Some ion-exchange resins have already been shown to release NDMA (formed as a minor byproduct of the manufacturing process) and California is proposing to regulate NDMA in drinking water (CA EPA – NDMA, 2014). Currently, an MCL has not been issued but a notification level of $10 \, \text{ng} \, \text{L}^{-1}$ (parts per trillion) has been designated so NDMA release from WBA resins is therefore a major potential issue which will affect their use in the removal of chromate from drinking water.

Recently, a relatively new resin has been proposed for chromate removal from drinking water that minimizes the possibility of NDMA formation. Dow has developed a WBA resin based upon a phenol formaldehyde polymer (PWA-7) that has been claimed to be effective for Cr(VI) removal. Since there are no amine functionalities to react with the adsorbed Cr(VI), the NDMA formation potential is greatly reduced. Further data is required to confirm the efficacy of this new resin.

23.4.2   *Addition of Fe(II) salts*

Chromate can be effectively removed from drinking water by reducing it to Cr(III) using an Fe(II) salt and precipitating it out in conjunction with $Fe(OH)_3$ precipitate formed following the oxidation of the Fe(II). At pHs greater than 6.5, the redox reaction that occurs between Cr(VI) and Fe(II) is as follows (Lee and Herring, 2003):

$$CrO_4^{2-} + 3Fe^{2+} + 8H_2O \rightarrow Cr(OH)_3 + 3Fe(OH)_3 + 4H^+ \qquad (23.2)$$

Thus, in theory, it takes a three-fold molar excess of iron over chromate to achieve the required reduction to Cr(III). In practice, however, significant Fe(II) is consumed by other side reactions so an iron addition considerably in excess of the stoichiometric amount is required. McGuire and coworkers showed that an Fe:Cr mass ratio of 10:1 was required for 98.5% Cr(VI) removal from water obtained from wells in Glendale, California. Unpublished data from the author's laboratory using a synthetic drinking water showed a similar excess of iron was required to the McGuire study and also that the optimum pH for the reaction was between 7.5 and 8.0.

The addition of iron salts to drinking water and the subsequent filtration technology required to remove the iron hydroxide precipitate is well known in the water industry and was described previously in this chapter where it was applied for As removal. Consequently, this approach should be readily acceptable to the drinking water industry because the nature of the waste residuals and the filtration technology is well understood and can thus easily be adapted for chromate removal.

23.5   CONCLUSIONS

The removal of As from drinking water is well understood and a number of different technologies and media have been proven to be effective in the USA, and worldwide, on a commercial scale to remove As to below the drinking water MCL. Consequently, there is little incentive or need to develop new technologies in the developed world where there is an adequate water infrastructure. However, there is still an urgent requirement to develop new As removal technologies that are suitable for the developing world in places such as West Bengal and Bangladesh where there is no infrastructure and insufficient funding to install, operate and maintain standard As water treatment systems. The necessity of removing Cr(VI) from drinking water is imminent and full scale Cr(VI) removal systems are likely to be installed in California in the very near future and, depending upon the passage of further legislation, later in the rest of the United States of America. Once commercial Cr(VI) removal plants have been commissioned and are operational, it will be possible to determine which is the most effective and economic remediation technique.

23.5.1   *Disclaimer*

The views in this article are a reflection of the personal views of the author. Any mention of specific materials, trade names or companies should not be taken as an endorsement of any specific technology, product, manufacturer or supplier.

ACKNOWLEDGEMENTS

I would like to thank Layne for giving me permission to use photographs of their arsenic removal plants and would also like to thank Professors Nalan Kabay and Marek Bryjak for inviting me to write this chapter.

REFERENCES

Aragon, M., Kottenstette, R., Dwyer, B., Aragon, A., Everett, R., Holub, W., Siegel, M. & Wright, J. (2007) *Arsenic Pilot Plant Operation and Results – Anthony, New Mexico*. Sandia Report SAND2007-6059.

Berdal, A., Verrie, D. & Zaganiaris, E. (2000) Removal of arsenic from potable water by ion exchange resins. In: Greig, J.A. (ed.) *Ion Exchange at the Millennium*. Imperial College Press, London, UK. p. 101.

California Environmental Protection Agency (2014) *Chromium 6*. Available from: http://www.waterboards.ca.gov/drinking_water/certlic/drinkingwater/Chromium6.shtml.

California Environmental Protection Agency (2014) *Groundwater information sheet: N-nitrosodimethyl-amine (NDMA)*. Available from: http://www.waterboards.ca.gov/water_issues/programs/gama/docs/ndma.pdf [accessed July 2015].

Davis, C.C., Chen, H.-W. & Edwards, M. (2002) Modeling silica sorption to iron hydroxide. *Environmental Science & Technology*, 36, 582–587.

Kapaj, P., Peterson, H., Liber, K. & Bhattacharya, P. (2006) Human health effects from chronic arsenic poisoning—a review. *Journal of Environment Science & Health* Part A, 41, 2399–2428.

Lee, G. & Hering, J.G. (2003) Removal of chromium (VI) from drinking water by redox-assisted coagulation with iron(II). *Journal of Water Supply: Research and Technology – Aqua*, 52, 319–332.

McGuire, M.J., Blute, N.K., Seidel, C., Qin, G. & Fong, L. (2006) Pilot scale studies of hexavalent chromium removal. *Journal AWWA*, 98, 134–143.

Meng, X., Bang, S. & Korfiatis, G.P. (2000) Effects of silicate, sulfate and carbonate on arsenic removal by ferric chloride. *Water Research*, 34, 1255–1261.

Möller, T. & Sylvester, P. (2008) Effect of silica and pH on arsenic uptake by resin/iron oxide hybrid media. *Water Research*, 42, 1760–1766.

SenGupta, A.K. & Cumbal, L.H. (2007) Hybrid anion exchanger for selective removal of contaminating ligands from fluids and method of manufacture thereof. US Patent No. 7,291,578.

Sharma, S.K., Petrusevski, B. & Amy, G. (2008) Chromium removal from water: a review. *Journal of Water Supply: Research and Technology – Aqua*, 57, 541–553.

Smith, A.H., Biggs, M.L., Moore, L., Haque, R., Steinmaus, C., Chung, J., Hernandez, A. & Lopipero, P. (1999) Cancer risks from arsenic in drinking water: implications for drinking water standards. In: Chappell, W.R., Abernathy, C.O. & Calderon, R.L. (eds.) *Arsenic exposure and health effects*. Elsevier Science. pp. 191–199.

Smith, S.D. & Edwards, M. (2005) The influence of silica and calcium on arsenate sorption to oxide surfaces. *Journal of Water Supply: Research and Technology*, 54, 201–211.

Steinmaus, C., Moore, L., Hopenhayn-Rich, C., Biggs, M.L. & Smith, A.H. (2000) Arsenic in drinking water and bladder cancer. *Cancer Investigation*, 18, 174–182.

Sutton, R. (2010) *Chromium-6 in US Tap Water*. Available from: http://www.ewg.org/chromium6-in-tap-water. [accessed July 2015].

Sylvester, P., Westerhoff, P., Möller, T., Badruzzaman, M. & Boyd, O. (2007) A hybrid sorbent utilizing nanoparticles of hydrous iron oxide for arsenic removal from drinking water. *Journal of Environmental Engineering Science*, 24, 104–112.

Wang, L., Chen, A.S.C., Tong, N. & Coonfare, C.T. (2007) Arsenic removal from drinking water by ion exchange. USEPA emonstration project at Fruitland, ID. Six month evaluation report. EPA/600/R-07/0-17.

Subject index

Sustainable Water Developments

Book Series Editor: Jochen Bundschuh

ISSN: 2373-7506

Publisher: CRC Press/Balkema, Taylor & Francis Group

1. Membrane Technologies for Water Treatment: Removal of Toxic Trace
   Elements with Emphasis on Arsenic, Fluoride and Uranium
   Editors: Alberto Figoli, Jan Hoinkis & Jochen Bundschuh
   2016
   ISBN: 978-1-138-02720-6 (Hbk)

2. Innovative Materials and Methods for Water Treatment:
   Solutions for Arsenic and Chromium Removal
   Editors: Marek Bryjak, Nalan Kabay, Bernabé L. Rivas & Jochen Bundschuh
   2016
   ISBN: 978-1-138-02749-7 (Hbk)